Bioplastics for Food Packaging

Bioplastics for Food Packaging

Materials, Procedures, and Applications

Edited by

Mehran Ghasemlou

Department of Chemical Engineering, The University of Melbourne, Parkville, VIC, Australia

Shima Jafarzadeh

School of Engineering, Deakin University, Geelong Waurn Ponds Campus, Geelong, VIC, Australia;
Center for Sustainable Bioproducts, Deakin University, Geelong Waurn Ponds Campus, Geelong, VIC, Australia

ELSEVIER

Elsevier
Radarweg 29, PO Box 211, 1000 AE Amsterdam, Netherlands
125 London Wall, London EC2Y 5AS, United Kingdom
50 Hampshire Street, 5th Floor, Cambridge, MA 02139, United States

ISBN: 978-0-443-41654-5

For Information on all Elsevier publications
visit our website at https://www.elsevier.com/books-and-journals

Publisher: Matthew Deans
Acquisitions Editor: Jinia Dasgupta
Editorial Project Manager: Abhinav Singh
Production Project Manager: Maria Shalini A.
Cover Designer: Bakyalakshmi S

Typeset by MPS Limited, Chennai, India

Dedication

To my parents, for their boundless support. Without their love and encouragement, the completion of this work would not have been possible.

—Mehran Ghasemlou

To my dear husband, Mohammad. Without his love and encouragement, the completion of this work would not have been possible.

—Shima Jafarzadeh

Contents

List of contributors

Ahmadreza Abedinia Department of Food Science and Technology, Da.C., Islamic Azad University, Damghan, Iran; Postgraduate School, Universitas Brawijaya, Malang, East Java, Indonesia

Benu Adhikari School of Science, STEM College, RMIT University, Melbourne, VIC, Australia; Centre for Advanced Materials and Industrial Chemistry (CAMIC), RMIT University, Melbourne, VIC, Australia

Sahar Akhavan-Mahdavi Department of Food Materials and Process Design Engineering, Gorgan University of Agricultural Science and Natural Resources, Gorgan, Golestan, Iran

Maryam Azizi-lalabadi Research Center of Oils and Fats, Health Technology Institute, Kermanshah University of Medical Sciences, Kermanshah, Iran

Colin J. Barrow Center for Sustainable Bioproducts, Deakin University, Geelong Waurn Ponds Campus, Geelong, VIC, Australia

Gaelle Anne Chantal Bihan Deakin Center for Sustainable Bioproducts, Deakin University, Geelong Waurn Ponds Campus, Geelong, VIC, Australia

Shaghayegh Khajehali Chaleshtori Department of Food Science and Technology, Faculty of Agriculture, Ferdowsi University of Mashhad, Mashhad, Iran; Member of Iran National Standards Organization, Tehran, Iran

Fugen Daver School of Engineering, RMIT University, Bundoora, VIC, Australia

Md. Forshed Dewan School of Life and Environmental Sciences, Deakin University, Geelong Waurn Ponds Campus, Geelong, VIC, Australia

Mohammed Ali Dheyab School of Physics, Universiti Sains Malaysia, Pulau Pinang, Penang, Malaysia; Nano-Biotechnology Research and Innovation (NanoBRI), Institute for Research in Molecular Medicine (INFORMM), Universiti Sains Malaysia, Pulau Pinang, Penang, Malaysia

Mina Dokouhaki School of Science, STEM College, RMIT University, Melbourne, VIC, Australia

Yasaman Esmaeili School of Engineering, Deakin University, Geelong Waurn Ponds Campus, Geelong, VIC, Australia

Bahareh Farasati Far Department of Chemistry, Iran University of Science and Technology, Tehran, Iran

Asgar Farahnaky School of Science, STEM College, RMIT University, Melbourne, VIC, Australia

Mehran Ghasemlou School of Science, STEM College, RMIT University, Melbourne, VIC, Australia; Center for Sustainable Bioproducts, Deakin University, Geelong Waurn Ponds Campus, Geelong, VIC, Australia; Department of Chemical Engineering, The University of Melbourne, Parkville, VIC, Australia

Morteza Ghobakhloo Division of Industrial Engineering and Management, Uppsala University, Uppsala, Sweden

P.D.S. Amanda Goonathilaka Deakin Center for Sustainable Bioproducts, Deakin University, Geelong Waurn Ponds Campus, Geelong, VIC, Australia

Simin Hedayatnia Department of Food Biotechnology, Faculty of Food Science and Technology, Universiti Putra Malaysia, Selangor, Malaysia

Thi Thuy Hang Hoang Deakin Center for Sustainable Bioproducts, Deakin University, Geelong Waurn Ponds Campus, Geelong, VIC, Australia

Mohammad Iranmanesh School of Hospitality, Tourism and Events, Centre for Asian Modernisation, Taylor's University, Selangor, Malaysia

Elena P. Ivanova School of Science, STEM College, RMIT University, Melbourne, VIC, Australia

Shima Jafarzadeh School of Engineering, Deakin University, Geelong Waurn Ponds Campus, Geelong, VIC, Australia; Center for Sustainable Bioproducts, Deakin University, Geelong Waurn Ponds Campus, Geelong, VIC, Australia

Mehdi Jahanbakhshi School of Chemical Engineering, University of Tehran, Tehran, Iran

Minoo Naebe Center for Sustainable Bioproducts, Deakin University, Geelong Waurn Ponds Campus, Geelong, VIC, Australia; Institute for Frontier Materials, Deakin University, Geelong Waurn Ponds Campus, Geelong, VIC, Australia

Ali Najafi Department of food Science and Technology , Da.C, Islamic Azad University, Damghan, Iran

Hoang Chinh Nguyen Centre for Sustainable Bioproducts, Deakin University, Geelong Waurn Ponds Campus, Geelong, VIC, Australia

Ali Nikmanesh Department of Food and Drug, Semnan University of Medical Sciences, Semnan, Iran

Nazila Oladzadabbasabadi School of Science, STEM College, RMIT University, Melbourne, VIC, Australia; School of Engineering, Deakin University, Geelong Waurn Ponds Campus, Geelong, VIC, Australia

Sachin Talekar Centre for Sustainable Bioproducts, Deakin University, Geelong Waurn Ponds Campus, Geelong, VIC, Australia; Department of Biological Sciences, BITS-Pilani, K.K. Birla Goa campus, Sancoale, Goa, India

Mohammad Tarahi School of Science, STEM College, RMIT University, Melbourne, VIC, Australia

Caixia Wan Department of Chemical and Biomedical Engineering, University of Missouri, Columbia, MO, United States

Hananeh Yazdanbakhsh Research Center of Oils and Fats, Health Technology Institute, Kermanshah University of Medical Sciences, Kermanshah, Iran

About the editors

Dr. Mehran Ghasemlou is currently a McKenzie Research Fellow in the Department of Chemical Engineering at the University of Melbourne, Australia, where he collaborates with a multidisciplinary team to develop innovative bioplastic products. Before this appointment, he was a Deakin University Postdoctoral Research Fellow (DUPRF) at the Centre for Sustainable Bioproducts (CSB), Deakin University, Australia. He worked as a chief scientist at MyEco Group (Melbourne, Australia) for several years, leading research and innovations in bioplastic materials. Dr. Ghasemlou received his PhD in materials science and engineering from RMIT University in 2021. He is currently supervising 8 PhD scholars and has secured thousands of dollars in research grants and industrial support as a principal investigator or coinvestigator. As a top innovator in polymer/material science, he has published nearly 100 scientific research papers, review articles, and book chapters in reputable national and international journals. His work is widely recognized and heavily cited across disciplines and countries, with his publications accumulating >8300 citations to date (as of April 16, 2026). His name has featured for five consecutive years (2021–25) in the World Ranking of Top 2% Scientists in the subject-wise analyses conducted by a team of scientists at Stanford University, United States. The outstanding quality, performance, excellence, significance, and innovative potential of his research have been recognized by many awards and honors, including the 2022 RMIT Higher Degree Research (HDR) Science Award, 2021 RMIT Vice-Chancellor's Research Impact Award, 2021 Publication Prize, 2013 USDA Food Safety Research Fellowship, the 2011 University of Tehran Medal for Excellence in Research Thesis, and 2011 Silver Medal Award for Young Scientists in Applied Research Category. His expertise lies in bio-inspired materials, food packaging, and bioplastic technology, and his work focuses on identifying ideal materials for the development of alternative bio-based packaging methods to help build sustainability in Australia. He also contributes to the development of new methods for biodegradation of a variety of biopolymers and conduct studies in renewable packaging materials. His mission is to leverage science and technology to eliminate litter and divert plastic waste into a renewable resource to support the growth of Australia's circular economy.

Dr. Shima Jafarzadeh is a research fellow at the Centre for Sustainable Bioproducts (CSB), Deakin University, Australia. Prior to joining Deakin, she worked as a research fellow at Edith Cowan University (ECU), where she contributed to research on bioplastics and sustainable polymer materials. Her work focuses on developing sustainable bioplastics, thin-film composites, and biopolymeric materials, particularly for environmentally friendly packaging applications. She specializes in converting food waste and biomass into high-performance biodegradable materials, helping to address global challenges related to plastic pollution while supporting the transition toward a circular economy. Dr. Jafarzadeh has received several prestigious awards, with the most recent including the Best Early Career Researcher Award at Deakin University (2025). She is also ranked among the World's Top 2% Scientists (2024–25) based on citation impact. As of April 2026, she has published more than 75 scholarly outputs, including peer-reviewed journal articles and book chapters, with over 5800 citations and an h-index of 42. A significant proportion of her publications appear in journals ranked within the top 10% globally, demonstrating the strong quality, visibility, and impact of her research. She has secured ~$1 million in national and international competitive research funding and has supervised more than 10 researchers, including PhD students, master's students, and research assistants. Through collaboration with industry and international partners, her work aims to develop practical and scalable solutions that transform waste into valuable sustainable materials. She has received awards for her quality publications and served as a guest editor for reputable journals in the field.

Preface

Plastics are indispensable in everyday life and industrial operations, but the environmental impact of plastic waste is a huge concern. Plastic pollution is pervasive and only expected to get worse with serious consequences for both the environment and human health. The vast majority of plastics are derived from fossil fuels, and every stage of the plastic life cycle (from extracting resources and refining chemicals, to recycling and incineration) pumps greenhouse gases into our atmosphere. The urgent need to move away from fossil-based plastics has provided fertile ground for the emergence of alternatives, collectively referred to as "bioplastics." The ever-rising consumer demand for bioplastics has fueled their production, which is projected to reach 5.73 million tonnes by 2029. As the demand for bioplastics in food packaging continues to rise, there is an increasing focus on studying their chemical structures, biodegradability, and functional properties. Against this backdrop, we present *Bioplastics for Food Packaging: Materials, Procedures, and Applications*, a comprehensive book that explores the use of diverse bioplastics in sustainable packaging technologies. We conceived this book in response to the rapidly evolving role of bioplastics in shaping sustainable food packaging systems. The idea for this book originated from numerous discussions with colleagues, researchers, and industrial experts who identified a significant gap between the theoretical development of bioplastics (whether renewable or fossil-based) and their practical application in food packaging. As editors, we felt a strong need to compile a resource that not only showcases the state of the art in material science but also acts as a practical guide for those striving to make food supply chains more circular, resilient, and sustainable. What sets this book apart is its focus on affordable and scalable methods designed to make bioplastics more competitive with conventional petroleum-based plastics. The market currently faces significant hurdles because existing packaging often fails to provide adequate protection against spoilage without leaving a lasting environmental footprint. Each chapter in this book has been rigorously peer-reviewed and reflects contributions by a diverse team of researchers, academics, and industry experts working across polymer science, materials science, food science, and environmental sciences. We worked carefully to curate the content to ensure it offers conceptual depth, empirical evidence, and actionable insights into the emerging applications of bioplastics in food packaging. We are confident that this book will become an essential reference and study guide for students, researchers, and food industry professionals. We extend our heartfelt congratulations to the authors for this exceptional contribution and urge the readers to delve into its pages. This is not merely a guide to technological advancements but a visionary roadmap to safer and more sustainable food packaging. We invite our readers to share their feedback and suggestions to help us improve the future editions of this work.

Mehran Ghasemlou
Shima Jafarzadeh

Acknowledgments

Writing a book is more complicated than we thought, but seeing it come to life has been more rewarding than we ever imagined. The completion of this work—or at least its existence in its current form—would not have been possible without the tireless and dedicated efforts of our esteemed contributing authors. The high quality of research presented in these chapters is a source of great pride for both editors. We, the editors, wish to express our profound gratitude to everyone who contributed to the creation of this work. More importantly, we thank our chapter authors for sharing their knowledge and insights. Their timely contributions and diligent revisions form the backbone of this book. While it is not possible to list every individual by name, we must acknowledge our students and postdoctoral researchers for their hard work behind the scenes that built the body of this work. We also thank our colleagues, mentors, and research supervisors for the insightful discussions that helped sharpen the focus of this book. We are grateful to the peer reviewers for their careful evaluations and constructive suggestions, which helped to increase the scientific quality and clarity of every chapter. A special note of thanks is reserved for our art editor, whose artistic talent and meticulous attention to detail in redrawing the figures and illustrations were critical in distinguishing this book from others in the field. We further thank Ms. Yasaman Esmaeili for her dedicated assistance and technical support in managing the submission of several chapters into the ELSA platform. We acknowledge the institutional support and resources provided by our respective universities and research groups, which established the necessary environment for this interdisciplinary collaboration. Our appreciation goes to the publishing team and editorial staff at Elsevier, particularly Ms. Ana Claudia Garcia and Mr. Abhinav Singh, for their continuous support and coordination. Their professional guidance and patience—from the initial proposal stage through the editing and final production—have been critical to the successful completion of this project. Finally, none of this would be possible without the support of our families and their sacrifices. We express our boundless gratitude for their guidance, encouragement, and understanding they showed during the many long hours required to bring this project to fruition. We conclude these acknowledgments with a heartfelt thanks to our family members for their unwavering support. The completion of this book was a massive project, and we may have missed mentioning some individuals who supported us. To all those we may have inadvertently omitted, we offer our sincere thanks, and we extend our apologies to all researchers and colleagues whose work we could not properly cite or acknowledge.

Dr. Mehran Ghasemlou
Dr. Shima Jafarzadeh

CHAPTER 1

Plastic waste and current food packaging materials

Thi Thuy Hang Hoang[1], Gaelle Anne Chantal Bihan[1], P.D.S. Amanda Goonathilaka[1], Mehran Ghasemlou[2,3,4], Hoang Chinh Nguyen[4], and Colin J. Barrow[4]

[1]Deakin Center for Sustainable Bioproducts, Deakin University, Geelong Waurn Ponds Campus, Geelong, VIC, Australia, [2]School of Science, STEM College, RMIT University, Melbourne, VIC, Australia, [3]Department of Chemical Engineering, The University of Melbourne, Parkville, VIC, Australia, [4]Centre for Sustainable Bioproducts, Deakin University, Geelong Waurn Ponds Campus, Geelong, VIC, Australia

1.1 Introduction

The 21st century is often described as the "Plastic Era," because plastic has replaced wood, glass, ceramic, and metal in almost every facet of our daily life. Plastics are ubiquitous in agriculture, construction, transportation, heat insulation, packaging, medical devices, electronics, furniture, children's toys, leisure items, and medicine and have become indispensable in modern society (Jambeck et al., 2015; Nayanathara Thathsarani Pilapitiya and Ratnayake, 2024). Due to their affordability, durability, corrosion resistance, good electrical insulation, low electrical/thermal conductivity and low density, there are no better alternative materials with similar features that can compete with plastics. Moreover, these outstanding characteristics are not limited to a particular field and product class. Plastic can contribute to durable and adaptable packaging products or constitute a part of an aircraft or car to reduce transportation costs and lower fuel consumption. In food applications, plastics plays a crucial role in extending the shelf life of perishable goods by acting as a barrier against microbial growth or cross-contamination, thereby helping to reduce or mitigate food spoilage and organic waste. Plastics act as a tool for safeguarding food safety and maintaining quality throughout the supply chain (Marsh and Bugusu, 2007).

However, these remarkable and unique features that have rendered plastics a material of choice are also responsible for their negative environmental impact. The global population increased from 2.5 billion in 1950 to over 8.2 billion in 2025, with a projected growth to 10 billion by 2050. Over the same period, annual plastic production rose from 1.7 million tons to 400 million tons largely driven by technological advances that demanded more-than-expected plastic production (OECD, 2022). Plastic packaging accounts for ~40% of this output, with an

Bioplastics for Food Packaging. DOI: https://doi.org/10.1016/B978-0-443-41654-5.00013-5

average service life of around 6 months, making it the largest end-use segment (OECD, 2022). Importantly, most of plastic packaging is designed for single use and is discarded soon after purchase. In Europe, over 60% of all plastic packaging waste comes from the food and beverage sectors, and in Australia, food packaging contributes to an estimated 900,000 tons of plastic packaging each year. Common plastic materials include polyethylene (PE) (low and high density), polypropylene (PP), poly(vinyl chloride) (PVC), polyethylene terephthalate (PET), and polystyrene (PS) (Fig. 1.1) (Selke et al., 2021).

Solving the food waste problem is becoming challenging when these materials are combined to make multilayer plastics. Consequently, around 300 million tons of plastic waste are discarded annually, which is equal to the weight of the entire human population. Yet, only 9%–10% of this plastic waste is recycled; the remainder is either incinerated (12%) or left to accumulate in landfills and the natural environment (~79%) (Hurley et al., 2020). Plastic waste mismanagement is increasingly evident in both freshwater and marine environments. Rivers are recognized as a significant source of plastic debris that eventually reach the ocean. Each year, more than 11 million tons of plastic enter the ocean. Alarmingly, current projections suggest that plastic leakage could triple by 2040 unless the plastic industry adopts a radical shift in its manufacturing practices (Pew Charitable Trust, 2020). These plastics can break down into tiny microplastics and even nanoplastics driven by environmental factors such as microorganisms, ultraviolet (UV) exposure, and physical or mechanical processes (wave or wind action). Several studies suggested that microplastics can be found in air, water,

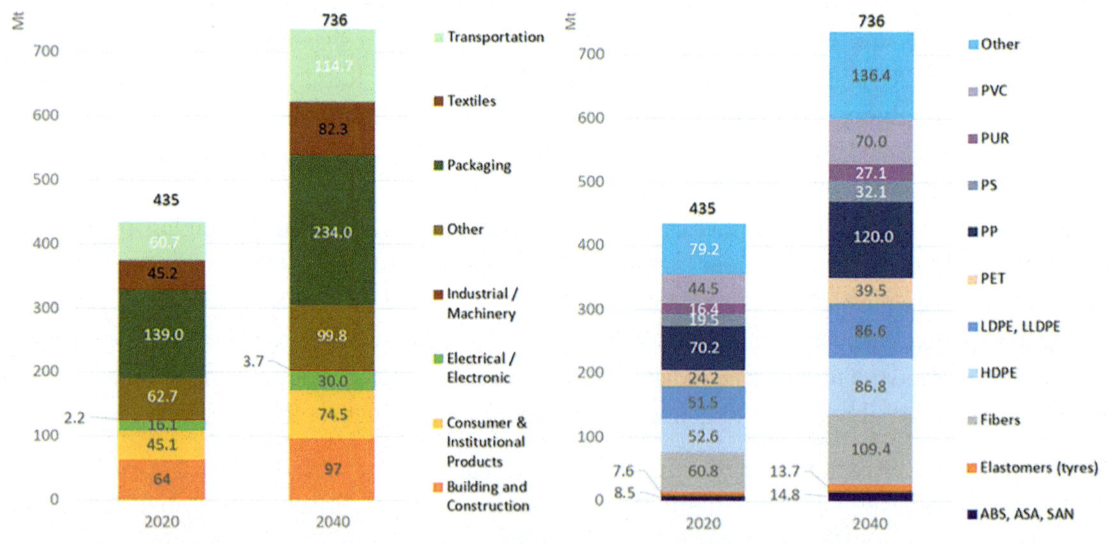

Figure 1.1

Global plastics production (measured in tons per year) by different industrial sectors (left-hand panel) and by type of polymer (right-hand panel) in 2020 and 2040.

The statistics are sourced from the OECD ENV-Linkages model (OECD, 2024).

soil, food, and even in human and animal blood and placentas (Kadac-Czapska et al., 2023). The bioaccumulation of plastics in the human body has raised increasing concern over potential long-term health effects since the 1980s. Despite the alarms sounded by scientists and the urgent need for radical changes to mitigate climate change and seek alternatives to fossil-based plastics, global responses to curb plastic pollution had been largely unsuccessful (OECD, 2024). Some developed countries have enacted bans or restrictions on single-use plastics, whereas others, particularly developing countries and conflict areas led by weaker infrastructures and policies or climate-skeptic administrations, have reversed or stalled such initiatives (Plastics Europe, 2024a; Zero Waste Europe, 2021). For instance, the U.S. government in 2025 rolled back an initiative that previously promoted the use of paper-based alternatives to single-use plastics. The COVID-19 pandemic intensified global plastic consumption, particularly due to increased demand for disposable food service items and sterilized products (Oliveira et al., 2021). To tackle these challenges, there is an urgent need to reduce dependence on conventional, petroleum-based plastics and facilitate transition toward recyclable, biodegradable, or sustainably sourced alternatives. However, this transition must carefully consider the need to maintain food safety standards, as risks of food contamination can limit the use of recycled plastics. Effective mitigation of plastic waste, especially in food packaging, ultimately requires a global coordinated policy effort, innovations in materials science, and systemic changes in production and consumption patterns.

In this chapter, we discuss environmental aspects of plastic waste. We begin our discussion by a comprehensive review of conventional plastics including their history, production, and applications. A detailed analysis of the global plastic waste crisis and its impact on the environment, wildlife, human health, and socioeconomic consequences is subsequently provided. We then highlight the important roles of plastics in food packaging in terms of function, materials, and practical challenges. Finally, this chapter discusses the future of food packaging by proposing some directions for the sustainable and responsible management of plastic waste.

1.2 Conventional plastics: Production, properties, and applications

Although natural polymers have always existed, such as polysaccharides (e.g., cellulose, starch, alginate, and carrageenan), polypeptides (e.g., collagen and gelatine) or DNA, artificial and fully synthetic polymers have only recently become part of everyday life. Artificial polymers, obtained by the chemical modification of natural polymers, emerged in the 19th century with various attempts to modify natural rubber and cellulose. One of the major discoveries was the vulcanization of rubber using sulfur, patented by both Thomas Hancock in 1843 in the United Kingdom and Charles Goodyear in the United States in 1844 (Gilbert, 2017). Both found out that increasing the amount of sulfur yielded a hard product variously called ebonite, vulcanite, or hard rubber. In 1846, Christian Friedrich Schönbein built on the work of Théophile-Jules Pelouze and established controlled conditions for the nitration of cellulose, which was used as an explosive or in the form of collodion used for wound dressing (Schönbein, 1911). Meanwhile, in 1866, Marcellin Berthelot explained the polymerization of styrene (Berthelot, 1911). However, the application of synthetic polymers did not become feasible until 1909. In that year, Fritz Hofmann successfully created synthetic rubber, and Leo Hendrik Baekeland developed the industrial production of a phenol-formaldehyde synthetic polymer commercialized as *Bakelite* (Gilbert, 2017). The following decades witnessed the boom of synthetic polymer development, such as five of today's major thermoplastics: PS, PVC, the polyolefins PE and PP, and the acrylic poly(methyl methacrylate).

At the bottom of the pyramid, we find thermoplastics named commodity plastics such as PP, PE, PVC, PS, and certain grades of PET. They represent the most cost-effective polymers produced in large volumes to support everyday applications, including food packaging. However, despite limited thermal and mechanical resistance, they have a large range of applications. Polymers can also be sorted by class, namely, thermoplastics, elastomers, and thermosets (Fig. 1.2). The heat response of each polymer governs both application promise and recyclability. Thermoplastic polymers such as PE, PP, PS, PET, and PVC soften when heated and are generally recyclable without losing their properties. Elastomers exhibit high elasticity and can stretch or deform under heat without breaking with; rubbers being the major category of elastomers. In contrast, thermoset polymers are heavily cross-linked, which prevent them from reprocessing and recycling. Epoxy and polyester resins and vulcanized rubber are typical thermosets. Amorphous thermoplastics are typically transparent and can be easily shaped by extrusion processes or injection molding over a broad temperature range. In

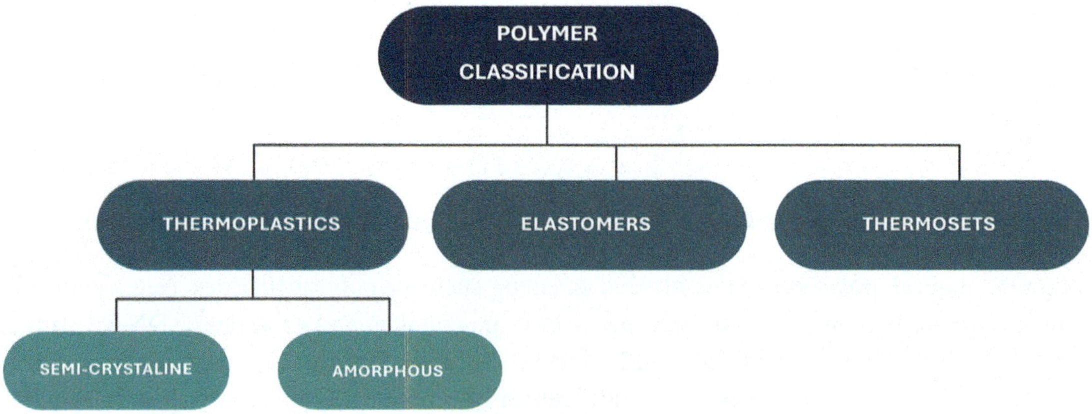

Figure 1.2

The classification of polymers into thermoplastics, elastomers, and thermosets. Thermoplastics can also be categorized into semicrystalline and amorphous polymers. A semicrystalline polymer has a mixture of orderly and random crystals regions, while an amorphous polymer has chains arranged completely at random, with no ordered crystal areas.

contrast, semicrystalline thermoplastics are typically opaque due to their highly ordered crystalline domains that act as a melting barrier and are more challenging to thermoform (Fig. 1.3).

1.2.1 Polyethylene

PE is by far the world's most widely used plastic primarily for packaging (plastic bags, films, containers, etc.). This polymer is grouped under the polyolefins family under the thermoplastic classification and is made from the polymerization of ethylene (C_2H_4) using Ziegler-Natta or Phillips catalysts (Table 1.1). Ethylene is produced from petrochemical sources or natural gas. PE chains can be linear or branched and, the variations of density, branching, crystallinity, and molecular weight, will have different mechanical and physical properties and thus diverse applications (Fig. 1.4).

1.2.2 Polypropylene

PP is a versatile thermoplastic polymer belonging to the polyolefin family, which is produced by the polymerization of propylene (C_3H_6) using Ziegler-Natta or metallocene catalysts (Fig. 1.5).

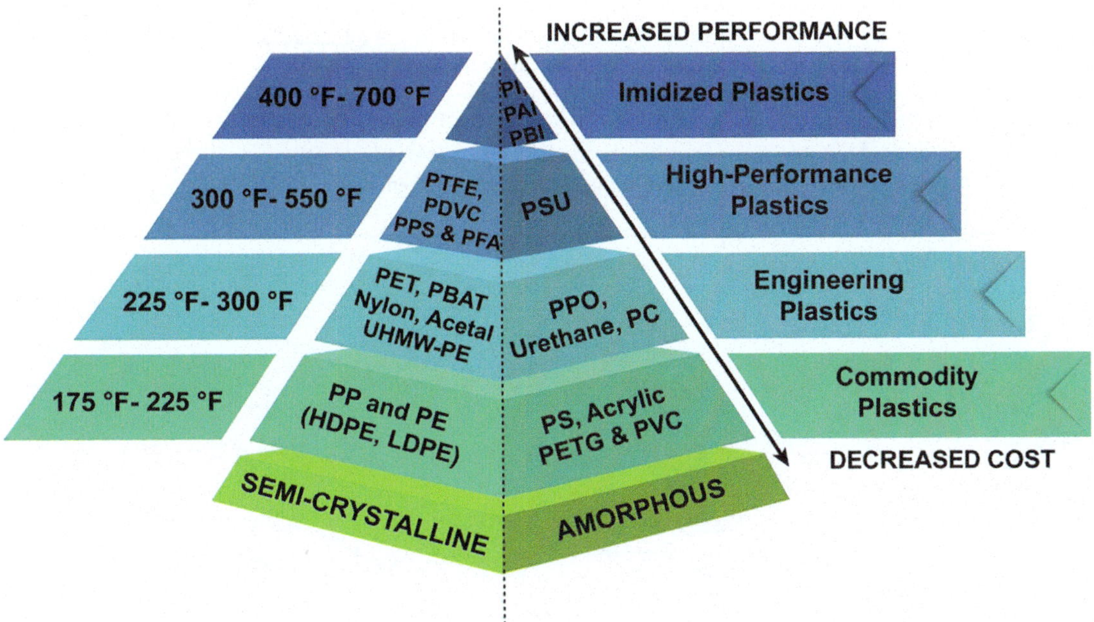

Figure 1.3
Thermoplastic classification by cost and thermal resistance from commodity to high-performance materials.

Table 1.1 Technical properties, features, potential applications, and proposed line structure of different types of polyethylene.

Type of PE	Technical properties	Features and potential application	Structure
LDPE: low-density PE	Density = 0.915–0.935 g/cm^3 melting point = 105°C–115°C semicrystalline thermoplastic	Flexible, low strength, good thermal insulation, cheap. It is used in plastic bags, film wraps, squeeze bottles.	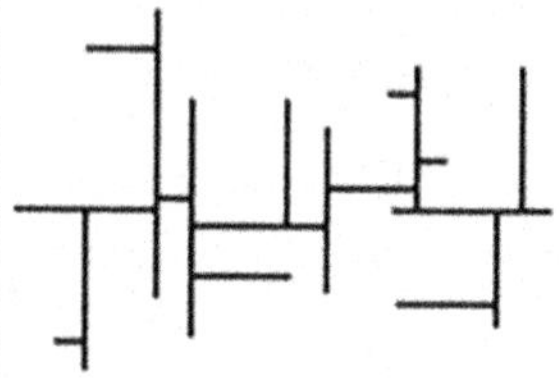
LLDPE: linear low-density PE	Density = 0.90–0.93 g/cm^3 melting point = 100°C–130°C	Similar to LDPE but with more strength/stretch. It is used in packaging.	
MDPE: middle-density PE	Density = 0.93–0.94 g/cm^3 melting point = 120°C–130°C	Less dense than HDPE and less notch sensitive. It is used in gas pipes and fittings, sacks, shrink film, packaging films, carrier bags, and screw closures.	
HDPE: high-density PE	Density = 0.94–0.97 g/cm^3 melting point = 128°C–136°C	Strong, stiff, chemical-resistant. It is used in bottles, milk jugs, pipes, and containers.	
UHMWPE: ultra-high-molecular weight PE	Density = 0.94–0.97 g/cm^3 melting point = 130°C–136°C	Extremely tough. It is used in bulletproof vests, joint implants, and industrial gears.	
PEX: cross-linked PE	Density = 0.94 g/cm^3 melting point = 121°C–127°C	Durable, heat-resistant. It is used in plumbing, and heating systems.	

PE, polyethylene.

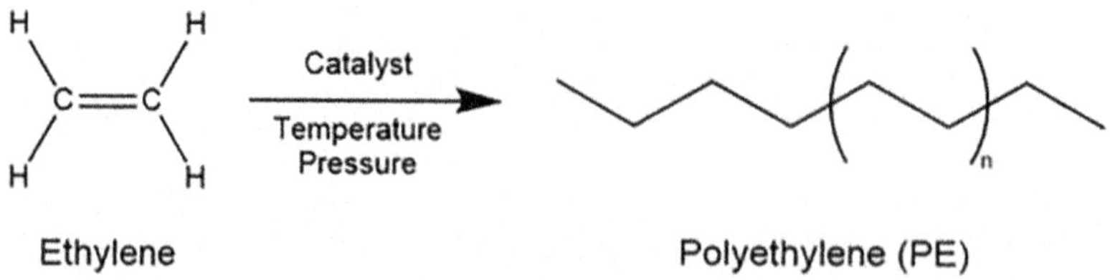

Figure 1.4
Chemical structure of ethylene and its polymerization to polyethylene.

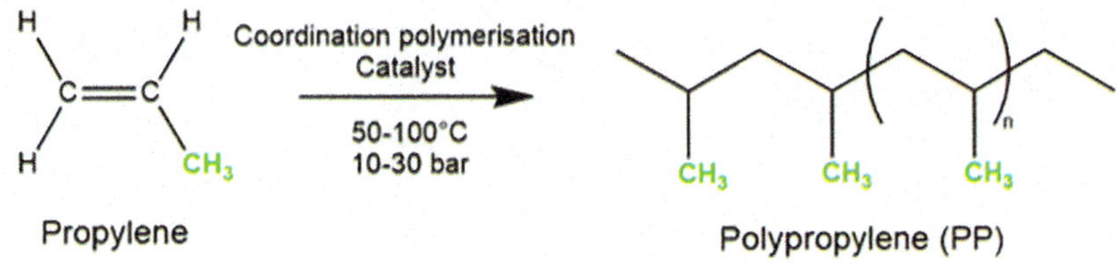

Figure 1.5
Chemical structure of propylene and its polymerization to polypropylene.

Depending on the catalyst used and the arrangement of methyl groups along the polymer chain, PP can be isotactic, syndiotactic, or atactic, with isotactic PP being the most commercially important due to its superior crystallinity and mechanical strength. PP exhibits excellent chemical resistance, fatigue resistance, and a relatively high melting point. Its low density and processability contribute to its popularity in both consumer and industrial goods such as packaging films, rigid containers, automotive parts, nonwovens textiles, medical devices, and household goods (Fig. 1.6).

1.2.3 Polyethylene terephthalate

PET is a thermoplastic polyester produced through the polycondensation of ethylene glycol and terephthalic acid. It is known for its excellent strength-to-weight ratio, transparency, effective barrier properties, and resistance to moisture and gases. PET is widely used in food and beverage packaging, particularly for bottles and containers, as well as in textile polyester fibers, thermoformed trays, and films. It can be amorphous or semicrystalline, depending on processing conditions. PET is highly recyclable, and its recycling stream is well established in many countries. Its balance of performance, cost, and recyclability potential makes it a key material in packaging and synthetic textiles (Fig. 1.7).

1.2.4 Poly(vinyl chloride)

PVC is a widely used synthetic thermoplastic polymer produced by the polymerization of vinyl chloride monomers (C_2H_3Cl). PVC can be rigid or flexible depending on the processing conditions. It can also exhibit tacticity—atactic, syndiotactic, or isotactic—but commercial PVC is generally atactic. Rigid PVC is known for its strength, chemical resistance, and durability, making it ideal for construction applications such as pipes, window frames, and siding. Flexible PVC is used in cables, medical tubing, flooring, and synthetic leather. The versatility and cost-effectiveness of PVC has made it one of the most widely produced plastics, although concerns about its difficult recyclability have driven interest in safer alternatives (Fig. 1.8).

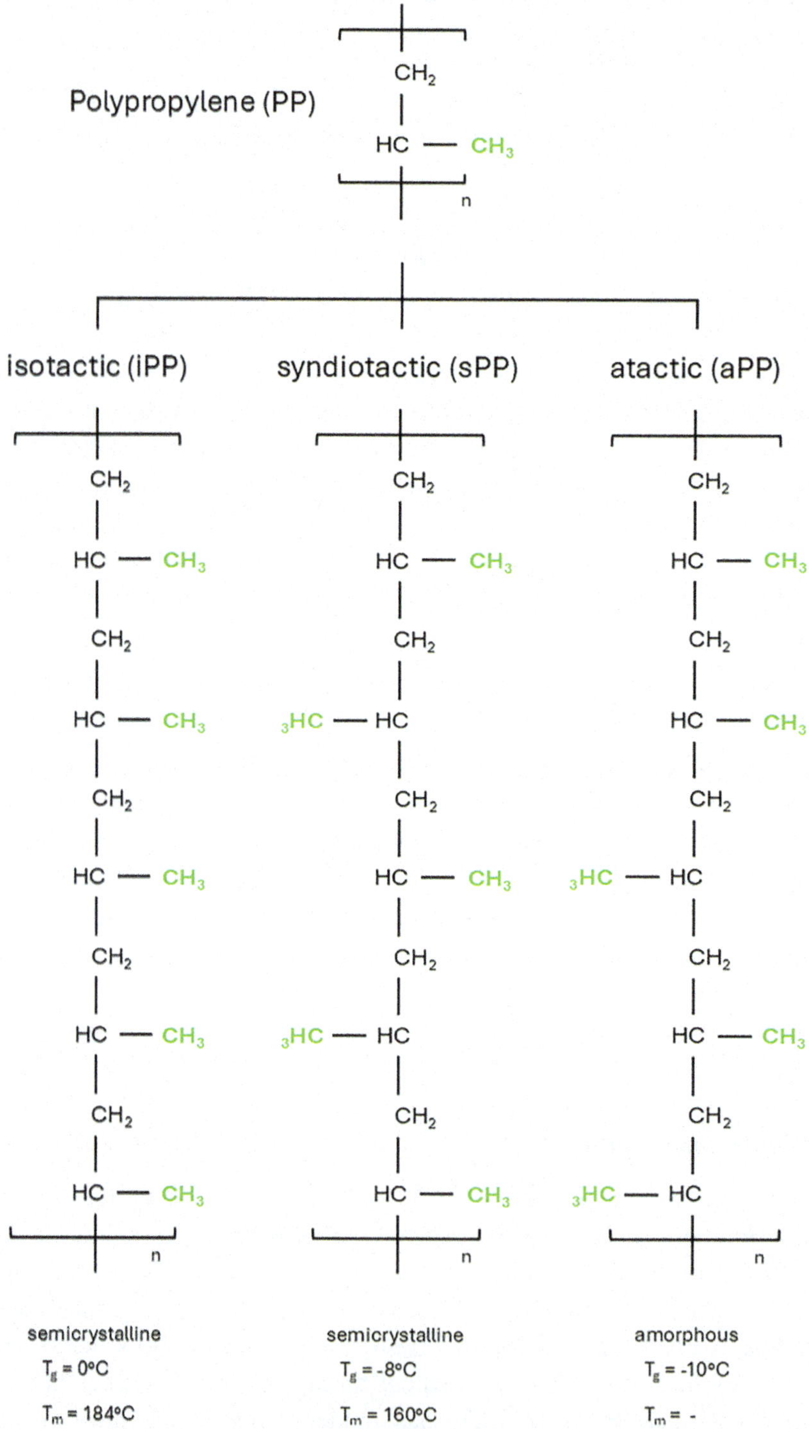

Figure 1.6
The stereochemical arrangement of the pendant-methyl groups along the polypropylene backbone divides them into three groups of atactic (aPP), syndiotactic (sPP), and isotactic (iPP).

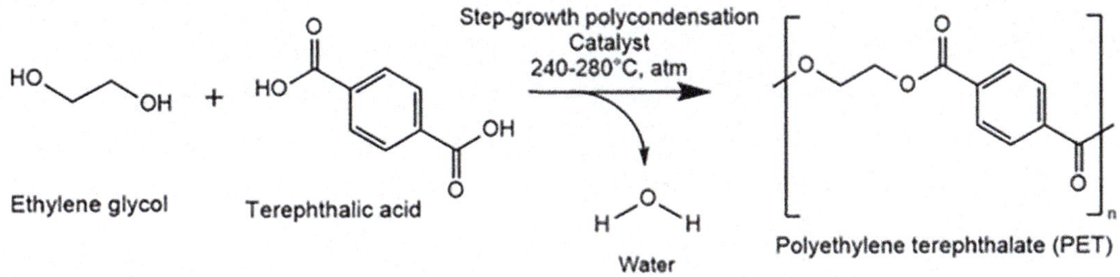

Figure 1.7
Chemical structure of ethylene glycol and terephthalic acid and its polycondensation to polyethylene terephthalate.

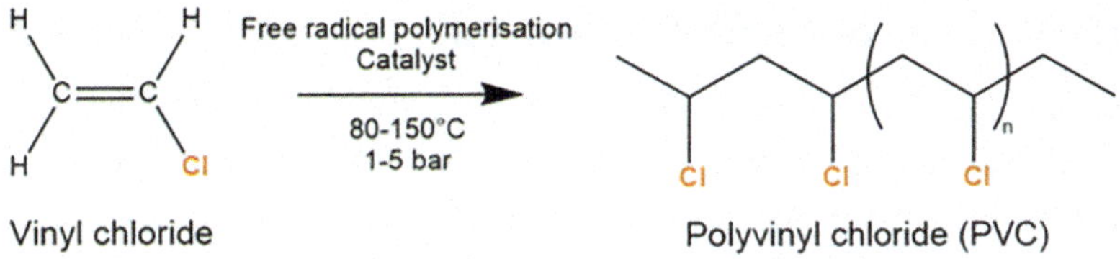

Figure 1.8
Chemical structure of vinyl chloride and its polymerization to poly(vinyl chloride).

1.2.5 Polystyrene

PS is an aromatic thermoplastic polymer synthesized from the polymerization of styrene monomers (C_8H_8), typically through free-radical initiation. PS can exist in atactic, isotactic, or syndiotactic forms; however, commercial PS is generally atactic and amorphous, which results in rigidity and brittleness. It is valued for its clarity, ease of molding, and low cost. PS is widely used in disposable food containers, cutlery, compact-disk cases, insulation materials (in its expanded form, EPS), and lightweight packaging. Despite its mechanically weak and brittle nature, PS had been a popular choice for single-use products and packaging due to its low production cost and favorable processing characteristics (Fig. 1.9).

1.3 Global plastic waste crisis and its impact

Plastic is synthetic organic polymer derived from fossil fuels, such as natural gas and petroleum. Plastic is used in nearly every consumer and industrial activity, from construction and transportation to electronics and agriculture. When plastics are improperly disposed, plastic pollution becomes a major driver of biodiversity loss and ecosystem degradation. It threatens human health, affects food and water safety, burdens economic activity, and worsens climate change.

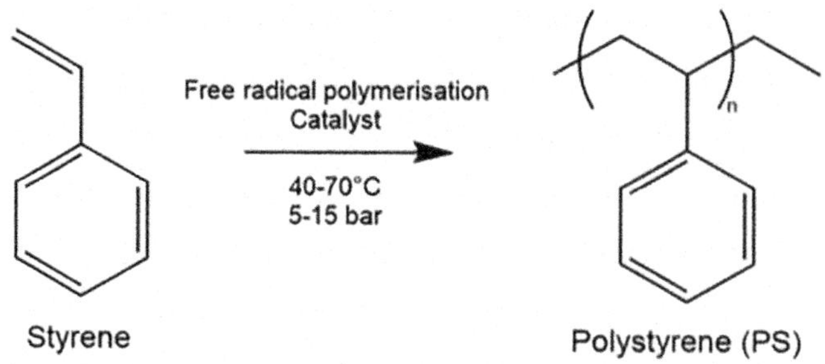

Figure 1.9
Chemical structure of styrene and its polymerization to polystyrene.

1.3.1 *Plastics in marine and terrestrial systems*

Terrestrial environments are increasingly recognized as major sinks for plastic pollution, largely because of plastic production, use, and disposal that predominantly occurs on land. Agricultural sectors contribute significantly to this issue, given widespread use of plastic materials such as mulch films, greenhouse coverings, wrappings, and packaging (Hurley et al., 2020). These items, once discarded, fragment into microplastics under environmental stressors such as UV radiation, temperature fluctuations, and mechanical abrasion. Microplastics can persist in soils for decades, posing risks to soil health, microbial biodiversity, and plant growth. Furthermore, improper waste management practices—particularly illegal dumping and open-air burning—exacerbate plastic accumulation in terrestrial landscapes. Transport mechanisms such as wind (aeolian processes), surface runoff, and human activities further redistribute plastic debris across diverse land ecosystems (Azevedo-Santos et al., 2021). In addition, plastics in terrestrial settings can leach toxic additives, including phthalates and bisphenol A (BPA), into surrounding soils and groundwater, raising concerns about long-term environmental and human health effects. These impacts underscore the need for improved waste regulation and sustainable alternatives in land-based sectors such as agriculture and urban development (Naqash et al., 2025). The marine environment receives an estimated 8 to 23 million tos of plastic waste annually (U.N. Environment Program, 2025). A significant proportion of this waste, around 80%, originates from land-based sources, primarily transported by rivers such as the Ganges, Indus, and Yangtze (Parker, 2021). In total, about 14 million tons of plastic are believed to enter the ocean each year, with one garbage truck's worth of plastic dumped into the sea every minute (Lau, 2021). Microplastic pollution further adds complexity to the plastic waste crisis. These particles arise from both primary sources, such as microbeads in cosmetics and synthetic fibers in textiles, and secondary sources generated through the fragmentation of larger plastic items. Microplastics typically enter marine environments through wastewater discharge and surface runoff. It is estimated that over 170 trillion plastic particles are currently circulating in the ocean, imposing an environmental cost of approximately US$13 billion per year (Geneva Environment Network, 2025).

Although larger plastic items—such as bags, bottle caps, and ropes—are more visible, microplastics are generally deemed more hazardous due to their extensive dispersal and potential to infiltrate marine ecosystems at the cellular level. Nevertheless, large plastic debris can pose its own ecological risks. It can entangle marine life, be mistakenly ingested, or become physically embedded in organisms, thereby disrupting feeding, mobility, and reproduction (Nelms et al., 2016; Thiel et al., 2018).

1.3.2 Plastics in the air pollution

Numerous studies have identified multiple sources of microplastics across diverse sectors such as packaging, construction, automotive, textile, medical, electronics, agriculture, and sports (Andrady, 2011; Brahney et al., 2021). Atmospheric microplastics consist primarily of carbon and oxygen, with minor trace elements (Gasperi et al., 2018). Synthetic polymer-based fibers, such as polyester, nylon, and acrylic constitute more than 90% of the airborne microplastics and fragment into microfibers more readily than natural fibers (Dris et al., 2016, 2017; Singh, 2024). Major emission pathways include waste-handling operations, clothing abrasion, industrial materials, tire wear, industrial emissions, combustion sources, and certain agricultural practices (Gasperi et al., 2018; Napper and Thompson, 2016; Sommer et al., 2018).

Recent research has shown that microplastics are not restricted to land and water, and they can be transported through the atmosphere. Atmospheric deposition of microplastics can contaminate remote and pristine environments such as the Himalayas (Yang et al., 2021) and the Western Italian Alps (Parolini et al., 2021). These records have shown that continental microplastics can travel over 1000 km into the atmosphere and influence the global atmospheric environment (Wang et al., 2020). In addition, the impact of this transport was highlighted by the discovery of significant fiber concentrations in outdoor places with heavy foot circulation on a university campus in coastal California. The results imply that tarpaulins and outdoor textiles play a major role in raising the concentration of certain microplastics, such PVC fibers in ambient air (Sridharan et al., 2022).

1.3.3 Impact of plastics in wildlife

Plastic pollution is one of the most urgent environmental challenges of our time. Beyond the visible accumulation of discarded plastic waste, a more profound and insidious threat lies in its pervasive impact on animal life across the planet. From microscopic plankton to large marine mammals such as whales, no group of wildlife remains unaffected by the plastic pollution. Across marine, freshwater, and terrestrial ecosystems, the consequences are even more serious and life-threatening. Many species ingest plastic debris, often mistaking it for food, which can result in internal injuries, gastrointestinal blockages, malnutrition, and, ultimately, death. Seabirds, turtles, whales, and fish are particularly vulnerable, with studies reporting that over 90% of seabirds have mistakenly ingested plastic (Wilcox et al., 2015). In addition to ingestion, entanglement in plastic waste—such as discarded fishing nets, six-pack rings, and packaging straps—poses a critical threat. These materials can cause suffocation, drowning, reduced mobility, and infections from open wounds (Gall and Thompson, 2015). Entanglement typically occurs when animals become trapped in loops or openings of plastic debris, preventing escape from predators or hindering natural behaviors such as migration.

Microplastics can bring additional risks to wildlife. These tiny plastic particles are easily ingested by lower-trophic organisms like plankton and subsequently accumulate up the food chain, affecting fish, marine mammals, and birds (Jeong et al., 2024). In addition, microplastics can also impair animal health. For example, in blue mussels, exposure to high-density PE (0–80 μm) caused histological changes and strong inflammatory responses, including granuloma formation and lysosomal membrane instability (Von Moos et al., 2012). In earthworms, PS particles (1–100 μm) induced oxidative stress and severe DNA damage (Xu et al., 2021). PVC microparticles (0.1–1000 μm) triggered oxidative damage in the brain and liver and decreased expression of growth-hormone receptor and binding-protein genes in juvenile goldfish (Romano et al., 2020). These outcomes underscore the extensive toll that plastic pollution takes on individual health, species survival, and the overall stability of ecosystems.

1.3.4 *Impact of plastics in human health*

Plastic waste, particularly in the form of microplastics and nanoplastics, poses a complex and emerging threat to human health due to its pervasive distribution and bioavailability across ecosystems through multiple exposure routes. Several reports provide evidence of human ingestion of microplastics through diet. Microplastics are now routinely found in terrestrial and aquatic environments and have entered the food chain, drinking water supplies, and even atmospheric dust (Cox et al., 2019; Gasperi et al., 2018). A recent study estimated that humans may ingest tens of thousands of microplastic particles annually, with higher exposure among individuals who consume bottled water or seafood (Senathirajah et al., 2021). More alarmingly, microplastics have been detected in human blood, lungs, feces, and placenta (Amato-Lourenço et al., 2021; Leslie et al., 2022; Ragusa et al., 2021; Schwabl et al., 2019), suggesting their widespread internal exposure and dangerous-level potential for bioaccumulation.

According to estimates, individuals following a typical American diet may ingest between 74,000 and 121,000 microplastic particles each year, and those who regularly drink bottled water could take in up to 90,000 additional particles annually (Goyal et al., 2023). Remarkably, brewing a single plastic teabag at 95°C can release approximately 11.6 billion microplastics and 3.1 billion nanoplastics per cup (Hernandez et al., 2019). In addition, microplastic contamination has been detected in alcoholic beverages—for example, white wine contained 2563–5857 particles per liter originating from bottle stoppers (Prata et al., 2020), while sparkling water showed contamination levels of 40 ± 19 particles per liter attributed to bottle packaging (Kankanige and Babel, 2020).

Health risks associated with plastics stem from both the physical characteristics and the chemical toxicity of these particles. Physically, micro- or nanoplastics may cause cellular damage, immune disruption, and inflammation by interacting with gut epithelium, alveolar tissue, and other organs (Wright and Kelly, 2017). Chemically, plastics contain various additives such as plasticizers (e.g., phthalates), brominated flame retardants, and stabilizers that can leach into biological systems. Many of these additives have known endocrine-disrupting chemistries that interfere with hormone function, potentially leading to metabolic disorders, infertility, developmental defects, and certain cancers (Talsness et al., 2009). Moreover, plastics are highly capable of absorbing persistent organic pollutants such as polychlorinated biphenyls, polycyclic aromatic hydrocarbons, and heavy metals from their surrounding environment, and these toxic substances may be released into human tissues when ingested (Teuten et al., 2009). Tiny fibers in the deep-lung tissues (Enyoh et al., 2019), and/or microplastics, can easily travel through the bloodstream to lymph nodes and lead to major health issues such lung cancer, asthma-like syndrome, coughing, chest tightness, recurrent fever, and body aches (Wang et al., 2021).

The small size of nanoplastics further increases their hazard potential. Unlike larger microplastics, nanoplastics can translocate across cell membranes, passage through the blood–brain barrier, and accumulate in neural tissues, raising serious concerns about neurotoxicity and cognitive impairment (Mattsson et al., 2017). Since the data on direct human exposure to microplastics is unavailable, scientists used animal models for this purpose and have shown behavioral changes, neuroinflammation, and oxidative stress after exposure to nanoplastics. In addition, plastics have the potential to interfere with the gut microbiome, which is essential for digestion, immunity, and mental well-being. Research involving mice has shown that exposure to nanoplastics can alter the composition and function of gut bacteria (Jin et al., 2018).

Vulnerable populations such as children, pregnant women, and workers in plastic manufacturing, waste management, and textiles face increased risks because of their developing biological systems and/or occupational exposures to airborne particles and plastic-associated chemicals (Prata, 2018). While human epidemiological studies are still in early stages, the toxicological evidence from in vitro and in vivo models underscores the need for urgent preventive action. Reducing plastic production, eliminating hazardous additives, promoting safer alternatives, and improving global waste management systems are critical steps toward mitigating health risks associated with plastic pollution.

1.3.5 Economic and esthetic impacts of plastics

Plastic pollution generates substantial economic costs and causes visible esthetic degradation, both of which affect communities, industries, and ecosystems on a global scale. Economically, plastic debris imposes direct financial burdens through increased waste management costs, reduced tourism revenue, and losses in fisheries. For instance, clean-up operations on European beaches cost governments millions of euros annually, with the United Kingdom alone spending an estimated £18 million per year on coastal cleanups (Newman et al., 2015). Marine plastic debris damages fishing gear, reduces catch quality, and fishing time, producing significant losses for the commercial fishing sector (Richardson et al., 2019). Abandoned, lost, or discarded fishing gear—often plastic-based—poses an additional threat through "ghost fishing," which leads to ongoing environmental damage and economic loss (Macfadyen et al., 2009). Plastic blockages in storm-water systems also cause urban flooding and costly infrastructure damage, particularly in low-income countries with inadequate waste disposal systems (Alam and Ahmade, 2013; Mwanza and Mbohwa, 2017). Seafood safety can be adversely affected when plastics contaminate marine resources, reducing consumer demand and leading to economic losses. For instance, more than 1750 tons of plastic were released into the Indian ocean by the 2002 MV X-Press Pearl shipwreck, and tainted plastic particles were discovered in the gills of seafood species (Perera et al., 2022). Sri Lanka suffered financial losses as a result of the sharp decline in seafood consumption (Ratnayake, 2023).

Beyond these economic impacts, the esthetic consequences of plastic pollution are pronounced and widespread. Accumulations of plastic litter in public parks, rivers, coastlines, and urban streets diminish the scenic and recreational value of these areas. Studies show that people are less likely to visit or feel satisfied with areas heavily affected by plastic litter, even when other natural features are present (Wyles et al., 2016). Littered environments have also been linked to a decrease in property values and overall life satisfaction among residents (Beaumont et al., 2019). In tourism-dependent economies, visual pollution from plastic debris directly affects visitor experience and perception, and revenue (Jang et al., 2014). These findings highlight not only the visible degradation caused by plastic waste but also its indirect effects on mental well-being, social identity, and economic opportunity. Together, the esthetic and economic impacts of plastic pollution reinforce the urgency of global policy interventions, sustainable product design, and improved waste governance to protect both environmental quality and human livelihoods.

1.4 Plastics in food packaging

1.4.1 Type of food packaging

Plastic remains fundamental to modern food preservation owing to its low permeability to gases and moisture, mechanical strength, and chemical resistance. There are four types of packaging products that are used to protect foodstuff. Commonly plastics are selected based on their specific performance characteristics in relation to food type and storage conditions (Velásquez et al., 2021). Recent advances have enabled the incorporation of active components—such as antimicrobials, oxygen scavengers, and ethylene absorbers—into the polymer matrix to inhibit spoilage and extend shelf life in perishable foods like meat, seafood, and fresh produce (Ceballos-Santos et al., 2024; Gaikwad et al., 2020). However, there is always a chance these functional substances migrate into food. Additives such as BPA (Khalili Sadrabad et al., 2023), phthalates (Alp and Yerlikaya, 2020), and nonintentionally added substances have been shown to leach from packaging into food matrices, particularly under conditions involving heat, acidity, or high fat content (Kato and Conte-Junior, 2021). Understanding the migration kinetics of these additives from plastics is essential for evaluating chronic dietary exposure and ensuring toxicological safety (Muzeza et al., 2023). Current research thus should prioritize the development of safer packaging formulations, advanced analytical detection methods, and substitution with bioinert or biodegradable alternatives (Jafarzadeh et al., 2024).

Glass packaging is widely regarded for its exceptional chemical stability and barrier properties. Due to its amorphous and inert silica-based structure, glass does not leach harmful substances or react with food contents. Such character makes it particularly suitable for the storage of acidic, alcoholic, or highly sensitive products such as fruit juices, fermented pickles, baby food, and wine. Glass is impermeable to oxygen, carbon dioxide, water vapor, and other volatile compounds and less likely to interact with the external environment, thereby preserving the organoleptic properties, nutritional value, and shelf life of food products. Furthermore, glass is highly resistant to thermal shock and can withstand sterilization processes at high temperatures, which make it ideal for pasteurized or retorted products (Jan et al., 2022). However, despite these advantages, glass is fragile and prone to breakage during handling and transport and heavy.

Metal packaging, such as aluminum and tinplate, is widely used for canned goods, carbonated beverages, and heat-treated products due to its robustness, opacity, and excellent barrier properties against light, oxygen, and microorganisms. Aluminum cans are lightweight and highly recyclable, whereas tin-coated steel is suited for preserving nutrient-dense foods under thermal processing (Balo and Sua, 2023). However, corrosion and potential metal leaching in acidic media necessitate internal coatings or lacquers, which can complicate recycling and raise health concerns if degraded (Akram et al., 2023).

Paper and cardboard packaging are commonly used for dry foods such as cereals, flour, and tea, as well as for outer cartons and labels. They are renewable, biodegradable, and cost-effective, and can be easily printed for branding and information. However, their inherent hydrophilicity limits their protective functions unless laminated or coated with synthetic barriers (Adibi et al., 2023). These coatings, often made of plastic or wax, reduce the recyclability and biodegradability of the final product. Despite this, paper-based packaging continues to play a major role in secondary and tertiary packaging, especially in combination with other materials.

1.4.2 Function of plastics for food packaging

Food packaging is a vital element of the modern food supply chain, fulfilling a range of technical functions including protection, containment, and preservation. It serves as a barrier against environmental and mechanical stressors such as oxygen, moisture, light, and microbial contamination, thereby maintaining food quality and extending shelf life (Han, 2005; Robertson, 2005). Innovations such as modified atmosphere packaging (MAP) and multilayer plastic films exemplify this role by reducing spoilage and minimizing exposure to contaminants (Coles et al., 2003). Containment ensures the structural integrity and hygienic transport of various food forms, particularly liquids and powders (Singh and Heldman, 2001). Beyond the passive protective role of plastics, active packaging systems have been developed to enhance microbial safety. For example, antimicrobial films containing compounds like sorbic acid and glycerol monooleate inhibit microbial growth while preserving nutritional value (Corrales et al., 2014). Additionally, UV radiation is a significant factor in the degradation of photosensitive products, contributing to color fading, oxidation, and nutrient loss. To address this, recent research has focused on UV-blocking packaging materials. Chitosan-lignin films have demonstrated strong potential for blocking UV radiation while increasing mechanical integrity (Izaguirre et al., 2020). PP films coated with benzotriazole-based polymer have been developed to meet industrial demands for durability and UV resistance (Cohen et al., 2017). Recent advances have also introduced biodegradable biopolymer films incorporating conjugated fatty acids and polyphenols, offering broad-spectrum UV absorption (Ezati et al., 2023). Similar breakthroughs to increase UV shielding and water resistance have been achieved by reinforcing chitosan films with lignin nanoparticles (Zhang et al., 2023). In addition, plant-derived UV absorbers are being explored as sustainable alternatives for packaging applications (Alizadeh Sani et al., 2024). Collectively, these innovations signal a growing shift toward multifunctional, environmentally responsible materials that support both food preservation and ecological sustainability.

The communication function of food packaging is critical for delivering information about product transparency, regulatory compliance, and consumer safety. Packaging serves as a medium through which essential details—such as product identity, nutritional composition, ingredient lists, allergen warnings, storage instructions, manufacturing and expiry dates, and traceability codes—are conveyed to both consumers and supply chain stakeholders (Marsh and Bugusu, 2007; Grunert et al., 2010). This information not only facilitates informed purchasing decisions but also plays a vital role in managing health risks and dietary preferences. In regulated markets, packaging must comply with strict labeling laws to support public health initiatives and reduce foodborne illnesses. Beyond mandatory labeling, modern packaging often includes smart communication technologies, including barcodes, QR codes, and RFID tags to enhance traceability and inventory management throughout the supply chain (Ahari and Soufiani, 2021). In the context of consumer psychology, visual and textual elements on packaging—including color, typography, imagery, and eco-labels—also influence brand perception, perceived product quality, and environmental values (Becker et al., 2011; Silayoi and Speece, 2007). As such, packaging functions not merely as a passive container but as an interactive platform that bridges producers, retailers, and consumers with accurate, timely, and persuasive information.

Recent innovations in smart food packaging have substantially advanced real-time product monitoring, microbial safety, and environmental sustainability. One notable development is a battery-free, autonomous system that incorporates gas sensors with near-field communication. This platform continuously assessed freshness and could extend the shelf life of fish products by up to 14 days through controlled release of active compounds (Douaki et al., 2025). Intelligent packaging systems equipped with sensors, indicators, and RFID technology have further enhanced spoilage detection by tracking environmental conditions and microbial activity (Mkhari et al., 2025). In parallel, the integration of biosensing capabilities into biodegradable, plant-based materials offer a dual benefit: reducing dependency on fossil-based plastics while maintaining effective freshness monitoring (Baneshi et al., 2024). Collectively, these developments illustrate a paradigm shift toward multifunctional, eco-intelligent packaging systems that are designed to enhance food safety, reduce waste, and are aligned with circular economy principles.

1.4.3 Food packaging waste

World plastic production reached 413.8 million tons in 2023, and packaging is the largest end-use market for plastics, with an approximately 40% share (Plastics Europe, 2024a). Within this, 35% of all plastic packaging is used in food applications. The number reflects the scale and importance of plastics in food packaging for preservation, transportation, and safety. Unfortunately, most food packing are designed to be single use. Some of the major single-use plastics in the food sector are drinking water bottles, bottle caps, food wrappers, grocery bags, lids, straws, stirrers, and take-away containers.

Consumer behaviors strongly influence this trend. Demand for convenience, time-saving meals, and ready-to-eat foods has driven the use of individually wrapped and single-serving packages (Ncube et al., 2021). Online delivery services contribute to plastic waste because of using these single-use items. Heightened hygiene concerns during the COVID-19 pandemic further accelerated the production of single-use packaging items (Oliveira et al., 2021). These patterns, reinforced by a fast-paced urban lifestyles and a growing middle class worldwide, contribute to the proliferation of packaging waste. Public schools in the United States generate about 42% of the daily solid waste as food packaging (Heiges et al., 2022). A study across educational levels from elementary to university led to the development of the WASTE tool, a standardized audit instrument designed to consistently assess food packaging waste and support efforts to reduce environmental impact.

A large share of household food waste contributes to plastic packaging waste in which most food items are purchased (Bishop et al., 2021). Many of these food packaging become waste in the same year as they entered the market. Today, 95% of plastic packaging material value worth US$80–120 billion is lost after a short first use (Ellen MacArthur Foundation and McKinsey & Company, 2016). The short service life, high production volume, and low recyclability of plastic packaging create major solid waste challenges worldwide (Kan and Miller, 2022).

The composition of plastics used in food packaging is highly complex, comprising both monomaterial plastics and multilayered structures that combine polymers, metals, and paper. According to European Analysis, almost 20% of flexible packaging consists of multilayer films. These multilayer laminates are often used for temperature control, extended shelf life, and product protection. They vary in recyclability and are either nonrecyclable or economically unfeasible to recover in most municipal systems (Zero Waste Europe, 2021). Thus, most of these packaging ends up in landfills or incinerators.

Another growing concern for plastic packaging is the contamination of packaging waste with food residues, which complicates collection, sorting, and processing (Hemachandra et al., 2024). Contaminated packaging is often considered unsuitable for recycling and is directed to landfilling or incineration. In many regions, particularly in developing countries, the absence of effective waste separation and recycling infrastructure exacerbates the issue. These operational barriers severely limit the potential for a circular economy in food packaging, despite increasing public awareness and policy pressure (Ellen MacArthur Foundation and McKinsey & Company, 2016).

1.5 Management of food packaging plastic waste

The waste hierarchy emphasizes reduction, reuse, and then recycling before other waste treatments and disposal. However, a major problem with plastic food packaging waste is its low recyclability. This leads to about 90% of the food packaging reaching landfills or being incinerated (Chakori et al., 2021). The estimated annual loss in the value of plastic packaging waste during sorting and processing alone is US$80–120 billion (Environment, 2021).

1.5.1 Recycling

Recycling is a cornerstone of plastic waste management strategies, but its effectiveness in the context of food packaging is limited. Even after 40 years since the launch of the first universal recycling symbol, only 14% of plastic packaging waste is collected for recycling. When losses during sorting and reprocessing are considered, just 5% of the material's value is retained for subsequent use. As a result, only 9%–10% of plastic waste is successfully recycled globally (OECD, 2022). There are two types of recycling methods: mechanical recycling and chemical recycling. Mechanical recycling is the most widely used method and requires clean and uncontaminated input streams that are shredded and melted into pellets for reuse in new plastic products (Hopewell et al., 2009). However, this criterion is rarely met by food packaging waste due to food residue, complex multilayer materials, and labeling adhesives.

Emerging technologies, such as chemical recycling and enzymatic depolymerization, offer better pathways for recovering value from hard-to-recycle plastics. In these methods, plastic waste is depolymerized under controlled temperatures into its monomeric components or converted into new chemicals (Damayanti et al., 2022). These approaches are still at early commercial stages and are often energy intensive. In many cases, the cost of recovering and processing packaging waste outweighs the market value of the recycled material, discouraging investment in infrastructure (Foundation, 2016).

Efforts to improve recycling rates include the introduction of extended producer responsibility (EPR) schemes and better design for recyclability, and clearer labels. In 2022, a total of 43.7 million tons of fossil-based plastics and 10.4 million tons of circular plastics, those made from recycled or renewable materials, were converted into new plastic products. This reflects a growing shift toward circular feedstocks and a more than 70% increase in the use of postconsumer recycled content since 2018 (Plastics Europe, 2024a). With continued progress supported by governmental policies, improved collection systems, contamination control, and improved economic incentives for recycling, the role of recycling in managing food packaging waste can be meaningfully strengthened.

1.5.2 Landfilling

Landfilling is the most common disposal route for plastic packaging waste worldwide, particularly in regions where recycling infrastructures are limited. Plastics that are contaminated with food, composed of multiple materials, or economically nonviable for recycling typically end up in landfills. In some countries, landfill remains the default waste management strategy due to lower upfront costs and limited regulatory restrictions

(Silpa Kaza et al., 2018). Landfilling prevents immediate environmental exposure, but it does not eliminate long-term risks. Plastics in landfills can take centuries to degrade, during which they may leach harmful additives or break down into microplastics that infiltrate soil and groundwater. The space burden of landfills, especially in urban areas, and their greenhouse gas emissions further diminish their sustainability (Huang et al., 2022; Pan et al., 2020). Some governments have implemented landfill taxes or bans on certain types of plastic packaging to encourage reduction and recovery, but such enforcements remain inconsistent (OECD, 2024). Without robust policy and investment, reliance on landfilling will continue to undermine sustainability goals.

1.5.3 Incineration

Incineration, including waste-to-energy systems, offers an alternative disposal pathway by converting waste into energy. This method reduces the volume of waste to 90% and yields useful energy (Roberts et al., 2022). However, combustion of plastic releases greenhouse gases and can emit hazardous pollutants such as dioxins and furans, especially when waste is improperly sorted or incinerators lack adequate filtration technology. In some regions, incineration is embraced as a pragmatic solution to manage nonrecyclable waste and generate power. Countries like Sweden and Japan use it extensively under strict environmental regulations (Roberts et al., 2022). However, in many other countries, incineration is criticized for discouraging recycling and perpetuating the production of disposable packaging (Europe Annual Report, 2021). From a food packaging standpoint, incineration may handle composite and contaminated materials more effectively than recycling, but it does not align with circular economy principles. Moreover, its infrastructure-intensive nature and potential negative health impacts mean it must be approached cautiously and transparently.

1.5.4 Emerging approaches

In response to the limitations of traditional waste management methods, several innovative and systemic strategies are being adopted to address plastic food packaging waste. There has been a clear shift toward systemic, collaborative, and innovation-driven approaches to redesign, reuse, and improve recycling of plastics, supported by stronger policy frameworks and industry commitments (Ellen MacArthur Foundation and McKinsey & Company, 2016). Emerging approaches such as zero waste models, circular economy initiatives, EPR, and deposit–return schemes are gaining attention globally (Europe Annual Report, 2021; OECD, 2022).

Zero waste models aim to redesign products and systems to eliminate waste altogether, focusing on reuse, refill, and closed-loop systems. Circular economy initiatives keep the materials in use for as long as possible through improved design, recycling, and resource recovery. EPR policies shift the financial and operational responsibility for postconsumer packaging waste onto producers, encouraging them to design more recyclable and less wasteful packaging. Deposit refund schemes incentivize consumers to return packaging such as bottles or containers by offering a small monetary refund, significantly improving return and recycling rates. These emerging approaches reflect a broader shift toward integrated, upstream solutions that target the root causes of packaging waste and promote long-term resource efficiency. In addition to systemic and regulatory measures, there are other approaches to recognize the role of individual behavior in reducing plastic packaging waste. Behavioral interventions including social norm messaging, habit-disrupting nudges, and contextual reminders can support upstream waste prevention and complement policies like EPR and deposit–return systems.

A key study by Salazar et al. (2021) tested various normative messages at a marine park in Portugal to reduce the demand for plastic straws. The intervention included descriptive, injunctive, moral, and combined messages, which were either positive or negative. Surprisingly, the most effective outcome was obtained through a negative injunctive message: "Choose not to use straws with your drink. The planet thanks you!" This message reduced straw use by 38% compared to a control group, although the result was not statistically significant. Other interventions focus on the timing and placement of reminders. For example, stickers placed on reusable bags at the point of action significantly increased the return rate of plastic bags by 83% compared to settings without reminders. Similarly, programs that provide students with reusable water bottles have been shown to reduce the use of disposable bottles and increase support for campus bans on bottled water.

These findings suggest that behaviorally informed strategies can help disrupt default behaviors associated with plastic packaging waste. While such interventions can be impactful on their own, their effectiveness is likely enhanced when combined with broader policy tools such as bans, levies, and incentives.

1.6 Future of plastic food packaging

The escalating plastic pollution crisis poses a significant threat to environmental and human health, primarily due to the persistent reliance on fossil-based polymers. Bioplastics are increasingly proposed as a sustainable alternative to conventional plastics, particularly in sectors like food packaging where single-use materials are dominant. However, the large-scale adoption of bioplastics faces several challenges. Many commercially available bioplastics require high-energy inputs for production and often degrade only under controlled industrial composting conditions. Without proper infrastructure, even biodegradable plastics can accumulate in the environment and generate microplastics. Additionally, bioplastics currently remain more costly than petroleum-based alternatives, which limits their competitiveness in the global market.

Nonetheless, advances in microbial fermentation, metabolic-pathway engineering, and the use of agricultural waste as feedstock have begun to address these limitations. Such innovations are improving the scalability, functionality, and environmental performance of next-generation bioplastics. As fossil fuel prices rise and environmental regulations tighten, the economic and ecological advantages of bioplastics are becoming more apparent. To support their responsible use in food systems, comprehensive research is required to assess bioplastics' environmental fate, food safety, and long-term toxicity. Effective integration also depends on systemic reform, including improved labeling, consumer education, and investment in composting and recycling infrastructure; however, with continued innovation, bioplastics have the potential to significantly reduce environmental burdens and support global climate goals.

1.7 Conclusion

Plastics remain the preferred material for consumer packaging solutions and are the most widely produced materials globally across various application markets. They are extensively used in the packaging industry, particularly in the food sector, where most packaging is disposable. After a short period of use, these plastics often end up in the environment, where they can harm both animals and humans by polluting land and water systems and even entering the food chain. In response, recent advancements in packaging science have focused on the development of innovative materials aimed at extending product shelf life and preserving nutritional quality while mitigating environmental impact. A radical shift is currently underway to transform from conventional plastics to biodegradable and bio-based alternatives. Despite progress, substantial technological and regulatory challenges persist in achieving packaging systems that effectively reduce food waste, enhance product safety, and align with global sustainability goals.

References

Adibi, A., Trinh, B.M., Mekonnen, T.H., 2023. Recent progress in sustainable barrier paper coating for food packaging applications. Progress in Organic Coatings 181, 107566. Available at https://doi.org/10.1016/j.porgcoat.2023.107566.

Ahari, H., Soufiani, S.P., 2021. Smart and active food packaging: insights in novel food packaging. Frontiers in Microbiology 12, 657233. Available at https://doi.org/10.3389/fmicb.2021.657233.

Akram, N., Saeed, M., Mansha, A., Bokhari, T.H., Ali, A., 2023. Metal packaging for food items advantages, disadvantages and applications. Green Sustainable Process for Chemical and Environmental Engineering and Science: Methods for Producing Smart Packaging. Elsevier, Pakistan, pp. 129–141.

Alam, P., Ahmade, K., 2013. Impact of solid waste on health and the environment. International Journal of Sustainable Development and Green Economics (IJSDGE) 2 (1), 165–168.

Alizadeh Sani, M., Khezerlou, A., Tavassoli, M., Abedini, A.H., McClements, D.J., 2024. Development of sustainable UV-screening food packaging materials: a review of recent advances. Trends in Food Science and Technology 145, 104366. Available at https://doi.org/10.1016/j.tifs.2024.104366.

Alp, A.C., Yerlikaya, P., 2020. Phthalate ester migration into food: effect of packaging material and time. European Food Research and Technology 246 (3), 425–435. Available at https://doi.org/10.1007/s00217-019-03412-y.

Amato-Lourenço, L.F., Carvalho-Oliveira, R., Júnior, G.R., dos Santos Galvão, L., Ando, R.A., Mauad, T., 2021. Presence of airborne microplastics in human lung tissue. Journal of Hazardous Materials 416, 126124. Available at https://doi.org/10.1016/j.jhazmat.2021.126124.

Andrady, A.L., 2011. Microplastics in the marine environment. Marine Pollution Bulletin 62 (8), 1596–1605. Available at https://doi.org/10.1016/j.marpolbul.2011.05.030.

Azevedo-Santos, V.M., Brito, M.F.G., Manoel, P.S., Perroca, J.F., Rodrigues-Filho, J.L., Paschoal, L.R.P., et al., 2021. Plastic pollution: a focus on freshwater biodiversity. Ambio 50 (7), 1313–1324. Available at https://doi.org/10.1007/s13280-020-01496-5.

Balo, F., Sua, L.S., 2023. Chapter 16 - Evaluating the sustainable metal packaging for cooked foods among food packaging materials. In: Inamuddin, Altalhi, T., Cruz, J.N. (Eds.), Green Sustainable Process for Chemical and Environmental Engineering and Science: Methods for Producing Smart Packaging. Elsevier, Turkey, pp. 283–302. Available at https://doi.org/10.1016/B978-0-323-95644-4.00004-8.

Baneshi, M., Aryee, A.N.A., English, M., Mkandawire, M., 2024. Designing plant-based smart food packaging solutions for prolonging consumable life of perishable foods. Food Chemistry Advances 5, 100769. Available at https://doi.org/10.1016/j.focha.2024.100769.

Beaumont, N.J., Aanesen, M., Austen, M.C., Börger, T., Clark, J.R., Cole, M., et al., 2019. Global ecological, social and economic impacts of marine plastic. Marine Pollution Bulletin 142, 189–195. Available at https://doi.org/10.1016/j.marpolbul.2019.03.022.

Becker, L., van Rompay, T.J.L., Schifferstein, H.N.J., Galetzka, M., 2011. Tough package, strong taste: the influence of packaging design on taste impressions and product evaluations. Food Quality and Preference 22 (1), 17–23. Available at https://doi.org/10.1016/j.foodqual.2010.06.007.

Berthelot, M.P.E., 1911. Encyclopædia Britannica 3, 811.

Bishop, G., Styles, D., Lens, P.N.L., 2021. Environmental performance of bioplastic packaging on fresh food produce: a consequential life cycle assessment. Journal of Cleaner Production 317, 128377. Available at https://doi.org/10.1016/j.jclepro.2021.128377.

Brahney, J., Mahowald, N., Prank, M., Cornwell, G., Klimont, Z., Matsui, H., et al., 2021. Constraining the atmospheric limb of the plastic cycle. Proceedings of the National Academy of Sciences of the United States of America 118 (16). Available at https://doi.org/10.1073/pnas.2020719118.

Ceballos-Santos, S., de Sousa, D.B., García, P.G., Laso, J., Margallo, M., Aldaco, R., 2024. Exploring the environmental impacts of plastic packaging: a comprehensive life cycle analysis for seafood distribution crates. Science of the Total Environment 951, 175452. Available at https://doi.org/10.1016/j.scitotenv.2024.175452.

Chakori, S., Aziz, A.A., Smith, C., Dargusch, P., 2021. Untangling the underlying drivers of the use of single-use food packaging. Ecological Economics 185, 107063. Available at https://doi.org/10.1016/j.ecolecon.2021.107063.

Cohen, S., Haham, H., Pellach, M., Margel, S., 2017. Design of UV-absorbing polypropylene films with polymeric benzotriaziole based nano- and microparticle coatings. ACS Applied Materials & Interfaces 9 (1), 868–875. Available at https://doi.org/10.1021/acsami.6b12821.

Coles, R., McDowell, D., Kirwan, M.J., 2003. Food Packaging Technology, Vol. 5. CRC Press, Boca Raton.

Corrales, M., Fernández, A., Han, J.H., 2014. Chapter 7 - Antimicrobial packaging systems. In: Han, J.H. (Ed.), Innovations in Food Packaging, second ed. Academic Press, pp. 133–170. Available at https://doi.org/10.1016/B978-0-12-394601-0.00007-2.

Cox, K.D., Covernton, G.A., Davies, H.L., Dower, J.F., Juanes, F., Dudas, S.E., 2019. Human consumption of microplastics. Environmental Science and Technology 53 (12), 7068–7074. Available at https://doi.org/10.1021/acs.est.9b01517.

Damayanti, D., Saputri, D.R., Marpaung, D.S.S., Yusupandi, F., Sanjaya, A., Simbolon, Y.M., et al., 2022. Current prospects for plastic waste treatment. Polymers 14 (15), 3133. Available at https://doi.org/10.3390/polym14153133.

Douaki, A., Ahmed, M., Longo, E., Windisch, G., Riaz, R., Inam, S., et al., 2025. Battery-free, stretchable, and autonomous smart packaging. Advanced Science 12 (22), e2417539. Available at https://doi.org/10.1002/advs.202417539.

Dris, R., Gasperi, J., Saad, M., Mirande, C., Tassin, B., 2016. Synthetic fibers in atmospheric fallout: a source of microplastics in the environment? Marine Pollution Bulletin 104 (1–2), 290–293. Available at https://doi.org/10.1016/j.marpolbul.2016.01.006.

Dris, R., Gasperi, J., Mirande, C., Mandin, C., Guerrouache, M., Langlois, V., et al., 2017. A first overview of textile fibers, including microplastics, in indoor and outdoor environments. Environmental Pollution 221, 453–458. Available at https://doi.org/10.1016/j.envpol.2016.12.013.

Ellen MacArthur Foundation and McKinsey & Company, 2016. The New Plastics Economy — Rethinking the future of plastics. World Economic Forum, Ellen MacArthur Foundation and McKinsey & Company.

Enyoh, C.E., Verla, A.W., Verla, E.N., Ibe, F.C., Amaobi, C.E., 2019. Airborne microplastics: a review study on method for analysis, occurrence, movement and risks. Environmental Monitoring and Assessment 191 (11), 668. Available at https://doi.org/10.1007/s10661-019-7842-0.

Ezati, P., Khan, A., Priyadarshi, R., Bhattacharya, T., Tammina, S.K., Rhim, J.W., 2023. Biopolymer-based UV protection functional films for food packaging. Food Hydrocolloids 142, 108771. Available at https://doi.org/10.1016/j.foodhyd.2023.108771.

Gaikwad, K.K., Singh, S., Negi, Y.S., 2020. Ethylene scavengers for active packaging of fresh food produce. Environmental Chemistry Letters 18, 269–284.

Gall, S.C., Thompson, R.C., 2015. The impact of debris on marine life. Marine Pollution Bulletin 92 (1–2), 170–179. Available at https://doi.org/10.1016/j.marpolbul.2014.12.041.

Gasperi, J., Wright, S.L., Dris, R., Collard, F., Mandin, C., Guerrouache, M., et al., 2018. Microplastics in air: are we breathing it in? Current Opinion in Environmental Science and Health 1, 1–5. Available at https://doi.org/10.1016/j.coesh.2017.10.002.

Geneva Environment Network, 2025. Plastics and Biodiversity. Available at https://www.genevaenvironment network.org/resources/updates/plastics-and-biodiversity.

Gilbert, M., 2017. Chapter 1 - Plastics Materials: Introduction and Historical Development. In: Gilbert, M. (Ed.), Brydson's Plastics Materials, eighth ed. Butterworth-Heinemann, pp. 1–18. Available at https://doi.org/10.1016/b978-0-323-35824-8.00001-3.

Goyal, T., Singh, S., Das Gupta, G., Verma, S.K., 2023. Microplastics in environment: a comprehension on sources, analytical detection, health concerns, and remediation. Environmental Science and Pollution Research 30 (54), 114707–114721. Available at https://doi.org/10.1007/s11356-023-30526-4.

Grunert, K.G., Wills, J.M., Fernández-Celemín, L., 2010. Nutrition knowledge, and use and understanding of nutrition information on food labels among consumers in the UK. Appetite 55 (2), 177–189. Available at https://doi.org/10.1016/j.appet.2010.05.045.

Han, J.H., 2005. Innovations in Food Packaging, second ed. Academic Press. Available at https://doi.org/10.1016/C2011-0-06876-X.

Heiges, J., Lee, D.L., Vollmer, L., Wobbekind, K., Thompson, H.R., Gosliner, W., et al., 2022. Evaluating food packaging waste in schools: a systematic literature review. International Journal of Environmental Research and Public Health 19 (9), 506. Available at https://doi.org/10.3390/ijerph19095607.

Hemachandra, S., Hadjikakou, M., Pettigrew, S., 2024. A scoping review of food packaging life cycle assessments that account for packaging-related food waste. The International Journal of Life Cycle Assessment 29 (10), 1899–1915. Available at https://doi.org/10.1007/s11367-024-02349-z.

Hernandez, L.M., Xu, E.G., Larsson, H.C.E., Tahara, R., Maisuria, V.B., Tufenkji, N., 2019. Plastic teabags release billions of microparticles and nanoparticles into tea. Environmental Science and Technology 53 (21), 12300–12310. Available at https://doi.org/10.1021/acs.est.9b02540.

Hopewell, J., Dvorak, R., Kosior, E., 2009. Plastics recycling: challenges and opportunities. Philosophical Transactions of the Royal Society B: Biological Sciences 364 (1526), 2115–2126. Available at https://doi.org/10.1098/rstb.2008.0311.

Huang, J., Veksha, A., Chan, W.P., Giannis, A., Lisak, G., 2022. Chemical recycling of plastic waste for sustainable material management: a prospective review on catalysts and processes. Renewable and Sustainable Energy Reviews 154, 111866. Available at https://doi.org/10.1016/j.rser.2021.111866.

Hurley, R., Horton, A., Lusher, A., Nizzetto, L., 2020. Chapter 7 - Plastic waste in the terrestrial environment. In: Letcher, T.M. (Ed.), Plastic Waste and Recycling: Environmental Impact, Societal Issues, Prevention, and Solutions. Academic Press, pp. 163–193. Available at https://doi.org/10.1016/B978-0-12-817880-5.00007-4.

Izaguirre, N., Gordobil, O., Robles, E., Labidi, J., 2020. Enhancement of UV absorbance and mechanical properties of chitosan films by the incorporation of solvolytically fractionated lignins. International Journal of Biological Macromolecules 155, 447–455. Available at https://doi.org/10.1016/j.ijbiomac.2020.03.162.

Jafarzadeh, S., Yildiz, Z., Yildiz, P., Strachowski, P., Forough, M., Esmaeili, Y., et al., 2024. Advanced technologies in biodegradable packaging using intelligent sensing to fight food waste. International Journal of Biological Macromolecules 261 (Pt 1), 129647. Available at https://doi.org/10.1016/j.ijbiomac.2024.129647.

Jambeck, J.R., Geyer, R., Wilcox, C., Siegler, T.R., Perryman, M., Andrady, A., et al., 2015. Plastic waste inputs from land into the ocean. Science 347 (6223), 768–771. Available at https://doi.org/10.1126/science.1260352.

Jan, B., Rizvi, Q.ul eain H., Shams, R., Dar, A.H., Majid, I., Khan, S.A., 2022. Containers for food packaging application. In: Parameswaranpillai, J., Salim, N.V., Pulikkalparambil, H., Mavinkere Rangappa, S., Suchart Siengchin, I.h. (Eds.), Containers for Food Packaging Application. Springer, Singapore, pp. 105–125. Available at https://doi.org/10.1007/978-981-16-8146-2_5.

Jang, Y.C., Hong, S., Lee, J., Lee, M.J., Shim, W.J., 2014. Estimation of lost tourism revenue in Geoje Island from the 2011 marine debris pollution event in South Korea. Marine Pollution Bulletin 81 (1), 49–54. Available at https://doi.org/10.1016/j.marpolbul.2014.02.021.

Jeong, E., Lee, J.Y., Redwan, M., 2024. Animal exposure to microplastics and health effects: a review. Emerging Contaminants 10 (4), 100369. Available at https://doi.org/10.1016/j.emcon.2024.100369.

Jin, Y., Xia, J., Pan, Z., Yang, J., Wang, W., Fu, Z., 2018. Polystyrene microplastics induce microbiota dysbiosis and inflammation in the gut of adult zebrafish. Environmental Pollution 235, 322–329. Available at https://doi.org/10.1016/j.envpol.2017.12.088.

Kadac-Czapska, K., Knez, E., Gierszewska, M., Olewnik-Kruszkowska, E., Grembecka, M., 2023. Microplastics derived from food packaging waste—their origin and health risks. Materials 16 (2), 674. Available at https://doi.org/10.3390/ma16020674.

Kan, M., Miller, S.A., 2022. Environmental impacts of plastic packaging of food products. Resources, Conservation and Recycling 180, 106156. Available at https://doi.org/10.1016/j.resconrec.2022.106156.

Kankanige, D., Babel, S., 2020. Smaller-sized micro-plastics (MPs) contamination in single-use PET-bottled water in Thailand. Science of the Total Environment 717, 137232. Available at https://doi.org/10.1016/j.scitotenv.2020.137232.

Kato, L.S., Conte-Junior, C.A., 2021. Safety of plastic food packaging: the challenges about non-intentionally added substances (NIAS) discovery, identification and risk assessment. Polymers 13 (13), 2077. Available at https://doi.org/10.3390/polym13132077.

Khalili Sadrabad, E., Hashemi, S.A., Nadjarzadeh, A., Askari, E., Akrami Mohajeri, F., Ramroudi, F., 2023. Bisphenol A release from food and beverage containers – a review. Food Science & Nutrition 11 (7), 3718–3728. Available at https://doi.org/10.1002/fsn3.3398.

Lau, W., 2021. Plastic Pollution Affects Sea Life. Pew. Available from: https://www.pew.org/en/trust/archive/winter-2021/ocean-plastic-pollution-is-a-huge-but-solvable-problem.

Leslie, H.A., van Velzen, M.J.M., Brandsma, S.H., Vethaak, A.D., Garcia-Vallejo, J.J., Lamoree, M.H., 2022. Discovery and quantification of plastic particle pollution in human blood. Environment International 163, 107199. Available at https://doi.org/10.1016/j.envint.2022.107199.

Macfadyen, G., Huntington, T., Cappell, R., 2009. Abandoned, Lost or Otherwise Discarded Fishing Gear, Vol. 523. United Nations Environment Programme Nairobi.

Marsh, K., Bugusu, B., 2007. Food packaging? Roles, materials, and environmental issues. Journal of Food Science 72 (3), R39. Available at https://doi.org/10.1111/j.1750-3841.2007.00301.x.

Mattsson, K., Johnson, E.V., Malmendal, A., Linse, S., Hansson, L.A., Cedervall, T., 2017. Brain damage and behavioural disorders in fish induced by plastic nanoparticles delivered through the food chain. Scientific Reports 7 (1), 11452. Available at https://doi.org/10.1038/s41598-017-10813-0.

Mkhari, T., Adeyemi, J.O., Fawole, O.A., 2025. Recent advances in the fabrication of intelligent packaging for food preservation: a review. Processes 13 (2), 539. Available at https://doi.org/10.3390/pr13020539.

Muzeza, C., Ngole-Jeme, V., Msagati, T.A.M., 2023. The mechanisms of plastic food-packaging monomers' migration into food matrix and the implications on human health. Foods 12 (18), 3364.

Mwanza, B.G., Mbohwa, C., 2017. Drivers to sustainable plastic solid waste recycling: a review. Procedia Manufacturing 8, 649–656. Available at https://doi.org/10.1016/j.promfg.2017.02.083.

Napper, I.E., Thompson, R.C., 2016. Release of synthetic microplastic plastic fibres from domestic washing machines: effects of fabric type and washing conditions. Marine Pollution Bulletin 112 (1–2), 39–45. Available at https://doi.org/10.1016/j.marpolbul.2016.09.025.

Naqash, N., Yadav, K.K., Shaik, A.S., Alam, M.W., Djajadi, D., Sunarto, D.A., et al., 2025. Microplastic pollution in terrestrial systems: sources and implications for soil functioning and plant performance. Water, Air, and Soil Pollution 236 (3), 172. Available at https://doi.org/10.1007/s11270-025-07818-1.

Nayanathara Thathsarani Pilapitiya, P.G.C., Ratnayake, A.S., 2024. The world of plastic waste: a review. Cleaner Materials 11, 100220. Available at https://doi.org/10.1016/j.clema.2024.100220.

Ncube, L.K., Ude, A.U., Ogunmuyiwa, E.N., Zulkifli, R., Beas, I.N., 2021. An overview of plastic waste generation and management in food packaging industries. Recycling 6 (1), 1–25. Available at https://doi.org/10.3390/recycling6010012.

Nelms, S.E., Duncan, E.M., Broderick, A.C., Galloway, T.S., Godfrey, M.H., Hamann, M., et al., 2016. Plastic and marine turtles: a review and call for research. ICES Journal of Marine Science 73 (2), 165–181. Available at https://doi.org/10.1093/icesjms/fsv165.

Newman, S., Watkins, E., Farmer, A., Brink, P.T., Schweitzer, J.P., 2015. The economics of marine litter. In: Bergmann, M., Gutow, L., Klages, M. (Eds.), Marine Anthropogenic Litter. Springer International Publishing, Cham, pp. 367–394. Available at https://doi.org/10.1007/978-3-319-16510-3.

OECD, 2022. Global Plastics Outlook: Policy Scenarios to 2060. OECD Publishing, Paris. Available at https://doi.org/10.1787/aa1edf33-en.

OECD, 2024. Policy Scenarios for Eliminating Plastic Pollution by 2040. OECD Publishing, Paris. Available at https://doi.org/10.1787/76400890-en.

Oliveira, W.Qd, Azeredo, H.M.Cd, Neri-Numa, I.A., Pastore, G.M., 2021. Food packaging wastes amid the COVID-19 pandemic: trends and challenges. Trends in Food Science and Technology 116, 1195–1199. Available at https://doi.org/10.1016/j.tifs.2021.05.027.

Pan, D., Su, F., Liu, C., Guo, Z., 2020. Research progress for plastic waste management and manufacture of value-added products. Advanced Composites and Hybrid Materials 3 (4), 443–461. Available at https://doi.org/10.1007/s42114-020-00190-0.

Parker, L., 2021. Plastic gets to the oceans through over 1,000 rivers. Available at https://www.nationalgeographic.com/environment/article/plastic-gets-to-oceans-through-over-1000-rivers.

Parolini, M., Antonioli, D., Borgogno, F., Gibellino, M.C., Fresta, J., Albonico, C., et al., 2021. Microplastic contamination in snow from western Italian Alps. International Journal of Environmental Research and Public Health 18 (2), 768. Available at https://doi.org/10.3390/ijerph18020768.

Perera, U.L.H.P., Subasinghe, H.C.S., Ratnayake, A.S., Weerasingha, W.A.D.B., Wijewardhana, T.D.U., 2022. Maritime pollution in the Indian Ocean after the MV X-Press Pearl accident. Marine Pollution Bulletin 185 (Pt A), 114301. Available at https://doi.org/10.1016/j.marpolbul.2022.114301.

Pew Charitable Trust, 2020. Breaking the Plastic Wave.

Plastics Europe, 2024a. The Circular Economy for Plastics: A European Analysis.

Prata, J.C., 2018. Airborne microplastics: consequences to human health? Environmental Pollution 234, 115–126. Available at https://doi.org/10.1016/j.envpol.2017.11.043.

Prata, J.C., Paço, A., Reis, V., da Costa, J.P., Fernandes, A.J.S., da Costa, F.M., et al., 2020. Identification of microplastics in white wines capped with polyethylene stoppers using micro-Raman spectroscopy. Food Chemistry 331, 127323. Available at https://doi.org/10.1016/j.foodchem.2020.127323.

Ragusa, A., Svelato, A., Santacroce, C., Catalano, P., Notarstefano, V., Carnevali, O., et al., 2021. Plasticenta: first evidence of microplastics in human placenta. Environment International 146, 106274. Available at https://doi.org/10.1016/j.envint.2020.106274.

Ratnayake, A.S., 2023. Marine pollution after the MV X-Press Pearl maritime disaster: societal impacts based on research outcomes. Societal Impacts 1 (1–2), 100003. Available at https://doi.org/10.1016/j.socimp.2023.100003.

Richardson, K., Hardesty, B.D., Wilcox, C., 2019. Estimates of fishing gear loss rates at a global scale: a literature review and meta-analysis. Fish and Fisheries 20 (6), 1218–1231.

Roberts, D., Edo, M., Johansson, I., Hoffman, B., Becidan, M., Ciceri, G., Murphy, F., Trois, C., Curran, T.P., 2022. Material and Energy Valorisation of Waste in a Circular Economy. IEA Bioenergy: Task 36.

Robertson, G.L., 2005. Food Packaging: Principles and Practice, 3rd ed. CRC Press, Boca Raton.

Romano, N., Renukdas, N., Fischer, H., Shrivastava, J., Baruah, K., Egnew, N., et al., 2020. Differential modulation of oxidative stress, antioxidant defense, histomorphology, ion-regulation and growth marker gene expression in goldfish (Carassius auratus) following exposure to different dose of virgin microplastics. Comparative Biochemistry and Physiology Part C: Toxicology & Pharmacology 238, 108862. Available at https://doi.org/10.1016/j.cbpc.2020.108862.

Salazar, G., Neves, J., Alves, V., Silva, B., Giger, J.-C., Veríssimo, D., et al., 2021. The effectiveness and efficiency of using normative messages to reduce waste: a real world experiment. PLOS One 16 (12), e0261734. Available at https://doi.org/10.1371/journal.pone.0261734.

Schönbein, C.F., 1911. The Encyclopaedia Britannica - Vol.24 - 11th ed. University Press, Cambridge.

Schwabl, P., Köppel, S., Königshofer, P., Bucsics, T., Trauner, M., Reiberger, T., et al., 2019. Detection of various microplastics in human stool. Annals of Internal Medicine 171 (7), 453–457. Available at https://doi.org/10.7326/m19-0618.

Selke, S.E.M., Culter, J.D., Auras, R.A., Rabnawaz, M., 2021. Major Plastics in Packaging. Plastics Packaging. Carl Hanser Verlag GmbH & Co. KG, pp. 105–164. Available at https://doi.org/10.1007/978-1-56990-823-5_4.

Senathirajah, K., Attwood, S., Bhagwat, G., Carbery, M., Wilson, S., Palanisami, T., 2021. Estimation of the mass of microplastics ingested – a pivotal first step towards human health risk assessment. Journal of Hazardous Materials 404 (Pt B), 124004. Available at https://doi.org/10.1016/j.jhazmat.2020.124004.

Silayoi, P., Speece, M., 2007. The importance of packaging attributes: a conjoint analysis approach. European Journal of Marketing 41 (11/12), 1495–1517. Available at https://doi.org/10.1108/03090560710821279.

Silpa Kaza, L.Y., Bhada-Tata, P., Van Woerden, F., 2018. What a Waste 2.0: A Global Snapshot of Solid Waste Management to 2050. World Bank.

Singh, B.P., 2024. Understanding the dynamics and implications of airborne microplastics in atmosphere. Air Quality, Atmosphere and Health 17 (11), 2661–2674. Available at https://doi.org/10.1007/s11869-024-01593-4.

Singh, R.P., Heldman, D.R., 2001. Introduction to Food Engineering.

Sommer, F., Dietze, V., Baum, A., Sauer, J., Gilge, S., Maschowski, C., et al., 2018. Tire abrasion as a major source of microplastics in the environment. Aerosol and Air Quality Research 18 (8), 2014–2028. Available at https://doi.org/10.4209/aaqr.2018.03.0099.

Sridharan, S., Kumar, M., Saha, M., Kirkham, M.B., Singh, L., Bolan, N.S., 2022. The polymers and their additives in particulate plastics: what makes them hazardous to the fauna? Science of the Total Environment 824, 153828. Available at https://doi.org/10.1016/j.scitotenv.2022.153828.

Talsness, C.E., Andrade, A.J.M., Kuriyama, S.N., Taylor, J.A., Saal, F.S.V., 2009. Components of plastic: experimental studies in animals and relevance for human health. Philosophical Transactions of the Royal Society B: Biological Sciences 364 (1526), 2079–2096. Available at https://doi.org/10.1098/rstb.2008.0281.

Teuten, E.L., Saquing, J.M., Knappe, D.R.U., Barlaz, M.A., Jonsson, S., Björn, A., et al., 2009. Transport and release of chemicals from plastics to the environment and to wildlife. Philosophical Transactions of the Royal Society B: Biological Sciences 364 (1526), 2027–2045. Available at https://doi.org/10.1098/rstb.2008.0284.

Thiel, M., Luna-Jorquera, G., álvarez-Varas, R., Gallardo, C., Hinojosa, I.A., Luna, N., et al., 2018. Impacts of marine plastic pollution from continental coasts to subtropical gyres—fish, seabirds, and other vertebrates in the SE Pacific. Frontiers in Marine Science 5, 359654. Available at https://doi.org/10.3389/fmars.2018.00238.

U.N. Environment Program, 2025. Plastic pollution. Available from: https://www.unep.org/plastic-pollution.

U.N. Environment Program, 2021. From Pollution to Solution: A Global Assessment of Marine Litter and Plastic Pollution. Available at https://www.unep.org/resources/pollution-solution-global-assessment-marine-litter-and-plastic-pollution.

Velásquez, E., Guerrero Correa, M., Garrido, L., Guarda, A., Galotto, M.J., López de Dicastillo, C., Parameswaranpillai, J., Mavinkere Rangappa, S., Gulihonnehalli Rajkumar, A., Siengchin, S., 2021. Food packaging plastics: identification and recycling. In: Parameswaranpillai, J., Mavinkere Rangappa, S., Gulihonnehalli Rajkumar, A., Siengchin, S. (Eds.), Recent Developments in Plastic Recycling. Composites Science and Technology. Springer, Singapore, pp. 311–343. Available at https://doi.org/10.1007/978-981-16-3627-1_14.

Von Moos, N., Burkhardt-Holm, P., Köhler, A., 2012. Uptake and effects of microplastics on cells and tissue of the blue mussel Mytilus edulis L. after an experimental exposure. Environmental Science and Technology 46 (20), 11327–11335. Available at https://doi.org/10.1021/es302332w.

Wang, X., Li, C., Liu, K., Zhu, L., Song, Z., Li, D., 2020. Atmospheric microplastic over the South China Sea and East Indian Ocean: abundance, distribution and source. Journal of Hazardous Materials 389, 121846. Available at https://doi.org/10.1016/j.jhazmat.2019.121846.

Wang, Y., Huang, J., Zhu, F., Zhou, S., 2021. Airborne microplastics: a review on the occurrence, migration and risks to humans. Bulletin of Environmental Contamination and Toxicology 107 (4), 657–664. Available at https://doi.org/10.1007/s00128-021-03180-0.

Wright, S.L., Kelly, F.J., 2017. Plastic and human health: a micro issue? Environmental Science and Technology 51 (12), 6634–6647. Available at https://doi.org/10.1021/acs.est.7b00423.

Wilcox, C., Van Sebille, E., Hardesty, B.D., 2015. Threat of plastic pollution to seabirds is global, pervasive, and increasing. Proceedings of the National Academy of Sciences of the United States of America 112 (38), 11899–11904.

Wyles, K.J., Pahl, S., Thomas, K., Thompson, R.C., 2016. Factors that can undermine the psychological benefits of coastal environments: exploring the effect of tidal state, presence, and type of litter. Environment and Behavior 48 (9), 1095–1126. Available from. Available at https://doi.org/10.1177/0013916515592177, http://eab.sagepub.com/archive/.

Xu, G., Liu, Y., Song, X., Li, M., Yu, Y., 2021. Size effects of microplastics on accumulation and elimination of phenanthrene in earthworms. Journal of Hazardous Materials 403, 123966. Available at https://doi.org/10.1016/j.jhazmat.2020.123966.

Yang, H., He, Y., Yan, Y., Junaid, M., Wang, J., 2021. Characteristics, toxic effects, and analytical methods of microplastics in the atmosphere. Nanomaterials 11 (10), 2747. Available at https://doi.org/10.3390/nano11102747.

Zero Waste Europe, 2021. Zero Waste Europe Annual Report. Available at: https://zerowasteeurope.eu/wp-content/uploads/2022/06/ZWE-Annual-Report-2021.pdf.

Zhang, W., Gao, P., Jiang, Q., Xia, W., 2023. Green fabrication of lignin nanoparticles/chitosan films for refrigerated fish preservation application. Food Hydrocolloids 139, 108548. Available at https://doi.org/10.1016/j.foodhyd.2023.108548.

Further reading

Meeker, J.D., Sathyanarayana, S., Swan, S.H., 2009. Phthalates and other additives in plastics: human exposure and associated health outcomes. Philosophical Transactions of the Royal Society B: Biological Sciences 364 (1526), 2097–2113.

Plastics Europe, 2024b. Plastics – The Fast Facts.

The ocean. Marine pollution bulletin.

Yam, K.L., Takhistov, P.T., Miltz, J., 2005. Intelligent packaging: concepts and applications. Journal of food science 70 (1), R1–R10.

CHAPTER 2

Understanding green packaging consumers and managerial strategies

Mohammad Iranmanesh[1] and Morteza Ghobakhloo[2]

[1]School of Hospitality, Tourism and Events, Centre for Asian Modernisation, Taylor's University, Selangor, Malaysia, [2]Division of Industrial Engineering and Management, Uppsala University, Uppsala, Sweden

2.1 Introduction

Global consumption patterns have led to a sharp increase in waste generation, placing growing pressure on natural ecosystems. Packaging plays a major role in this development, as it is often used for a very short period but has persisted in the environment for many years (Mahmoud et al., 2022). In many developed countries, packaging waste represents between 30% and 35% of total municipal solid waste (Mudgal et al., 2024). In the European Union, each individual discards around 150 kg of packaging waste every year (Duarte et al., 2024). At the global level, the World Bank estimates that total solid waste will reach 3.88 billion tons by 2050 (De Canio, 2023). These figures highlight the scale of the environmental challenge associated with packaging and underline the urgency of reducing its negative impact. Plastic packaging is a particularly serious concern. Plastic waste generation remains high across regions, reaching about 221 kg per person in the United States and 114 kg per person in the European Union (OECD, 2022). Large quantities of this waste are not properly managed. In Europe, around 40% of packaging waste is sent to landfills, while approximately 32% leaks into land and marine environments (Mahmoud et al., 2022). Globally, only about 9% of plastic waste is recycled (Suhartanto et al., 2021). In 2019 alone, an estimated 1.7 million tons of plastic waste were dumped into seas, with a further 6.1 million tons entering rivers and other aquatic systems (Mahmoud et al., 2022). Projections suggest that by 2030, packaging waste will increase by 19%, while plastic waste will rise by 46% (European Commission, 2022). These trends raise serious concerns about long-term environmental damage, particularly to marine ecosystems.

In response to these challenges, green packaging has been promoted as a key approach to reducing the environmental footprint of consumption. Green packaging refers to packaging that uses environmentally friendly materials, is safe for human health, and performs its function effectively throughout the product life cycle. Its development is closely linked to

Bioplastics for Food Packaging. DOI: https://doi.org/10.1016/B978-0-443-41654-5.00006-8

global sustainability objectives, including efforts to encourage responsible consumption and protect natural ecosystems. Reflecting this shift, the global sustainable packaging market was valued at approximately US$258 billion in 2020 and is expected to reach US$385 billion by 2028 (De Canio, 2023). In addition, the market for packaging made from recycled materials is projected to reach more than US$440 billion by 2025 (Popovic et al., 2019). These figures indicate strong industry interest and investment in sustainable packaging solutions. However, despite increasing environmental awareness and strong market growth forecasts, the actual adoption of green packaging by consumers remains limited. Green products are estimated to account for less than 4% of the global market (Ketelsen et al., 2020). Many consumers express concern about environmental issues, yet this concern does not consistently translate into purchasing decisions. Price remains a major barrier, as green packaging is often more expensive than conventional alternatives (Duarte et al., 2024). Consumers are also frequently uncertain about the environmental benefits of different materials and how they should be disposed of. In addition, sustainability is rarely prioritized when it is perceived to reduce convenience, product protection, or visual appeal (Granato et al., 2022). As a result, conventional packaging continues to dominate consumer markets.

For managers and firms, increasing the adoption of green packaging is both a social responsibility and a strategic challenge. Green packaging can enhance brand reputation and provide opportunities for market differentiation (Popovic et al., 2019). At the same time, firms must address consumer skepticism, concerns about misleading environmental claims, and strong expectations regarding quality and price. Achieving wider adoption therefore requires a clear understanding of consumer behavior and effective communication of environmental value without compromising functional performance. This chapter addresses three key questions. First, it examines who the green packaging consumer is. Second, it explores why consumers choose to purchase green packaging. Third, it considers how managers can encourage wider adoption of green packaging in consumer markets. By answering these questions, the chapter aims to support both academic research and managerial practice in advancing more sustainable packaging systems.

2.2 Who is the green packaging consumer?

Consumer responses to green packaging are not uniform across the marketplace. Instead, they vary according to differences in consumers' socioeconomic conditions, environmental knowledge, household characteristics, and cultural norms. This heterogeneity indicates that environmentally responsible packaging is understood, valued, and acted upon differently across consumer segments. Recognizing these variations is essential, as it helps identify the appropriate target markets for green packaging. From a managerial perspective, understanding which consumers have greater propensity to buy, which face financial or informational constraints, and which require additional support allows for more effective segmentation and more credible strategies.

Gender has been one of the most frequently examined demographic dimensions in green consumption research. A substantial body of evidence suggests that women tend to display stronger pro-environmental behaviors and higher purchase intentions toward green packaging than men (Wandosell et al., 2021). This pattern is often attributed to women's greater involvement in household shopping decisions and waste management activities, which increases their exposure to packaging-related environmental consequences (Boz et al., 2020). Women are also more likely to engage in behaviors such as recycling, reuse, and careful product comparison, which align closely with the objectives of sustainable packaging.

Age and generational cohort further differentiate green packaging responses by shaping both environmental values and economic capacity. Younger consumers, particularly those belonging to Generation Z and the Millennial cohort, are frequently characterized by greater environmental awareness and stronger normative support for sustainable consumption (Hao et al., 2019). They are often among the earliest adopters of green packaging concepts and tend to express optimism about the long-term benefits of packaging innovation (Noëth et al., 2024). Nevertheless, high awareness does not always translate into higher willingness to pay. Limited income and greater price sensitivity can constrain younger consumers' ability to support green packaging financially, leading to a gap between attitudes and purchasing behavior (Ketelsen et al., 2020). In contrast, older consumers, including Baby Boomers, have been found to exhibit a higher willingness to pay price premiums for environmentally friendly packaging. Greater financial stability and greater consumption experience may enable these consumers to prioritize long-term environmental considerations over short-term cost savings (Ketelsen et al., 2020).

Educational level plays a consistently positive role in shaping responses to green packaging. Consumers with higher levels of education generally demonstrate more favorable attitudes and stronger purchase intentions toward environmentally responsible packaging options (Güngördü Belbağ, 2025). Education enhances the ability to interpret eco-labels, sustainability claims, and technical information related to recyclability or material composition. As a result, educated consumers are better positioned to distinguish between substantive environmental improvements and superficial green claims. This cognitive advantage reduces uncertainty and increases trust in green packaging initiatives, thereby facilitating adoption (Boz et al., 2020). In contrast, lower levels of education may limit engagement not because of low concern, but due to difficulties in processing complex environmental information. Income level further conditions green packaging behavior by influencing affordability and price tolerance. Higher-income consumers are more likely to purchase products with eco-friendly packaging and to accept associated price premiums (Duarte et al., 2024; Güngördü Belbağ, 2025). For these consumers, green packaging is often seen as a quality attribute that fits with their broader lifestyle values. Lower-income consumers, by comparison, tend to prioritize immediate economic considerations, making them more sensitive to even modest price increases. This does not imply a lack of environmental concern but rather reflects external barriers that limit the translation of concern into action. Without supportive pricing strategies or incentives, green packaging may remain inaccessible to large segments of the population.

Geographical and cultural context also shapes how consumers evaluate and respond to green packaging. In developing markets such as India and China, consumers often report relatively high willingness to pay for environmentally friendly packaging, driven by rising environmental awareness and visible pollution challenges (Hao et al., 2019; Prakash et al., 2019). At the same time, these consumers may lack detailed technical knowledge about specific packaging benefits, relying more on general cues and brand trust. In developed markets such as Germany and the United States, consumers tend to evaluate green packaging through more differentiated criteria, with preferences shaped by local waste management infrastructure and regulatory frameworks (Herbes et al., 2018). For example, recyclability may be prioritized over biodegradability in contexts where recycling systems are well established, whereas the opposite may occur elsewhere. These cross-national differences highlight that green packaging preferences are embedded within broader institutional and cultural environments rather than driven solely by individual attitudes.

2.3 Why do consumers purchase green packaging?

Consumer intention to purchase products with green packaging is shaped by a combination of internal motivations, product-related evaluations, social influences, emotional responses, and cognitive assessments. Rather than operating in isolation, these determinants interact to form a comprehensive decision-making process that guides environmentally responsible consumption. The key drivers are organized into five interrelated categories.

2.3.1 Value orientation and moral motivation

Value orientation and moral motivation explain why consumers care about green packaging and why they decide to choose it. These factors shape how people understand environmental issues and how they judge their own buying decisions. Values influence whether consumers view green packaging as meaningful or unnecessary. They also affect how strongly people feel about acting in an environmentally responsible way during everyday shopping. As a result, value orientation helps explain differences in consumer behavior even when people have similar levels of environmental awareness.

Some consumers are guided by concern for other people and for the natural environment. These individuals choose green packaging because they believe it helps protect communities, reduces harm, and supports a healthier future (Prakash et al., 2019). They see environmentally friendly packaging as a shared responsibility rather than a personal sacrifice. This motivation becomes stronger when environmental problems such as waste and pollution are clearly visible in daily life. In these cases, green packaging is viewed as a practical way to reduce negative environmental impact. Other consumers are motivated mainly by personal benefits. They may prefer green packaging because they believe it is safer, healthier, or better for their own well-being (Nuojua et al., 2024). Many people connect sustainable packaging with lower exposure to harmful chemicals and improved product quality. Green packaging can also improve social image, especially when it is visible to others, such as reusable bags or cups (Magnier and Crié, 2015). These personal rewards often create positive feelings, such as pride or satisfaction, which encourage repeat purchasing (Ahmed et al., 2023). However, when green packaging is seen as too costly or inconvenient, personal benefits may no longer support the decision (Steenis et al., 2018).

Values are more likely to lead to action when consumers feel personally responsible for environmental problems. When people believe their daily choices contribute to waste and pollution, they become more willing to change their behavior. This sense of responsibility creates an inner pressure to act in line with personal moral standards. Consumers then avoid packaging options that conflict with their beliefs and choose alternatives that allow them to feel comfortable with their decisions (Chen et al., 2017). This moral pressure can remain strong even when green options require extra effort. Some consumers also think beyond the present when making packaging choices. They consider how today's actions may affect future generations and the long-term condition of the environment. For these individuals, green packaging represents a way to reduce future environmental damage. They see their purchasing decisions as part of a wider effort to protect resources for others (Ketelsen et al., 2020). Viewing consumption from this future-focused perspective adds meaning to sustainable choices and strengthens ongoing support for green packaging.

2.3.2 Self-identity and psychological attachment

Self-identity and psychological attachment explain how consumers' views about themselves influence their choices of green packaging. Buying decisions are not based only on price or usefulness; they are also shaped by how people see themselves and what their choices mean to them. When green packaging fits a person's self-image, the decision seems more meaningful. These internal factors help turn general environmental awareness into actual purchasing behavior. They also explain why some consumers act on their concerns while others do not.

Green self-identity refers to whether people see themselves as environmentally responsible. When individuals think of themselves as caring about the environment, they try to behave in ways that match this view. Simply knowing about environmental problems is not enough to change behavior unless it becomes part of how people define themselves (Moshood et al., 2022). Buying products with green packaging allows consumers to stay consistent with their personal standards and avoid feeling uncomfortable about their choices (Su et al., 2021). Green packaging also sends a message to others, as visible consumption choices reflect personal values. When people do not strongly identify as environmentally responsible, practical issues such as price or convenience often take priority. Psychological ownership describes a deeper personal connection to green products. It refers to the feeling that a product or idea matters personally, even if it is not legally owned. When consumers feel connected to green packaging, they care more about its success and are more willing to support it. This connection can make people more accepting of higher prices or small inconveniences linked to sustainable options (Chang, 2023). Psychological ownership grows when consumers understand the product, feel familiar with it, and believe their choices matter. As a result, consumers move from simply noticing green information to actively supporting environmentally friendly products (Chang et al., 2021).

Using green packaging also affects how consumers feel about themselves. Making sustainable choices helps people see themselves as responsible and morally aware. This positive self-view acts as an inner reward and encourages continued environmentally friendly behavior (Chang, 2023). Consumers often feel reassured when their actions match their ethical beliefs (Steenis et al., 2018). When traditional packaging is seen as harmful to the environment, this feeling becomes stronger. Choosing green alternatives helps consumers feel comfortable with their decisions and supports positive self-evaluation, even when other product features are similar. Self-identity is also shaped by lifestyle choices, especially for consumers who value health and sustainability in daily life. For these individuals, green packaging is not an occasional choice but part of how they live (McCarthy and Wang, 2022). They often believe that protecting the environment also supports personal health and safety (De Canio, 2023). This view makes them more open to new packaging solutions and less concerned about unfamiliar formats. Because green packaging fits their everyday habits and beliefs, they are more likely to choose it regularly and remain supportive over time.

2.3.3 Affective drivers

Emotional factors play an important role in shaping consumer responses to green packaging. Even when price, quality, and usefulness are acceptable, emotions often determine whether consumers actually choose environmentally friendly options. These emotions influence how people feel about their decisions and whether those decisions feel right or satisfying. Feelings related to moral comfort, self-evaluation, and care for the natural environment strongly affect green purchasing behavior. As a result, emotional responses often work alongside rational thinking rather than replacing it. A key emotional factor is the positive feeling people experience when they make environmentally responsible choices (Güngördü Belbağ, 2025). Many consumers feel satisfied when they buy products with green packaging because the decision matches their personal beliefs about protecting the environment. This positive feeling provides reassurance and helps consumers feel comfortable with their choices. Over time, such experiences can improve overall feelings about daily consumption. Green packaging also allows consumers to express their values to themselves, which strengthens their sense of identity and makes the decision more meaningful (Ahmed et al., 2023).

Negative emotions, especially guilt and discomfort, also influence behavior. Consumers may expect to feel guilty if they choose packaging that harms the environment, and this expectation encourages them to avoid such options (Enginkaya and Hayrun Sağlam, 2025). These feelings usually arise when there is a mismatch between what people believe is right and what they actually buy. Instead of discouraging action, this emotional discomfort often leads consumers to think more carefully about their choices. Guilt is particularly strong when people throw away plastic packaging, as disposal highlights waste and environmental harm. These experiences often influence later decisions by increasing attention to packaging materials and waste outcomes (Chen et al., 2017). Positive emotions such as pride also support green purchasing decisions. Consumers may feel proud when they choose environmentally friendly packaging, especially if it requires extra effort or higher cost. This feeling strengthens self-respect and reinforces the idea that they are acting responsibly (Mahmoud et al., 2022). Pride also has a social aspect, as visible green packaging can show others that the consumer cares about the environment (Magnier and Crié, 2015). In settings where sustainability is valued, such choices can lead to positive social feedback. This makes green packaging more attractive even when other options are easier or cheaper. Emotional connection to nature further shapes responses to green packaging. Consumers who feel close to nature, or to places such as oceans and forests, are more sensitive to environmental harm caused by packaging waste. Seeing pollution in natural environments often strengthens emotional reactions and concern (Nuojua et al., 2024). These feelings encourage consumers to prefer packaging that can be recycled or produces less waste. Even simple images of nature on packaging can trigger these emotional responses and influence attitudes (Ahmed et al., 2023). Such emotional links make environmental issues feel more personal and relevant. Beyond single emotional reactions, green satisfaction reflects a broader sense of fulfillment from making environmentally responsible choices. This satisfaction occurs when consumers believe that packaging meets their expectations for environmental performance (Pan et al., 2021). Feeling satisfied increases confidence in the product and the brand. It also makes consumers more willing to repeat the behavior in the future (Suhartanto et al., 2021). Over time, this emotional fulfillment supports stable and regular green purchasing rather than one-time decisions.

2.3.4 Cognitive evaluations

Cognitive evaluations explain how consumers think through their decisions about green packaging. These evaluations involve careful consideration rather than emotional reactions.

Consumers assess whether their actions matter, whether green options are easy to use, whether information can be trusted, and whether the effort or cost is reasonable. Even when people care about the environment, these thought processes strongly affect whether they actually buy products with green packaging. In this way, rational thinking often determines whether good intentions lead to real behavior. One important factor is whether consumers believe their individual actions can make a difference. Some people feel that choosing green packaging helps reduce environmental problems, even in a small way. When consumers hold this belief, they are more willing to change their buying habits and support sustainable options. This belief is especially important when green packaging is new or unfamiliar, as it gives consumers confidence that their choice is worthwhile (Güngördü Belbağ, 2025; Ischen et al., 2022). On the other hand, when people think that environmental problems are too large to be influenced by individual action, motivation decreases. In such cases, concern for the environment alone is often not enough to encourage green purchasing (Gao and Shao, 2022). Another key factor is how easy consumers believe it is to buy green packaging. This includes whether they feel confident about finding it, understanding it, and paying for it. When consumers believe that green packaging is within their control, they are more likely to intend to purchase it (Gao and Shao, 2022). Feeling capable reduces hesitation and makes change seem manageable. However, when green options appear difficult to locate, confusing, or too expensive, intention weakens. This can happen even when consumers hold positive views about environmental protection (Leong et al., 2024).

Practical barriers such as price and availability also influence decision-making. Many consumers compare the benefits of green packaging with the extra money or effort required. Higher prices often discourage purchase, especially when the difference between green and conventional packaging is not noticeable (Moshood et al., 2022). For consumers with limited budgets, affordability becomes a priority over environmental concerns (Wandosell et al., 2021). Availability plays a similar role. When green packaging is sold in regular stores and placed alongside standard products, it feels easier to choose (Moshood et al., 2025). When it requires extra searching or special trips, consumers may decide it is not worth the effort (Leong et al., 2024). Trust in environmental information strongly affects how consumers judge green packaging. When claims are unclear or seem exaggerated, consumers may doubt their accuracy. This doubt reduces the value of green features and can lead to negative reactions (Pan et al., 2021; Güngördü Belbağ, 2025). Problems also arise when packaging looks excessive or over-packaged but claims to be environmentally friendly. Such mixed signals can confuse consumers and reduce confidence in the brand (Chen et al., 2017). Trust is therefore essential for helping consumers accept environmental messages and feel comfortable with their choices.

Knowledge and familiarity also shape how deeply consumers think about green packaging. Consumers who know more about environmental issues are better able to judge claims and understand differences between materials. They rely less on surface features and make more informed decisions (Moshood et al., 2022). Consumers with limited knowledge often depend on simple cues such as colors, symbols, or basic labels (De Canio, 2023; Steenis et al., 2017). This makes them more open to misunderstanding or misleading information. As a result, gaps between real environmental impact and consumer beliefs remain a challenge in encouraging wider use of green packaging.

2.3.5 Packaging attributes and social influence

Consumer decisions about green packaging are shaped by more than environmental concern alone. People consider the product itself, the signals sent by packaging design and information, and the social setting in which the purchase takes place. These elements work together to influence whether positive views about sustainability lead to real buying behavior. Even consumers who care about the environment may hesitate if these factors do not support their decision. At a basic level, green packaging must meet everyday expectations related to quality, safety, and ease of use. Most consumers are not willing to accept packaging that feels weak, unsafe, or inconvenient, even if it is environmentally friendly. If green packaging is seen as less able to protect the product or keep it fresh, interest quickly decreases (Granato et al., 2022). Price also plays an important role. While some consumers are open to paying more for sustainable options, higher prices often discourage purchase, especially among those with limited budgets (Moshood et al., 2022). Many consumers feel that companies should cover the extra cost of sustainability (Wandosell et al., 2021). However, when green packaging is viewed as durable or reusable, it can be seen as better value, which makes higher prices easier to accept.

Visual appearance strongly influences how consumers judge green packaging. Before reading any information, people notice materials, colors, and textures. Materials such as paper, cardboard, and glass are often seen as more environmentally friendly, while plastic is commonly viewed negatively because of its link to pollution (Ischen et al., 2022; Steenis et al., 2017). Design features also matter. Earth tones (beige, green, and brown), simple and minimalist graphics, and nonglossy finishes are often linked to sustainability, while overly complex or layered packaging may suggest waste (Scott and Vigar-Ellis, 2014; Magnier and Crié, 2015). Packaging that looks environmentally friendly can also shape how consumers view the product inside. Even without changes to the product itself, consumers may believe it is healthier or higher in quality based on the packaging alone (Granato et al., 2022).

Clear information is needed to support these visual impressions. Labels and symbols help consumers understand whether packaging is truly environmentally friendly, especially when they lack detailed knowledge (Ischen et al., 2022; Nguyen-Viet et al., 2024). Simple and clear labels make decisions easier and reduce confusion. However, information is only effective when it is trusted. Many consumers are doubtful of environmental claims because of unclear or exaggerated messages. When trust is low, green features lose their influence and may even cause negative reactions (Chang, 2023; Boz et al., 2020). Independent certifications are often seen as more reliable than company claims and help rebuild confidence in green packaging (Moshood et al., 2022). Social influence and the retail setting also shape consumer choices. People often consider what others expect from them when making purchasing decisions. If environmentally responsible behavior is supported by family, friends, or society, consumers are more likely to choose green packaging (Leong et al., 2024). These choices can also bring personal satisfaction, as consumers feel they are acting in line with shared values (Ahmed et al., 2023). Because packaging is visible in daily life, it can signal personal beliefs to others. Retailers play an important role by making green options easy to find and clearly displayed (De Canio, 2023). Stores with a strong environmental image increase consumer trust and make sustainable choices feel more normal and achievable (Su et al., 2021). By doing so, retailers help turn positive attitudes toward sustainability into actual purchasing behavior.

2.4 How can managers promote green packaging adoption?

To successfully drive the adoption of green packaging, managers must move beyond treating sustainability as a mere compliance issue and instead integrate it into the core marketing and operational strategy. The literature suggests a multidimensional approach involving supply chain collaboration, design optimization, trust building, and internal employee engagement.

2.4.1 Prioritize functional performance

Managers should begin by understanding that sustainability cannot come at the expense of basic packaging performance. For most consumers, packaging is expected to protect the product, keep it hygienic, and make everyday use simple and reliable. Environmental benefits are often appreciated, but they rarely compensate for functional failure (Ketelsen et al., 2020; Granato et al., 2022). If a package leaks, breaks, or causes the product to spoil, consumers respond with frustration and negative feelings toward the brand. This reaction occurs even among environmentally concerned buyers (Sonck-Rautio et al., 2024). As a result, managers should treat functional performance as a fixed baseline and ensure that any sustainable packaging solution performs at least as well as conventional options under real conditions such as transport, storage, and repeated use. Product protection and preservation are especially critical for food packaging, where the environmental impact of food waste is usually far greater than that of the package itself (Ketelsen et al., 2020; López-Miguens et al., 2025). Packaging that extends shelf life, slows spoilage, or protects food from contamination can make a meaningful contribution to sustainability, even if it uses slightly more material. Consumers place high value on freshness and safety, and they expect packaging to deliver these outcomes consistently. Managers should therefore evaluate green packaging not only by material choice but also by how effectively it reduces food waste and maintains product quality throughout the supply chain and at home (López-Miguens et al., 2025).

Material choice and design must also account for consumer perceptions of quality and trust. Many consumers associate transparent packaging with freshness and honesty, particularly for fresh products such as meat, fruits, and vegetables (Oloyede and Lignou, 2021). However, several sustainable materials are naturally opaque or cloudy, which can reduce confidence if not handled carefully. At the same time, there is a common belief that environmentally friendly packaging is weaker or less durable. Managers need to invest in research, testing, and thoughtful design to ensure that sustainable materials feel strong, resist moisture and temperature changes, and meet daily handling expectations (McCarthy and Wang, 2022; Granato et al., 2022). When new solutions such as edible packaging are introduced, concerns about hygiene and safety are even stronger, making clear standards, testing, and communication essential to avoid rejection. Efforts to reduce environmental impact through material reduction or reuse must not undermine convenience. Lightweight packaging can fail if it becomes too flexible, hard to reseal, or uncomfortable to handle, leading consumers to see it as poorly designed (Ketelsen et al., 2020; Popovic et al., 2019). Convenience strongly influences purchase decisions, so sustainable packaging should remain easy to open, close, empty, and store (Sonck-Rautio et al., 2024). Reusable packaging systems are more likely to succeed in product categories where handling feels simple and hygiene concerns are low, such as dry foods or bread, while categories like frozen or raw foods may face greater resistance (Noëth et al., 2024). Finally, disposal should be treated as part of functionality. Consumers prefer packaging that is easy to sort and recycle without complex steps (Sonck-Rautio et al., 2024). Simple structures and clear disposal instructions reduce frustration, support correct behavior, and improve acceptance of green packaging overall.

2.4.2 Leverage both implicit and explicit design cues

Managers should understand that green packaging communicates both through written messages and through its visual and physical appearance. Because many purchase decisions are made quickly and with limited attention, consumers often rely on what they see and feel rather than taking time to read detailed text. Features such as material, texture, color, and shape help consumers form an immediate judgment about whether a package is environmentally friendly. For this reason, sustainability should be clearly visible at first glance, while written information plays a supporting role for those who want confirmation. If the package does not look green, written claims alone are unlikely to be convincing.

Material choice is one of the most influential signals in shaping these first impressions. Consumers commonly view paper, cardboard, and glass as more environmentally friendly than plastic, even when scientific assessments sometimes suggest otherwise (Herbes et al., 2018; Ketelsen et al., 2020). A natural or cardboard-like appearance is often associated with biodegradability, simplicity, and care for the environment, which can increase trust and attention (Ischen et al., 2022). Color and surface finish reinforce these impressions. Soft, natural colors such as brown, green, and beige, along with matte or slightly rough textures, are generally seen as more natural and less artificial than bright colors or glossy surfaces (Magnier and Crié, 2015; Scott and Vigar-Ellis, 2014). While the color green can support an eco-friendly image, it is not sufficient on its own and may even raise suspicion if it is not supported by material and texture cues (Boz et al., 2020). Managers should therefore ensure that material, color, and finish work together to communicate a consistent and believable sustainability message.

Design simplicity further strengthens this message. Consumers tend to associate clean and minimal designs with honesty, modern values, and environmental responsibility, while busy graphics and complex structures can signal over-packaging and waste (Akturan et al., 2025). Reducing visual clutter, avoiding unnecessary layers, and eliminating oversized or secondary packaging are often interpreted as clear signs of genuine environmental commitment. Simple nature-related imagery, such as leaves or hand-drawn elements, can support these perceptions without making the design feel excessive or artificial (Magnier and Crié, 2015). Overall, simplicity in both structure and appearance helps align the package with consumers' expectations of what "green" should look like. Written messages on packaging should be chosen with care. In most shopping environments, especially supermarkets, consumers are distracted and unlikely to process detailed environmental information. Short, emotionally appealing messages about protecting nature or contributing to a better future are usually more effective than technical explanations. Detailed or highly specific claims are better suited to smaller groups of highly informed consumers or to low-distraction settings such as specialty stores or online platforms (Aagerup et al., 2019). Regardless of the message style, credibility is essential. Emotional appeals should be supported by clear and recognized proof, such as trusted eco-labels or certifications (Ischen et al., 2022; Magnier and Crié, 2015). This combination allows packaging to attract attention through design and emotion while also reassuring consumers that the environmental claims are genuine and trustworthy.

2.4.3 Build trust and eliminate "greenwashing" risks

Managers should recognize that trust is one of the biggest challenges in green packaging because many consumers have become skeptical after repeated exposure to unclear or exaggerated environmental claims (Chang, 2023). Sustainability cannot be treated as a marketing layer added to the final stage. It must be real, visible, and consistent throughout the packaging design. When consumers notice a gap between what the package claims and what it actually looks like or does, they quickly interpret this as greenwashing. Building trust therefore requires that environmental promises are supported by clear structural decisions, simple design, and actions that consumers can easily observe. Packaging volume is one of the most powerful and immediate signals of environmental intent. Consumers often interpret large boxes, multiple layers, and unnecessary inserts as clear signs of waste, even if the materials used are recyclable or biodegradable (Oloyede and Lignou, 2021). Excessive packaging can damage a brand's green image, weaken emotional attachment, and in some cases turn the package into what consumers see as "branded litter," where the brand becomes associated with the waste problem rather than the solution (Chen et al., 2017). Over-packaging also creates practical and psychological resistance. Consumers may feel they are paying for waste instead of value, and they may react negatively if they feel manipulated or pressured by unnecessary materials (Chang et al., 2021). Reducing packaging volume by removing nonessential layers not only lowers material use but also improves convenience and sends a strong signal of honesty and environmental care (Magnier and Crié, 2015).

Credible eco-labels are another essential tool for building trust and reducing skepticism. Most consumers lack the technical knowledge needed to judge packaging sustainability on their own and are often confused by terms such as bio-based, biodegradable, or recyclable (Friedrich, 2021). Recognized third-party certifications help close this knowledge gap and provide reassurance that environmental claims are verified. These labels are particularly important for consumers with limited or moderate environmental knowledge, while still reinforcing confidence among more informed consumers (De Canio, 2023). However, managers should avoid vague wording or self-created symbols, which are often viewed with suspicion. Even trusted labels need to be clearly visible and supported by communication that explains their meaning, as many shoppers overlook them in busy retail environments (Scott and Vigar-Ellis, 2014). Ultimately, trust depends on consistency between what the package says and what it shows. Consumers rely heavily on how packaging looks and feels when judging environmental responsibility. If a package promotes sustainability but uses heavy, glossy, or plastic-heavy materials, this mismatch is quickly perceived as misleading and can trigger strong greenwashing reactions (Steenis et al., 2017). A holistic design approach is therefore essential, where material choice, texture, color, structure, and written claims all point in the same direction (Ischen et al., 2022). When visual cues and verbal messages support each other, they reinforce credibility and strengthen the perception that the brand's environmental commitment is genuine.

2.4.4 Integrate supply chain and retail context

Managers should approach green packaging as a system-wide effort rather than a standalone design decision. Packaging choices affect, and are affected by, manufacturers, logistics providers, retailers, and even waste management actors. Treating packaging only as a product feature limits its impact and increases the risk of failure. A coordinated approach across the supply chain allows sustainability goals to be aligned with functional, economic, and commercial requirements. When all actors work toward shared objectives, green packaging becomes easier to implement, more credible, and more effective.

Retailers play a particularly important role because they control what consumers see and how products are presented at the point of sale (De Canio, 2023). Their environmental reputation strongly influences consumer trust and willingness to choose green options. When consumers shop at retailers known for sustainability, they are more likely to believe environmental claims and act in environmentally responsible ways (De Canio, 2023; Nguyen-Viet et al., 2024). Managers should therefore prioritize collaboration with retailers that have strong social and environmental commitments, as this context amplifies the credibility of green packaging. Joint efforts such as shared messaging, shelf placement, and in-store signs can gently guide consumers toward sustainable choices without requiring them to actively search for them. Placing green-packaged products in visible and familiar locations or highlighting their benefits at the shelf can help overcome convenience barriers that often prevent sustainable choices (Ketelsen et al., 2020). Beyond retail collaboration, effective green packaging requires deep integration across the entire supply chain. Decisions about materials, design, logistics, and communication should not be made in isolation. Collaboration with suppliers helps ensure that materials are responsibly sourced and suitable for recycling or reuse, while coordination with retailers and waste handlers supports proper disposal. Within the firm, marketing, logistics, and environmental teams must work together. Marketing may seek distinctive designs, logistics may prioritize strength and stackability, and environmental teams may push for material reduction. Without integration, these goals can conflict and weaken overall sustainability outcomes (Afif et al., 2022).

An integrated approach also helps managers manage trade-offs and share costs and benefits fairly. Reducing material weight, for example, may lower environmental impact but increase the risk of damage during transport if logistics needs are ignored, leading to more waste overall (Afif et al., 2022). Similarly, sustainable packaging often requires upfront investments in research, new materials, or reverse logistics systems. Integration allows stakeholders to see long-term gains such as reduced waste handling, improved efficiency, or increased sales from stronger consumer trust. This shared-value perspective is especially important for reusable packaging systems, where consumers become part of the supply chain by returning empty packages (Noëth et al., 2024). In such cases, reverse logistics and waste management must be treated as core operations rather than external issues, ensuring that the system truly supports circular use (Wandosell et al., 2021).

2.4.5 Strategic pricing and value perception

Managers should accept that price remains one of the strongest barriers to the widespread adoption of green packaging, even as environmental awareness continues to grow. Many consumers support sustainability in principle but are unwilling or unable to pay significantly more at the shelf. This means pricing decisions cannot be based only on higher material or production costs. Instead, managers need to shift toward value-based pricing by strengthening how consumers perceive the overall benefits of green packaging. The goal is to increase the sense that the product delivers more value in total, not just that it is "more expensive because it is green."

To justify a price premium, managers must expand the perceived value of green packaging beyond its environmental benefits (Suhartanto et al., 2021). Consumers evaluate value across several dimensions at once, including functional performance, emotional satisfaction, and social meaning (Ahmed et al., 2023). Green packaging can create emotional rewards, such as feeling good about making a responsible choice or experiencing a sense of moral satisfaction. It can also serve as a social signal, allowing consumers to express their values and feel proud of the products they use and display (Ahmed et al., 2023; Mahmoud et al., 2022). However, these psychological and social benefits only matter if the package also delivers high functional quality. Consumers are not willing to accept weaker protection, lower durability, or higher risk in exchange for sustainability. If green packaging is seen as less reliable or lower in quality, its perceived value drops quickly, and so does purchase intention. Managers must therefore ensure and clearly communicate that sustainable packaging performs at least as well as conventional alternatives. Because willingness to pay varies widely, a single pricing approach rarely works for all consumers. Managers should focus early efforts on segments that are more open to paying a premium, such as consumers whose lifestyles strongly align with health, sustainability, and social responsibility. These consumers are often more willing to try new solutions, including biodegradable or reusable packaging, because such choices fit their identity and values (McCarthy and Wang, 2022). For people with a strong sense of environmental self-identity, paying more is not simply a financial decision but a way to act consistently with who they believe they are (Kesgin et al., 2025). At the same time, managers should be cautious when targeting younger consumers. While many younger shoppers express strong environmental concern, their actual spending power may be limited, although this pattern can differ across countries and income contexts (Martinho et al., 2015; Popovic et al., 2019).

For the broader mass market, where price sensitivity is high, managers need to reduce the perception of extra cost. One effective approach is to highlight economic benefits that come with green packaging, especially in reusable systems. By communicating long-term savings, such as lower cost per use, managers can frame sustainability as financially sensible rather than costly (Scott and Vigar-Ellis, 2014). Another strategy is to combine environmental benefits with personal benefits that matter to consumers, such as health and safety. Packaging that is positioned as both environmentally friendly and safer for personal use can justify a higher price more easily than sustainability claims alone (Prakash et al., 2019). Finally, clear communication and the use of trusted eco-labels can reduce skepticism around higher prices. When consumers understand and trust the value behind the price, they are more likely to accept and support green packaging choices (Güngördü Belbağ, 2025).

2.5 Conclusion

This chapter explored green packaging within the context of rising environmental pressure, growing waste volumes, and limited consumer adoption. Packaging waste continues to represent a large share of municipal waste, while plastic pollution remains a serious threat to land and marine ecosystems. Although markets for sustainable packaging are expanding and environmental concern is widely expressed, green packaging still accounts for a small share of consumer purchases. This gap highlights the need to better understand both consumer behavior and managerial decision-making.

The chapter first addressed who the green packaging consumer is (Fig. 2.1). The analysis showed that adoption varies across gender, age, education, income, and cultural context. Women, highly educated consumers, and higher-income groups are generally more supportive of green packaging, although financial constraints limit adoption for many consumers who still express environmental concern. Cultural and institutional differences further shape how packaging is evaluated, showing that green packaging preferences are influenced by local waste systems and social norms rather than individual attitudes alone.

The chapter then explained why consumers choose green packaging. Purchase decisions were shown to result from the interaction of moral values, self-identity, emotional responses, and rational evaluations. Consumers are motivated by concern for the environment, personal well-being, and social image, but these motivations only lead to action when green packaging is seen as effective, trustworthy, and easy to use. Emotions such as pride, satisfaction, and guilt

Figure 2.1
Green packaging—consumer motivations and marketing pathways.

reinforce behavior, while beliefs about personal impact, price fairness, and information credibility strongly influence final decisions.

Finally, the chapter discussed how managers can promote green packaging adoption. The findings emphasize that sustainability must not compromise functional performance, especially in food packaging where product protection and waste reduction are critical. Managers should use clear design cues, reduce unnecessary packaging, and rely on credible labels to build trust and avoid greenwashing perceptions. Collaboration across the supply chain and with retailers is essential to ensure consistency, visibility, and proper disposal. Pricing strategies should focus on value perception by combining environmental benefits with functional, emotional, and economic values.

References

Aagerup, U., Frank, A.S., Hultqvist, E., 2019. The persuasive effects of emotional green packaging claims. British Food Journal 121 (12), 3233–3246. Available at https://doi.org/10.1108/BFJ-08-2019-0652.

Afif, K., Rebolledo, C., Roy, J., 2022. Drivers, barriers and performance outcomes of sustainable packaging: a systematic literature review. British Food Journal 124 (3), 915–935. Available at https://doi.org/10.1108/BFJ-02-2021-0150.

Ahmed, R.R., Streimikiene, D., Qadir, H., Streimikis, J., 2023. Effect of green marketing mix, green customer value, and attitude on green purchase intention: evidence from the USA. Environmental Science and Pollution Research 30 (5), 11473–11495. Available at https://doi.org/10.1007/s11356-022-22944-7.

Akturan, U., Kuter, D., Gökkut, B.S., 2025. Mirror, mirror on the wall, who is the most convincing of them all? A systematic literature review on packaging. Journal of Business Research 193, 115355. Available at https://doi.org/10.1016/j.jbusres.2025.115355.

Boz, Z., Korhonen, V., Koelsch Sand, C., 2020. Consumer considerations for the implementation of sustainable packaging: a review. Sustainability 12 (6), 2192. Available at https://doi.org/10.3390/su12062192.

Chang, T.-W., 2023. Double-edged sword effect of packaging: antecedents and consumer consequences of a company's green packaging design. Journal of Cleaner Production 406, 137037. Available at https://doi.org/10.1016/j.jclepro.2023.137037.

Chang, T.W., Chen, Y.S., Yeh, Y.L., Li, H.X., 2021. Sustainable consumption models for customers: investigating the significant antecedents of green purchase behavior from the perspective of information asymmetry. Journal of Environmental Planning and Management 64 (9), 1668–1688. Available at https://doi.org/10.1080/09640568.2020.1837087.

Chen, Y.S., Hung, S.T., Wang, T.Y., Huang, A.F., Liao, Y.W., 2017. The influence of excessive product packaging on green brand attachment: the mediation roles of green brand attitude and green brand image. Sustainability (Switzerland) 9 (4), 654. Available at https://doi.org/10.3390/su9040654.

De Canio, F., 2023. Consumer willingness to pay more for pro-environmental packages: the moderating role of familiarity. Journal of Environmental Management 339, 117828. Available at https://doi.org/10.1016/j.jenvman.2023.117828.

Duarte, P., Silva, S.C., Roza, A.S., Dias, J.C., 2024. Enhancing consumer purchase intentions for sustainable packaging products: an in-depth analysis of key determinants and strategic insights. Sustainable Futures 7, 100193. Available at https://doi.org/10.1016/j.sftr.2024.100193.

Enginkaya, E., Hayrun Sağlam, M., 2025. Navigating sustainability: the role of consumer psychology in shaping sustainable behavior. SAGE Open 15 (4). Available at https://doi.org/10.1177/21582440251376481.

European Commission, 2022. European Green Deal: putting an end to wasteful packaging, boosting reuse and recycling. https://ec.europa.eu/commission/presscorner/detail/en/ip_22_7155.

Friedrich, D., 2021. Market and business-related key factors supporting the use of compostable bioplastics in the apparel industry: a cross-sector analysis. Journal of Cleaner Production 297, 126716. Available at https://doi.org/10.1016/j.jclepro.2021.126716.

Gao, S., Shao, B., 2022. Why do consumers switch to biodegradable plastic consumption? The effect of push, pull and mooring on the plastic consumption intention of young consumers. Sustainability 14 (23), 15819. Available at https://doi.org/10.3390/su142315819.

Granato, G., Fischer, A.R.H., van Trijp, H.C.M., 2022. The price of sustainability: how consumers trade-off conventional packaging benefits against sustainability. Journal of Cleaner Production 365, 132739. Available at https://doi.org/10.1016/j.jclepro.2022.132739.

Güngördü Belbağ, A., 2025. Consumer behavior and sustainability: what role does eco-friendly packaging play in emerging markets? Journal of Economic and Administrative Sciences. Available at https://doi.org/10.1108/jeas-12-2024-0529 [Epub ahead of print].

Hao, Y., Liu, H., Chen, H., Sha, Y., Ji, H., Fan, J., 2019. What affect consumers' willingness to pay for green packaging? Evidence from China. Resources, Conservation and Recycling 141, 21–29. Available at https://doi.org/10.1016/j.resconrec.2018.10.001.

Herbes, C., Beuthner, C., Ramme, I., 2018. Consumer attitudes towards biobased packaging – a cross-cultural comparative study. Journal of Cleaner Production 194, 203–218. Available at https://doi.org/10.1016/j.jclepro.2018.05.106.

Ischen, C., Meijers, M.H.C., Vandeberg, L., Smit, E.G., 2022. Seen as green? Assessing the salience and greenness of environmentally friendly packaging cues. Journal of Food Products Marketing 28 (1), 31–48. Available at https://doi.org/10.1080/10454446.2022.2038757.

Kesgin, M., Can, A.S., Ding, L., Legg, M., Schuler, D., 2025. Legacy matters: encouraging willingness to pay a premium for environmentally friendly off-premises food packaging. International Journal of Hospitality Management 126, 104037. Available at https://doi.org/10.1016/j.ijhm.2024.104037.

Ketelsen, M., Janssen, M., Hamm, U., 2020. Consumers' response to environmentally-friendly food packaging - a systematic review. Journal of Cleaner Production 254, 120123. Available at https://doi.org/10.1016/j.jclepro.2020.120123.

Leong, C.M., She, L., Lim, T.Y., Wong, N.M., 2024. To buy or not to buy? Green packaging, gender differences and the intention to purchase. International Journal of Social Economics 51 (11), 1353–1369. Available at https://doi.org/10.1108/IJSE-04-2023-0249.

López-Miguens, M.J., Álvarez-González, P., Dopico-Parada, A., 2025. Can active and intelligent packaging support sustainability in food sector? Insights from a consumer's viewpoint. Waste Management 205, 115018. Available at https://doi.org/10.1016/j.wasman.2025.115018.

Magnier, L., Crié, D., 2015. Communicating packaging eco-friendliness: an exploration of consumers' perceptions of eco-designed packaging. International Journal of Retail and Distribution Management 43 (4–5), 350–366. Available at https://doi.org/10.1108/IJRDM-04-2014-0048.

Mahmoud, M.A., Tsetse, E.K.K., Tulasi, E.E., Muddey, D.K., 2022. Green packaging, environmental awareness, willingness to pay and consumers' purchase decisions. Sustainability (Switzerland) 14 (23), 16091. Available at https://doi.org/10.3390/su142316091.

Martinho, G., Pires, A., Portela, G., Fonseca, M., 2015. Factors affecting consumers' choices concerning sustainable packaging during product purchase and recycling. Resources, Conservation and Recycling 103, 58–68. Available at https://doi.org/10.1016/j.resconrec.2015.07.012.

McCarthy, B., Wang, P., 2022. Shades of sustainability: who are the buyers and non-buyers of sustainable packaging? Journal of Food Products Marketing 28 (4), 153–178. Available at https://doi.org/10.1080/10454446.2022.2064202.

Moshood, T.D., Nawanir, G., Mahmud, F., Mohamad, F., Ahmad, M.H., AbdulGhani, A., 2022. Why do consumers purchase biodegradable plastic? The impact of hedonics and environmental motivations on switching intention from synthetic to biodegradable plastic among the young consumers. Journal of Retailing and Consumer Services 64, 102807. Available at https://doi.org/10.1016/j.jretconser.2021.102807.

Moshood, T.D., Mahmud, F., Nawanir, G., Ahmad, M.H., Mohamad, F., AbdulGhani, A., 2025. Determinants driving the uptake of biodegradable plastics: a sequential exploratory mixed-methods research approach. Microplastics 4 (1), 7. Available at https://doi.org/10.3390/microplastics4010007.

Mudgal, D, Pagone, E, Salonitis, K, 2024. Selecting sustainable packaging materials and strategies: A holistic approach considering whole life cycle and customer preferences. Journal of Cleaner Production 48, 144133.

Nguyen-Viet, B., Nguyet-Phuong Vo, H., Nguyen-Trang Pham, T., 2024. Modeling consumers' purchase intention for environment-friendly packaged products: an empirical study in Vietnam. Journal of International Food & Agribusiness Marketing 36 (4), 739–762. Available at https://doi.org/10.1080/08974438.2023.2192215.

Noëth, E., Van Opstal, W., Du Bois, E., 2024. Introducing reusable food packaging: customer preferences and design implications for successful market entry. Business Strategy and the Environment 33 (7), 6507–6532. Available at https://doi.org/10.1002/bse.3820.

Nuojua, S., Pahl, S., Thompson, R.C., 2024. Plastic alternatives and substitutes in the packaging sector – a UK consumer perspective. Sustainable Production and Consumption 46, 68–81. Available at https://doi.org/10.1016/j.spc.2024.02.019.

Oloyede, O.O., Lignou, S., 2021. Sustainable paper-based packaging: a consumer's perspective. Foods 10 (5), 1035. Available at https://doi.org/10.3390/foods10051035.

Pan, C., Lei, Y., Wu, J., Wang, Y., 2021. The influence of green packaging on consumers' green purchase intention in the context of online-to-offline commerce. Journal of Systems and Information Technology 23 (2), 133–153. Available at https://doi.org/10.1108/jsit-11-2019-0242.

Popovic, I., Bossink, B.A.G., van der Sijde, P.C., 2019. Factors influencing consumers' decision to purchase food in environmentally friendly packaging: what do we know and where do we go from here? Sustainability 11 (24), 7197. Available at https://doi.org/10.3390/su11247197.

Prakash, G., Choudhary, S., Kumar, A., Garza-Reyes, J.A., Khan, S.A.R., Panda, T.K., 2019. Do altruistic and egoistic values influence consumers' attitudes and purchase intentions towards eco-friendly packaged products? An empirical investigation. Journal of Retailing and Consumer Services 50, 163–169. Available at https://doi.org/10.1016/j.jretconser.2019.05.011.

Scott, L., Vigar-Ellis, D., 2014. Consumer understanding, perceptions and behaviours with regard to environmentally friendly packaging in a developing nation. International Journal of Consumer Studies 38 (6), 642–649. Available at https://doi.org/10.1111/ijcs.12136.

Sonck-Rautio, K., Lahtinen, T., Tynkkynen, N., 2024. Consumer meaning-making of packaging functions for sustainable food packaging–Insights from qualitative research in Finland. Current Research in Environmental Sustainability 7, 100259.

Steenis, N.D., van Herpen, E., van der Lans, I.A., Ligthart, T.N., van Trijp, H.C.M., 2017. Consumer response to packaging design: the role of packaging materials and graphics in sustainability perceptions and product evaluations. Journal of Cleaner Production 162, 286–298. Available at https://doi.org/10.1016/j.jclepro.2017.06.036.

Steenis, N.D., van der Lans, I.A., van Herpen, E., van Trijp, H.C.M., 2018. Effects of sustainable design strategies on consumer preferences for redesigned packaging. Journal of Cleaner Production 205, 854–865. Available at https://doi.org/10.1016/j.jclepro.2018.09.137.

Su, D.N., Duong, T.H., Thanh Tran Dinh, M., Nguyen-Phuoc, D.Q., Johnson, L.W., 2021. Behavior towards shopping at retailers practicing sustainable grocery packaging: the influences of intra-personal and retailer-based contextual factors. Journal of Cleaner Production 279, 123683. Available at https://doi.org/10.1016/j.jclepro.2020.123683.

Suhartanto, D., Kartikasari, A., Hapsari, R., Budianto, B.S., Najib, M., Astor, Y., 2021. Predicting young customers' intention to repurchase green plastic products: incorporating trust model into purchase intention model. Journal of Asia Business Studies 15 (3), 441–456. Available at https://doi.org/10.1108/JABS-04-2020-0150.

Wandosell, G., Parra-Meroño, M.C., Alcayde, A., Baños, R., 2021. Green packaging from consumer and business perspectives. Sustainability 13 (3), 1356. Available at https://doi.org/10.3390/su13031356.

CHAPTER 3

Introduction to the use of bioplastics in food packaging

Mohammad Tarahi[1], Asgar Farahnaky[1], Mehran Ghasemlou[1,2,3], and Mina Dokouhaki[1]

[1]School of Science, STEM College, RMIT University, Melbourne, VIC, Australia, [2]Center for Sustainable Bioproducts, Deakin University, Geelong Waurn Ponds Campus, Geelong, VIC, Australia, [3]Department of Chemical Engineering, The University of Melbourne, Parkville, VIC, Australia

3.1 Introduction

Food packaging plays a crucial role in the supply chain, providing essential services to food products, including protection, preservation, containment, convenience, and communication. Such functionalities are designed to preserve food quality, extend shelf life, enable safe transportation, and deliver comprehensive product information to consumers (Harnkarnsujarit et al., 2021). Among the various packaging materials, plastics have become an automatic choice as a packaging medium, largely due to their versatility, flexibility in processing, low cost, lightweight, and excellent water/gas barrier properties (Doanh et al., 2014). Large-scale production of plastics began in the early 1950s, and our use of plastics witnessed phenomenal growth from 2 million metric tons per year in 1950 to over 415 million tons in 2019, and is projected to reach 1.2 billion tons per year by 2060 (Jayakumar et al., 2023; Rahardiyan et al., 2023). Food packaging is the largest market use of plastics and contributes nearly 30% of all global plastic materials (Westlake et al., 2023). This trend is primarily attributed to a global shift from reusable to disposable plastic containers, driven by rising urbanization, lifestyle changes, and the globalization of food supply chains (Jayakumar et al., 2023). This consumer-oriented desire for convenience has also led to devastating global environmental issues.

Plastics are primarily derived from nonrenewable fossil-based sources and are known to be highly resistant to natural degradation. A recent study estimated that around 79% of plastics have accumulated in landfills/open dumps or the natural environment, while only ~9% are recycled (Westlake et al., 2023). It is also estimated that around 13 million metric tons of plastics enter the oceans every year, threatening marine ecosystems and breaking up into microplastics and nanoplastics (Kong et al., 2023). During the COVID-19 pandemic, plastic production and consumption have been intensified because the use of single-use plastics such

Bioplastics for Food Packaging. DOI: https://doi.org/10.1016/B978-0-443-41654-5.00014-7

as disposable personal protective equipment and takeout containers was encouraged, underscoring the need for innovative and sustainable packaging solutions (Ghasemlou et al., 2024).

In response to increasing environmental concerns and regulatory pressures, academics, manufacturers, and policymakers have turned their attention to bioplastics, which include plastics that are either bio-based, biodegradable, or feature both properties. Some bioplastics can be derived from renewable and biological sources such as starch, cellulose, proteins, and microbial biomass (Kong et al., 2023). Certain types can degrade into biogases and biomass (mostly carbon dioxide and water) under aerobic conditions and methane under anaerobic conditions, thereby solving the end-of-life issues associated with conventional plastics (Rahardiyan et al., 2023). Bioplastics can be engineered to mimic the properties of conventional plastics and tailored to obtain the desired shape for specific applications in the food industry, such as films, trays, semirigid cups, tubs, and containers with similar or even better functionality (Negrete-Bolagay and Guerrero, 2024). Importantly, bioplastics represent a diverse class of materials with varying chemical structures, degradation behavior, and functional characteristics. For instance, some bio-based bioplastics, such as bio-based polyethylene (bio-PE) and bio-based polyethylene terephthalate (bio-PET), are not biodegradable and are designed as sustainable replacements for conventional plastics, while other bioplastics, such as polylactic acid (PLA) and polyhydroxyalkanoates (PHAs), belong to an entirely novel bio-based and biodegradable family that can support the promotion of a circular lifecycle (Rosenboom et al., 2022; Siracusa and Blanco, 2020). However, despite their enormous advantages, bioplastics face several technological, economic, and policy-related hurdles, including limited mechanical strength, poor barrier properties, and higher production costs, which make them less competitive than conventional plastics.

This chapter provides an in-depth overview of bioplastics as emerging materials in the transition toward more circular and sustainable food packaging. It highlights their historical development, classification, biodegradability, and marketability, along with key manufacturing techniques and potential applications across various food systems. The chapter also seeks to provide an overview of current trends and issues related to regulatory standards and addresses the critical challenges and innovations. The insights presented here aim to support policymakers and packaging manufacturers in advancing the adoption of bioplastics toward achieving environmentally responsible packaging solutions.

3.2 Overview of bioplastics development

3.2.1 History of plastics/bioplastics

The origins of bioplastics trace back to ancient times, when natural polymers such as plant gums, shellac, and starch were used in construction, waterproofing, and textile applications (Fig. 3.1). For example, natural latex extracted from rubber trees was used by indigenous

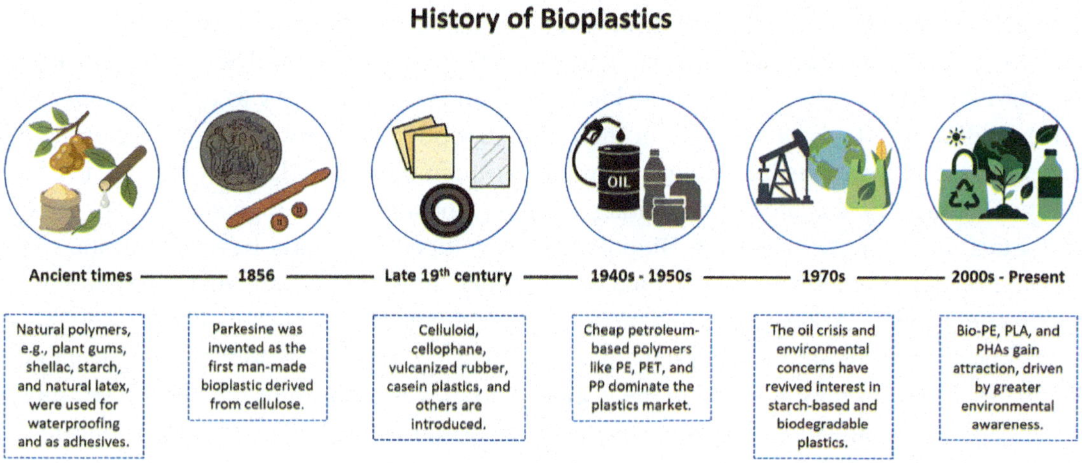

Figure 3.1
History of bioplastics.
The historical development of bioplastics from the ancient use of natural polymers to today's modern bio-based plastics.

people long before European exploration. An early use of biopolymer was in 1495, when Christopher Columbus landed on the island of Haiti and saw its natives playing games with elastic rubber balls (Lackner, 2015). Over many centuries, starch was used as a natural adhesive for paper and fabric, and it was substituted for natural rubber in many waterproof materials. Although these early materials were not similar to the current plastics we use, they mark humanity's initial use of bio-based materials.

The first bioplastics in the modern era are believed to have been discovered during the 19th century. The first man-made bioplastic derived from cellulose (Parkesine) was invented in 1856 by English metallurgist Alexander Parkes. The evolution of bioplastics was followed by celluloid and cellophane, which are both based on modified cellulose and found broad commercial use. In 1907, Leo Hendrik Bakelite invented the first fully synthetic plastic by reacting phenol and formaldehyde, marking the onset of fossil-based plastics (Lackner, 2015). During World War I and II, growing shortages of natural rubber and nylon supplies led to rapid and sustained growth of synthetic plastics, and nitrocellulose bioplastics gradually lost their position (Doanh et al., 2014). With the rising supply of cheap crude oil in the 1940s and 1950s, mass production of petroleum-based plastics such as PE, PET, and polypropylene surged. These materials became the dominant component of every product because of their lightweight, low cost, high performance, and scalability. As a result, biofuels and bioplastics eventually vanished from the markets.

Interest in bioplastics re-emerged during the 1970s crude oil crisis. Indeed, rising oil prices, environmental problems, and favorable legislation by several global governments allowed bioplastics to take over the plastic business. Early efforts were to make a blend of starch and synthetic polymers to enhance biodegradability. Although these hybridized materials offered some environmental benefits, their drawbacks, including their inability to degrade under real-world landfilling, played a key role in their slow adoption. Despite the lack of commercial success, these innovations laid the groundwork for the modern bioplastics industry (Lackner, 2015). The late 1980s and 1990s witnessed many business developments in the area of bioplastics. In 1983, Marlborough Biopolymers developed a bacteria-based bioplastic called biopal. An Italian company, Novamont, was founded in 1990 and remains a key bioplastic manufacturer today. At the same time, major chemical companies, such as BASF and Cargill, began investing in bio-based materials due to increasing public pressure and environmental awareness. Littering plastic waste in the oceans, which was realized in the late 1990s, further fueled capital investment and market changes for biodegradable and compostable plastics. In the 2000s, social media advertising was a critical player in promoting organic and eco-friendly products. In response to growing consumer demand for bioplastics, large companies such as Coca-Cola, Pepsi, and Nestlé began to integrate bioplastics into their production processes. Coca-Cola launched its PlantBottle, which was the first recyclable plastic bottle made partially from bio-based PET. Meanwhile, many biotech companies such as Corbion and Avantium developed and commercialized bioplastic technologies through strategic alliances and partnerships with food and beverage companies (Cooke and Pomeroy, 2023). Braskem in Brazil is the only commercial-scale producer of bio-PE from sugarcane, and NatureWorks in the United States is the leading company of PLA from corn (Lackner, 2015).

Despite the growing demand for bioplastics, a main issue is to keep production costs down. Today, the driving forces behind the viability of bioplastics are rising global awareness of plastic pollution, government regulations, and the growing demand of consumers for environmentally friendly products (Cooke and Pomeroy, 2023). Indeed, consumer behavior for high-quality sustainable/green products is shaping the plastic market, and encouraging businesses to adopt bioplastics not only for food packaging and shopping bags but also for medical devices and clothing.

3.2.2 Classification of bioplastics

According to the European Bioplastics Committee, a material is considered bioplastic if it is either bio-based or biodegradable, or both. This ill-defined terminology is confusing, and the terms "bio-based" and "biodegradable" are used interchangeably by the general public and even industrial plastic experts (Lackner et al., 2023). Bioplastics can be broadly classified based on two fundamental criteria: the source of raw materials and their biodegradability. In terms of source, bioplastics are either bio-based, derived wholly or partly from renewable biological resources (e.g., starch, cellulose, chitin, or keratin protein) or fossil-based, derived from petrochemical feedstocks (Ghasemlou et al., 2024). Regarding biodegradability, bioplastics can be biodegradable, breaking down into water, carbon dioxide, and biomass through aerobic or anaerobic microbial activities, or nonbiodegradable, meaning they persist in the natural environments and exhibit similar behavior to conventional plastics (Rahardiyan et al., 2023). This dual-axis framework leads to four main categories of bioplastics: (1) bio-based and biodegradable, (2) bio-based and nonbiodegradable, (3) fossil-based and biodegradable, and (4) fossil-based and nonbiodegradable (which is conventional plastics) (Fig. 3.2). Importantly, not all bio-based plastics are biodegradable. For example, bio-PE is chemically identical to its fossil-derived counterpart and is not biodegradable, although its monomers are made from renewable resources (Kong et al., 2023). Some other bioplastics, such as poly(butylene adipate-co-terephthalate) (PBAT) and polycaprolactone (PCL), are nonrenewable from fossil resources but biodegradable.

In addition to the source and degradation behavior, bioplastics can also be categorized by their polymer origin into three groups: (1) bioplastics with natural polymers derived from biomass, such as starch, cellulose, and chitosan; (2) bioplastics with chemically synthesized polymers from renewable monomers, such as PLA; and (3) bioplastics with microbial polymers produced by microbial fermentation, such as PHAs (Rahardiyan et al., 2023). Each group has distinct advantages and disadvantages in terms of mechanical strength, thermal resistance, and processability, all of which should be carefully evaluated to select the most appropriate bioplastic for specific food packaging applications.

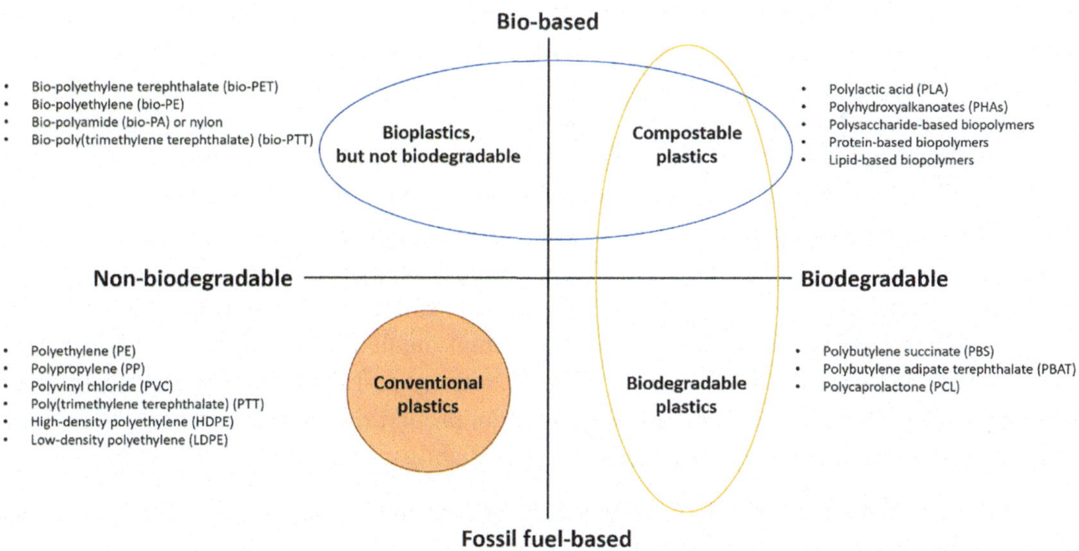

Figure 3.2
Classification of bioplastics.
Classification of bioplastics based on raw material source (bio-based versus fossil-based) and biodegradability (biodegradable versus nonbiodegradable).

3.2.3 Biodegradability and compostability

Biodegradability and compostability are two essential tests in evaluating the end-of-life of bioplastics. Unlike petroleum-based plastics, which can stay for up to a thousand years in the marine environment, some bioplastics are designed to break down rapidly after disposal under specific conditions. With global plastic waste projected to reach 121 million tons by 2050, the urgency to find bioplastics that undergo fast degradation has never been greater (Pottinger et al., 2024). Biodegradation is a process in which microorganisms, such as bacteria and fungi, metabolize polymer chains into smaller molecules, such as water, carbon dioxide, methane, and organic biomass. Compostability is a subset of biodegradability in which the material does not break down naturally but decomposes under human-driven conditions in industrial composting facilities. Compostable bioplastics are further classified into home composting (temperature of ~28°C) and industrial composting (temperature of ~58°C), based on degradation conditions (Cruz et al., 2022; Lackner, 2015).

The biodegradation process involves several distinct stages: biodeterioration, biofragmentation, depolymerization, assimilation, and mineralization. In the initial stage, microorganisms adhere to the surface of the plastic and produce extracellular polymeric substances (EPS), forming biofilms that initiate physical and chemical modifications. These EPS can generate microscopic fissures and pores and change the pH of the surface, facilitating the penetration of enzymes. In subsequent stages, extracellular enzymes such as depolymerases, lipases, esterases, and oxygenases cleave the polymer chains into oligomers and monomers. The enzymatic action in this stage also constitutes the biofragmentation and depolymerization phases. In the assimilation stage, the resulting small molecules are then absorbed by the microbial cells and metabolized mainly into carbon dioxide and water (Rahardiyan et al., 2023). Polymer characteristics such as molecular weight, chain flexibility, hydrophobicity, and crystallinity can significantly impact the effectiveness and speed of the biodegradation. For example, polymers with lower molecular weight and less-ordered amorphous regions are more accessible to enzymatic attack than those with higher molecular weight and ordered crystalline domains (Thakur et al., 2018). Besides, environmental factors, including temperature, moisture, pH, oxygen availability, and salinity, can determine the microbial diversity. Several studies have shown that microbial-mediated degradation is faster under warm and humid conditions that favor microbial activity and growth (Negrete-Bolagay and Guerrero, 2024). Therefore, optimizing both material properties and environmental conditions is essential to enhance the overall efficiency and reliability of bioplastic degradation.

Despite their immense potential, biodegradable and compostable plastics still face critical challenges, and their degradation efficiency varies significantly across different environments. For instance, many bioplastics tend to degrade very slowly in aquatic and oceanic systems, and they show little to no disintegration after several years. Another major concern about the use of biodegradable bioplastics is the formation of micro- and nanoplastics from incomplete degradation, which can accumulate in marine organisms and pose risks to biodiversity and functional ecosystems. Aerobic decomposition of bioplastics typically yields carbon dioxide and biomass; anaerobic decomposition in landfills yields methane. Although compositing is considered a more sustainable alternative to landfilling, the production of greenhouse gases such as methane, with the potential to contribute to global warming, is much higher than that of carbon dioxide. These issues highlight the need for definitive end-of-life options and stricter environmental guidelines for biodegradable plastics to fulfill their intended ecological benefits.

3.2.4 Market size

The global market for bioplastics, particularly in the food packaging sector, has demonstrated substantial growth over the past two decades and is poised to drive further expansion. This growth is driven by increasing environmental awareness among consumers, regulatory pressures against conventional plastics, and ongoing advancements in material science and industrial biotechnology. According to European Bioplastics, global bioplastics production capacity reached approximately 2.4 million tons in 2021 and is expected to surge to 7.6 million tons by 2026 (Rahardiyan et al., 2023). In terms of market value, the sector was worth US$9.17 billion in 2020, and it will grow to over US$26 billion by 2026. This surge of interest represents a compound annual growth rate of more than 10%, which can lead to robust and sustained global economic expansion. Despite this rapid growth, bioplastics still represent only about 0.5%–1% of global plastic production, indicating many opportunities for market penetration and product development. The food packaging sector had about a 47% share of the bioplastic market in 2020 (Cooke and Pomeroy, 2023). This is due to the sector's massive consumption of single-use products, such as trays, food wrappers, drinking bottles, bottle caps, and grocery bags. With the global bans on single-use plastics and clearer legislation for compostable plastic products, food packaging is expected to remain the key growth engine for the bioplastics market. Consumers' attitudes and behaviors toward using bioplastic products are stronger than years ago, and they are willing to pay more for disposable containers or jackets made of bioplastics. An earlier study reported positive attitudes and perceptions of consumers in 42 countries toward replacing petroleum-based plastic-based cutlery for food packaging and serving with those made from biodegradable and bio-based ingredients (Filho et al., 2022). Cost, however, is a major obstacle to widespread adoption of bioplastics. Bioplastics typically cost US$2–6 per kilogram, compared to US$1–2 per kilogram for conventional plastics (Cooke and Pomeroy, 2023). This price gap is largely due to the higher cost of raw materials, lack of large-scale and low-cost production, and ongoing investment in Research and Development. The adoption of bioplastics is particularly more pronounced in cost-sensitive industries such as fast-moving consumer goods, where food packaging constitutes a substantial portion of total costs (Harnkarnsujarit et al., 2021). Although bioplastics are still not at par in price with petroleum-based plastics, increased production capacity and technological advances are expected to make them more price-competitive over the years. From a regional perspective, Asia and Europe are anticipated to dominate the global bioplastics market by 2026, driven by government subsidies and incentives, intense industrial investment, and growing consumer awareness and demand for green products. Historical data shows that Europe held 40% of the global bioplastics market in 2008, and Asian countries significantly increased their market share to 35% in 2018 (Byun and Kim, 2014). Although North America has been a significant player for years, its contribution has dropped due to faster growth rates of the Asian market. In terms of material types, PLA and starch blends have been leading contributors to the biodegradable bioplastics market. Some market analyses, such as a survey by Freedonia and Pira International, had projected PLA to capture around 40%–42.5% of the bioplastics market in the 2010s, with starch blends holding 20%–30%. The fastest-growing segment in the following years was bio-based and nonbiodegradable plastics, such as bio-PET and bio-PE, which together made up over 50% of bioplastics production capacity in 2011 (Byun and Kim, 2014). The high compatibility with the existing plastic processing and less reliance on fossil fuels make them particularly attractive for large-scale industrial applications.

3.3 Techniques to manufacture bioplastics for food packaging

The manufacture of bioplastics for food packaging relies on a range of processes that convert raw materials into functional packaging components. Each technique influences the structural and barrier properties of the final product and has different scalability, biodegradability, and environmental footprints. This section provides an overview of the key techniques used to produce bioplastic-based packaging materials and explores their diverse applications in preserving food quality and safety. Fig. 3.3 presents key techniques for manufacturing bioplastics based on their processing principles.

3.3.1 Polymerization techniques

Polymerization is a chemical synthesis method by which monomers derived from renewable sources are converted into macromolecular bioplastics. One of the most well-known examples is the conversion of lactic acid into PLA via ring-opening polymerization. This method has an important role in enabling the production of biodegradable plastics with high molecular weight and distribution, which are essential for achieving the desired mechanical and barrier properties (Yu et al., 2023). In addition to PLA, other bio-based materials, such as whey protein, have been polymerized to develop novel biodegradable plastics (Chalermthai et al., 2021). However, polymerization processes have their own challenges. The presence of residual unreacted monomers or unwanted by-products in the bioplastic can adversely affect their mechanical performance, thermal stability, and safety. Additionally, the overall economic viability of polymerization for bioplastics production is influenced by the

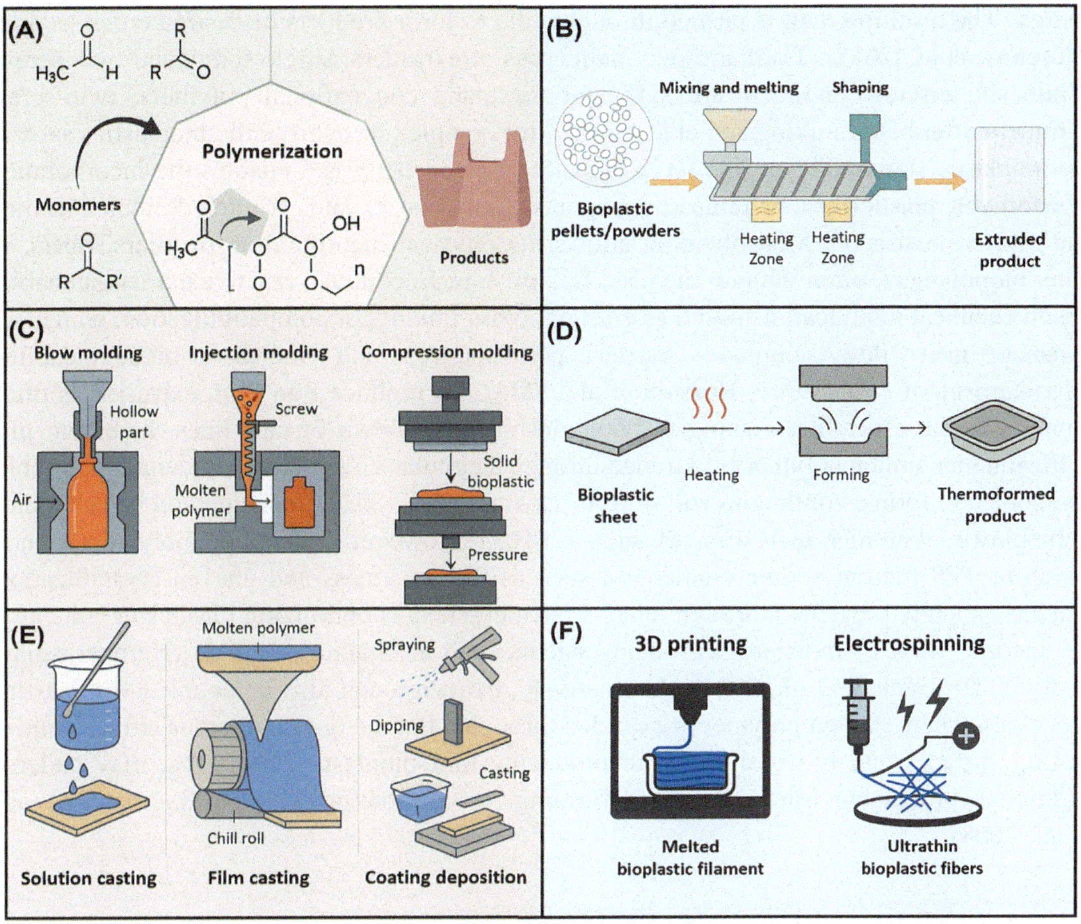

Figure 3.3
Overview of key techniques to manufacture bioplastics for food packaging applications. (A) Polymerization of renewable monomers, (B) extrusion-based processing, (C) molding methods (blow, injection, and compression molding), (D) thermoforming of heated sheets, (E) casting and coating techniques, and (E) advanced methods such as three-dimensional printing and electrospinning.

availability and cost of renewable monomer feedstocks. In food packaging applications, the purity, nontoxicity, and safety of the products or ingredients must be carefully assessed to ensure consumer safety.

3.3.2 Extrusion-based techniques

Extrusion is a dominant and highly versatile technique used in the continuous production of bioplastic films, sheets, pellets, and tubes. It involves feeding a biopolymer, often in pellet or powder form, into a screw-driven barrel, where it is subjected to thermal and mechanical energy. The resulting melt is pushed through a die to form products of desired cross sections (Rosenow et al., 2025). There are two main types of extruders: single screw and twin screw. While single-screw extruders are sufficient for many conventional polymers, twin-screw extruders offer better mixing and are preferred for complex or hydrophilic bioplastics such as thermoplastic starch (TPS). This setup enhances homogeneity and enables the incorporation of additives, plasticizers, or reinforcing agents. Extrusion extends to include melt blending and reactive extrusion. Melt blending allows the physical incorporation of fillers, fibers, or other biopolymers, often without the need for solvents. In contrast, reactive extrusion enables in situ chemical modifications, such as grafting, cross-linking, or compatibilization, which can enhance melt flow, improve barrier performance, and increase biodegradability (Harnkarnsujarit et al., 2021; Hossain et al., 2024). To produce thin film, extrusion is often combined with other film-making technologies such as blown or cast lines. In blown film extrusion, the molten polymer is extruded through a circular die, inflated into a tubular bubble, and cooled to form a continuous roll of film (Hossain et al., 2024). This method is well-suited to bioplastics with high melt strength, such as PBAT. However, hydrophilic polymers such as starch or TPS present certain challenges, such as film tackiness and uneven crystallization, that are typically prevented through many strategies, such as optimizing plasticizer content or the incorporation of moisture-scavenging agents such as titanium dioxide (Harnkarnsujarit et al., 2021; Siddiqui et al., 2024). Alternatively, extrusion can also be coupled with casting processes, where molten polymer is extruded through a flat die onto numerous chilled rolls or casting surfaces and is widely used in producing thin bioplastic sheets that may undergo additional processing such as thermoforming or lamination (Li et al., 2015; Wang et al., 2008).

3.3.3 Molding techniques

Molding techniques are indispensable for shaping bioplastics into solid packaging items and offer unique advantages in processing speed, design flexibility, and mechanical reliability. Among these, injection molding is the most widely adopted. In this process, melted bioplastic pellets are injected into a closed cavity mold under high pressure and subsequently cooled to form three-dimensional (3D) components, such as caps, closures, and food containers. Screw-type injection systems offer better melt mixing and material homogeneity (Cruz et al., 2022; Harnkarnsujarit et al., 2021). Compression molding is a more cost-effective solution for producing bulkier components or materials with low flowability. In this process, the polymer is placed in a preheated mold cavity and shaped under compressive force. This method is especially well-suited for natural fiber-reinforced composites, such as those with bamboo or cellulose pulp (Siddiqui et al., 2024). It also generates less material waste, which is useful when working with expensive or thermally sensitive materials. Blow molding is another technique to produce hollow packaging products such as bottles. It often includes extrusion and injection blow molding. In extrusion blow molding, a molten tube (i.e., parison) is extruded and inflated inside a mold, while in injection blow molding, the parison is first formed by injection and then transferred to a second mold for inflation. Bioplastics such as PLA and bio-PET are commonly used for this purpose (Harnkarnsujarit et al., 2021). Critical parameters, including wall thickness, crystallization behavior, and cooling efficiency, strongly influence the structural and mechanical performance of the final product.

3.3.4 Thermoforming techniques

Thermoforming is a widely used technique in the food packaging industry to convert pre-extruded bioplastic sheets into 3D forms such as trays, lids, and cups. The process typically begins with the production of flat sheets, typically using cast extrusion. These sheets are then heated above their glass transition temperature (T_g) or slightly below their melting point (T_m). At this stage, the material softens but does not melt. The softened sheet is clamped and molded using vacuum or air pressure to achieve the desired shape (Subramanian, 2024). Unlike molding techniques that shape molten bioplastics directly, thermoforming works with solid sheets, making it a secondary shaping process. This provides several advantages. The method is less complex and more cost-effective, particularly for mid-volume production, and it enables high throughput with minimal material waste. However, thermoforming has limited design flexibility and lower dimensional accuracy compared to injection molding (Cruz et al., 2022; Harnkarnsujarit et al., 2021). Materials like PLA and TPS/PLA blends are commonly used in this process. For instance, trays fabricated from TPS/PLA blends using combined twin-screw extrusion and vacuum thermoforming have demonstrated excellent performance in freeze–thaw cycles and under humid storage conditions. Moreover, incorporating fillers like zeolites has been shown to improve gas barrier properties and odor retention (Harnkarnsujarit et al., 2021). However, excessive filler loading may cause processing defects such as pinholes, which can compromise the integrity of the packaging.

3.3.5 Casting and coating techniques

Casting and coating techniques are primarily used for producing thin films and functional packaging surfaces. In solution casting, biopolymers such as starch, proteins, and chitosan are dissolved in a suitable solvent and then spread onto a flat substrate. A uniform film is formed upon evaporation of the solvent. This method is especially suitable for laboratory-scale studies and for making bioactive films containing antimicrobial or antioxidant agents (Shah et al., 2025a, 2025b). Film casting, although closely related, may also involve melt casting, where the polymer is applied in molten form rather than a solution. This approach is frequently used to fabricate starch-based, agar-based, or composite films containing fillers like natural fibers or nanoparticles and enables greater flexibility in both formulation and structure (Hossain et al., 2024). Coating deposition techniques such as spraying, dipping, and casting are used to enhance the surface properties of packaging materials. These coatings may act as oxygen and moisture barriers or deliver targeted bioactive functionalities, such as antimicrobial activity (Siddiqui et al., 2024; Zhang et al., 2023). For instance, a chitosan-based coating enriched with combined orange essential oil and potassium permanganate demonstrated improved surface uniformity and significantly increased water barrier performance (Shah et al., 2025a, 2025b). Despite their advantages, casting/coating techniques face limitations such as uneven film thickness, prolonged drying times, and poor scalability. Additionally, solvent recovery and environmental impact also remain significant concerns for industrial-scale production.

3.3.6 Three-dimensional printing and electrospinning techniques

Advanced fabrication methods, such as 3D printing and electrospinning, offer precise control over structure and functionality, making them highly promising for next-generation bioplastic applications. 3D printing, particularly fused deposition modeling, fabricates objects layer-by-layer from melted bioplastic filaments such as PLA and its blends. This method allows for rapid prototyping and the development of highly customized packaging designs, including smart packaging systems (Ali et al., 2024). However, its successful implementation requires precise calibration of printing parameters and may involve postprocessing steps, such as surface finishing or assembly, to achieve the desired functionality and appearance. On the other hand, electrospinning uses a high-voltage electric field to produce ultrafine bioplastic fibers from a polymer solution. In this process, the polymer solution is drawn into a jet and stretched into nanofibers, which are collected as a nonwoven mat. These mats have high surface area, excellent mechanical properties, and the ability to encapsulate bioactive compounds without thermal degradation. Such features make them valuable for food packaging applications. Electrospun layers have also been used as interlayers in multilayer bioplastic films to improve barrier performance (Akhouy et al., 2023; Siddiqui et al., 2024). While both methods hold significant potential, they remain resource- and equipment-intensive. As such, they are currently more applicable to laboratory research or high-value niche applications rather than large-scale commercial production.

3.4 Emerging applications of bioplastics in food packaging

3.4.1 Fruits and vegetables

Fruits and vegetables are highly perishable commodities that continue their metabolic activities postharvest, making them particularly vulnerable to rapid spoilage. Physiological and biochemical deterioration, such as water loss, tissue softening, oxidative browning, and microbial decay, can lead to a staggering global food loss of approximately 500 million tons annually. With 20%–30% of postharvest fruits and vegetables wasted, the need for effective and sustainable packaging materials is critical (Hadidi et al., 2024). As summarized in Table 3.1, bioplastics, particularly those incorporated with active or intelligent agents, offer

Table 3.1 Emerging applications of bioplastics to preserve a wide range of fruits and vegetables.

Bioplastic matrix	Additive(s)	Food product(s)	Packaging type	Packaging properties	Food preservation	References
PLA	ZnO NPs and cinnamaldehyde	Apple	Films	• PLA nanoblend film with 6 wt.% cinnamaldehyde and ZnO NPs showed lower OP and higher WVP than pure PLA • Elongation at break and elastic modulus increased, while TS decreased with ZnO addition	– Nanoblend packaging retained firmness, total phenolics, and color of fresh-cut apple during 14-day storage at $4°C \pm 1°C$ – Microbial growth was significantly inhibited, enhancing food safety and extending shelf life	Li et al. (2017)
PLA	BEO, TiO_2 NPs, and Ag NPs	Mango	Films	• PLA nanocomposite films showed lower weight loss than PLA and PLA/BEO films between days 9 and 12 • Films incorporating BEO and NPs improved barrier and antimicrobial properties	– Delayed firmness loss, color change, and vitamin C degradation over 15 days at room temperature – Better control of total acidity and microbial growth than non-nanocomposite PLA-based films – Extended postharvest shelf life of mangoes up to 15 days without compromising sensory acceptance	Chi et al. (2019)
PLA and CMC	-	Mango	Films	• PLA films blended with up to 4% CMC showed the highest moisture absorption • The 75 μm thick PLA/CMC films effectively controlled relative humidity inside the package	– Mangoes packed with 4% CMC film had a shelf life of 42 days, compared to ~21 days for unpacked controls – Maintained better firmness, color, TSS, and acidity throughout export storage conditions	Kamthai and Magaraphan (2019)
PLA	-	Peach and cherry tomato	Bags	• Three-layer PLA films with high water vapor permeability were used for EMAP • PLA films showed higher WV permeability than conventional OPP films	– EMAP maintained the freshness, color, and aroma of fresh produce – Optimized gas composition within the package helped preserve sensory and physicochemical quality – Bio-based EMAP extended shelf life without compromising produce quality	Mistriotis et al. (2016)

(*Continued*)

Table 3.1 (Continued)

Bioplastic matrix	Additive(s)	Food product(s)	Packaging type	Packaging properties	Food preservation	References
PLA	-	Spinach	Trays	• Maintained comparable shelf life to PET for both fresh-cut and cooked spinach • Require improvement to address water accumulation issues	– PLA trays preserved the flavor of fresh spinach longer than PET	Botondi et al. (2015)
PLA and PBAT	ZnO NPs and GSE	Onion, cabbage, and carrot	Films	• PLA/PBAT film showed moderate antioxidant capacity, 36.3% (ABTS) and 21.9% (DPPH)	– Combined treatment with UV-C LED and antimicrobial film reduced Escherichia coli by up to 2.82 log CFU/g and *Staphylococcus aureus* by up to 1.77 log CFU/g after 7 days at 10°C – Synergistic antibacterial effect achieved across different vegetables	Kim et al. (2022)
PLA and PBAT	-	Broccoli	Films	• Nature Fresh film (PLA/PBAT blend) is biodegradable and designed as a stretch-cling alternative to petroleum-based films	– Broccoli packaged with Nature Fresh had similar gas composition and pigment retention to LLDPE-wrapped samples – Maintained acceptable sensory quality of broccoli for at least 21 days at 2°C	Paulsen et al. (2022)
PLA and PS	-	Cherry tomato	Films	• Bilayer PS/PLA films improved toughness, thermal stability, and barrier properties compared to PLA alone • PLA layer enhanced water resistance and tensile strength over monolayer PS film	– Reduced weight loss and better retention of organic acids and vitamin C in packaged cherry tomatoes – Bilayer films helped extend shelf life and maintain the nutritional quality of fresh produce	Zhou et al. (2019)
PLA and PBS	OEO	Lettuce	Films	• Films with OEO exhibited antimicrobial and antioxidant activity, and enhancing bioactive packaging functionality	– PUFAs and phytosterols were preserved up to 8 days in films with 5%–10% OEO, but decreased without OEO – Nutritional quality of lettuce remained stable with OEO-enriched packaging during cold storage	Llana-Ruíz-Cabello et al. (2019)

PBS and PBAT	-	Mango	Films	• 20 wt.% PBAT improved PBS film's elongation by 373%, while maintaining 27 MPa TS and 262 MPa Young's modulus	– Color, TSS, and weight loss remained stable during 33 days of storage	Winotapun et al. (2024)
PBAT, TPS, and CNW	-	Lettuce	Films	• TPS increased water vapor permeability, while CNW led to porous surface structures	– Films demonstrated antimicrobial activity within 10 days against molds, yeasts, and psychrotrophic bacteria	da Silva et al. (2019)
PHA and CS	Cinnamaldehyde	Strawberry	Films	• Increasing the PHA:CS ratio increased particle size and shifted surface charge from negative to positive • Microspheres showed high encapsulation efficiency and biodegradability, with physical interaction confirmed by FTIR	– Cinnamaldehyde-loaded microspheres effectively inhibited bacterial growth, helping preserve strawberries during storage	Yang et al. (2020)
CMC and GG	Ag^0NC	Strawberry	Films	• Structural and functional characterization confirmed nanoparticle dispersion and thermal stability • Films showed strong antimicrobial activity against 10 foodborne pathogens via zone of inhibition assays	– Extended shelf life and preserved freshness of strawberries, better than conventional packaging – Minimized weight loss and maintained visual quality during storage	Kanikireddy et al. (2020)

ABTS: 2,2'-azino-bis(3-ethylbenzothiazoline-6-sulfonic acid); Ag^0NC: silver nanocomposite; BEO: bergamot essential oil; CMC: carboxymethyl cellulose; CNW: cellulose nanowhiskers; CS: chitosan; DPPH: 2,2-diphenyl-1-picrylhydrazyl; EMAP: equilibrium-modified atmosphere packaging; FTIR: Fourier transform infrared spectroscopy; GG: guar gum; GSE: grapefruit seed extract; LLDPE: linear low-density polyethylene; NPs: nanoparticles; OEO: oregano (*Origanum vulgare*) essential oil; OP: oxygen permeability; PBAT: poly(butylene adipate-co-terephthalate); PBS: poly(butylene succinate); PET: polyethylene terephthalate; PHA: polyhydroxyalkanoate; PLA: polylactic acid; PS: pea starch; PUFAs: polyunsaturated fatty acids; TPS: thermoplastic starch; TS: tensile strength; TSS: total soluble solid; WVP: water vapor permeability.

innovative solutions to extend the shelf life of fresh products while supporting global sustainability goals.

One of the most widely studied bioplastics for fruit and vegetable packaging is PLA, which is valued for its compatibility with natural additives and nanomaterials. For example, PLA/ZnO nanofilms containing 6 wt.% cinnamaldehyde effectively preserved fresh-cut apples stored at 4°C ± 1°C for 14 days. These films improved firmness, phenolic content, and microbial safety while reducing polyphenol oxidase activity and browning (Li et al., 2017). Similarly, PLA nanocomposite films with bergamot essential oil, nano-TiO_2, and nano-Ag extended the shelf life of mangoes to 15 days by reducing weight loss from day 9 to day 12, and maintaining vitamin C, color, and total acidity (Chi et al., 2019). Another promising application involves PLA/carboxymethyl cellulose (CMC) blends also effectively reduced respiration and ethylene production in "Nam Dok Mai" mangoes (*Mangifera indica* Linn). At 4% CMC concentration, this packaging significantly extended the mango shelf life up to 42 days under export conditions (13°C ± 1°C, 90% ± 5% RH), outperforming unpackaged controls. Such bioplastics not only provide a protective barrier but also offer active functions, such as humidity regulation, that are essential for preserving moisture-sensitive fruits (Kamthai and Magaraphan, 2019). Besides, equilibrium-modified atmosphere packaging systems using microperforated multilayer PLA can further improve the preservation of perishable products such as peaches and cherry tomatoes (Mistriotis et al., 2016). In leafy vegetables such as spinach, PLA hinged trays better preserved aroma and color than PET trays over 16 days at 4°C (Botondi et al., 2015). Likewise, Nature Fresh stretch films (PLA/PBAT blend) maintained the quality of broccoli heads for at least 21 days at 2°C, comparable to conventional linear low-density polyethylene while providing environmental benefits (Paulsen et al., 2022). In lettuce, PLA/polybutylene succinate (PBS) (95:5%) films containing 5%–10% oregano (*Origanum vulgare*) essential oil (OEO) effectively preserved polyunsaturated fatty acids (PUFAs) and phytosterols (PHSTs) for up to 8 days, while films without OEO showed significant degradation (Llana-Ruíz-Cabello et al., 2019). Recent advances have focused on antimicrobial and controlled-release systems in bio-based and biodegradable matrices. For instance, PHA/chitosan porous microspheres loaded with cinnamaldehyde exhibited adjustable release profiles from 7 to 60 days, significantly extending the shelf life of strawberries and inhibiting microbial growth (Yang et al., 2020). Similarly, CMC/guar gum/silver nanocomposite films showed strong antimicrobial activity against 10 foodborne pathogens, effectively reducing weight loss and maintaining strawberry freshness during storage (Kanikireddy et al., 2020).

3.4.2 Bakeries and pastas

Bakery and pasta products are staple foods worldwide but are highly susceptible to spoilage and staling. According to Goryńska-Goldmann et al. (2021), bakery companies report average daily losses ranging from 9.7% to 14.4% of their total production, with bread and fresh pastries being the most affected items. In response, bioplastic packaging offers a sustainable solution that not only reduces environmental impact but also enhances product preservation through the use of active and intelligent agents (Table 3.2).

In bakery applications, bioplastics have shown remarkable performance and effectiveness in preserving freshness and extending shelf life. For instance, electrospun composite fibers made from zein and PLA, containing 5%–20% carvacrol, demonstrated antioxidant activity ranging from 62% to 75% for zein and 53% to 65% for PLA. These films exhibited a controlled-release profile and effectively delayed spoilage in whole wheat bread (Altan et al., 2018). Similarly, gelatin-based films enriched with black pepper (*Piper nigrum*) oleoresin significantly reduced water vapor and oil permeability while improving mechanical flexibility, which contributed to maintaining bread texture and sensory attributes (Figueroa-Lopez et al., 2018). In another study, bio-nanocomposite films composed of corn starch, chitosan, and nanoclay, plasticized with sorbitol and reinforced with grapefruit seed extract, showed outstanding antifungal properties. These films prevented fungal growth on bread for up to 20 days at 25°C and 59% RH, compared to only 6 days for synthetic plastic films (Jha, 2020). In a further study, incorporation of 2% zinc oxide nanoparticles into chitosan/CMC/oleic acid films significantly decreased water vapor permeability from 8.27×10^{-7} to 1.96×10^{-7} g/m·h·Pa. These films also increased the microbial shelf life of sliced wheat bread from 3 days to 35 days and helped retain moisture and reduce firmness loss during storage (Noshirvani et al., 2017). Cakes, including butter and sponge types, have also benefited from active bioplastic packaging systems. Bovine gelatin nanocomposite-emulsion films significantly enhanced barrier properties, reducing water vapor permeability from 9.680×10^{-1} to 6.067×10^{-10} g/m/s·Pa, and oxygen permeability from 39.262 to 29.645 $cm^3 \cdot \mu m/m^2 \cdot d \cdot kPa$. These improvements contributed to extended freshness and reduced fungal growth in sponge cakes without preservatives (Sahraee et al., 2020). In another study, acetylated cassava starch films with 1%–10% maltol delayed mold growth on butter cakes by up to 6 times (maximum 19 days), while preserving aroma and sensory qualities (Promhuad et al., 2022). In addition, chitosan films incorporating potassium sorbate or vanillin showed no visible mold for 8 days, whereas mold appeared by day 4 and day 7 in cakes wrapped with commercial stretch film and plain chitosan film, respectively (Sangsuwan et al., 2015).

In the context of pasta packaging, bioplastics have shown potential for both dry and fresh variants. For dry pasta, biodegradable films made from triticale flour maintained physicochemical stability over 45 days, with no significant changes in moisture, breaking force, or cooking quality compared to commercial packaging. Although antioxidant activity (2,2'-azino-bis(3-ethylbenzothiazoline-6-sulfonic acid) assay) declined slightly, total phenolic content and ferric-reducing ability of plasma-based antioxidant activity remained stable during storage (Aranibar et al., 2020). Bioplastics packaging solution for fresh pasta has also shown promising results. Extruded films composed of rice flour, PBAT, glycerol,

Table 3.2 Emerging applications of bioplastics to preserve bakery and pasta products.

Bioplastic matrix	Additive(s)	Food product	Packaging type	Packaging properties	Food preservation	References
Zein and PLA	Carvacrol	Bread	Fibrous films	• Antioxidant activity increased with carvacrol content; up to 75% for zein and 65% for PLA • Carvacrol altered fiber morphology and enhanced functional properties of the films	– Films helped preserve bread freshness through sustained antioxidant release – Suitable for active packaging of bakery products with minimal quality loss	Altan et al. (2018)
Gelatin	Clove, nutmeg, and black pepper oleoresins	Bread	Films	• Oleoresin incorporation improved film elongation and reduced water vapor and oil permeability • Black pepper oleoresin provided the most favorable barrier and structural properties	– Films with black pepper oleoresin best preserve bread quality during storage	Figueroa-Lopez et al. (2018)
CS, chitosan, and nanoclay	KS and GFSE	Bread	Films	• CS/chitosan/nanoclay/SOR/GFSE films had the highest crystallinity and thermal stability • SOR-plasticized films showed the highest TS and the lowest WVP • Hydrogen bonding confirmed via FTIR as the main interaction; enhanced barrier performance	– No fungal growth on bread packed in GFSE films for 20 days (vs. 6 days for plastic) – GFSE films showed strong antifungal activity (max. inhibition against *Aspergillus niger*) – Films maintained bread safety and extended shelf life at 25°C, 59% RH	Jha (2020)
CMC, chitosan, and OA	ZnO NPs	Bread	Films	• WVP decreased from 8.27×10^{-7} to 1.96×10^{-7} g/m·h·Pa with 2% ZnO NPs • ZnO NPs and OA enhanced the moisture barrier and film structure	– No fungal growth observed for 15 days with 1%–2% ZnO NP coatings – Microbial shelf life extended from 3 to 35 days with 2% ZnO NPs coating	Noshirvani et al. (2017)
CG, gelatin, and FA	*Cymbopogon citratus* EO	Bread	Films	• Elongation at break increased from 2% to 140% with EO incorporation • Solubility significantly decreased due to FA-induced cross-linking • EO-containing films had lower rupture stress but maintained smooth morphology	– CG/gelatin/EO films extended bread shelf life from 3 to 6 days vs. commercial film – Bread packed in CG/gelatin/EO films showed increased harshness due to higher permeability	Oliveira et al. (2020)

Gelatin, GN, GE, or GNE	-	Sponge cake	Films	• Nanoparticles reduced WVP from 9.680 ± 0.460 to 6.067 ± 0.337 × 10^{-10} g·m/s·m^2·Pa • OP decreased from 39.262 to 29.645 cm^3·μm/m^2·d·kPa with nanocomposite films • GNE films showed the highest overall barrier performance among all tested materials	– GNE films significantly reduced fungal growth in preservative-free sponge cake – Cakes in GNE films had better texture and higher organoleptic acceptance during storage – Acidity and peroxide values remained lower in cakes packaged with GNE, indicating better stability	Sahraee et al. (2020)
Cassava starch	Maltol	Butter cake	Films	• Incorporation of 1%–10% maltol reduced film strength and flexibility • Increased maltol modified film permeability to water vapor and oxygen	– Butter cake wrapped in films with 10% maltol showed no mold for up to 19 days – Fungal growth was delayed 2.7× with 1%–5% maltol in films	Promhuad et al. (2022)
Chitosan	KS or vanillin	Butter cake	Films	• KS reduced transparency and strength but increased flexibility and OP • Vanillin decreased film strength and permeability and increased yellow coloration • Chitosan films offered improved barrier and antimicrobial properties vs. commercial film	– Mold appeared on cake by day 4 (stretch film) and day 7 (plain chitosan film) – No mold observed for 8 days with chitosan films containing vanillin or sorbate	Sangsuwan et al. (2015)
Triticale flour	-	Pasta	Bags	• Biodegradable triticale flour bags showed stable performance over 45 days	– No microbial growth, moisture changes, or breaking force differences observed – Cooking quality remained unaffected by packaging type or storage time – Total phenolic content and FRAP antioxidant activity stayed stable throughout	Aranibar et al. (2020)
Rice flour and PBAT	KS	Pasta	Films	• Sorbate-containing films reduced slime formation on the pasta surface	– Microbial counts were lower in pasta with migrated sorbic acid from films	Sousa et al. (2016)

(*Continued*)

Table 3.2 (Continued)

Bioplastic matrix	Additive(s)	Food product	Packaging type	Packaging properties	Food preservation	References
Chitosan	Chestnut extract	Pasta	Sachets	• Effective migration allowed preservation while keeping final sorbic acid levels low • Films showed adequate mechanical strength for use as sachets in pasta packaging	– Addition of 3 g/100 g KS extended shelf life effectively – Microbial growth was completely inhibited on fresh pasta for 60 days at 8°C – Moisture migration during the first 9 days led to starch retrogradation	Korge et al. (2020)

CG: cashew gum; CMC: carboxymethyl cellulose; CS: corn starch; CG: cashew gum; EO: essential oil; FA: ferulic acid; FRAP: ferric-reducing ability of plasma; FTIR: Fourier transform infrared spectroscopy; GE: gelatin emulsion; GFSE: grapefruit seed extract; GN: gelatin nanocomposite; GNE: two-layer gelatin nanocomposite and gelatin emulsion; KS: potassium sorbate; NPs: nanoparticles; OA: oleic acid; OP: oxygen permeability; PBAT: poly(butylene adipate-co-terephthalate); PLA: polylactic acid; SOR: sorbitol; TS: tensile strength; WVP: water vapor permeability.

and potassium sorbate effectively reduced microbial growth and surface slime while maintaining low sorbic acid migration into the product (Sousa et al., 2016). Additionally, chitosan-based films incorporating tannin-rich chestnut extract preserved filled fresh pasta for 60 days under refrigerated conditions (8°C, 60% RH). These films prevented microbial growth and delayed retrogradation, particularly during the first 9 days of storage (Korge et al., 2020). Overall, these studies underscore the strong potential of bioplastics to replace conventional plastics and improve the preservation, safety, and sustainability of bakery and pasta products.

3.4.3 Meat, poultry, and seafood

Muscle-based food products are highly perishable due to their rich nutritional content, high moisture levels, and susceptibility to microbial growth and oxidative degradation, even under refrigerated conditions. Traditional plastic packaging offers a physical barrier to external contamination, but it poses serious environmental concerns due to poor recyclability and persistence in landfills (Hadidi et al., 2024; Zhang et al., 2023). In this respect, bioplastics have emerged as a sustainable alternative to extend shelf life, enhance food safety, and maintain sensory quality in meat, poultry, and seafood products (Table 3.3). In red meat packaging, PLA composite films enriched with *Mentha piperita* (MEO) and *Bunium persicum* (BPO) essential oils, along with nanocellulose, significantly enhanced microbial stability and extended the shelf life of ground beef from 4 to 7 days. These films also maintained acceptable sensory qualities during 12 days of refrigerated storage (Talebi et al., 2018). Similarly, corn starch-based films embedded with *Zataria multiflora* essential oil (ZEO) and cinnamaldehyde in nanoemulsion form achieved improved oxidative stability, with low peroxide values of 3.70 meq/kg lipid and thiobarbituric acid reactive substances (TBARS) of 1.03 mg MDA/kg (Amiri et al., 2019). In pork packaging, Chinese yam starch films containing 3% eugenol exhibited controlled-release behavior and effectively reduced microbial growth. These films extended the shelf life of pork by over 50%, showcasing the practical utility of natural antimicrobials embedded in biodegradable matrices (Cheng et al., 2019). Processed meats like ham and sausages also benefit from bioplastic packaging. For instance, polyhydroxybutyrate/polycaprolactone (PHB/PCL) films activated with 4000 IU/cm^3 of nisin inhibited *Lactobacillus plantarum* and extended the shelf life of sliced ham (Correa et al., 2017). Starch films developed with subcritical water technology and loaded with chitosan and carvacrol showed complete inhibition of *Listeria monocytogenes* over 28 days at 4°C, although native microbiota was only partially inhibited (Zhao et al., 2019). In the case of dry sausage, PLA films containing up to 40% propolis extract exhibited concentration-dependent antimicrobial activity. These films were more effective against *Staphylococcus aureus* and retained high phenolic content that was gradually released into the food matrix (Safaei and Roosta Azad, 2020).

Table 3.3 Emerging applications of bioplastics to preserve meat, poultry, and seafood products.

Bioplastic matrix	Additive(s)	Food product(s)	Packaging type	Packaging properties	Food preservation	References
PLA	MEO, BPO, and NC	Ground beef	Films	• WVP decreased as EO concentration increased, improving barrier properties • Addition of EOs reduced lightness and increased yellowness of PLA composite films	– Active films with 1% MEO or BPO extend ground beef shelf life from 4 to 7 days at 4°C – PLA/0.5% MEO/NC film yielded the best sensory acceptance among treatments during 12-day storage	Talebi et al. (2018)
Corn starch	ZEO and CIN	Ground beef patties	Films	• Nanoemulsified essential oils improved film flexibility and performance • NZEOC showed no improvement over NZEO + CIN in antioxidant performance	– NZEO film provided the best oxidative stability; PV = 3.70 meq/kg, TBARS = 1.03 mg MDA/kg, and carbonyl = 0.83 nmol/mg – NZEO film yielded the highest sensory acceptability score after 20 days at 4°C ± 1°C	Amiri et al. (2019)
Yam starch	Eugenol	Pork	Films	• D-sorbitol plasticized yam starch films showed enhanced compatibility for eugenol incorporation • Eugenol release was effectively controlled via diffusion and erosion mechanisms	– Pork wrapped in films with 3% eugenol had a shelf life extended by more than 50% – Antibacterial activity was dose-dependent and most effective against *Escherichia coli* > *Listeria monocytogenes* > *Staphylococcus aureus*	Cheng et al. (2019)
PHB and PCL	Organo-clays + nisin	Ham	Films	• Organo-clays (Cloisite 30B & 10 A) improved tensile strength, thermal stability, and water vapor barrier • Nisin adsorption was not hindered by clay presence, allowing a dual-function nanocomposite design	– Nisin-activated PHB/PCL films effectively suppressed *Lactiplantibacillus plantarum* growth on sliced ham	Correa et al. (2017)

Cassava starch	Gallic acid, chitosan, and/or carvacrol	Ham	Films	• Chitosan and carvacrol enhanced antimicrobial activity, while gallic acid had a minimal effect • Films maintained physical integrity throughout storage and supported bioactive release	– Films with chitosan and carvacrol fully inhibited *L. monocytogenes* on ham during 28 days at 4°C – Growth of native meat microbiota was delayed by 1–2 weeks by chitosan-based films	Zhao et al. (2019)
PLA	Propolis extract	Dry meat sausage	Films	• Incorporation of 10%–40% w/w propolis extract into PLA reduced tensile strength and flexibility	– Phenolic content in sausage slices increased with PE concentration, enhancing antioxidant protection – PE-containing films preserved the oxidative stability of sausages for up to 4 days under refrigeration	Safaei and Roosta Azad (2020)
Chitosan, PCL, and NC	GSE	Chicken breast fillet	Bilayer films	• Bilayer films with 15% GSE and 2% NC showed enhanced total color change • Films with GSE and NC significantly reduced lipid oxidation	– Bilayer films with GSE maintained the pH stability of chicken breast fillets over 15 days – TBARS values were lower in samples with active films compared to the control – TMAB and TCB counts were significantly lower than the control throughout storage, indicating better microbiological quality	Sogut and Seydim (2019)
CMC and OM	ZnO NPs	Chicken breast meat	Films	• CMC films enriched with 50% OM and ZnO NPs showed the strongest barrier and antimicrobial performance	– TVC, *S. aureus*, and LAB counts were significantly lower in samples packed with OM/ZnO films – CMC/OM50%/ZnO-packaged samples received the highest sensory scores after 12 days at 4°C	Mohammadi et al. (2019)

(*Continued*)

Table 3.3 (Continued)

Bioplastic matrix	Additive(s)	Food product(s)	Packaging type	Packaging properties	Food preservation	References
PLA and nanochitosan	PIE	Chicken fillet	Films	• PLA films incorporated with nanochitosan and 1.5% PIE improved barrier functionality • Wrapped samples had lower peroxide values (<2 meq/1000 g lipid) vs. control (2.01 meq/1000 g lipid)	– Active PLA films extended the refrigerated shelf life of chicken fillets beyond 10 days – pH remained stable in samples wrapped with PLA/nanochitosan/PIE films, unlike control and PLA-only samples	Javaherzadeh et al. (2020)
PBS and PBSA	-	Raw and smoked poultry meats	Films	• PBS-based films showed higher WVT (20.9–21.0 g/m^2/day) vs. PA/PE (2.1 g/m^2/day) • OP was higher in PBSA (115.4) vs. PA/PE (74.7 ml/m^2/day/0.1 MPa) • PBS films had much lower light transmittance (6.9%–15.7%) than PA/PE (82.6%), offering better light protection	– Packaged poultry showed stable pH and water activity throughout storage with PBS-based films – No significant microbiological deterioration observed in raw or smoked meats over shelf life – Volatile compound profiles and color remained largely unchanged during storage, indicating freshness	Vytejčková et al. (2017)
PBAT	OEO	Fish fillet	Films	• OEO incorporation reduced tensile strength, elongation at break, and elastic modulus • SEM revealed increased surface roughness in OEO-loaded films	– Active films significantly reduced *S. aureus*, total coliforms, and psychrotrophic bacteria on fish fillets	Cardoso et al. (2017)
PLA	OEO	Rainbow trout	Films	• Addition of 0.5% OEO increased PLA film flexibility, enhancing elongation at break from 2.82% to 16.78% (6-fold) • OEO acted as a plasticizer, reducing T_g and altering tensile properties and surface roughness	– PLA films with 1.5% OEO significantly reduced microbial growth in rainbow trout during chilled storage – Antimicrobial activity was highest against *S. aureus* and least effective against *E. coli*	Javidi et al. (2016)
CMC and SC	Probiotic bacteria	Trout fillet	Films	• Films showed stable mechanical and morphological performance during application	– Probiotic films significantly reduced microbial loads (Enterobacteriaceae, TVC, and	Mozaffarzogh et al. (2020)

				• Viability of probiotics declined over time	psychrotrophs) compared to the control – Chemical spoilage indicators (TVB-N, TBARS, peroxide, carbonyls, biogenic amines) remained lower in treated fillets – Shelf life of trout fillets extended up to 14 days while delivering probiotic health benefits	
Agar	PH and/ or CEO	*Paralichthys orbignyanus* fillet	Films	• PH (A20) incorporation increased film solubility, water vapor permeability, and elongation at break • CEO films were less transparent due to the oil content • A20 hydrolysate showed better antimicrobial activity than P20 due to lower MW and higher hydrophobic amino acids	– CEO films significantly delayed the growth of H_2S-producing microorganisms during 15 days at 5°C – CEO-based films extended shelf life more effectively than PH or control films	da Rocha et al. (2018)
PLA	ZnO NPs + ZEO + MEO	*Otolithes ruber* fillet	Films	• PLA films with 1.5% ZnO + 1.5% ZEO or MEO showed strong antibacterial zones; 691 mm^2 (ZEO) and 513.33 mm^2 (MEO) against *S. aureus* • Antioxidant activity reached 69.14% for ZEO and 49.08% for MEO films	– Active films extended the shelf life of *Otolithes ruber* fillets from 8 to 16 days at 4°C ± 1°C – Lowest TBARS (0.8 mg MA/kg) and TVB-N (21.23 mg/100 g) were observed in fillets wrapped with PLA/ZnO/ZEO	Heydari-Majd et al. (2019)
Cassava starch and PVA	RPPE	Shrimp	Intelligent films	• Adding 1.00 wt.% RPPE improved film compactness and water vapor/UV barrier properties • FTIR confirmed hydrogen bonding between betacyanins and film matrix, enhancing structural integrity	– Films with 1.00 wt.% extract visually changed color in response to volatile nitrogen from shrimp spoilage – Intelligent films enabled real-time monitoring of shrimp freshness via colorimetric response to spoilage	Qin et al. (2020)

(*Continued*)

Table 3.3 (Continued)

Bioplastic matrix	Additive(s)	Food product(s)	Packaging type	Packaging properties	Food preservation	References
Chitosan and PVA	Anthocyanins from *Syzygium cumini*	Shrimp	Intelligent films	• Incorporation of anthocyanins significantly affected film thickness and optical properties • Anthocyanins were well dispersed in the chitosan/PVA matrix, improving film uniformity	– Films changed color from red to blue in response to shrimp spoilage, indicating freshness status – Films functioned effectively as freshness indicators across −20°C to 20°C storage temperatures	Merz et al. (2020)
PLA	Nanochitosan	Indian white prawn	Films	• Nanochitosan addition imparted antimicrobial activity against aerobic microorganisms • Solvent-casted films had stable structural integrity and functional performance for seafood packaging	– PLA/nanochitosan films preserved microbial and biochemical quality of *Fenneropenaeus indicus* for up to 18 days at chilled conditions – Prawn samples showed delayed spoilage and maintained freshness during storage	Fathima et al. (2018)
PLA and PHB	FEN oil	Oyster	Films	• PLA/PHB/FEN film had comparable mechanical strength to EVOH but lower oxygen barrier capacity • FEN oil released up to 38.82 μg/mL in fat simulant (65% ethanol), ideal for oily food packaging	– PLA/PHB/FEN reduced *E. coli* and *S. aureus* by 1.2 and 1.3 log CFU/g, respectively – Aerobic and anaerobic bacterial loads in oysters reduced by ~1 log CFU/g with PLA/PHB/FEN film – Shelf life of oysters extended by 2–3 days while maintaining pH, TVB-N, texture, and amino acid profile	Miao et al. (2019)

BPO: ***Bunium percicum*** essential oil; CEO clove essential oil; CIN: cinnamaldehyde; CMC: carboxymethyl cellulose; EVOH: ethylene vinyl alcohol; FEN: fennel; FTIR: Fourier transform infrared spectroscopy; GSE: grape seed extract; LAB: lactic acid bacteria; MEO: ***Menthe piperita*** L. essential oil; MW molecular weight; NC: nanocellulose; NPs: nanoparticles; OEO: oregano (***Origanum vulgare***) essential oil; OM: okra mucilage; OP: oxygen permeability; PA/PE: polyamide/polyethylene; PBAT: poly(butylene adipate-co-terephthalate); PBS: polybutylene succinate; PBSA: polybutylene succinate-co-adipate; PCL: polycaprolactone; PH: protein hydrolysate; PHB: polyhydroxybutyrate; PIE: ***Polylophium involucratum*** essential oil; PLA: polylactic acid; PV: peroxide value; PVA: polyvinyl alcohol; RPPE: red pitaya (Hylocereus polyrhizus) peel extract; SC: sodium caseinate; SEM: scanning electron microscope; TBARS: thiobarbituric acid reactive substances; TCB: total coliform bacteria; TMAB: total mesophilic aerobic bacteria; TVB-N: total volatile base nitrogen; TVC: total viable counts; WVP: water vapor permeability; WVT: water vapor transmission; ZEO: ***Zataria multiflora*** Boiss. essential oil.

Poultry products like chicken and turkey meat have also been effectively preserved using bioplastic technologies. Bilayer films composed of chitosan and PCL and embedded with 15% grape seed extract and 2% nanocellulose reduced total mesophilic aerobic bacteria and total coliform bacteria in chicken breast fillets over 15 days of refrigerated storage. These films stabilized pH and reduced TBARS values, indicating lower lipid oxidation (Sogut and Seydim, 2019). Another nanocomposite film made from CMC, okra mucilage, and zinc oxide nanoparticles showed the strongest inhibition of microbial growth and lipid oxidation and improved sensory properties, and maintained color (higher L^* values) (Mohammadi et al., 2019). Furthermore, PLA films containing 1.5% *Polylophium involucratum* essential oil kept total volatile base nitrogen (TVB-N) levels below 23 mg/100 g and peroxide values to less than 2 meq/1000 g lipid, compared to 27.9 mg/100 g and 2.01 meq/1000 g in the control, over 10 days of storage (Javaherzadeh et al., 2020). Bioplastics based on PBS and polybutylene succinate-co-adipate were evaluated against conventional polyamide/polyethylene films for vacuum-packaged raw chicken and turkey meat. While PBS-based films had higher water vapor transmission rates (20.9–21.0 vs. 2.1 g/m^2 per day) and oxygen permeability (85.1–115.4 vs. 74.7 ml/m^2 per day per 0.1 MPa), they maintained the color, pH, and microbial stability throughout the storage period (Vytejčková et al., 2017).

Seafood packaging poses unique challenges due to the high PUFA content and moisture levels. Bioplastics have provided meaningful improvements in extending seafood shelf life while maintaining nutritional and sensory quality (Table 3.3). For example, PBAT films incorporated with OEO effectively suppressed total coliforms, *S. aureus*, and psychrotrophic bacteria in fish fillets, while higher OEO concentrations enhanced antioxidant activity (Cardoso et al., 2017). Similarly, PLA films with 1.5% OEO increased elongation at break from 2.82% to 16.78% and suppressed microbial growth in rainbow trout during chilled storage (Javidi et al., 2016). In trout packaging, films composed of CMC and sodium caseinate, enriched with probiotics such as *Lactobacillus acidophilus* and *Bifidobacterium bifidum*, extended shelf life by at least 14 days. These films significantly reduced biogenic amine formation and oxidative spoilage compared to control samples (Mozaffarzogh et al., 2020). Agar-based films embedded with protein hydrolysates or clove essential oil also improved flounder fillet stability by reducing TVB-N and pH and inhibiting H_2S-producing bacteria (da Rocha et al., 2018). PLA/ZnO films combined with ZEO and *Mentha piperita* essential oils extended the shelf life of *Otolithes ruber* fillets from 8 to 16 days at 4°C. These films demonstrated large antibacterial zone areas (up to 691 mm^2 against *S. aureus*) and strong antioxidant effects (up to 69.14% for ZEO-containing films), with TBARS and TVB-N values of 0.8 mg MA/kg and 21.23 mg/100 g, respectively (Heydari-Majd et al., 2019). In addition to active protection, intelligent bioplastics have been developed to monitor the freshness of seafood products. Films formulated with jambolan fruit and red pitaya peel changed color in response to spoilage-associated volatile nitrogen compounds in shrimp, providing real-time visual cues for detecting spoilage (Merz et al., 2020; Qin et al., 2020).

3.5 Regulatory frameworks governing bioplastics in food packaging

The global shift toward sustainable packaging has catalyzed the development of comprehensive regulatory frameworks. These frameworks govern the production, labeling, application, and disposal of bioplastics and are critical to ensuring product safety, environmental protection, market transparency, and consumer confidence. Regulations differ by region, but there is growing convergence toward principles that are aligned with the circular economy and the emerging bioeconomy.

3.5.1 European Union

The European Union (EU) is a global leader in the regulation of bioplastics and the development of bioeconomy policies. The 2012 European Bioeconomy Strategy laid a foundational and comprehensive roadmap for transitioning toward a circular economy based on renewable biological resources. This approach integrates multiple sectors such as sustainable agriculture, forestry, fisheries, food, and bio-based products, including bioplastics (2018). The European "green deal and circular economy action plan" further strengthens this strategy by mandating a reduction in plastic use, promoting reusable and recyclable materials, and encouraging investments in bio-based industries (Rosenboom et al., 2022). A key regulatory milestone is the Single-Use Plastics Directive (EU 2019/904), which bans the use of specific single-use plastic items, including oxo-degradable plastics that fragment into microplastics without undergoing full biodegradation (European Union, 2019). In 2021, the EU introduced a €800 tax per ton on nonrecycled plastic waste, further incentivizing relevant industries to adopt compostable and recyclable materials (Rosenboom et al., 2022). Ongoing policy efforts to revise and harmonize biodegradation standards are underway to ensure alignment with real-world conditions and to mitigate greenwashing practices (Painuli et al., 2024). Bioplastics are at the nexus of environmental and industrial policy and are expected to play a strategic role in reducing the EU's reliance on fossil-based plastics.

3.5.2 The United States

The development of the bioplastics sector in the United States is closely tied to federal initiatives aimed at fostering bio-based innovation. The Biomass Research and Development Act of 2000 and the United States Department of Agriculture (USDA)'s Program for the cornerstone of this regulatory landscape. The latter mandates federal agencies to prioritize the procurement of certified bio-based products and supports a standardized labeling system to promote market transparency and consumer confidence. In parallel, the National Bioeconomy

Blueprint identified biological research and innovation as critical drivers of sustainable economic growth and national competitiveness (Gong et al., 2024). At the state level, regulatory momentum is gaining traction through a diverse patchwork of legislation. States such as California, New York, and Oregon have implemented bans or restrictions on single-use plastics, mandating the use of biodegradable or compostable alternatives (Rosenboom et al., 2022). More broadly, the proposed "Break Free from Plastic Pollution Act" outlines comprehensive reforms, including extended producer responsibility frameworks, restrictions on nonrecyclable plastics, and enhanced labeling requirements (Lowenthal, 2020). These federal and state-level efforts signal an evolving governmental policy that seeks to accelerate the integration of bioplastics into mainstream packaging systems through a combination of regulatory mandates and market incentives.

3.5.3 Asia Pacific region

Asian countries are increasingly becoming key contributors to the global bioplastics value chain, driven by both environmental concerns and strategic industrial policies. In China, a series of regulatory measures have been enacted since 2020 to progressively ban single-use plastics and establish national biodegradability standards such as GB/T 41010-2021. These actions have triggered substantial investment in domestic bioplastics manufacturing, particularly in key polymers such as PLA, PBAT, and PBS (Jayakumar et al., 2023; Rosenboom et al., 2022). Japan has demonstrated sustained commitment through initiatives such as the Biomass Nippon Strategy and the Plastic Resource Circulation Act 2021, both designed to promote the usage of bioplastics and mitigate marine plastic littering (Gong et al., 2024; Suwa, 2022). Concurrently, countries like South Korea, Thailand, and Singapore are implementing a range of incentive mechanisms such as tax relief, subsidies, and preferential procurement policies to stimulate bioplastic research and commercialization. For example, Thailand classified bioplastics as a key industry under its National Innovation Strategy released in 2008, thereby reinforcing institutional support to increase domestic capacity for market development (Sirirat, 2017).

3.6 Global standards and certification systems

Given that bioplastics are often visually indistinguishable from conventional plastics, standardized testing and certification are essential to validate environmental claims and ensure appropriate waste handling. International standardization organizations, such as the International Organization for Standardization (ISO), the American Society for Testing and Materials (ASTM), and the European Committee for Standardization (CEN), have developed rigorous protocols to assess compostability, biodegradability, and bio-based content. For instance, ISO 17088 specifies criteria for compostable plastics such as biodegradability ($\geq$60% carbon dioxide evolution within 180 days), disintegration (<10% residual mass after 84 days), and adherence to ecotoxicological limits to ensure that the resulting compost does not negatively impact plant growth or soil health, and that concentrations of heavy metals or harmful substances remain below defined safety thresholds. Similarly, ASTM D6400 specifies that plastics must biodegrade under industrial composting environments, with defined thresholds for carbon dioxide generation, physical disintegration, and absence of harmful residues (Painuli et al., 2024). In the CEN, the EN 13432 remains the benchmark for certifying industrial compostability, demanding ≥90% biodegradation within 6 months and disintegration within 12 weeks. Additionally, it requires that the resulting compost should pass seed germination testing and comply with heavy metal limits. To address bio-based content claims, EN 16640 and EN 16785 provide standardized methodologies (e.g., radiocarbon analysis) for quantifying the proportion of renewable carbon in a given material (Painuli et al., 2024). Third-party certification bodies such as DIN Certco and TÜV AUSTRIA (formerly Vincotte) have internationally recognized ecolabels, such as "Seedling," "OK Compost Industrial/Home," and "OK Biodegradable Soil/Water," which facilitate product differentiation and enhance consumer trust. In the United States, the Biodegradable Products Institute (BPI) certifies products under ASTM guidelines, while in Japan, the Japan Bioplastic Association (JBPA) issues the "GreenPla" certification based on ISO protocols and additional ecotoxicity thresholds for aquatic ecosystems (Jayakumar et al., 2023).

3.7 Current challenges and future perspectives

Bioplastics are a promising pathway toward mitigating the global plastic crisis, particularly in the context of food packaging. However, their widespread adoption currently faces several challenges, including technical shortcomings, high production costs, limited consumer awareness, insufficient regulatory frameworks, and unresolved environmental impacts. Addressing these challenges requires coordinated advances across research, industry, and policy.

One of the most pressing technical issues is the limited functional performance of certain widely used biopolymers such as TPS. Although TPS is attractive due to its biodegradability and low cost, it suffers from poor mechanical strength, low water resistance, and weak barrier properties. These shortcomings become particularly evident in high-moisture or fluctuating-temperature conditions, such as in refrigerated food storage followed by thawing. The hydrophilic nature of starch leads to moisture absorption, swelling, and structural breakdown, reducing packaging integrity (García-Guzmán et al., 2022; Othman and Fadzil, 2021). This not only compromises food safety and shelf life but also limits TPS's use in advanced applications like vacuum packaging (Rahardiyan et al., 2023). Emerging strategies such as incorporating nanofillers, optimizing plasticizer content, and employing multilayer casting techniques have shown promise in enhancing the performance of TPS, but further research is needed to validate these modifications under commercial conditions.

Economic feasibility remains another substantial barrier. Currently, most bioplastics are more expensive than their fossil-based counterparts due to higher feedstock costs, underdeveloped manufacturing technologies, and limited production scale. This cost differential, often referred to as the "green premium," is a critical deterrent for food manufacturers operating on tight margins. Although surveys suggest younger consumers are more willing to pay a premium for sustainable products, price sensitivity in mass markets continues to restrict adoption (Cooke and Pomeroy, 2023). To overcome this barrier, targeted investments in infrastructure, research into low-cost feedstocks (e.g., algae, agricultural waste, hemp, and seaweed), and the optimization of production processes are essential. Moreover, policy mechanisms such as subsidies for green materials, tax credits, and reduced import tariffs for renewable raw materials, technologies, and equipment could accelerate the scale-up and reduce production costs over time.

Feedstock-related challenges also raise concerns about sustainability and ethics. The current reliance on first-generation feedstocks, such as sugarcane, corn, and cassava, poses a dual threat; competition with food crops and potential land-use changes that may offset environmental gains. This has sparked criticism regarding the true sustainability of certain bioplastics, particularly when grown in regions where deforestation or monoculture practices occur (Jayakumar et al., 2023). Transitioning to second-generation feedstocks like agricultural residues, invasive plant species, and nonfood biomass is a key direction for future development. Even more promising are third-generation feedstocks such as microalgae and seaweed, which offer higher productivity with minimal land and freshwater use, making them ideal candidates for climate-resilient, circular production models. Besides, widespread misconceptions about bioplastics further contribute to environmental mismanagement, particularly the false belief that all bioplastics are biodegradable in any environmental condition. This has led to improper disposal practices and contamination of existing recycling streams. The degradability of bioplastics like PLA is highly dependent on specific industrial composting conditions that include high temperatures (>55°C), controlled humidity, and specific microbial consortia. In marine or natural soil environments, such conditions are rarely met, resulting in long-term persistence and potential harm to ecosystems. Additionally, mislabeling bioplastics as "compostable" without specifying the required conditions not only misleads consumers but also undermines efforts to build a circular plastic economy (Ghasemlou et al., 2024). There is a growing call for standardized, transparent, and enforceable labeling systems that communicate biodegradability, recyclability, and end-of-life pathways clearly to all stakeholders. On the other hand, methane emissions from anaerobic degradation in landfills and the formation of micro- and nanomaterials from incomplete decomposition present new ecological risks. The long-term effects of these degradation products on soil, water, and food chains remain insufficiently studied. Furthermore, the migration of bioplastic additives or breakdown intermediates into food and their interactions with biological macromolecules is a largely unexplored field, raising potential food safety concerns (Jayakumar et al., 2023). As such, future research should include lifecycle assessments, ecotoxicity assays, and migration studies to ensure bioplastics are not only functionally viable but also safe for human and environmental health.

Looking forward, the development of next-generation bioplastics will likely depend on innovation in materials science, manufacturing processes, and smart packaging technologies. The integration of nanomaterials such as nanoclays, cellulose nanocrystals, and bio-based antioxidants can significantly improve the mechanical, barrier, and active properties of bioplastics (Jha, 2020; Sogut and Seydim, 2019). Smart functionalities, including indicators for spoilage or temperature changes, can further enhance consumer value and drive market differentiation. Additionally, multilayer film structures that combine different biopolymers for synergistic effects have gained increasing attention, particularly for demanding applications like fresh produce packaging or high-fat food storage (Qin et al., 2020; Winotapun et al., 2024). At a systemic level, the successful development of a robust bioplastics industry also requires regulatory and logistical reforms. Governments must invest in waste management systems that can handle compostable and biodegradable packaging materials separately from traditional plastics. This includes standardized waste collection systems, industrial composting facilities, and anaerobic digestion technologies. Clearer policies and targeted incentives can encourage corporations toward eco-conscious product design and effective end-of-life management. At the same time, public education initiatives are critical to correcting consumer misconceptions and fostering responsible behavior.

3.8 Conclusion

The transition from fossil-based plastics to bioplastics in food packaging is a cornerstone in advancing circular economy goals and reducing long-term ecological impact. As this chapter demonstrates, bioplastics sourced from renewable feedstocks can be tailored for specific packaging functions, and often support biodegradability or compostability. From packaging fresh produce and bakery goods to highly perishable meat and seafood products, bioplastics have shown measurable benefits in preserving food quality and extending shelf life when reinforced with bioactive agents or nanocomposites. Nonetheless, the adoption of bioplastics is restricted by several challenges. Material limitations, including low mechanical strength, poor barrier characteristics, and limited thermal stability, along with variable degradation profiles and high production costs, represent significant obstacles to their widespread adoption. The continued use of first-generation feedstocks intensifies concerns surrounding food security and land use, while public misconceptions and inconsistent labeling complicate responsible end-of-life disposal. To realize the full potential of bioplastics, a multifaceted and coordinated approach is essential. Innovations in materials science, including multilayer film design, nanotechnology integration, and adoption of third-generation feedstocks such as algae, will be vital to improving performance and scalability. At the same time, public education, clearer labeling standards, and investment in building extra industrial composting infrastructure are critical to maximizing the environmental potential of bioplastics. Ultimately, bioplastics are not a single bullet for plastic waste problems, but their strategic development and responsible integration into food packaging systems can reduce the environmental footprint of the agri-food sector.

References

Akhouy, G., Aziz, K., Gebrati, L., El Achaby, M., Akgul, Y., Yap, P.S., et al., 2023. Recent applications on biopolymers electrospinning: strategies, challenges and way forwards. Polymer-Plastics Technology and Materials 62 (13), 1754–1775. Available at https://doi.org/10.1080/25740881.2023.2234459.

Ali, S., Deiab, I., Pervaiz, S., 2024. State-of-the-art review on fused deposition modeling (FDM) for 3D printing of polymer blends and composites: innovations, challenges, and applications. The International Journal of Advanced Manufacturing Technology 135 (11-12), 5085–5113. Available at https://doi.org/10.1007/s00170-024-14820-0.

Altan, A., Aytac, Z., Uyar, T., 2018. Carvacrol loaded electrospun fibrous films from zein and poly(lactic acid) for active food packaging. Food Hydrocolloids 81, 48–59. Available at https://doi.org/10.1016/j.foodhyd.2018.02.028.

Amiri, E., Aminzare, M., Azar, H.H., Mehrasbi, M.R., 2019. Combined antioxidant and sensory effects of corn starch films with nanoemulsion of Zataria multiflora essential oil fortified with cinnamaldehyde on fresh ground beef patties. Meat Science 153, 66–74. Available at https://doi.org/10.1016/j.meatsci.2019.03.004.

Aranibar, C., Pedrotti, F., Archilla, M., Vázquez, C., Borneo, R., Aguirre, A., 2020. Storage and preservation of dry pasta into biodegradable packaging made from triticale flour. Journal of Food Science and Technology 57 (2), 693–701. Available at https://doi.org/10.1007/s13197-019-04102-z.

Botondi, R., Bartoloni, S., Baccelloni, S., Mencarelli, F., 2015. Biodegradable PLA (polylactic acid) hinged trays keep quality of fresh-cut and cooked spinach. Journal of Food Science and Technology 52 (9), 5938–5945. Available at https://doi.org/10.1007/s13197-014-1695-x.

Byun, Y., Kim, Y.T., 2014. Chapter 15 - Utilization of bioplastics for food packaging industry. In: Han, J.H. (Ed.), Innovations in Food Packaging, second ed. Academic Press, pp. 369–390. Available at https://doi.org/10.1016/B978-0-12-394601-0.00015-1.

Cardoso, L.G., Pereira Santos, J.C., Camilloto, G.P., Miranda, A.L., Druzian, J.I., Guimarães, A.G., 2017. Development of active films poly (butylene adipate co-terephthalate) – PBAT incorporated with oregano essential oil and application in fish fillet preservation. Industrial Crops and Products 108, 388–397. Available at https://doi.org/10.1016/j.indcrop.2017.06.058.

Chalermthai, B., Giwa, A., Schmidt, J.E., Taher, H.A.B., 2021. Life cycle assessment of bioplastic production from whey protein obtained from dairy residues. Bioresource Technology Reports 15, 100695. Available at https://doi.org/10.1016/j.biteb.2021.100695.

Cheng, J., Wang, H., Kang, S., Xia, L., Jiang, S., Chen, M., et al., 2019. An active packaging film based on yam starch with eugenol and its application for pork preservation. Food Hydrocolloids 96, 546–554. Available at https://doi.org/10.1016/j.foodhyd.2019.06.007.

Chi, H., Song, S., Luo, M., Zhang, C., Li, W., Li, L., et al., 2019. Effect of PLA nanocomposite films containing bergamot essential oil, TiO2 nanoparticles, and Ag nanoparticles on shelf life of mangoes. Scientia Horticulturae 249, 192–198. Available at https://doi.org/10.1016/j.scienta.2019.01.059.

Cooke, T., Pomeroy, R.S., 2023. The bioplastics market: history, commercialization trends, and the new eco-consumer. In: Pomeroy, R.S. (Ed.), Rethinking Polyester Polyurethanes: Algae Based Renewable, Sustainable, Biodegradable and Recyclable Materials. Elsevier, pp. 261–280. Available at https://doi.org/10.1016/B978-0-323-99982-3.00003-1.

Correa, J.P., Molina, V., Sanchez, M., Kainz, C., Eisenberg, P., Massani, M.B., 2017. Improving ham shelf life with a polyhydroxybutyrate/polycaprolactone biodegradable film activated with nisin. Food Packaging and Shelf Life 11, 31–39. Available at https://doi.org/10.1016/j.fpsl.2016.11.004.

Cruz, R.M.S., Krauter, V., Krauter, S., Agriopoulou, S., Weinrich, R., Herbes, C., et al., 2022. Bioplastics for food packaging: environmental impact, trends and regulatory aspects. Foods 11 (19), 3087. Available at https://doi.org/10.3390/foods11193087.

da Rocha, M., Alemán, A., Romani, V.P., López-Caballero, M.E., Gómez-Guillén, M.C., Montero, P., et al., 2018. Effects of agar films incorporated with fish protein hydrolysate or clove essential oil on flounder (Paralichthys orbignyanus) fillets shelf-life. Food Hydrocolloids 81, 351–363. Available at https://doi.org/10.1016/j.foodhyd.2018.03.017.

da Silva, J.B.A., Santana, J.S., de Almeida Lucas, A., Passador, F.R., de Sousa Costa, L.A., Pereira, F.V., et al., 2019. PBAT/TPS-nanowhiskers blends preparation and application as food packaging. Journal of Applied Polymer Science 136 (26), 47699. Available at https://doi.org/10.1002/app.47699.

Doanh, V.H., Vu, D.C., Thien, Q.T., Son, N.P., Sy, T.N., Giang, N., et al., 2014. A new rapid neutron activation analysis system at Dalat nuclear research reactor. Nuclear Science and Technology 4 (1), 84–90. Available at https://doi.org/10.53747/jnst.v4i1.218.

European Commission, 2018. Plastic waste: an European strategy to protect the planet, defend our citizens and empower our industries. Available from: https://ec.europa.eu/commission/presscorner/detail/en/IP_18_5.

European Union, 2019. Directive (EU) 2019/904 of the European Parliament and of the Council of 5 June 2019 on the reduction of the impact of certain plastic products on the environment. Available from: https://eur-lex.europa.eu/eli/dir/2019/904/oj.

Fathima, P.E., Panda, S.K., Ashraf, P.M., Varghese, T.O., Bindu, J., 2018. Polylactic acid/chitosan films for packaging of Indian white prawn (Fenneropenaeus indicus). International Journal of Biological Macromolecules 117, 1002–1010. Available at https://doi.org/10.1016/j.ijbiomac.2018.05.214.

Figueroa-Lopez, K.J., Andrade-Mahecha, M.M., Torres-Vargas, O.L., 2018. Development of antimicrobial biocomposite films to preserve the quality of bread. Molecules (Basel, Switzerland) 23 (1), 212. Available at https://doi.org/10.3390/molecules23010212.

Filho, W.L., Barbir, J., Rimi Abubakar, I., Paço, A., Stasiskiene, Z., Hornbogen, M., et al., 2022. Consumer attitudes and concerns with bioplastics use: an international study. PLoS One 17 (4), e0266918. Available at https://doi.org/10.1371/journal.pone.0266918.

García-Guzmán, L., Cabrera-Barjas, G., Soria-Hernández, C.G., Castaño, J., Guadarrama-Lezama, A.Y., Rodríguez Llamazares, S., 2022. Progress in starch-based materials for food packaging applications. Polysaccharides 3 (1), 136–177. Available at https://doi.org/10.3390/polysaccharides3010007.

Ghasemlou, M., Barrow, C.J., Adhikari, B., 2024. The future of bioplastics in food packaging: an industrial perspective. Food Packaging and Shelf Life 43, 101279. Available at https://doi.org/10.1016/j.fpsl.2024.101279.

Gong, L., Passari, A.K., Yin, C., Kumar Thakur, V., Newbold, J., Clark, W., et al., 2024. Sustainable utilization of fruit and vegetable waste bioresources for bioplastics production. Critical Reviews in Biotechnology 44 (2), 236–254. Available at https://doi.org/10.1080/07388551.2022.2157241.

Goryńska-Goldmann, E., Gazdecki, M., Rejman, K., Łaba, S., Kobus-Cisowska, J., Szczepański, K., 2021. Magnitude, causes and scope for reducing food losses in the baking and confectionery industry—a multi-method approach. Agriculture (Switzerland) 11 (10), 936. Available at https://doi.org/10.3390/agriculture11100936.

Hadidi, M., Liñán-Atero, R., Tarahi, M., Christodoulou, M.C., Aghababaei, F., 2024. The potential health benefits of gallic acid: therapeutic and food applications. Antioxidants 13 (8), 1001. Available at https://doi.org/10.3390/antiox13081001.

Harnkarnsujarit, N., Wongphan, P., Chatkitanan, T., Laorenza, Y., Srisa, A., 2021. Bioplastic for sustainable food packaging. In: Galanakis, C.M. (Ed.), Sustainable Food Processing and Engineering Challenges. Academic Press, pp. 203–277. Available at https://doi.org/10.1016/B978-0-12-822714-5.00007-3.

Heydari-Majd, M., Ghanbarzadeh, B., Shahidi-Noghabi, M., Najafi, M.A., Hosseini, M., 2019. A new active nanocomposite film based on PLA/ZnO nanoparticle/essential oils for the preservation of refrigerated Otolithes ruber fillets. Food Packaging and Shelf Life 19, 94–103. Available at https://doi.org/10.1016/j.fpsl.2018.12.002.

Hossain, M.T., Shahid, M.A., Akter, S., Ferdous, J., Afroz, K., Refat, K.R.I., et al., 2024. Cellulose and starch-based bioplastics: a review of advances and challenges for sustainability. Polymer-Plastics Technology and Materials. 63 (10), 1329–1349. Available at https://doi.org/10.1080/25740881.2024.2329980.

Javaherzadeh, R., Tabatabaee Bafroee, A.S., Kanjari, A., 2020. Preservation effect of Polylophium involucratum essential oil incorporated poly lactic acid/ nanochitosan composite film on shelf life and sensory properties of chicken fillets at refrigeration temperature. LWT 118, 108783. Available at https://doi.org/10.1016/j.lwt.2019.108783.

Javidi, Z., Hosseini, S.F., Rezaei, M., 2016. Development of flexible bactericidal films based on poly(lactic acid) and essential oil and its effectiveness to reduce microbial growth of refrigerated rainbow trout. LWT - Food Science and Technology 72, 251–260. Available at https://doi.org/10.1016/j.lwt.2016.04.052.

Jayakumar, A., Radoor, S., Siengchin, S., Shin, G.H., Kim, J.T., 2023. Recent progress of bioplastics in their properties, standards, certifications and regulations: a review. Science of the Total Environment 878, 163156. Available at https://doi.org/10.1016/j.scitotenv.2023.163156.

Jha, P., 2020. Effect of plasticizer and antimicrobial agents on functional properties of bionanocomposite films based on corn starch-chitosan for food packaging applications. International Journal of Biological Macromolecules 160, 571–582. Available at https://doi.org/10.1016/j.ijbiomac.2020.05.242.

Kamthai, S., Magaraphan, R., 2019. Development of an active polylactic acid (PLA) packaging film by adding bleached bagasse carboxymethyl cellulose (CMC B) for mango storage life extension. Packaging Technology and Science 32 (2), 103–116. Available at https://doi.org/10.1002/pts.2420.

Kanikireddy, V., Varaprasad, K., Sandhya Rani, M., Venkataswamy, P., Jagan Mohan Reddy, B., Vithal, M., 2020. Biosynthesis of CMC-Guar gum-Ag0 nanocomposites for inactivation of food pathogenic microbes and its effect on the shelf life of strawberries. Carbohydrate Polymers 236, 116053. Available at https://doi.org/10.1016/j.carbpol.2020.116053.

Kim, H.J., Lee, C.L., Yoon, K.S., Rhim, J.W., 2022. Synergistic effect of UV-C LED irradiation and PLA/PBAT-based antimicrobial packaging film on fresh-cut vegetables. Food Control 138, 109027. Available at https://doi.org/10.1016/j.foodcont.2022.109027.

Kong, U., Mohammad Rawi, N.F., Tay, G.S., 2023. The potential applications of reinforced bioplastics in various industries: a review. Polymers 15 (10), 2399. Available at https://doi.org/10.3390/polym15102399.

Korge, K., Bajić, M., Likozar, B., Novak, U., 2020. Active chitosan–chestnut extract films used for packaging and storage of fresh pasta. International Journal of Food Science and Technology 55 (8), 3043–3052.

Lackner, M., 2015. Bioplastics - Biobased plastics as renewable and/or biodegradable alternatives to petroplastics. In: Othmer, K. (Ed.), Kirk-Othmer Encyclopedia of Chemical Technology. Wiley, pp. 1–41. Available at https://doi.org/10.1002/0471238961.koe00006.

Lackner, M., Mukherjee, A., Koller, M., 2023. What are "bioplastics"? defining renewability, biosynthesis, biodegradability, and biocompatibility. Polymers 15 (24), 4695. Available at https://doi.org/10.3390/polym15244695.

Li, G., Shankar, S., Rhim, J.W., Oh, B.Y., 2015. Effects of preparation method on properties of poly(butylene adipate-co-terephthalate) films. Food Science and Biotechnology 24 (5), 1679–1685. Available at https://doi.org/10.1007/s10068-015-0218-5.

Li, W., Li, L., Cao, Y., Lan, T., Chen, H., Qin, Y., 2017. Effects of PLA film incorporated with ZnO nanoparticle on the quality attributes of fresh-cut apple. Nanomaterials 7 (8), 207. Available at https://doi.org/10.3390/nano7080207.

Llana-Ruíz-Cabello, M., Puerto, M., Pichardo, S., Jiménez-Morillo, N.T., Bermúdez, J.M., Aucejo, S., et al., 2019. Preservation of phytosterol and PUFA during ready-to-eat lettuce shelf-life in active bio-package. Food Packaging and Shelf Life 22, 100410. Available at https://doi.org/10.1016/j.fpsl.2019.100410.

Lowenthal, A.S., 2020. H.R.5845 - Break Free From Plastic Pollution Act of 2020. Available from: https://www.congress.gov/bill/116th-congress/house-bill/5845.

Merz, B., Capello, C., Leandro, G.C., Moritz, D.E., Monteiro, A.R., Valencia, G.A., 2020. A novel colorimetric indicator film based on chitosan, polyvinyl alcohol and anthocyanins from jambolan (Syzygium cumini) fruit for monitoring shrimp freshness. International Journal of Biological Macromolecules 153, 625–632. Available at https://doi.org/10.1016/j.ijbiomac.2020.03.048.

Miao, L., Walton, W.C., Wang, L., Li, L., Wang, Y., 2019. Characterization of polylactic acids-polyhydroxybutyrate based packaging film with fennel oil, and its application on oysters. Food Packaging and Shelf Life 22, 100388. Available at https://doi.org/10.1016/j.fpsl.2019.100388.

Mistriotis, A., Briassoulis, D., Giannoulis, A., D'Aquino, S., 2016. Design of biodegradable bio-based equilibrium modified atmosphere packaging (EMAP) for fresh fruits and vegetables by using micro-perforated poly-lactic acid (PLA) films. Postharvest Biology and Technology 111, 380–389. Available at https://doi.org/10.1016/j.postharvbio.2015.09.022.

Mohammadi, H., Kamkar, A., Misaghi, A., Zunabovic-Pichler, M., Fatehi, S., 2019. Nanocomposite films with CMC, okra mucilage, and ZnO nanoparticles: extending the shelf-life of chicken breast meat. Food Packaging and Shelf Life 21, 100330. Available at https://doi.org/10.1016/j.fpsl.2019.100330.

Mozaffarzogh, M., Misaghi, A., Shahbazi, Y., Kamkar, A., 2020. Evaluation of probiotic carboxymethyl cellulose-sodium caseinate films and their application in extending shelf life quality of fresh trout fillets. LWT 126, 109305. Available at https://doi.org/10.1016/j.lwt.2020.109305.

Negrete-Bolagay, D., Guerrero, V.H., 2024. Opportunities and challenges in the application of bioplastics: perspectives from formulation, processing, and performance. Polymers 16 (18), 2561. Available at https://doi.org/10.3390/polym16182561.

Noshirvani, N., Ghanbarzadeh, B., Rezaei Mokarram, R., Hashemi, M., 2017. Novel active packaging based on carboxymethyl cellulose-chitosan-ZnO NPs nanocomposite for increasing the shelf life of bread. Food Packaging and Shelf Life 11, 106–114. Available at https://doi.org/10.1016/j.fpsl.2017.01.010.

Oliveira, M.A., Gonzaga, M.L.C., Bastos, M.S.R., Magalhães, H.C.R., Benevides, S.D., Furtado, R.F., et al., 2020. Packaging with cashew gum/gelatin/essential oil for bread: release potential of the citral. Food Packaging and Shelf Life 23, 100431. Available at https://doi.org/10.1016/j.fpsl.2019.100431.

Othman, S.A., Fadzil, N.F., 2021. Implementation of nanofiller for bioplastic reinforcement: a review. International Journal of Advanced Research in Technology and Innovation 3 (2), 19–27.

Painuli, R., Raghav, S., Kumar, D., 2024. Standards issues toward bioplastics. In: Mishra, A.K., Hussain, C.M. (Eds.), Bioplastics for Sustainability: Manufacture, Technologies, and Environment. Elsevier, India, pp. 143–159. Available at https://doi.org/10.1016/B978-0-323-95199-9.00010-X.

Paulsen, E., Lema, P., Martínez-Romero, D., García-Viguera, C., 2022. Use of PLA/PBAT stretch-cling film as an ecofriendly alternative for individual wrapping of broccoli heads. Scientia Horticulturae 304, 111260. Available at https://doi.org/10.1016/j.scienta.2022.111260.

Pottinger, A.S., Geyer, R., Biyani, N., Martinez, C.C., Nathan, N., Morse, M.R., et al., 2024. Pathways to reduce global plastic waste mismanagement and greenhouse gas emissions by 2050. Science (New York, N.Y.) 386 (6726), 1168–1173. Available at https://doi.org/10.1126/science.adr3837.

Promhuad, K., Bumbudsanpharoke, N., Wadaugsorn, K., Sonchaeng, U., Harnkarnsujarit, N., 2022. Maltol-incorporated acetylated cassava starch films for shelf-life-extension packaging of bakery products. Polymers 14 (24), 5342. Available at https://doi.org/10.3390/polym14245342.

Qin, Y., Liu, Y., Zhang, X., Liu, J., 2020. Development of active and intelligent packaging by incorporating betalains from red pitaya (Hylocereus polyrhizus) peel into starch/polyvinyl alcohol films. Food Hydrocolloids 100, 105410. Available at https://doi.org/10.1016/j.foodhyd.2019.105410.

Rahardiyan, D., Moko, E.M., Tan, J.S., Lee, C.K., 2023. Thermoplastic starch (TPS) bioplastic, the green solution for single-use petroleum plastic food packaging – a review. Enzyme and Microbial Technology 168, 110260. Available at https://doi.org/10.1016/j.enzmictec.2023.110260.

Rosenboom, J.G., Langer, R., Traverso, G., 2022. Bioplastics for a circular economy. Nature Reviews Materials 7 (2), 117–137. Available at https://doi.org/10.1038/s41578-021-00407-8.

Rosenow, P., Fernández-Ayuso, C., López-García, P., Minguez-Enkovaara, L.F., 2025. Design, new materials, and production challenges of bioplastics-based food packaging. Materials 18 (3), 673. Available at https://doi.org/10.3390/ma18030673.

Safaei, M., Roosta Azad, R., 2020. Preparation and characterization of poly-lactic acid based films containing propolis ethanolic extract to be used in dry meat sausage packaging. Journal of Food Science and Technology 57 (4), 1242–1250. Available at https://doi.org/10.1007/s13197-019-04156-z.

Sahraee, S., Milani, J.M., Ghanbarzadeh, B., Hamishehkar, H., 2020. Development of emulsion films based on bovine gelatin-nano chitin-nano ZnO for cake packaging. Food Science & Nutrition 8 (2), 1303–1312. Available at https://doi.org/10.1002/fsn3.1424.

Sangsuwan, J., Rattanapanone, N., Pongsirikul, I., 2015. Development of active chitosan films incorporating potassium sorbate or vanillin to extend the shelf life of butter cake. International Journal of Food Science and Technology 50 (2), 323–330. Available at https://doi.org/10.1111/ijfs.12631.

Shah, Y.A., Bhatia, S., Al-Harrasi, A., Tarahi, M., Khan, M.R., 2025. Chitosan-orange (Citrus sinensis) essential oil coatings with combined treatment of KMnO4 to enhance the shelf-life and storage quality of Ercolini pears. Food Science & Nutrition 13 (3), e70045.

Shah, Y.A., Bhatia, S., Al-Harrasi, A., Tarahi, M., Khan, T.S., Alam, T., et al., 2025. Comparative study of pectin and kappa-carrageenan-based films loaded with boswellic acid and acetyl-11-keto-beta-boswellic acid for active packaging. Journal of Food Quality 2025 (1), 3191264. Available at https://doi.org/10.1155/jfq/3191264.

Siddiqui, S.A., Yang, X., Deshmukh, R.K., Gaikwad, K.K., Bahmid, N.A., Castro-Muñoz, R., 2024. Recent advances in reinforced bioplastics for food packaging – a critical review. International Journal of Biological Macromolecules 263, 130399. Available at https://doi.org/10.1016/j.ijbiomac.2024.130399.

Siracusa, V., Blanco, I., 2020. Bio-polyethylene (Bio-PE), bio-polypropylene (Bio-PP) and bio-poly(ethylene terephthalate) (Bio-PET): recent developments in bio-based polymers analogous to petroleum-derived ones for packaging and engineering applications. Polymers 12 (8), 1641. Available at https://doi.org/10.3390/polym12081641.

Sirirat, P., 2017. Sustainable management strategies: Thailand's sustainable plastic-management leading the Thai plastic industry towards becoming a Bio-Plastic Hub. Available at https://openrepository.aut.ac.nz/server/api/core/bitstreams/b08e0b18-25d5-450e-a50a-cada56ccff39/content.

Sogut, E., Seydim, A.C., 2019. The effects of chitosan- and polycaprolactone-based bilayer films incorporated with grape seed extract and nanocellulose on the quality of chicken breast fillets. LWT 101, 799–805. Available at https://doi.org/10.1016/j.lwt.2018.11.097.

Sousa, G.M., Yamashita, F., Soares Júnior, M.S., 2016. Application of biodegradable films made from rice flour, poly(butylene adipate-co-terphthalate), glycerol and potassium sorbate in the preservation of fresh food pastas. LWT 65, 39–45. Available at https://doi.org/10.1016/j.lwt.2015.07.054.

Subramanian, M.N., 2024. Thermoforming: Processing and Technology. Scrivener Publishing LLC. Available at https://doi.org/10.1002/9781119555889.

Suwa, Y., 2022. Recent progress on the promotion and standardization of bioplastics in Japan. European Bioplastics.

Talebi, F., Misaghi, A., Khanjari, A., Kamkar, A., Gandomi, H., Rezaeigolestani, M., 2018. Incorporation of spice essential oils into poly-lactic acid film matrix with the aim of extending microbiological and sensorial shelf life of ground beef. LWT 96, 482–490. Available at https://doi.org/10.1016/j.lwt.2018.05.067.

Thakur, S., Chaudhary, J., Sharma, B., Verma, A., Tamulevicius, S., Thakur, V.K., 2018. Sustainability of bioplastics: opportunities and challenges. Current Opinion in Green and Sustainable Chemistry 13, 68–75. Available at https://doi.org/10.1016/j.cogsc.2018.04.013.

Vytejčková, S., Vápenka, L., Hradecký, J., Dobiáš, J., Hajšlová, J., Loriot, C., et al., 2017. Testing of polybutylene succinate based films for poultry meat packaging. Polymer Testing 60, 357–364. Available at https://doi.org/10.1016/j.polymertesting.2017.04.018.

Wang, J., Lu, Y., Yuan, H., Dou, P., 2008. Crystallization, orientation morphology, and mechanical properties of biaxially oriented starch/polyvinyl alcohol films. Journal of Applied Polymer Science 110 (1), 523–530. Available at https://doi.org/10.1002/app.28646.

Westlake, J.R., Tran, M.W., Jiang, Y., Zhang, X., Burrows, A.D., Xie, M., 2023. Biodegradable biopolymers for active packaging: demand, development and directions. Sustainable Food Technology 1 (1), 50–72. Available at https://doi.org/10.1039/d2fb00004k.

Winotapun, C., Tameesrisuk, M., Sirirutbunkajal, P., Sungdech, P., Leelaphiwat, P., 2024. Enhancing gas transmission rate of PBS/PBAT composite films: a study on microperforated film solutions for mango storage. ACS Omega 9 (3), 3469–3479. Available at https://doi.org/10.1021/acsomega.3c06999.

Yang, S., Miao, Q., Huang, Y., Jian, P., Wang, X., Tu, M., 2020. Preparation of cinnamaldehyde-loaded polyhydroxyalkanoate/chitosan porous microspheres with adjustable controlled-release property and its application in fruit preservation. Food Packaging and Shelf Life 26, 100596. Available at https://doi.org/10.1016/j.fpsl.2020.100596.

Yu, J., Xu, S., Liu, B., Wang, H., Qiao, F., Ren, X., et al., 2023. PLA bioplastic production: from monomer to the polymer. European Polymer Journal 193, 112076. Available at https://doi.org/10.1016/j.eurpolymj.2023.112076.

Zhang, W., Hadidi, M., Karaca, A.C., Hedayati, S., Tarahi, M., Assadpour, E., et al., 2023. Chitosan-grafted phenolic acids as an efficient biopolymer for food packaging films/coatings. Carbohydrate Polymers 314, 120901. Available at https://doi.org/10.1016/j.carbpol.2023.120901.

Zhao, Y., Teixeira, J.S., Saldaña, M.D.A., Gänzle, M.G., 2019. Antimicrobial activity of bioactive starch packaging films against Listeria monocytogenes and reconstituted meat microbiota on ham. International Journal of Food Microbiology 305, 108253. Available at https://doi.org/10.1016/j.ijfoodmicro.2019.108253.

Zhou, X., Yang, R., Wang, B., Chen, K., 2019. Development and characterization of bilayer films based on pea starch/polylactic acid and use in the cherry tomatoes packaging. Carbohydrate Polymers 222, 114912. Available at https://doi.org/10.1016/j.carbpol.2019.05.042.

CHAPTER 4

Bio-based and biodegradable bioplastics in food packaging

Md. Forshed Dewan[1], Hoang Chinh Nguyen[2], Sachin Talekar[2,3], Colin J. Barrow[2], and Mehran Ghasemlou[2,4,5]

[1]School of Life and Environmental Sciences, Deakin University, Geelong Waurn Ponds Campus, Geelong, VIC, Australia, [2]Centre for Sustainable Bioproducts, Deakin University, Geelong Waurn Ponds Campus, Geelong, VIC, Australia, [3]Department of Biological Sciences, BITS-Pilani, K.K. Birla Goa campus, Sancoale, Goa, India, [4]School of Science, STEM College, RMIT University, Melbourne, VIC, Australia, [5]Department of Chemical Engineering, The University of Melbourne, Parkville, VIC, Australia

4.1 Introduction

Food packaging is essential not only for containment but also for maintaining food quality, ensuring safety, and enabling distribution. Packaging provides a protective barrier against physical, biological, and chemical factors that can extend shelf life and preserve nutritional value (Garba, 2023). Additionally, packaging serves as a communication tool, providing essential information such as ingredient lists, expiration dates, and brand identity, which builds consumer trust. However, extensive reliance on plastics in this sector has raised substantial environmental concerns (Rani et al., 2024).

Plastic pollution from food packaging is primarily driven by the overproduction and poor end-of-life management of conventional petroleum-based plastics. The food sector is the largest consumer of plastics, with widespread use of single-use plastic items (Gündoğdu et al., 2024). The persistent accumulation of these plastic items in landfills and aquatic environments accelerates the formation of microplastics, which are now pervasive across ecosystems, and are ingested by living organisms throughout the food chain (Song, 2024). Moreover, plastics often contain harmful additives that can leach into the soil and water, posing risks to aquatic, and terrestrial ecosystems, and ultimately to human health (Gündoğdu et al., 2024). Based on this argument, a transition toward environmentally sustainable packaging materials is needed to reduce environmental impact, improve recyclability, and respond to the growing consumer demand for sustainable practices (Rani et al., 2024).

Bio-based and biodegradable plastics (BBPs) are increasingly recognized as viable candidates to address these challenges. Substituting conventional plastics with BBPs is considered a good

Bioplastics for Food Packaging. DOI: https://doi.org/10.1016/B978-0-443-41654-5.00001-9

solution to mitigate plastic pollution and reduce greenhouse gas emissions (Medina et al., 2025). BBPs offer significant environmental benefits over conventional plastics, primarily due to their renewable origins and their potential for complete or partial degradation at end of life. These bioplastics are often sourced from renewable resources, such as agricultural crops or by-products, and fermentation-derived products and reduce dependence on nonrenewable fossil resources (Samuel et al., 2024). As a result, BBPs can mitigate the environmental impact associated with persistent plastic waste. The development of BBP is aligned with the principles of green chemistry and the United Nations Sustainable Development Goals. Thus, they are a credible alternative to traditional plastics in food packaging applications (Moshood et al., 2022a).

This chapter provides an in-depth overview of the sources, properties, processing strategies, and applications of BBPs in the food packaging sector. It will also discuss their environmental implications, current market trends and regulatory considerations. Finally, future perspectives are discussed with emphasis on improving performance and enabling cost-effective scale-up of BBPs within circular economy.

4.2 Overview and classification of bioplastics

Bioplastics are broadly classified into three main categories based on their end-of-life behavior: (i) biodegradable and bio-based; (ii) nonbiodegradable and bio-based; and (iii) biodegradable and fossil-based (Ibrahim et al., 2021) (Fig. 4.1). Biodegradability of plastics/ bioplastics refer to their capacity to undergo microbial assimilation either under natural conditions (biodegradation) or controlled conditions (composting) (Ghasemlou et al., 2024).

4.2.1 Common misconceptions surrounding bioplastics

The terminology surrounding bioplastics is often confusing for consumers, policymakers and industrial specialists. A common misconception is that "bio-based" and "biodegradable" are interchangeable. In some literature, these terms are used synonymously, assuming they have the same meaning. However, they have different meanings. According to the European Bioplastics definition, "bio-based" refers to materials that are wholly or partly derived from renewable biomass, whereas "biodegradable" refers to materials that can be assimilated by microorganisms in soil or water and converted into carbon dioxide and water. Bio-based plastics are not inherently biodegradable. An excellent example of this type of bioplastic is bio-derived polyethylene (bio-PE). bio-PE is bio-based because its monomers are made from renewable feedstocks such as maize. bio-PE is chemically identical to conventional PE and does not biodegrade under environmental conditions. Some biodegradable bioplastics are entirely from fossil-based (non-bio-based) resources. A well-known example is polycaprolactone. With this context, the umbrella term "bioplastics" can be deceptive and is likely to

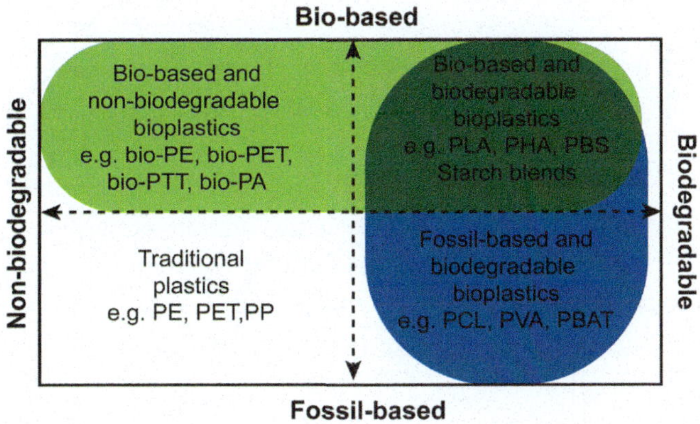

Figure 4.1

Classification of bioplastics based on their end-of-life behavior into three main groups: (i) bio-based and biodegradable, which are produced from renewable feedstocks (e.g., starch, cellulose, and polylactic acid) and can biodegrade; (ii) bio-based and nonbiodegradable, which are derived from renewable resources (e.,g bio-polyethylene, bio-PET) but are not degradable under environmental conditions; and (iii) fossil-based and biodegradable, which are produced from fossil feedstocks (e.g., polybutylene adipate terephthalate (PBAT)) and can biodegrade under either natural or controlled conditions. This figure explains the classification of bioplastics based on their end-of-life behavior.

cause confusion. To reduce ambiguity, the International Union of Pure and Applied Chemistry (IUPAC) recommended in 2015 that the term "bioplastics" be replaced with the more precise descriptors *bio-based plastics or biodegradable plastics* (Ghasemlou et al., 2024).

Another common source of confusion arises from the terms *biodegradable* and *compostable*, which are often used interchangeably (Lavagnolo et al., 2024). According to the American Society for Testing and Materials (ASTM), biodegradability is defined as microbial assimilation of a material into carbon dioxide, methane, and water, but with no specification of timeframe or environmental conditions. Since there is no fixed timeframe, all materials are theoretically biodegradable, although degradation may take months or years. However, a compostable plastic must disintegrate and mineralize within a defined period under controlled composting conditions, producing carbon dioxide, methane, water, and inorganic compounds (Lavagnolo et al., 2024). While all compostable plastics are biodegradable, not all biodegradable plastics are compostable (Cruz et al., 2022).

4.2.2 Definition and classification of biopolymer-based bioplastics

The term "biopolymer" refers to any polymer that is bio-based and/or biodegradable. In practice, the term bioplastic is often used instead of biopolymer to describe plastics that are designed to replace conventional ones made from oil or gas. This can be misleading because

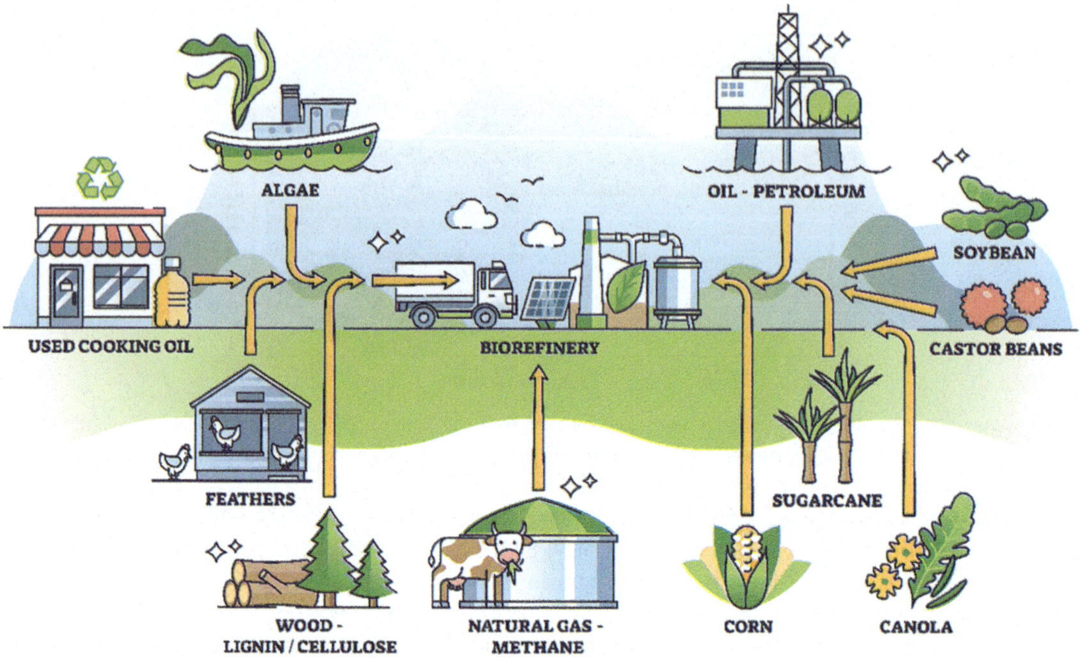

Figure 4.2

Diverse feedstocks for the production of bio-based polymers. Bio-based polymers can be derived from a broad spectrum of resources, including agricultural crops (corn, sugarcane, soybean, canola, and castor beans), lignocellulosic biomass, algae, animal by-products (feathers), waste streams (used cooking oil), and methane from natural gas. Image is licensed from istockphotoes.com. This figure explains the feedstocks that can be used to produce bio-based polymers.

not all biopolymers are considered plastics. Natural substances like collagen and cellulose are renewable and bio-based, but they are not usually described as bioplastics. Biopolymers can come from many different sources, including plants (e.g., starch, cellulose, and lignin), animals (e.g., collagen and chitosan), and microorganisms (e.g., polyhydroxyalkanoates). They can also be synthesized from renewable monomers (e.g., polylactic acid (PLA), or even from fossil feedstocks (e.g., polybutylene adipate terephthalate)) (Mehrpouya and Vahabi, 2023). From this perspective, classification of biopolymers requires a framework that considers both origin and degradability. Fig. 4.2 shows the raw materials that can be used to make BBPs.

4.3 Production of bio-based and biodegradable bioplastics for food packaging

Pulp-based BBPs are mainly composed of cellulose fibers. Pulp can be obtained directly from renewable resources, including wood, agricultural crops, and their associated residue or recovered from secondary streams including kraft and recycled paper. On an industrial scale, cellulose fibers are obtained through chemical and mechanical pulping. Mechanical pulping achieved high fiber yields but produces fibers with limited strength owing to lignin retention and hornification. Conversely, chemical pulping removes lignin and yields fibers of superior strength (Zhao et al., 2023). In addition to pulp, BBPs are derived from biopolymers naturally embedded in renewable biomass (Phadke and Rawtani, 2023). Current bioplastic production is dominated by carbohydrate-rich crops such as corn, sugarcane, potato, cassava, soybean, wheat, and so on, which are commonly classified as first-generation feedstocks. The first-generation feedstocks clearly compete with food production and raise concerns regarding land use and water consumption. To address these limitations, increasing attention is turned on second-generation feedstocks such as lignocellulosic biomass, municipal solid waste, and food waste, as well as third-generation feedstocks including algae, carbon dioxide, and methane, which have higher productivity and lower environmental impacts (Thomas et al., 2023). Conversion of feedstocks into BBPs can proceed through several technological routes. Microbial fermentation is most widely used, producing monomers that are subsequently polymerized into polymers. Alternative strategies include chemical polymerization of small monomers into large polymer precursors, direct extraction of naturally occurring biopolymers such as cellulose and starch, or the conversion of biomass into polymer precursors, followed by recovery (Jõgi and Bhat, 2020).

4.3.1 Microbial biosynthesis and chemical polymerization

The production of BBPs via bottom-up strategies begins with the generation of monomers, followed by their polymerization. In this approach, polysaccharide-rich feedstocks are fermented under controlled conditions using microbes such as lactic acid bacteria to yield building-block monomers. These monomers are subsequently polymerized into longer molecular chains through chemical or biotechnological routes (Haq et al., 2025; Zhao et al., 2023). A minimum of 100 monomer units are generally required to produce polymers with unique physical properties including tensile strength, flexibility, or fiber-forming ability, which differentiate them from materials composed of low-molecular-weight compounds (Ribeiro et al., 2025). Two common approaches are used for the synthesis of BBPs: ring-opening polymerization (ROP) and condensation polymerization. ROP is widely used for PLA by polymerizing lactide monomers derived from lactic acid. Condensation polymerization is an alternative method in which lactic acid monomers or oligomers are directly linked with the release of water (Haq et al., 2025; Zhao et al., 2023).

4.3.2 Extraction

Extraction is the first step of any top-down approach in producing BBPs. In this process, specific compounds are separated from the plant or animal-derived biomass, or from microbial biomass after fermentation. The aim is to recover naturally occurring polymers or biopolymers such as cellulose, starch, lactic acid, or other components, which then serve as building blocks for further processing. Extraction techniques are generally divided into conventional and advanced categories. Conventional methods including heat-assisted solvent extraction, Soxhlet extraction, and alkali and acid treatment are widely used in laboratory and industrial practice, but they often require high-energy inputs and large volumes of solvent. On the other hand, advanced "green" technologies such as enzymatic treatment, microwave-assisted extraction, ultrasound-assisted extraction, pulse-electric field, and rapid solvent systems are designed to improve efficiency while lowering environmental impact (Krishnan et al., 2024).

4.3.3 Blending and compounding

Blending and compounding are two important steps in the production of BBPs. These strategies have emerged as effective ways to improve the performance of biopolymers and overcome their shortcomings such as poor mechanical strength, limited barrier properties, and high production costs (Elliot, 2021; Silviya and Varma, 2009).

Blending involves combining two or more polymers to produce a new material with improved or complementary characteristics (Nyamweya, 2021). For instance, starch can be blended with chitosan or polyvinyl alcohol to improve mechanical strength and barrier properties while retaining biodegradability. Similarly, the mechanical properties of starch-based bioplastics have been shown to improve after reinforcement with natural fillers such as chitosan or algal proteins (Haq et al., 2025). In contrast, compounding refers to the incorporation of additives, fillers or fibers to generate tailored composites. A compounding step is necessary for producing homogeneous biocomposites from blended polymers prior to processing. For instance, cellulose fibers can be added into PLA to improve mechanical qualities, whereas nanoclays can be introduced into starch-based polymers to reduce oxygen and water vapor permeability. Compounding is typically carried out using a twin-screw extruder, followed by pelletizing, after which pellets are converted into finished or semifinished products (Fitch-Vargas et al., 2023; Silviya and Varma, 2009).

4.3.4 Processing techniques

After compounding or blending, pellets are converted into BBPs with defined shapes, sizes, colors, and properties tailored to application requirements. Different processing techniques are used for the production of BBPs. The techniques include extrusion (blown, co-extrusion, and reactive extrusion), injection molding, compression molding, blow molding, thermoforming, and casting methods such as solution casting and tape casting (Haq et al., 2025; Sid et al., 2021). Processing routes are generally divided into wet and dry techniques. Wet processes such as solution casting or electrospinning require the dissolution of the raw material in appropriate solvents, while dry processes such as compression molding, extrusion, or injection molding typically require the incorporation of plasticizers to improve flexibility and processability. These techniques often need to be redesigned for BBPs because the biopolymers have different properties from conventional plastics (Álvarez-Castillo et al., 2021). Selection of the most suitable processing method depends on both the intrinsic properties of the raw material (whether pulp-derived or polymer-based) and the performance criteria of the targeted product.

4.4 Functional properties and performance of bio-based and biodegradable bioplastics

4.4.1 Mechanical properties

The mechanical performance of BBPs is influenced by multiple factors including crystallinity, molecular weight, and degradation mechanisms. Key parameters such as Young's modulus, tensile strength, and elongation at break are crucial for packaging applications where they determine material selection, stress tolerance, and quality control. Due to their brittle character, many BBPs are less adaptable than conventional plastics. For instance, starch-derived BBPs generally exhibit inferior mechanical strength and flexibility, whereas chitin-based polymers have shown remarkable mechanical performance compared to starch biopolymers (Jha et al., 2024).

Several strategies have been developed to enhance the mechanical properties of BBPs. Reinforcing with metal oxide nanoparticles has been shown to improve tensile strength, elasticity, and elongation properties of BBPs (Anugrahwidya et al., 2021). Combining different BBPs (such as PLA–polyhydroxybutyrate (PHB) blends) can also result in materials with increased flexibility and/or impact resistance (Armentano et al., 2015). In addition, processing techniques such as melt extrusion and injection molding can be used to modify polymer structure and optimize mechanical properties. For example, PLA samples made using these methods showed up to a 73% increase in Young's modulus (Barkhad et al., 2020).

4.4.2 Barrier properties

The suitability of BBPs for food packaging is primarily determined by their barrier properties to gases and moisture. Barrier qualities are associated with the capacity of polymer to regulate the permeation (i.e., the transport of low-molecular-weight compounds across the packaging matrix). Effective food packaging requires resistance to the transfer of oxygen, nitrogen, carbon dioxide and other volatiles to protect food from outside environment. The permeability of these compounds depends on characteristics of the polymer materials such as thickness, size, and configuration, as well as processing conditions such as pressure and temperature (Robertson, 2009).

In general, BBPs have a hydrophilic character and provide adequate barrier properties against both oxygen and carbon dioxide. However, many BBPs have poor resistance to water vapor, and moisture can substantially alter their gas permeability (Donkor et al., 2023). Indeed, relative humidity (RH) strongly affects the oxygen transmission rate (OTR) of hydrophilic polymers. For the majority of BBPs, increasing RH increases OTR due to the plasticizing effect of water molecules on the polymer matrix, which expands free volume and facilitates the diffusion of oxygen molecules. Some BBPs were found to show no effect of RH on OTR and in certain cases even a reduction. For example, the OTR of PLA, PHB, and poly(3-hydroxybutyrate-co-3-hydroxyvalerate) (PHBV) remained unchanged or reduced with increasing RH (de Beukelaer et al., 2022). To manage or control the moisture sensitivity of BBPs, polymers can be cross-linked with inorganic fillers, modified with surface coatings, or blended with hydrophobic materials to reduce water uptake (Donkor et al., 2023).

4.4.3 Thermal properties

When considering the processing and utilization of BBPs in the food sector, their thermal properties are important because they determine how packaging materials respond to variations in temperature. To prevent deformation or deterioration at high temperatures, materials must have adequate thermal stability (Donkor et al., 2023). Polymers incorporating aromatic backbones or cross-linked networks generally display improved thermal resistance, whereas those containing oxygenated groups or double bonds in the main chain are less stable under elevated temperatures (Shaikh et al., 2021).

Glass transition temperature (T_g) represents the point where a BBP shifts from rigid and brittle state to soft and rubber-like behavior. Melting temperature (T_m) is the transition from solid to a liquid. The thermal decomposition temperature (T_d) is the point at which the polymer starts to chemically degrade. Most BBPs have relatively low T_g values, which limits their suitability for food packaging applications. It is also quite challenging to produce BBPs because of the narrow gaps between T_m and T_d. These limitations can be resolved by tailoring the thermal characteristics of BBPs. Common methods are composite formation through the addition of fillers, blending, copolymerization, or multilayer film manufacturing. Chemical modifications such as cross-linking and the incorporation of plasticizers are frequently used to adjust thermal parameters and improve processability (Gaur et al., 2020).

4.4.4 Optical properties

Optical performance is a crucial parameter for the practical use of BBPs in food packaging (Donkor et al., 2023). Transparent packaging is particularly desirable for perishable products (with shorter shelf life) because it enables customers to visually assess the quality of the food inside its package The transparency of a polymer is largely determined by its morphology, which can be adjusted through processing. Increasing crystallinity reduces clarity by enhancing the scattering of visible light, as crystalline structures fall within the wavelength range of visible light and act as scattering centers. Even transparent polymers do not transmit all wavelengths equally, and higher crystallinity consistently results in greater light scattering and lower transparency (Gaur et al., 2020).

Optical transparency is most often characterized by transmittance in the visible range (400–800 nm). Packaging materials with ~80% transmittance are considered transparent and those with a transmittance of >90% are highly transparent. Opaque materials have transmittance <10%, and translucency is considered for materials with transmittance of 10%–80%. BBPs such as carboxymethyl cellulose, PHBV, and regenerated cellulose have transmittance of 80%–90%, whereas amorphous cellulose and chitosan are highly transparent with transmittance of >90% (Guzman-Puyol et al., 2022).

Opacity is the second most common parameter for determining the optical performance of a plastic film. A high opacity values suggests that the sample is not very transparent. Opacity can be defined as the ratio of sample thickness to absorbance. Opacity has no upper limit. For food packaging, transparent materials typically have opacity values less than 5. Bio-based polymers such as cellulose acetate, calcium alginate, carboxymethyl cellulose, amaranth protein isolate, and carrageenan typically fall within this range. Highly transparent polymers such as gelatin and chitosan exhibited opacity values below 1, whereas sodium alginate showed opacity value close to the threshold of 5 (Guzman-Puyol et al., 2022).

4.4.5 Antimicrobial and antioxidant properties

In the food sector, BBPs are increasingly explored for use in active packaging, which is a recent and most innovative approach to food preservation. The functionality of active packaging is determined either from the inherent properties of the polymer or from the additives incorporated into the packaging system. BBP-based active packaging can provide moisture absorption, antibacterial, and antioxidant activities or support modified atmosphere conditions. For example, chitosan has inherent antibacterial and antifungal activity, mediated by electrostatic interactions between its amine groups and the negatively charged surfaces of microbial cells. This results in changing the cell permeability and its ultimate death. Additional functionality can also be introduced by adding active monomers into the polymer backbone (Salgado et al., 2021).

Instead of being applied directly to food, active agents can be incorporated into the packaging material, coated on its surface, or embedded within specific package components such as bags, labels, pads, and bottle caps. This strategy enables lower concentrations of active agents to achieve the desired effect while reducing undesirable flavors and odors in food. When compliant with current regulations, active chemicals may be added into the packaging systems to either release or absorb compounds to or from food or its surrounding environment, thereby modifying composition or sensory characteristics. Depending on the design, migration can occur partially or fully into the food or headspace through gradual diffusion, or activity may be restricted to direct contact between the packaging and the product (Nilsen-Nygaard et al., 2021).

4.5 Applications of bio-based and biodegradable bioplastics in food packaging

Plastics dominate the global packaging market, accounting for 49% of all material used. Over the past few decades, there have been notable scientific and technological advances in the use of plastic packaging, especially in the food sector. Common commodity plastics such as nylon, PE, polypropylene (PP), polyethylene terephthalate (PET), polystyrene (PS), and polycarbonate (PC) are nonbiodegradable, and their widespread use poses serious environmental risks and contributes significantly to municipal waste and microplastic pollution. As a result, waste management initiatives, including recycling and the adoption of BBPs are strongly recommended. Companies are increasingly required to report on sustainability in accordance with market and governmental requirements. This demand for environmentally responsible reporting is imposing additional obligations on industries (Bhagabati et al., 2020).

BBPs are estimated to replace almost 31% of conventional polymers (Nizamuddin et al., 2024). These materials are already used in global food packaging. Asia is a leading production hub, followed by Europe, which is rapidly expanding its production capacity. The European Union (EU) has implemented a policy framework that includes mandatory biowaste collection. China has imposed bans on certain single-use plastics. India has also prohibited specific single-use items. Globally, there is a trend toward using BBPs derived from renewable sources such as corn starch, sugarcane, and algae, which are intended to replace fossil-based plastics (Bauer et al., 2025; Moshood et al., 2022b).

Currently, there are a number of biodegradable polymers and plastics that are commercially available. Each material has the potential to contribute to the overarching goal of sustainability. A growing number of companies or start-ups manufactured different types of BBPs, and these products are widely marketed for food and nonfood applications (Table 4.1.). These products can be manufactured in different shapes including trays, plates, bottles, cups, lids, cutlery, cling films, bags, and flexible or rigid packaging forms. All these BBPs are either biodegradable, compostable, or recyclable, thereby potentially reducing the persistence of microplastics in the environment. One of the earliest commercial applications of bio-based plastics was their use in biodegradable and compostable coffee capsules (Phadke and Rawtani, 2023).

4.5.1 Applications of polylactic acid in food packaging

PLA is a bio-based polymer synthesized from lactic acid derived from renewable resources. Most lactic acids are synthesized through microbial fermentation and subsequently

Table 4.1 A summary of commercially available bio-based and biodegradable bioplastics in the market that are used for food packaging.

Name of Company or product	Type of BBPs	End of life behavior	Food packaging application
Paper Water Bottle	Sugarcane and bamboo pulp based	Biodegradable and compostable	Bottles
NatureFlex	Cellulose based	Compostable	Food packaging for coffee, tea, snacks, dried food, chocolate, confectionary, bakery, fresh produce, and custom food service
MATER-BI	Starch based and cellulose based	Biodegradable and compostable	Bags for fruits and vegetables, dry and fresh food packing, food serving plates, cups, cutlery, bowls, ice cream cups and scoops, and cup lids
Bio-Plastobag	Starch based	Compostable	Cling film, cutlery, bags for grocery, fruits and vegetables, and food packaging laminates
Ecovio	PLA and PBAT-based	Compostable	Coffee capsules, trays, plates, cups, lids, cutlery, fruit and vegetable bags, and paper packaging
Luminy	PLA based	Compostable and recyclable	Rigid food packaging, food serviceware (forks, spoons, knives, cups, plates, bowls, straws, trays, lids, and more) and flexible packaging (fresh food films, snack and confectionary packaging, bread bags, tapes, and grocery bags)
Ecolife	PLA based	Biodegradable and compostable	Cutlery and paper cups
Ingeo	PLA based	Compostable	Food and beverage packaging (dairy and dessert, coffee capsules, infusion, produce, dry containers, DELI containers, bakery, meat trays, and temper bands) and food serviceware (clear cups, cutlery, deli containers. hot cup lids, plates, takeout trays, and straws)
BOPLA packaging film	PLA based	Biodegradable and compostable	Food packaging film and fresh packaging film
InNature Pack	Plant fiber based	Biodegradable	Plates, bowls, cutlery, egg trays, drinkware, and accessories (coffee cup, lids, and cup carriers) and produce packaging (trays and baskets)

PBAT, polybutylene adipate terephthalate.

polymerized to produce PLA (Malhotra et al., 2023). The US Food and Drug Administration (FDA) has approved lactic acid, a safe food preservative. This has further stimulated interest in adopting PLA for food and biomedical applications. Migration of lactic acid monomers from PLA packaging into food occurs at negligible levels, supporting its use for food-contact applications. PLA also has several advantageous features that are suitable for food packaging. PLA has a high degree of transparency. In comparison to conventional fossil-derived polymers, PLA has lower permeability to carbon dioxide and oxygen, which is an important consideration for food packaging applications that require effective oxygen barriers. The oxygen barrier capability of PLA is comparable or slightly higher than that of PS. PLA also demonstrates favorable mechanical strength and water resistance. The PLA surface is highly resistant to penetration by oils and fats, which supports its suitability for packaging fatty foods. In terms of end-of-life behavior, PLA can be recycled, incinerated, or biodegraded. Under hydrothermal conditions such as boiling water or steam, PLA undergoes hydrolysis in lactic acid, enabling molecular recycling of the packaging material (Swetha et al., 2023).

4.5.2 Applications of polyhydroxyalkanoates in food packaging

Polyhydroxyalkanoates (PHAs) are polyesters naturally synthesized by bacterial fermentation and stored as intracellular granules in both Gram-positive and Gram-negative species (Swetha et al., 2023). A wide variety of PHA homopolymers and copolymers can mimic the performance of petroleum-based plastics, offering a safe, eco-friendly, and biodegradable substitutes. Nowadays, different PHA monomers are now being used to replace fossil-derived polymers in packaging and other consumer goods. The adaptability of PHAs, including their tunable transparency, strength, and flexibility, makes them useful for applications ranging from packaging to agriculture and healthcare (Park et al., 2024). Because of its greater thermal stability, universal biodegradation, and biocompatibility, PHA shows considerable promise as replacements for petroleum-based polymers. Demand for PHAs in disposable packaging including antibacterial food containers is increasing due to their favorable functional properties. The use of PHA-based packaging provides the dual benefit of lowering environmental impact while extending the shelf life of packaged foods. In the food sector, PHA is a preferred material that can substitute oil-based plastics due to its bio-based origin and marine biodegradability (Atarés et al., 2024).

4.5.3 Applications of starch-based bioplastics in food packaging

Starch is a polysaccharide found in large quantities in staple crops such as corn, wheat, rice, and potatoes. Starch is one of the most widely used biodegradable biopolymers in the bioplastics industry. The structure of starch consists of glucose units linked by glycosidic bonds. As starch is made from agricultural feedstocks rather than fossil resources, it is a sustainable substitute that reduces reliance on nonrenewable polymers. A key advantage of starch is its biodegradability. Under suitable environmental conditions, starch-based packaging can break down into natural compounds, helping to mitigate or reduce environmental contamination and the accumulation of waste. Starch-based bioplastics (either alone or hybridized form) are currently used as replacements for PS. They are found in bottles, coffee machine capsules, bowls, plates, cups, egg trays, disposable cutlery, food wrapping cling films, and thermoplastics packaging. Starch-rich crops are widely available and abundant, making them a scalable and affordable raw material for making bioplastics. Overall, starch combines natural availability, biodegradability, and renewability, making it a desirable option for the development of sustainable packaging materials (Pei et al., 2024; Sid et al., 2021).

4.5.4 Applications of cellulose-based bioplastics in food packaging

Cellulose and its derivatives are among the most abundant natural substances on earth. Their wide availability makes them both affordable and accessible. In food packaging, cellulose offers several advantages over other polysaccharides: it is inexpensive, widely available, easy to process and store, recyclable, and capable of carrying branding or customer information. Additionally, cellulose also imparts barrier properties that limit the migration of gases, lipids, moisture, and solutes. Bioplastics composed entirely of cellulose are mechanically fragile because the strong interactions between polymer chains restrict flexibility. This limitation can be addressed by adding a plasticizer to reduce intermolecular bonding. The addition of plasticizers also enhances flexibility, mechanical handling, and maintains structural integrity. The strength of the pulp-based product is due to hydrogen bonding between the cellulose fibers. Cellulose is also utilized in multilayer packaging systems, where it is combined with other nonedible or nonbiodegradable polymer films to improve performance. Because of its high tensile and compressive strength, cellulose is frequently used as substitute for PS in containers, trays, and egg cartons. In addition, cellulose-based films serve as carriers for food additives, helping to inhibit microbial spoilage and extend the shelf life of package products (Liu et al., 2021; Sid et al., 2021).

4.5.5 Applications of chitosan-based bioplastics in food packaging

Chitosan is a linear amino-functionalized polysaccharide derived from chitin. Chitin is the second most abundant natural polysaccharide after cellulose and can be sourced from crustaceans, insects, and fungi through chemical treatments that remove acetyl groups (Nguyen et al., 2024). Chitin, chitosan, and their derivatives have attracted increasing attention in food packaging because of their unique structures and important features, including antimicrobial activity, biodegradability, biocompatibility, and nontoxicity. Chitosan films demonstrated strong antimicrobial activity, which contributes to extending the shelf life of packaged food products. They also offer selective barrier properties, which is crucial for fresh food packaging. The incorporation of natural extracts and essential oils further enhanced the antioxidant activities of chitosan films, further strengthening their role in food preservation. Chitin, chitosan, and their derivatives have been widely combined with proteins, carbohydrates, synthetic polymers, and inorganic nanoparticles to develop composites tailored for food packaging applications (Olunusi et al., 2025; Zubair et al., 2020).

4.5.6 Applications of protein-based bioplastics in food packaging

Numerous plant- and animal-derived proteins, including wheat gluten, soy protein, whey protein, zein, casein, collagen, and gelatin, have been explored as feedstocks for making bioplastics. From a sustainability standpoint, however, the use of proteins in packaging must be carefully considered, as these are essential components of the human diet. Proteins are organized into fibrous or globular structure through hydrogen, covalent, and ionic interactions Their networks are stabilized by disulfide bonds, hydrogen bonds, and hydrophobic interactions (Acar et al., 2026; Nanda et al., 2022). Owing to diverse amino acid compositions and the potential for extensive intermolecular interactions, proteins provide a greater range of functional capabilities than lipids or polysaccharides. Protein-based packaging not only offers several advantages but also faces limitations, including poor mechanical strength, low moisture resistance, and susceptibility to microbial contamination. To address these challenges, researchers are investigating strategies such as blending with reinforcing materials, utilizing chemical or physical cross-links, and optimizing processing conditions. These approaches aim to improve the mechanical characteristics, water resistance, and overall durability, with the broader goal of making protein-based bioplastics as viable competitive with conventional plastics (Qazanfarzadeh and Kumaravel, 2023).

4.5.7 Application of seaweed-based bioplastics in food packaging

Seaweed, once overlooked as a low-value resource, is now getting attention as a promising feedstock for bioplastic production (Ayala et al., 2024). Seaweed-based bioplastics can be made by extracting biopolymers (like alginate, carrageenan, agar, and ulvan) or using whole biomass without separating the biomolecules or turning biomass into polymer precursors to synthesize PLA and PHA (Torrejon et al., 2025). Research has revealed that seaweed-based bioplastics can extend the shelf life of food products up to 40%, maintaining nutritional and sensory properties (Kajla et al., 2024). The bioplastics produced from seaweed offer various benefits for food packaging, including biodegradability, antioxidant and antimicrobial properties, and adjustable mechanical, thermal, and barrier characteristics. Additionally, these bioplastics can be customized to be edible, compostable, or water-soluble, and it is possible to enhance their physical properties, such as moisture content, tensile strength, and gas barrier performance for specific applications like withstanding microwave heating and freezer conditions. In the upcoming years, seaweed polymers' cost-effectiveness and cost-efficiency are critical concerns that need to be resolved to ensure their sustainable use in the packaging sector. A betterment of cultural conditions for cultivation and the advancement of cultivation technologies, alongside genetic and metabolic engineering, can significantly enhance the yield of specific polymers in seaweeds. Consequently, a commercially viable supply of seaweed polymers to address the financial challenges in seaweed-based bioplastic production can be ensured (Duarah et al., 2025).

4.6 Life cycle assessment and environmental impacts of bio-based and biodegradable bioplastics

Life cycle assessment (LCA) is a structured method to assess the environmental impacts of a product or process. When LCA is correctly applied, it can reveal both positive and negative effects, many of which are not immediately visible (Walker and Rothman, 2020). The most common approaches to perform LCA are cradle-to-grave and cradle-to-gate (Jha et al., 2024). Cradle-to-gate considers impacts from raw material extraction to the point when a product leaves the manufacturer, whereas cradle-to-grave extends the analysis to include end-of-life stages such as recycling, reuse, or disposal (Walker and Rothman, 2020). Practically, LCA cannot capture every detail of a real process. Its accuracy and reliability depend on the number of processes and the levels of detail used to describe inputs, outputs, and related operations. The quality of data, the way impacts are allocated among co-products, and the models used to calculate environmental effects also play a decisive role. Typical impact categories include energy use, greenhouse gas emissions, water and land use, acidification, and eutrophication. Careful selection of categories and methods is essential to ensure reliable and meaningful outcomes (Walker and Rothman, 2020). LCA is also frequently used to compare the environmental impact of bioplastics with petroleum-based plastics. However, conducting LCAs for bioplastics is challenging due to many uncertainties across the process. Key uncertainties arise from limited knowledge of industrial-scale production processes, variation in product performance, consumers behavior, and in the fate of bioplastics in different waste management systems. Much of available data is derived from laboratory or pilot-scale studies, which often fail to reflect real-world production and end-of-life conditions. As a result, comprehensive LCAs that cover all life cycle stages of bioplastics are still rare and uncommon (Molina-Besch, 2022). A circular life cycle of bio-based and biodegradable polymers is illustrated in Fig. 4.3.

BBPs such as PHB, PHA, and PLA hold strong potential to reduce greenhouse gas emissions and dependence on fossil resources. However, agricultural practices and the energy intensity of processing can result in higher acidification and eutrophication impacts. Studies have shown that integrating renewable energy into production significantly improves their environmental profile. For example, PHB and PLA can outperform some fossil-derived plastics when manufactured using renewable energy and composted at the end-of-life. In contrast, the use phase showed little difference between BBPs and conventional plastics. The large-scale adoption of biopolymers also raises concerns related to land-use change and resource competition, where even a modest increase in demand could offset some of their

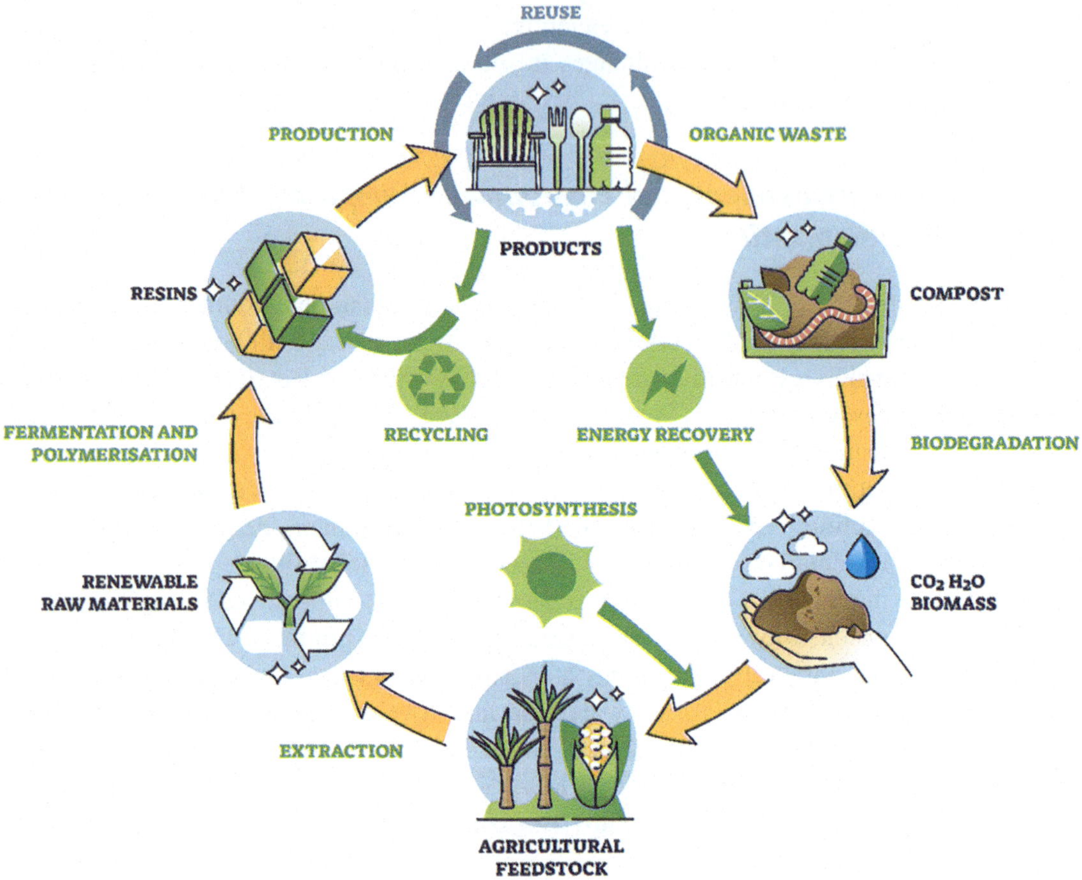

Figure 4.3

Circular life cycle of bio-based and biodegradable polymers. Agricultural feedstocks provide renewable raw materials that are extracted and converted into resins through fermentation and polymerization. These resins are processed into products, which can be reused, recycled, composted, or directed to energy recovery. At end-of-life, biodegradation generates carbon dioxide, water, and biomass, which are reintegrated into the biosphere through photosynthesis, thereby closing the loop. Image is licensed from istockphotoes.com. This figure explains the circular life cycle of bio-based and biodegradable polymers.

environmental benefits. To overcome this, second-generation feedstocks sourced from waste or nonfood biomass are increasingly seen as a more sustainable alternative. BBPs and their fossil-based counterparts showed almost similar results in the case of the use phase. Here, no major difference was observed (Jha et al., 2024; Molina-Besch, 2022).

BBPs can deliver environmental advantages through the utilization of renewable resources and the mitigation of greenhouse gas emissions, but they also carry risks. Indeed, if not adequately managed postconsumption, they can contribute to soil and water pollution, persistent microplastic formation, and competition for arable land needed for food production. Their overall environmental impact depends strongly on disposal conditions, including access to industrial composting facilities, renewable energy inputs in manufacturing, and the prevention of contamination in ecosystems where degradation may not incomplete (Atiwesh et al., 2021; Huang et al., 2025; Islam et al., 2024).

The end-of-life stage is a crucial aspect in evaluating the sustainability of BBPs. Commonly modeled options include composting, anaerobic digestion, recycling, incineration, and landfilling. Many LCAs rely on limited or generic data for these scenarios, which can lead to uncertainty in the results. In most assessments, recycling is often identified as the most favorable route, but this conclusion is not always supported by robust material-specific evidence. With growing environmental concern about plastic waste leakage into ecosystems, improving the quality and specificity of end-of-life data is vital. Industrial composting and well-managed recycling systems have the potential to substantially enhance the sustainability profile of BBPs if they are supported by accurate data and infrastructure (Molina-Besch, 2022). Overall, with renewable energy inputs and effective end-of-life strategies, BBPs represent a viable pathway toward more sustainable packaging systems.

4.7 Regulatory and market aspects of bio-based and biodegradable bioplastics

The global market for BBPs is expected to expand significantly between 2025 and 2035. Due to increasing concern over plastic pollution, governments are implementing strict regulations that ban or restrict conventional plastics and create strong demand for BBPs. Market Forecasts estimate the growth from US$6.3 billion in 2025 to US$15.6 billion by 2035, representing a compound annual growth rate of 9.5%. The food packaging sector is expected to dominated this expansion as companies adopt biodegradable films, compostable packaging, and renewable plastics to achieve sustainability goals (Future Market Insights, 2025).

Consumer behavior is also shifting in favor of BBPs. Rising environmental awareness is making customers willing to pay more for certified BBP products. Proper certification and labeling are therefore essential, particularly in food and personal care packaging. Misleading claims or incorrect certifications risk undermining customer's trust. To address this, producers much provide transparent information on product composition and disposal, supported by internationally recognized labeling requirements and educational initiatives (Nazareth et al., 2022).

The development and commercialization of BBPs for food packaging must comply with complex regulatory frameworks to ensure safety, consumer protection, and environmental sustainability. These frameworks vary across jurisdictions. Regulatory agencies such as the International Organization for Standardization (ISO), the European Bioplastics Association, European Food Safety Authority (EFSA) and the FDA provide standards, guidelines, and testing procedures. ISO develops standardized standards for environmental performance and biodegradability testing, while FDA and EFSA concentrate primarily on consumer safety and migration limits. Manufacturers seeking global market access must comply with these frameworks.

In the United States, the Food Contact Notification (FCN) program, established under the Food and Drug Administration Modernization Act (FDAMA) of 1997, provides a regulatory pathway to market for food-contact materials. Title 21 of the Code of Federal Regulations (21 CFR) specifies approved substances, use conditions, and migration limits. FCN submission to FDA includes information on identity, production methods, molecular weight profile, physicochemical properties, intended use, exposure conditions, and environmental impacts. Approval depends on rigorous demonstration of consumer safety (Dages, 2020).

In the European Union, BBPs used in food-contact applications are subject to the same stringent requirements as conventional plastics. Regulation (EU) No. 10/2011 requires migration testing to ensure that only approved substances are used in multicomponent materials. Certification under EN 13432 also requires ecotoxicity testing to verify compostability. In several cases, BBPs must undergo even more extensive testing than conventional plastics to ensure consumer and environmental safety (European Bioplastics, 2023).

ISO 17088 outlines methods for plastics, intended for industrial composting, including criteria for disintegration, aerobic biodegradation, chemical composition, and absence of harmful effects on terrestrial organisms. Labeling materials as “compostable” rather than simply “biodegradable” is recommended to reflect compliance with these criteria. ISO 17088 aligns closely with ASTM D6400 and the EN 13432 standards for industrial compostability (Nizamuddin et al., 2024).

4.8 Challenges and proposed solutions

Like other emerging materials, BBPs face a range of challenges that must be addressed to facilitate their wider adoption. A key challenge lies in meeting the target product properties expected by clients, consumers, and regulatory authorities. To replace conventional plastics, BBPs must demonstrate comparable affordability, stability, availability, processability, and user performance. In most cases, the only acceptable difference is improved end-of-life performance. This means that product specification will need to evolve to include sustainability criteria while considering recycling and other end-of-life limitations (Börner and Zinn, 2024).

Feedstock sustainability is another major challenge. Current production of BBPs relies heavily on first-generation feedstocks such as starches and plant oils. Continued dependence on these resources risks intensifying competition for agricultural land, with potential impacts on food security and biodiversity. Moving toward second-generation feedstocks such as organic waste or nonfood biomass is therefore essential. Advanced metabolic engineering and microbial mixed cultures, which are currently used on the laboratory scale, provide viable strategies to maintain productivity while reducing reliance on food-based resources (Börner and Zinn, 2024; Thomas et al., 2023).

The high production cost of BBPs is a significant barrier to commercialization. At present, they are estimated to be 2–10 times more expensive than fossil-based plastics. Reducing costs will require improvements in production technology, downstream processing (DSP), methods, and microbial strain development. In addition, reducing waste, valorizing residual cell biomass, and ensuring a sustainable feedstock supply with effective pretreatment are essential to lower the cost of BBP production. Continuous fermentation is an effective approach, as it enables the production of higher yields of high-quality PHAs while simplifying DSP process and generating less waste than fed-batch systems. The development of energy-efficient fermenters with enhanced mass transfer rates and gas recycling can further reduce production cost and improve the economic feasibility of BBPs (Börner and Zinn, 2024).

Strengthening the BBP value chain is also essential to ensuring consistent quality, reducing costs, and supporting regulatory compliance. Effective collaboration between academia, industry, and policymakers will be required to meet these goals. Academic research often lacks technical expertise in polymer processing, while industrial projects may not have the research capacity to address fundamental scientific questions. Precompetitive partnerships between academia and industry can help define clear technical targets, foster communication across the value chain, and improve alignment with both policymakers and consumers (Börner and Zinn, 2024).

Concerns regarding biodegradability must also be addressed. Although BBPs are designed to degrade in soil or compost, many can still fragment into microplastics during breakdown. These fragments are typically less persistent than those from conventional plastics, but complete degradation requires specific environmental conditions such as suitable temperature, moisture, and microbial activity. Such conditions are not always present in natural environments like oceans or landfills. The extent of microplastic formation depends on the polymer type, disposal pathways, and environmental conditions. To ensure that BBP deliver on their sustainability promise, compliance with biodegradability standards must be verified under realistic environmental conditions (Medina et al., 2025; Pilapitiya and Ratnayake, 2024).

Pulp-based BBPs face additional challenges. Secondary fibers obtained from recycled sources such as newspapers and books are widely used in molded pulp packaging but often contain residual inks and chemicals, which raises major concern for food-contact applications. Nevertheless, producing one ton of recycled pulp requires 27% less energy than the virgin pulp, and thus it is environmentally preferable. Substituting virgin fibers would undermine sustainability goals while increasing energy use. Further research is required to optimize pulp-based BBPs for safe and effective use in packaging (Zhao et al., 2023).

4.9 Conclusion

Among the different categories of bioplastics, BBPs are considered the most sustainable choice because they combine renewable origin with the capacity for biodegradation. The development and application of BBPs in food packaging present a promising pathway toward reducing reliance on fossil resources and lowering the environmental impacts of plastic waste. BBPs derived from starch, cellulose, PLA, and PHA have already demonstrated immense potential, but their widespread adoption is restricted by technical, economic, and regulatory challenges. Future progress will depend on continued improvements in material and ensure compliance with biodegradability standards. Without clear alignment to these standards, there is a risk that BBPs could be blended with bio-based polymers that are not biodegradable, and this will weaken their sustainability impact. BBPs are expected to become more cost-competitive with conventional plastic due to the adoption of second- and third-generation feedstocks and their integration into industrial-scale production. These developments will also enable more robust LCA, which is essential for determining optimal end-of-life pathways. Integration of BBPs within the frameworks of circular economy will be essential to accelerate their adoption and to achieve their immense potential.

References

Acar, E.G., Sezer, B., Gunes, G., 2026. 8.18 - Bioplastics for Food Packaging Applications. Comprehensive Polymer Science (Second Edition). Elsevier, pp. 398–424. Available at https://doi.org/10.1016/B978-0-323-95486-0.00052-1.

Anugrahwidya, R., Armynah, B., Tahir, D., 2021. Bioplastics starch-based with additional fiber and nanoparticle: characteristics and biodegradation performance: a review. Journal of Polymers and the Environment 29 (11), 3459–3476. Available at https://doi.org/10.1007/s10924-021-02152-z.

Armentano, I., Fortunati, E., Burgos, N., Dominici, F., Luzi, F., Fiori, S., et al., 2015. Bio-based PLA_PHB plasticized blend films: processing and structural characterization. LWT - Food Science and Technology 64 (2), 980–988. Available at https://doi.org/10.1016/j.lwt.2015.06.032.

Atarés, L., Chiralt, A., González-Martínez, C., Vargas, M., 2024. Production of polyhydroxyalkanoates for biodegradable food packaging applications using haloferax mediterranei and agrifood wastes. Foods 13 (6), 950. Available at https://doi.org/10.3390/foods13060950.

Atiwesh, G., Mikhael, A., Parrish, C.C., Banoub, J., Le, T.A.T., 2021. Environmental impact of bioplastic use: A review. Heliyon 7 (9), e07918. Available at https://doi.org/10.1016/j.heliyon.2021.e07918.

Ayala, M., Goosen, N., Michalak, L., Thomsen, M., Pizzol, M., 2024. Prospective LCA of brown seaweed-based bioplastic: upscaling from pilot to industrial scale. Sustainable Production and Consumption 52, 416–426. Available at https://doi.org/10.1016/j.spc.2024.11.020.

Álvarez-Castillo, E., Bengoechea, C., Felix, M., Guerrero, A., 2021. Protein-based bioplastics from biowastes: sources, processing, properties and applications. Bioplastics for Sustainable Development. Springer, Singapore, Spain, pp. 137–176. Available at http://doi.org/10.1007/978-981-16-1823-9_5.

Barkhad, M.S., Abu-Jdayil, B., Mourad, A.H.I., Iqbal, M.Z., 2020. Thermal insulation and mechanical properties of polylactic acid (PLA) at different processing conditions. Emirates Polymers 12 (9), 2091. Available at https://doi.org/10.3390/POLYM12092091.

Bauer, M.G., Henkel, F., Gürer, U., Lieleg, O., 2025. Bio-based and degradable food packaging materials: where are they? Advanced Materials Interfaces 12 (6), 2400645. Available at https://doi.org/10.1002/admi.202400645.

Bhagabati, P., Bhardwaj, U., Katiyar, V., 2020. Biobased and biodegradable polymers for food packaging: commercial status. Sustainable Polymers for Food Packaging: An Introduction. De Gruyter, India, pp. 15–22. Available at http://doi.org/10.1515/9783110648034-002.

Börner, T., Zinn, M., 2024. Key challenges in the advancement and industrialization of biobased and biodegradable plastics: a value chain overarching perspective. Frontiers in Bioengineering and Biotechnology 12, 1406278. Available at https://doi.org/10.3389/fbioe.2024.1406278.

Cruz, R.M.S., Krauter, V., Krauter, S., Agriopoulou, S., Weinrich, R., Herbes, C., et al., 2022. Bioplastics for food packaging: environmental impact, trends and regulatory aspects. Foods 11 (19), 3087. Available at https://doi.org/10.3390/foods11193087.

Dages, J.M., 2020. FDA authorized? Establishing a suitable food-contact legal status for bioplastics in the USA. Bioplastics Magazine 15, 48–49. Available at https://www.steptoe.com/a/web/212982/FDA-authorized-BioMAG-2021.pdf.

de Beukelaer, H., Hilhorst, M., Workala, Y., Maaskant, E., Post, W., 2022. Overview of the mechanical, thermal and barrier properties of biobased and/or biodegradable thermoplastic materials. Polymer Testing 116, 107803. Available at https://doi.org/10.1016/j.polymertesting.2022.107803.

Donkor, L., Kontoh, G., Yaya, A., Bediako, J.K., Apalangya, V., 2023. Bio-based and sustainable food packaging systems: relevance, challenges, and prospects. Applied Food Research 3 (2), 100356. Available at https://doi.org/10.1016/j.afres.2023.100356.

Duarah, R., Hazarika, P., Boruah, P., Lakshmi, D.S., Hazarika, S., 2025. Macroalgal (seaweed) derived bioplastic as a sustainable food packaging material – a critical review. International Journal of Biological Macromolecules 323 (Pt 2), 147094. Available at https://doi.org/10.1016/j.ijbiomac.2025.147094.

Elliot, D., 2021. Sponsored Content Plastics Technology Biopolymer Compounding Done Right. Available from https://www.ptonline.com/articles/biopolymer-compounding-done-right.

European Bioplastics, 2023. How bioplastics can contribute to the sustainability of food contact materials. Available from https://www.european-bioplastics.org/how-bioplastics-can-contribute-to-the-sustainability-of-food-contact-materials/.

Fitch-Vargas, P.R., Camacho-Hernández, I.L., Rodríguez-González, F.J., Martínez-Bustos, F., Calderón-Castro, A., J.d.J. Zazueta-Morales, E., et al., 2023. Effect of compounding and plastic processing methods on the development of bioplastics based on acetylated starch reinforced with sugarcane bagasse cellulose fibers. Industrial Crops and Products 192, 116084. Available at https://doi.org/10.1016/j.indcrop.2022.116084.

Future Market Insights, 2025. Biobased Biodegradable Plastic Market, Biobased Biodegradable Plastic Market Growth - Trends & Forecast 2025 to 2035. Available at https://www.futuremarketinsights.com/reports/global-biobased-biodegradable-plastics-market.

Gaur, S.S., Ghosh, T., Katiyar, V., 2020. General material property requirement for food packaging applications. Sustainable Polymers for Food Packaging: An Introduction. De Gruyter, India, pp. 93–126. Available at http://doi.org/10.1515/9783110648034-006.

Garba, A.I., 2023. Food preservation packaging. In: Tumuluru, J.S. (Ed.), Food Processing and Packaging Technologies: Recent Advances. IntechOpen. Available at http://doi.org/10.5772/INTECHOPEN.110043.

Ghasemlou, M., Barrow, C.J., Adhikari, B., 2024. The future of bioplastics in food packaging: an industrial perspective. Food Packaging and Shelf Life 43, 101279. Available at https://doi.org/10.1016/j.fpsl.2024.101279.

Guzman-Puyol, S., Benítez, J.J., Heredia-Guerrero, J.A., 2022. Transparency of polymeric food packaging materials. Food Research International 161, 111792. Available at https://doi.org/10.1016/j.foodres.2022.111792.

Gündoğdu, S., Bour, A., Köşker, A.R., Walther, B.A., Napierska, D., Mihai, F.C., et al., 2024. Review of microplastics and chemical risk posed by plastic packaging on the marine environment to inform the Global Plastics Treaty. Science of the Total Environment 946, 174000. Available at https://doi.org/10.1016/j.scitotenv.2024.174000.

Haq, F., Kiran, M., Khan, I.A., Mehmood, S., Aziz, T., Haroon, M., 2025. Exploring the pathways to sustainability: a comprehensive review of biodegradable plastics in the circular economy. Materials Today Sustainability 29, 101067. Available at https://doi.org/10.1016/j.mtsust.2024.101067.

Huang, S., Dong, Q., Che, S., Li, R., Tang, K.H.D., 2025. Bioplastics and biodegradable plastics: a review of recent advances, feasibility and cleaner production. Science of the Total Environment 969, 178911. Available at http://doi.org/10.1016/j.scitotenv.2025.178911.

Ibrahim, N.I., Shahar, F.S., Hameed Sultan, M.T., Md Shah, A.U., Azrie Safri, S.N., Yazik, M.H.M., 2021. Overview of bioplastic introduction and its applications in product packaging. Coatings 11 (11), 1423. Available at https://doi.org/10.3390/COATINGS11111423.

Islam, M., Xayachak, T., Haque, N., Lau, D., Bhuiyan, M., Pramanik, B.K., 2024. Impact of bioplastics on environment from its production to end-of-life. Process Safety and Environmental Protection 188, 151–166. Available at https://doi.org/10.1016/j.psep.2024.05.113.

Jha, S., Akula, B., Enyioma, H.H., Novak, M., Amin, V., Liang, H., 2024. Biodegradable biobased polymers: a review of the state of the art, challenges, and future directions. Polymers 16 (16), 2262. Available at https://doi.org/10.3390/polym16162262.

Jõgi, K., Bhat, R., 2020. Valorization of food processing wastes and by-products for bioplastic production. Sustainable Chemistry and Pharmacy 18, 100326. Available at https://doi.org/10.1016/j.scp.2020.100326.

Kajla, P., Chaudhary, V., Dewan, A., Bangar, S.P., Ramniwas, S., Rustagi, S., et al., 2024. Seaweed-based biopolymers for food packaging: a sustainable approach for a cleaner tomorrow. International Journal of Biological Macromolecules 274 (Pt 1), 133166. Available at https://doi.org/10.1016/j.ijbiomac.2024.133166.

Krishnan, L., Ravi, N., Kumar Mondal, A., Akter, F., Kumar, M., Ralph, P., et al., 2024. Seaweed-based polysaccharides – review of extraction, characterization, and bioplastic application. Green Chemistry 26 (10), 5790–5823. Available at https://doi.org/10.1039/d3gc04009g.

Lavagnolo, M.C., Poli, V., Zampini, A.M., Grossule, V., 2024. Biodegradability of bioplastics in different aquatic environments: a systematic review. Journal of Environmental Sciences 142, 169–181. Available at https://doi.org/10.1016/j.jes.2023.06.013.

Liu, Y., Ahmed, S., Sameen, D.E., Wang, Y., Lu, R., Dai, J., et al., 2021. A review of cellulose and its derivatives in biopolymer-based for food packaging application. Trends in Food Science and Technology 112, 532–546. Available at https://doi.org/10.1016/j.tifs.2021.04.016.

Malhotra, M., Garg, N., Chand, P., Jakhete, A., 2023. Bio-based bioplastics: current and future developments. Valorization of Biomass to Bioproducts: Biochemicals and Biomaterials. Elsevier, India, pp. 475–504. Available at http://doi.org/10.1016/B978-0-12-822887-6.00020-6.

Medina, A.A., Hassan, S., Dekiwadia, C., Chen, C., Ball, A.S., 2025. Bio-based plastics – a sustainable solution to plastic pollution. Microbiology Australia 46 (2), 96–101. Available at https://doi.org/10.1071/ma25027.

Mehrpouya, M., Vahabi, H., 2023. Chapter 1 - Additive manufacturing of biopolymers. In: Mehrpouya, M., Vahabi, H. (Eds.), Additive Manufacturing of Biopolymers: Handbook of Materials, Techniques, and Applications. Elsevier, The Netherlands, pp. 1–10. Available at https://doi.org/10.1016/B978-0-323-95151-7.00002-8.

Molina-Besch, K., 2022. Use phase and end-of-life modeling of biobased biodegradable plastics in life cycle assessment: a review. Clean Technologies and Environmental Policy 24 (10), 3253–3272. Available at https://doi.org/10.1007/s10098-022-02373-3.

Moshood, T.D., Nawanir, G., Mahmud, F., Mohamad, F., Ahmad, M.H., AbdulGhani, A., 2022. Sustainability of biodegradable plastics: new problem or solution to solve the global plastic pollution? Current Research in Green and Sustainable Chemistry 5, 100273. Available at https://doi.org/10.1016/j.crgsc.2022.100273.

Moshood, T.D., Nawanir, G., Mahmud, F., Mohamad, F., Ahmad, M.H., AbdulGhani, A., 2022. Biodegradable plastic applications towards sustainability: a recent innovations in the green product. Cleaner Engineering and Technology 6, 100404. Available at https://doi.org/10.1016/j.clet.2022.100404.

Nanda, S., Patra, B.R., Patel, R., Bakos, J., Dalai, A.K., 2022. Innovations in applications and prospects of bioplastics and biopolymers: a review. Environmental Chemistry Letters 20 (1), 379–395. Available at https://doi.org/10.1007/s10311-021-01334-4.

Nazareth, M.C., Marques, M.R.C., Pinheiro, L.M., Castro, Í.B., 2022. Key issues for bio-based, biodegradable and compostable plastics governance. Journal of Environmental Management 322, 116074. Available at https://doi.org/10.1016/j.jenvman.2022.116074.

Nguyen, M.T.P., Escribà-Gelonch, M., Hessel, V., Coad, B.R., 2024. A review of the current and future prospects for producing bioplastic films made from starch and chitosan. ACS Sustainable Chemistry and Engineering 12 (5), 1750–1768. Available at https://doi.org/10.1021/acssuschemeng.3c06094.

Nilsen-Nygaard, J., Fernández, E.N., Radusin, T., Rotabakk, B.T., Sarfraz, J., Sharmin, N., et al., 2021. Current status of biobased and biodegradable food packaging materials: impact on food quality and effect of innovative processing technologies. Comprehensive Reviews in Food Science and Food Safety 20 (2), 1333–1380. Available at https://doi.org/10.1111/1541-4337.12715.

Nizamuddin, S., Baloch, A.J., Chen, C., Arif, M., Mubarak, N.M., 2024. Bio-based plastics, biodegradable plastics, and compostable plastics: biodegradation mechanism, biodegradability standards and environmental stratagem. International Biodeterioration and Biodegradation 195, 105887. Available at https://doi.org/10.1016/j.ibiod.2024.105887.

Nyamweya, N.N., 2021. Applications of polymer blends in drug delivery. Future Journal of Pharmaceutical Sciences 7, 18. Available at https://doi.org/10.1186/s43094-020-00167-2.

Olunusi, S.O., Ramli, N.H., Adam, F., Hossain, E., Kim, T.H., 2025. Chitosan as a sustainable alternative for fresh food packaging: structural insights, modification strategies, and innovations for commercial viability. International Journal of Biological Macromolecules 318, 145065. Available at https://doi.org/10.1016/j.ijbiomac.2025.145065.

Park, H., He, H., Yan, X., Liu, X., Scrutton, N.S., Chen, G.Q., 2024. PHA is not just a bioplastic!. Biotechnology Advances 71, 108320. Available at https://doi.org/10.1016/j.biotechadv.2024.108320.

Pei, J., Palanisamy, C.P., Srinivasan, G.P., Panagal, M., Kumar, S.S.D., Mironescu, M., 2024. A comprehensive review on starch-based sustainable edible films loaded with bioactive components for food packaging. International Journal of Biological Macromolecules 274, 133332. Available at https://doi.org/10.1016/j.ijbiomac.2024.133332.

Phadke, G., Rawtani, D., 2023. Bioplastics as polymeric building blocks: paving the way for greener and cleaner environment. European Polymer Journal 199, 112453. Available at https://doi.org/10.1016/j.eurpolymj.2023.112453.

Pilapitiya, P.G.C.N.T., Ratnayake, A.S., 2024. The world of plastic waste: a review. Cleaner Materials 11, 100220. Available at https://doi.org/10.1016/J.CLEMA.2024.100220.

Qazanfarzadeh, Z., Kumaravel, V., 2023. Hydrophobisation approaches of protein-based bioplastics. Trends in Food Science & Technology 138, 27–43. Available at https://doi.org/10.1016/j.tifs.2023.06.002.

Rani, R., Tuteja, J., Sand, A., 2024. Sustainable novel food packaging in current scenario: an analysis to environmental perspectives. In: Singh, J., Singh, R.P., Kaushik, A.K., Adetunji, C.O., Rb. Sing, K. (Eds.), Nanobiotechnology for Food Processing and Packaging. Academic Press, pp. 249–264. Available at http://doi.org/10.1016/B978-0-323-91749-0.00015-0.

Ribeiro, A.P.C., Martins, M.O., Figueiras, A.O., Martins, L.M.D.R.S., 2025. Depolymerization: Concepts, Progress, and Challenges. Vol. 1: Core Concepts and Fundamentals. American Chemical Society. Available at https://doi.org/10.1021/BK-2025-1498.CH001.

Robertson, G.L., 2009. In: Robertson, G.L. (Ed.), Food Packaging and Shelf Life: A Practical GuideFood Packaging and Shelf Life. CRC Press. Available at http://doi.org/10.1201/9781420078459.

Salgado, P.R., Di Giorgio, L., Musso, Y.S., Mauri, A.N., 2021. Recent developments in smart food packaging focused on biobased and biodegradable polymers. Frontiers in Sustainable Food Systems 5, 630393. Available at https://doi.org/10.3389/fsufs.2021.630393.

Samuel, H.S., Ekpan, F.-D.M., Ori, M.O., 2024. Biodegradable, recyclable, and renewable polymers as alternatives to traditional petroleum-based plastics. Asian Journal of Environmental Research 1 (3), 152–165. Available at https://doi.org/10.69930/ajer.v1i3.86.

Shaikh, S., Yaqoob, M., Aggarwal, P., 2021. An overview of biodegradable packaging in food industry. Current Research in Food Science 4, 503–520. Available at https://doi.org/10.1016/j.crfs.2021.07.005.

Sid, S., Mor, R.S., Kishore, A., Sharanagat, V.S., 2021. Bio-sourced polymers as alternatives to conventional food packaging materials: a review. Trends in Food Science and Technology 115, 87–104. Available at https://doi.org/10.1016/j.tifs.2021.06.026.

Silviya, E.K., Varma, S., Unnikrishnan, G., Thomas, S., 2009. Chapter 4 - Compounding and mixing of polymers. In: Thomas, SAbu, Weimin, Yang (Eds.), Advances in Polymer Processing From Macro- to Nano- Scales. Woodhead Publishing, Cambridge, England, pp. 71–105. Available at https://doi.org/10.1533/9781845696429.1.71.

Song, S., 2024. The impact of microplastic bioaccumulation on marine ecosystems. Theoretical and Natural Science 71 (1), 78–83. Available at https://doi.org/10.54254/2753-8818/2024.la18908.

Swetha, T.A., Bora, A., Mohanrasu, K., Balaji, P., Raja, R., et al., 2023. A comprehensive review on polylactic acid (PLA) – synthesis, processing and application in food packaging. International Journal of Biological Macromolecules 234, 123715. Available at https://doi.org/10.1016/j.ijbiomac.2023.123715.

Thomas, A.P., Kasa, V.P., Dubey, B.K., Sen, R., Sarmah, A.K., 2023. Synthesis and commercialization of bioplastics: organic waste as a sustainable feedstock. Science of the Total Environment 904, 167243. Available at https://doi.org/10.1016/j.scitotenv.2023.167243.

Torrejon, V.M., Fernando, S., Roy, U.K., Onwukwe, U., Anguilano, L., 2025. Transforming seaweed into bioplastics: a review of cultivation, harvesting and processing methods. Applied Phycology 6 (1), 52–73. Available at https://doi.org/10.1080/26388081.2024.2435035.

Walker, S., Rothman, R., 2020. Life cycle assessment of bio-based and fossil-based plastic: a review. Journal of Cleaner Production 261, 121158. Available at https://doi.org/10.1016/j.jclepro.2020.121158.

Zhao, X., Wang, Y., Chen, X., Yu, X., Li, W., Zhang, S., et al., 2023. Sustainable bioplastics derived from renewable natural resources for food packaging. Matter 6 (1), 97–127. Available at https://doi.org/10.1016/j.matt.2022.11.006.

Zubair, M., Arshad, M., Pradhan, R.A., Ullah, A., 2020. Chapter 20 - Chitosan/chitin-based composites for food packaging applications. In: Gopi, S., Thomas, S., Pius, A. (Eds.), Handbook of Chitin and Chitosan: Volume 3: Chitin and Chitosan based Polymer Materials for Various Applications. Elsevier, pp. 641–670. Available at https://doi.org/10.1016/B978-0-12-817966-6.00020-0.

Bio-based and non-biodegradable bioplastics in food packaging

Mehran Ghasemlou[1,2,7], Fugen Daver[3], Mohammed Ali Dheyab[4,5], Colin J. Barrow[1], and Minoo Naebe[1,6]

[1]Center for Sustainable Bioproducts, Deakin University, Geelong Waurn Ponds Campus, Geelong, VIC, Australia, [2]School of Science, STEM College, RMIT University, Melbourne, VIC, Australia, [3]School of Engineering, RMIT University, Bundoora, VIC, Australia, [4]School of Physics, Universiti Sains Malaysia, Pulau Pinang, Penang, Malaysia, [5]Nano-Biotechnology Research and Innovation (NanoBRI), Institute for Research in Molecular Medicine (INFORMM), Universiti Sains Malaysia, Pulau Pinang, Penang, Malaysia, [6]Institute for Frontier Materials, Deakin University, Geelong Waurn Ponds Campus, Geelong, VIC, Australia, [7]Department of Chemical Engineering, The University of Melbourne, Parkville, VIC, Australia

5.1 Introduction

The second half of the 20th century witnessed an unprecedented expansion in the production of synthetic fossil-derived plastics, and they have become indispensable pillars of modern life because of their low cost, durability, flexibility, and convenience. Fossil-based plastics have been mass-produced on an industrial scale, finding use in every facet of daily life, from food packaging and textiles to electronics and automobile parts (Ghasemlou et al., 2025). Today, plastic production exceeds 380 million tons per year, and the linear take–make–waste economic model of use has generated enormous volumes of waste (Borrelle et al., 2020). Much of this plastic debris accumulates in landfills and natural ecosystems, or fragments into microplastics, polluting oceans and food chains. The production and degradation of plastics can also contribute to greenhouse gas emissions. In response to these growing environmental pressures, researchers and industries have turned their attention to materials with more sustainable character. Bioplastics have emerged as a promising avenue to help transform the current linear economy of plastics into a circular economy with minimal waste and emissions. Life cycle assessments suggest that replacing approximately two-thirds of conventional plastics with bioplastics could reduce CO_2-equivalent emissions by 241–316 million tons per year (Spierling et al., 2018). Bioplastics are intended to retain the functional advantages of conventional plastics while reducing reliance on fossil feedstocks. Indeed, bioplastics can offer similar or even better functionality than conventional plastics, often with a lower carbon footprint. Bioplastics are not a single solution to the plastic crisis, but they can meaningfully

Bioplastics for Food Packaging. DOI: https://doi.org/10.1016/B978-0-443-41654-5.00012-3

contribute to the circular economy. Their success will depend not on bio-based content alone, but on life cycle benefits, robust waste management infrastructure, and clear regulation (Ghasemlou et al., 2024). If these conditions are fulfilled, bioplastics can expand beyond niche applications and play a meaningful role in building a circular and low-carbon future.

Driven by increased public awareness of plastic pollution and climate change, global bioplastic production has steadily grown in recent years, reaching ~2.4 million tons in 2021 and is projected to rise to ~7.6 million tons by 2026. Bioplastics are materials that are bio-based, biodegradable, or both according to the definition of European Bioplastics organization (Hao et al., 2024). To state this explicitly, 100% bio-based plastics may be non-biodegradable, and 100% fossil-based plastics can be biodegradable. A common misconception is that bio-based plastics are equivalent to biodegradable plastics. Some bio-based plastics, such as bio-polyethylene (bio-PE) and bio-polyethylene terephthalate (bio-PET), are chemically identical to their fossil counterparts and do not biodegrade under ambient conditions (Siracusa and Blanco, 2020).

This chapter reviews biologically derived bioplastics, discussing their limitations and recent food packaging applications. It also explores the commercial development and global market of leading bio-based plastics and provides an overview of the key industrial players advancing their adoption worldwide.

5.2 Biologically derived non-biodegradable bioplastics

To reduce the environmental impact of plastics, considerable efforts are focused on replacing conventional petroleum-based plastics with new, lower-carbon materials derived from renewable feedstocks. As the demand for sustainable materials grows, a new class of plastics/bioplastics has gained increasing attention. Non-biodegradable bio-based plastics are a technologically important group of bio-based polymers that include bio-PE, bio-PET, bio-based polyamide, poly(trimethylene terephthalate) (PTT), bio-based poly(propylene) (bio-PP), and the more recently developed bio-based poly(ethylene furanoate). These materials are partly or entirely derived from renewable feedstocks, and their chemical structure and performance are virtually identical to those of their petrochemical counterparts. This means they can be processed, recycled, and used in existing systems without modification (Chen and Patel, 2012).

Today, these so-called "green polymers" account for nearly 57% of all bioplastics produced worldwide. Within this group, bio-PE and bio-PET (each with roughly 30% bio-based content) are the main players. Among them, bio-PET is the most widely produced, accounting for around 26% of total production (Stoleru et al., 2021). They are widely used in areas such as food packaging, disposable service items, shopping bags, nonwoven fabrics, and agricultural films. bio-PE is widely used in flexible films, while bio-PET is used in rigid packaging, such as beverage bottles (van den Oever and Molenveld, 2017). However, most bio-based materials are inherently more hydrophilic than their petrochemical analogues, and humidity can affect their barrier performance. Because bio-based, non-biodegradable bioplastics (i.e., bio-PE and bio-PET) have the same chemical composition as fossil-based plastics (i.e., PE and PET), their permeability to water vapor, oxygen, and carbon dioxide is also the same. To improve barrier performance, manufacturers often use lamination techniques together with biaxial orientation (Rabnawaz et al., 2017). Biaxially oriented films of PE and PET (often called BOPE or BOPET) have a high degree of crystallinity and dense molecular packing, which effectively reduces gas transmission. To further improve oxygen and moisture resistance, the biaxially oriented bio-PE and bio-PET films can be metalized (Jariyasakoolroj et al., 2020).

5.3 Synthesis and production bio-polyethylene

PE is the most widely produced plastic in the world, accounting for more than 30% of total global demand. PE is manufactured by polymerization of ethylene, which is a monomer derived primarily from fossil resources. This polymerization route, together with the choice of catalysts, determines whether the resulting material is high-density polyethylene (HDPE) or low-density polyethylene (LDPE). These materials are found in nearly every part of daily life such as packaging, films, pipes, and household goods. The dominance of PE arises from its low cost, excellent mechanical strength, and chemical resistance, but these properties also contribute to its nondegradability. From a technological perspective, bio-PE has structural and functional characteristics identical to those of conventional PE and can be processed using the same industrial infrastructure, without the need for additional investment in equipment. An important advantage for industry is that bio-PE can be potentially recycled through existing PE recovery systems without any modification (Soo et al., 2024).

Two main strategies have been proposed to mitigate the environmental impact of PE waste. The first strategy is the transition to bio-PE, which is currently produced from biofeedstocks such as sugarcane, corn, maize, wheat, starch, or other biomass sources (Siracusa and Blanco, 2020). This approach aims to shift PE production from fossil carbon while maintaining the same molecular composition, thereby reducing overall greenhouse gas emissions. In practice, sugars extracted from biomass are fermented to ethanol, which is then dehydrated to bio-ethylene and polymerized under conventional conditions. The resulting bio-PE is chemically identical to petrochemical PE and can be seamlessly integrated into existing processing and recycling infrastructures (Wang et al., 2023).

The second strategy is to develop PE biocomposites by incorporating natural or bio-based fillers to increase degradability while retaining mechanical performance. Because of its hydrophobic backbone, high molecular weight, and lack of reactive functional groups, PE is highly resistant to penetration and biodegradation. Incorporating organic fillers derived from agricultural residues, crop wastes, or food by-products introduces hydrophilic sites that promote moisture uptake and microbial colonization (Webb et al., 2012). These organic components increase surface area, facilitate biofilm formation, and enhance enzymatic activity, leading to faster degradation. This can likely address the end-of-life problem of PE in the environment, especially when it is inappropriately disposed of.

bio-PE is derived from renewable biomass rather than fossil fuels. Its production generally follows three main technological routes: (1) alcohol dehydration, (2) syngas conversion, and (3) biosynthetic pathway. Each route produces bio-ethylene, which is subsequently polymerized into bio-PE. Alcohol dehydration involves the dehydration of bio-ethanol produced through the fermentation of biomass feedstocks such as sugarcane, corn, lignocellulosic materials, or algae. This route is one of the most established and industrially mature methods, with existing infrastructure already in place. Brazil and the United States are responsible for roughly 81% of global bio-ethanol production. Bio-ethanol was first produced from first-generation feedstocks, which include edible crops such as sugarcane and corn but currently the most widely used method is the fermentation of sucrose derived from sugarcane (Singh and Rangaiah, 2017). The process begins with treatment of the bio-feedstocks to extract the juice and bagasse. The sugar-rich juice is then anaerobically fermented to produce bio-ethanol. Through distillation, an azeotropic solution containing 95.5 vol.% bio-ethanol is obtained, which is subsequently dehydrated to yield bio-ethylene. This monomer is polymerized under conventional conditions to produce different grades of bio-PE (Soo et al., 2024).

Sugarcane is the most common feedstock for producing bio-PE due to its abundance, high sugar content, and established processing infrastructure. However, the use of first-generation crops competes with food production, prompting research into second-generation feedstocks derived from lignocellulosic residues such as sugarcane bagasse, corn stalk, rice husk, and third-generation feedstocks such as algae. Depending on the source material, fermentation can yield up to ~12% ethanol from first-generation feedstocks, while second- and third-generation feedstocks generally produce lower yields of ~5%. The limited availability of fermentable sugars in lignocellulosic materials has encouraged the development of pretreatment methods such as alkali, acid, intense heat, compressed hot water, and ionic liquids, which improve hemicellulose saccharification by 70%–90% and increase bio-ethanol yield.

Algal biomass requires enzymatic pretreatment to break down cell walls and enhance bio-ethanol yield for subsequent bio-ethylene and bio-PE production. In addition, algae contain lipids and proteins that can be converted into bio-based ethylene glycol (EG) and amino acids, which can then be dehydrated or fermented into bio-ethylene (Ramachandra and Hebbale, 2020).

The second route of biomass-to-syngas conversion involves pyrolysis, gasification, and Fischer–Tropsch synthesis and can allow precise control over syngas composition, potentially improving bio-ethylene yield (AlNouss et al., 2020). Typically, second-generation feedstocks such as agricultural waste, food waste, and woody biomass are used in this route because they do not compete with food production. However, these processes are energy-, water-, and resource-intensive and complex. They also require high capital costs and catalysts, which are often deactivated by tar formation and fouling (AlNouss et al., 2020). Regardless of the production route, the resulting bio-PE exhibits properties identical to those of petroleum-derived PE. Each method has its own advantages and challenges depending on feedstock availability, technological maturity, environmental considerations, and cost. Overall, bio-PE is a more sustainable and environmentally friendly version of PE and can be recycled or processed using existing technologies without major investment. At the end of its life cycle, bio-PE behaves identically to fossil-based PE, and it is not biodegradable under environmental conditions. In recycling systems, bio-PE is indistinguishable from petrochemical PE and can circulate without disruption. When incinerated, it releases biogenic carbon instead of fossil carbon and its overall emission profile is similar. Life cycle assessments that include incineration often assume carbon neutrality for the biogenic fraction, improving the overall balance. With these apparent advantages, the global bio-PE market is expected to grow with increasing awareness and preference for sustainable materials (Trucillo et al., 2025).

5.3.1 Commercialization and global market of bio-polyethylene

The rise in global bio-PE demand is underpinned by a convergence of environmental, economic, and social drivers. Increasing awareness of plastic pollution and carbon emissions has reshaped consumer preferences and encouraged companies to adopt renewable materials. Policies such as carbon taxes, bans on single-use plastics, and incentives for using bio-based or biodegradable alternatives have also accelerated this transition. At the same time, investors and shareholders increasingly require firms to meet environmental, social, and governance benchmarks as a condition of continued support. In 2022, Asia dominated the global bioplastics market, producing about 41.4% of the world's total, followed by Europe (26.5%), North America (18.9%), South America (12.6%), and Australia (0.5%). Bioplastics have entered many industrial sectors, and bio-PE is now mainly used in building and construction, as well as in both rigid and flexible packaging. bio-PE represented ~11.0% of the total global bioplastic production in 2024. However, its relative market share is expected to slightly decline to 8.9% by 2029 due to the increasing demand for other biodegradable plastics such as polylactic acid (PLA) and polyhydroxyalkanoate (PHA), and the growing production of bio-PP. Nevertheless, the total production of bio-PE is still expected to rise over the next few years (Rabnawaz et al., 2017).

Braskem was the first company to commercialize bio-PE and is currently the global leader in its production. The company manufactures an extensive portfolio of bio-based HDPE (bio-HDPE) and bio-based LDPE (bio-LDPE) resins, supplying diverse sectors including food packaging, personal care, cleaning, cosmetics, and automotive industries. Currently, Braskem has several partners including Coca-Cola, Danone, Procter & Gamble, TetraPack, Johnson & Johnson, Acinplas, Shiseido, Toyota Tsusho, PetroPack, and NobelPack that use bio-derived PE for their products (Byun and Kim, 2014). For example, in 2011, the probiotic yogurt brand Actimel (Danone) has used bottles made from 95% or more of a bio-based PE and 5% or less of a non-bio-based plastic. Although these bottles are not biodegradable, they are fully recyclable (Fig. 5.1).

In the late 2000s, Braskem demonstrated that PE could be produced entirely from renewable feedstock without compromising product quality. Brazil was chosen for its facilities because of its proximity to agricultural resources and the efficiency of its sugarcane industry. In 2010, Braskem opened its large-scale plant in Triunfo, producing 200,000 tons of bio-PE annually. Over time, the capacity was expanded to around 260,000 tons per year in response to growing market demand. The company has also explored new regions for expansion and partnerships. A joint venture with SCG Chemicals in Thailand advanced plans to construct a new bio-ethylene plant that use local crops such as sugarcane and cassava, with a projected annual output of ~200,000 tons of bio-PE. Braskem also developed renewable-sourced PE waxes for use in coatings, adhesives, consumer care, and other applications (Martínez Bello et al., 2025).

The commercial manufacturing of bio-PE is based on a well-established pathway of fermenting sugarcane into ethanol, dehydrating ethanol into ethylene, and polymerizing

Figure 5.1
Actimel probiotic yogurt drink packaged in bottles made from over 95% bio-based polyethylene from Braskem. Image courtesy of Danone.

ethylene into bio-PE. The bio-PE resins are marketed under the brand "I'm green." This bio-PE product is reported to reduce energy consumption by up to ~80% compared to traditional PE production methods. This milestone significantly shifted the plastics industry toward renewable materials, encouraging other companies to enter the bioplastic market. For instance, LyondellBasell and Neste have a long-term collaboration to produce bioplastics. Neste's RE is a bioplastic manufacture that use biofeedstock such as fats and waste and residue oils to produce packaging materials. Other major companies such as Dow Chemical Company also uses bio-based feedstock to produce AFFINITY RE for packaging applications. Kuraray Co uses bio-PE for its PLANTIC products. Tetra Pak now uses bio-PE in bottle caps and tops and in carton packaging, and Procter & Gamble has introduced bio-PE bottles for several products such as Pantene. With the increasing emphasis on sustainable and bio-based materials, manufacturers and consumers throughout the bio-PE supply chain are anticipated to continue driving market growth in the coming years (Soo et al., 2024).

5.3.2 Applications of bio-polyethylene in food packaging

LDPE are commonly used as film layers and sealant layers in packaging due to its low melting point and excellent heat-seal properties, whereas HDPE has higher stiffness and is used in caps, bottles, and thicker containers. These roles can be translated directly to bio-PE grades. For instance, bio-HDPE provides impact resistance suitable for durable bottles and closures, while bio-LDPE yields soft and extensible films. bio-PE exhibits low oxygen barrier properties and is often combined with other layers (e.g. bio-PET or EVOH layer) when packaging oxygen-sensitive foods.

bio-PE has seen rapid uptake in food and beverage packaging, often as a means for brand owners to increase the renewable content of their packaging. Tetra Pak, which is the global leader in liquid food packaging, introduced bio-HDPE caps derived from sugarcane in 2011 and became the first in the carton industry to use plant-derived plastic closures. These caps, marketed as "LightCap 30," were identical to traditional caps except for a small leaf logo to indicate their renewable origin. Building on this success, Tetra Pak also began using bio-LDPE for the laminate film layers of cartons. A 2014 trial conducted in Brazil produced ~13 billion carton packages containing up to 82% renewable content by replacing fossil-based PE with Braskem's bio-LDPE in the package lining. These packages combined paperboard with bio-PE layers and caps, significantly cutting the fossil share of the material. Major food and beverage companies like Nestlé and Coca-Cola were among the early adopters, integrating Tetra Pak's sugarcane-derived StreamCap 1000 on products such as Nestlé's milk brands in Brazil.

Flexible packaging is another area for bio-PE. Sugarcane-based PE is used in films for bags, pouches, and wraps, especially for organic or sustainable food brands with low-carbon emission. For example, sugar-derived PE films and liners have appeared in snack packaging, bakery bags, and cereal liners. Braskem made a partnership with LyondellBasell and Neste in 2019 to commercially produce bio-PE (with >30% renewable content) for food contact, which Cofresco (Melitta Group) uses for food-storage bags and wraps (e.g. the Toppits brand in Europe). Likewise, Nestlé Waters and Danone have explored bio-PE and other bio-based plastics for bottled waters and dairy packaging. bio-PE has rarely been used as a coating in food packaging applications and has not been commercialized in this role (Pant et al., 2017).

5.4 Synthesis and production of bio-based poly(ethylene terephthalate)

PET is one of the most widely used polymers and is synthesized from EG and terephthalic acid (TPA) molecules. PET belongs to the family of polyesters and dominates the packaging industry, accounting for 20.7% of the total plastics used for packaging. EG monomers can be produced through two main strategies. The first strategy entails the synthesis of EG from bio-ethanol obtained through fermentation or hydrogenolysis of C5/C6 sugars such as glucose and sorbitol (Sheldon, 2014). The second strategy converts cellulose directly into EG using a carbon-supported tungsten carbide catalyst (Fig. 5.2) (Ji et al., 2008).

Although EG can be derived from renewable raw materials (e.g., bio-ethanol), an economically feasible route for large-scale production of bio-based TPA is not yet available. Therefore, bio-PET is only about 30% bio-based. Three major approaches have been developed for producing bio-based TPA from bio-derived p-xylene. First, C6 sugars are converted through an aqueous phase-reforming process into intermediate compounds, which are then transformed via dehydrocyclization into a BTX mixture (benzene, toluene, and xylenes). From this mixture, *p*-xylene is isolated and subsequently oxidized to produce TPA. Second, bio-

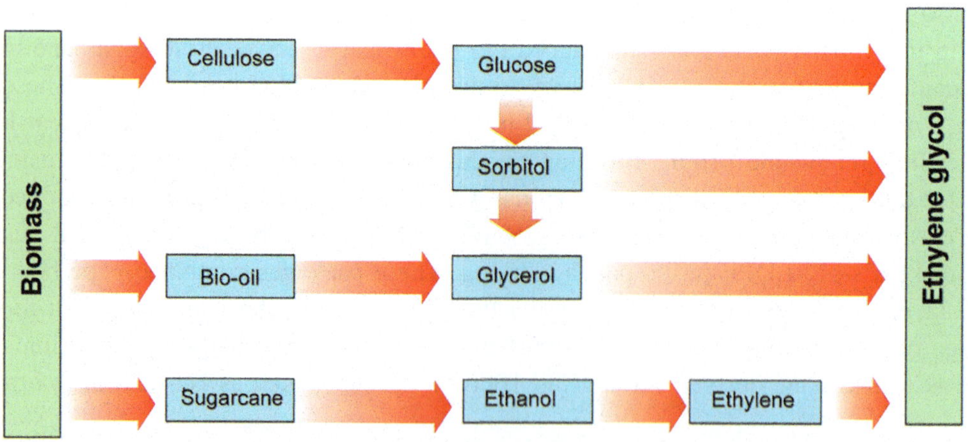

Figure 5.2
Pathways for the bio-based production of ethylene glycol (a component for bio-polyethylene terephthalate production) from renewable biomass sources.

isobutanol is converted to *p*-xylene through the Prins reaction. Third route involves the Diels–Alder addition of bio-ethylene (or acrolein) to biomass-derived 2,5-dimethylfuran, followed by elimination of water (or carbon dioxide) to yield *p*-xylene. Among these, the BTX-based route has been reported to offer the lowest environmental footprint. bio-PET has the same structural, physical, and mechanical properties as its petroleum-derived counterpart (Fig. 5.3) (Rabnawaz et al., 2017).

PET (or bio-PET) has a glass transition temperature (T_g) of 73°C and a melting temperature (T_m) of 245°C, which enables it to withstand relatively high temperatures. PET is a semicrystalline polymer due to its linear backbone, π–π interactions, and strong dipole–dipole interactions, which promote tighter molecular packing. This behavior strongly depends on processing history, including melting and quenching conditions. The inherently slow crystallization rate of PET results in small crystallites rather than large spherulites. These small crystallites do not scatter light effectively and allow PET bottles to have good optical clarity despite even at high degree of crystallinity. Ideally, the optical properties of PET can block damaging UV-B (315–280 nm) and UV-A (400–315 nm) light and are optically clear in the visible region: these are highly desirable for food packaging because they protect food

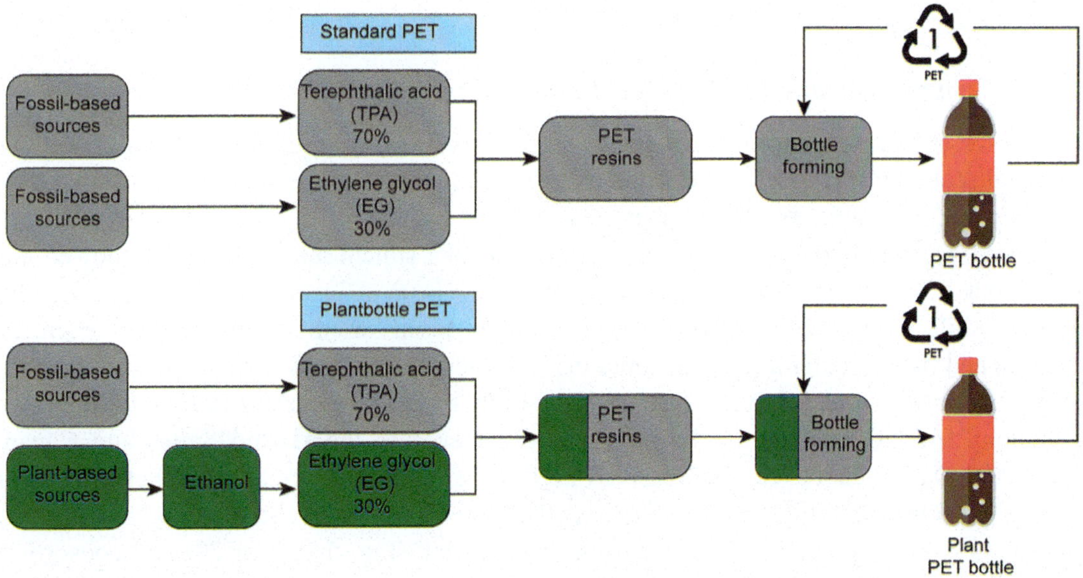

Figure 5.3
Comparison of production pathways for conventional and bio-based polyethylene terephthalate. Standard polyethylene terephthalate is entirely derived from fossil-based feedstocks and both terephthalic acid (70%) and ethylene glycol (30%) are derived from petroleum sources. PlantBottle polyethylene terephthalate replaces the fossil-derived ethylene glycol with a bio-based counterpart obtained from renewable ethanol, while maintaining the same molecular structure, processing, and recyclability as conventional polyethylene terephthalate.

products from photo-oxidation and their optical clarity in the visible region allows consumers to view the product in the container at the time of purchase (Shirvanimoghaddam et al., 2024). The degree of crystallization can be tuned using nucleating agents, which accelerate crystallization and promote the formation of spherulite. Indeed, translucent PET products are obtained when micron-sized spherulites are formed and are mainly used for high-temperature applications such as microwaveable containers. In addition, fully amorphous PET copolymer, which is named poly(ethylene terephthalate)-glycol (PET-G), are produced by incorporating diethylene glycol alongside EG during synthesis. PET-G is optically transparent and flexible and is suitable for films and flexible bottles. However, compared to standard PET, PET-G is more sensitive to UV exposure and surface scratching and is not suitable for autoclave applications (Dupaix and Boyce, 2005).

Despite all these benefits, the very low degradability of PET poses environmental concerns, particularly when used in short-lifetime products such as food (Siracusa and Blanco, 2020). Despite being non-biodegradable, bio-PET is recyclable, and its life cycle greenhouse gas emissions are estimated to be 25%–58% lower than those of petroleum-based PET. It is important to note that repeated mechanical recycling of PET bottles may lead to the generation of microplastic particles (Lee et al., 2025).

5.4.1 Commercialization and global market of bio-polyethylene terephthalate

bio-PET is already produced as A resin, which can be further processed into bottles at industrial scale. bio-PET (often marketed as a "green plastic") was commercialized by Coca-Cola in 2009 through its PlantBottle, made from bio-based EG and fossil-based TPA (Fig. 5.4a). Despite being advertised as fully renewable, stoichiometric analysis showed that only about 30% of the carbon content in bio-PET (from EG) originates from biological sources, while the remaining portion (from TPA) is fossil-based. It has the same chemical structure and functionality as conventional PET plastics, and its main advantage is to maintain the performance and recyclability of traditional PET. In 2011, Heinz licensed the PlantBottle technology from Coca-Cola for its ketchup bottles sold in the United States and Canada (Fig. 5.4b). Dasani also used plant-derived PET for water bottles, creating a package known as the "HybridBottle" (Fig. 5.4c). Similarly, the juice maker Odwalla has also begun using juice bottles that comprise 96% or more bio-based PET (Fig. 5.4d). In 2012, a consortium comprising Coca-Cola, Ford, Heinz, Nike, and Procter & Gamble was established to accelerate research and technological progress toward developing commercially viable 100% plant-based PET, with the goal of reducing fossil-fuel use and improving sustainability. This strategic working group was named the Plant PET Technology Collaborative (PTC) (Byun and Kim, 2014). In the same year, PepsiCo successfully produced the world's first pilot-scale 100% bio-PET made from renewable sources such as switchgrass, pine bark, and corn husks (Fig. 5.5). PepsiCo will commercialize 100% bio-based PET when it achieved

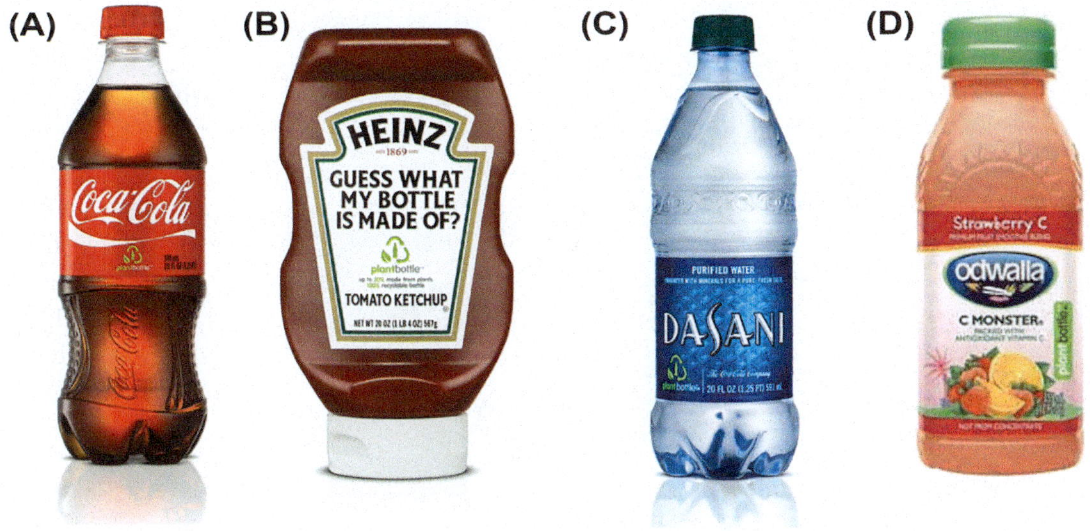

Figure 5.4

Commercial packaging applications using bio-based polyethylene terephthalate. (A) Coca-Cola's PlantBottle for carbonated beverages. Image courtesy of The Coca-Cola Company. (B) Heinz ketchup bottle produced under license from Coca-Cola's PlantBottle technology. Image courtesy of Heinz, Pittsburgh, PA. (C) Dasani water bottle incorporating partially bio-based polyethylene terephthalate. Image courtesy of The Coca-Cola Company. (D) Odwalla juice bottle made from plant-derived polyethylene terephthalate. Image courtesy of The Coca-Cola Company.

successful results from its pilot product. Ongoing efforts are directed toward producing bio-PET from other renewable resources, including orange peels, oat hulls, and potato scraps.

The PET market for soda and water bottles is enormous. By 2013, ~8% of Coca-Cola's PET bottles used PlantBottle technology, and this is equivalent to ~4 billion bottles or ~210 million pounds of PET. By 2015, Coca-Cola had distributed more than 35 billion PlantBottle packages in nearly 40 countries, demonstrating the scalability of bio-PET in mainstream products. The global TPA market exceeds 50 million tons per year, while PET bottles represent ~15 million tons annually, which indicates the vast potential for bio-based alternatives. Currently, PlantBottle packaging is available in nine countries, including Denmark, the United States, Canada, Japan, Brazil, Mexico, Norway, Sweden, and Chile. In Canada, PlantBottles are used for packaging Dasani, Coca-Cola, Diet Coca-Cola, Coca-Cola Zero, Sprite, Fresca, and Barq's. Adoption of PlantBottle packaging results in a reduction of CO_2 emissions by up to 19%. As discussed earlier, only the EG component of Coca-Cola's PlantBottle is bio-based and the TPA is still fossil-derived. If a bio-based TPA can be produced at similar quality and cost to fossil-based TPA, it would have an immediate market for making 100% bio-based PET. There is a strong drive toward making 100% bio-PET by synthesizing renewable TPA. Several chemical companies have developed processes

Figure 5.5
Prototype of PepsiCo's 100% bio-based polyethylene terephthalate bottle produced from renewable feedstocks such as orange peels, potato scraps, oat hulls, and other agricultural residues. The bottle is chemically identical to conventional polyethylene terephthalate but fully derived from plant-based sources.
Image courtesy of PepsiCo.

to make bio-based *p*-xylene (and hence TPA) from plant-derived feedstocks. Coca-Cola had worked with companies such as Virent and Gevo to produce bio-paraxylene from corn sugar or other biomass sources. In 2015, Coca-Cola unveiled its 100% bio-PET prototype made entirely from plant materials. This bottle used bio-based EG (from sugarcane) and bio-based TPA (from a bio-paraxylene derived partially from corn waste). Although full commercial scale of 100% bio-PET is still being established, the PlantBottle program proved the viability of the concept. Concurrently, consortiums like the NaturALL Bottle Alliance (Origin Materials with Nestlé Waters, Danone, and PepsiCo) have pursued bio-PET from nonfood biomass such as cardboard and wood scraps. As of the mid-2020s, these efforts are moving toward commercialization of 100% renewable PET resins (often termed "Plant PET"). Coca-Cola's vision is to ultimately convert all PET packaging to either recycled or plant-based materials.

5.4.2 *Applications of bio-polyethylene terephthalate in food packaging*

PET is well suited for food and beverage packaging applications. Owing to its excellent strength and stability, PET (or bio-PET) is used in both fibers (~65%) and packaging (~35%) including bottles (76%), films (13%), and containers (11%). Partially bio-based PET represents roughly 58% of all rigid bioplastic packaging materials (Rabnawaz et al., 2017). PET is used in a wide variety of applications, including water and soft-drink bottles, peanut butter jars, frozen food trays, and boil-in-bag packaging (Nikles and Farahat, 2005). Many suppliers have obtained medical or food-grade approvals for using bio-PET. For instance, the resin "BPF00002" from good natured products (used by Jones Healthcare) is a pharma-grade bio-PET with 30% plant-based content and does not have hazardous additives such as bisphenol A (BPA) or phthalates. Bottles made with bio-PET maintain their structural integrity under carbonation pressure (for sodas) and resist deformation. Clamshell food containers or thermoformed trays made from bio-PET provide the same crack-resistance and load-bearing capability as fossil-based PET. PET or bio-PET have a good barrier to oxygen and carbon dioxide. For example, A PET or bio-PET soda bottle can retain carbonation for months, whereas a similar PE-made bottle would allow gas to escape much faster. However, even the excellent gas-barrier properties of PET can pose challenges for products that require extended shelf life. PET can handle moderately high temperatures and hot-filled PET bottles can resist up to ~85°C with appropriate crystallization. bio-PET has the same thermal behavior and can withstand temperatures up to ~65°C in its amorphous form without deformation, which is adequate for warm filling and storage in hot climates. Therefore, bio-PET containers can be used for hot-fill juices or teas just like conventional PET containers, with no loss in performance.

PET is inherently flavor-neutral and safe for food contact and does not leach any hazardous substance. bio-PET has the same nonreactivity and contains no BPA or plasticizer content. It is resistant to attack by water and many chemicals, ensuring that food or beverage taste is not affected by packaging. bio-PET has been mostly used in the beverage industry. Bottles made from bio-PET are used for Coca-Cola and other soft-drink brands such as Dasani water and Minute Maid juice. Nestlé Waters (owner of brands such as Vittel and Pure Life) has shown interest in using bio-PET for its bottles to reach sustainability targets. Beyond beverages, food packaging trays and films have also seen some adoption of bio-PET. For example, PET is commonly used for clamshell containers (salads and fruits boxes), food blister packs, or trays that require durability and clarity. In Japan, several drink companies (e.g., Suntory) have introduced bottles using bio-PET for teas and waters. Toyota Tsusho has produced and used bio-PET in the context of making eco-friendly PET fibers for bottles and textiles. Oriented PET films are commonly used in cheese, meat, and soft drink packaging due to their excellent gas-barrier properties. Coextruded multilayer PET/EVOH/PET, where EVOH stands for ethylene vinyl alcohol, are used for oxygen-sensitive products such as ketchup and mayonnaise. In such systems, PET shields the water-sensitive EVOH layer from direct contact with moisture, while EVOH provides superior oxygen barrier performance (Lange and Wyser, 2003)

5.5 Synthesis and production of bio-based polypropylene

Polypropylene (PP) is one of the world's most widely used plastics, traditionally synthesized by polymerizing propylene derived from fossil hydrocarbons. bio-based polypropylene (bio-

PP) is PP produced from renewable, biological feedstocks instead of fossil feedstocks. From a chemical perspective, bio-PP is identical to conventional PP. There are several pathways to synthesize PP from biomass. The ethanol-to-propylene (ETP) route is a leading approach for large-scale production of bio-ethanol. In this route, biomass (e.g., sugarcane, corn, sugar beets, etc.) is fermented to ethanol, which is then catalytically converted to propylene. This conversion typically involves multiple steps including dehydration of ethanol to ethylene, oligomerization of ethylene to form butenes, and metathesis (or cracking) to yield propylene. In 2023, Braskem announced a major project to evaluate the production of bio-PP in the United States via ETP (Rahman and Bhoi, 2021).

Bio-naphtha is another strategy to introduce renewable feedstocks into the existing PP infrastructure. Companies such as Neste produce bio-naphtha and bio-propane from renewable sources such as used cooking oil, animal-fat waste, or nonfood vegetable oils. These biofeedstocks can be fed into steam crackers or propane dehydrogenation units to yield bio-based propylene. This approach has been rapidly commercialized. In 2019–2020, Borealis in Europe replaced a portion of fossil propane with Neste's renewable propane at its plant in Belgium and successfully produced the first volumes of certified renewable PP. Similarly, LyondellBasell (a major PP producer) began offering a bio-based grade PP using Neste's bio-naphtha method in 2021. LyondellBasell's process involves feeding bio-naphtha alongside fossil feed in a cracker, producing ethylene and propylene with a measurable C14 content (a marker of renewable origin). The propylene is then polymerized into PP. This co-feeding strategy is attractive because it does not require building new standalone bio-propylene plants and utilizes existing assets, and producers can scale the biocontent by adjusting the ratio of biofeed. There are also experimental routes, such as deriving propanol from biomass and dehydrating it to propylene. For instance, glycerol (a by-product of biodiesel production) can be converted to acetol and then to propylene or fermented to 1-propanol. Although not yet commercial, research in biochemistry and catalysis is exploring such routes, which could tap into waste streams like crude glycerol (from biodiesel) or syngas from biomass gasification (which could be catalytically transformed into light olefins). Global Bioenergies (a French company) demonstrated the fermentation of sugars to isobutene (a related C4 olefin). Isobutene can be converted to isooctane for fuels, but also potentially to propylene via cracking. While Global Bioenergies has focused more on fuels and isobutene derivatives, their technology shows the trend of bioengineering microbes to produce hydrocarbon building blocks (Niyogi et al., 2025).

When bio-propylene is obtained by any of these routes, it is polymerized using the same catalysts (e.g., Ziegler-Natta or metallocenes) and processes (bulk slurry or gas-phase polymerization) as fossil-based PP. A key advantage of bio-PP is that it is chemically identical to fossil-based PP and has the same physical, thermal, and mechanical properties. PP is a semicrystalline polymer and typically crystallizes in the isotactic form. The degree of crystallinity of PP depends on its thermal history and whether it is a homopolymer or a copolymer. Homopolymer PP has a crystallinity of ~50%–60%, a density of ~0.90–0.91 g/cm^3 and is translucent to opaque. Random copolymer PP has slightly lower crystallinity, better clarity, and greater impact resistance at low temperatures. All these variants can be produced as bio-PP using the appropriate co-monomers and catalysts. PP and bio-PP (isotactic homopolymer) have T_m of ~160°C–165°C and T_g of about −10°C. PP can typically withstand short-term exposure to ~100°C (boiling water) and is suitable for microwaveable containers and hot-fill applications. PP is also appreciated in packaging for its good balance of stiffness and toughness. For a typical PP homopolymer, Young's modulus is ~1.5–1.8 GPa, tensile strength is ~30–35 MPa, and elongation at break >200%. bio-PP with identical molecular weight and tacticity shows the same mechanical profile. PP is generally known for its excellent moisture barrier and moderate gas barrier properties. Its water vapor permeation rate is very low and is commonly used for snack food packaging. Therefore, PP films often need additional barrier layers (such as aluminum metallization or EVOH) for products that require long shelf life. Oriented bio-PP films have the same oxygen and carbon dioxide transmission rates as their fossil-based counterparts. A major advantage of bio-PP is that it does not disrupt recycling systems and packaging made from bio-PP can be recycled alongside conventional PP. bio-PP is not either biodegradable or compostable, and a bio-PP container will persist in the environment like a normal PP container.

5.5.1 Commercialization and global market of bio-based poly(propylene)

The drive to commercialize bio-PP has accelerated in recent years and numerous companies have announced production milestones and product launches. Braskem is the largest PP producer in the Americas and a pioneer in bio-based plastics. In 2024, Braskem announced the launch of its "WENEW" bio-circular PP in the United States. This product is manufactured using a mass balance approach with used cooking oil as the renewable feedstock. The bio-PP is intended for use in service packaging, snack food packaging, and other consumer goods. Borealis (Austria) has been a front-runner in integrating renewable feedstocks into polyolefins and began producing renewable PP in 2020 at its Belgium facility. Henkel started using Borealis renewable PP for packaging of its major detergent brands. LyondellBasell was also among the first to produce bio-PP using Neste's biofeed. In 2025, LyondellBasell collaborated with Shiseido, Futamura, and Iwatani in Japan to supply bio-PP for sustainable cosmetics packaging films. Mitsui Chemicals & Prime Polymer (Japan) has also been active in developing bio-PP. In 2023, Mitsui produced bio-PP used in packaging of seaweed snacks. SABIC (Saudi Arabia) is another major producer of certified renewable polyolefins. This company produced bio-PE and bio-PP from second-generation renewable feedstock and supplied some to Unilever in 2019 for use in Magnum ice cream tubs. Sinopec (China) also delivered its first bulk order of bio-PP for cosmetics packaging. As of mid-2020s, actual production volumes of bio-PP remain relatively small. The global bio-PP market size was estimated to be between US$100–US$250 million in 2024. The packaging sector is expected to remain the largest consumer of bio-PP, accounting for ~40%–50% of the market share during this period.

5.5.2 Applications of bio-based poly(propylene) in food packaging

PP is a workhorse material in food packaging and is used in almost every category of food packaging products. The introduction of bio-PP allows these applications to continue with a reduced carbon footprint, without requiring redesign of the packaging. For example, PP and bio-PP are widely used for injection-molded tubs, cups, and trays such as yogurt and dairy containers, margarine tubs, and microwaveable ready-meal trays. Several consumer brands that used bio-PP containers. For instance, Itochu began using bio-PP food containers in FamilyMart convenience stores in 2021. The bio-PP used in these applications is sourced from Borealis (and Borouge) via a mass balance approach. The low cost and moldability of PP make it a common choice for drink cups (especially cold drink cups) and their lids. For example, the lids of hot coffee cups are typically made from PP. A bio-PP lid has identical function and can withstand the heat. Screw caps for bottles (water, soda, juice, etc.) and flip-top caps on condiment bottles (ketchup, mayonnaise) are also commonly made from PP. Replacing these with bio-PP is straightforward and has already begun in some sectors. For example, Coca-Cola in Europe has trialed bio-based PP for some bottle caps using SABIC's mass-balanced renewable PP. PP is also heavily used in flexible packaging. Biaxially oriented polypropylene films are widely used for snack foods, bakery items, and other dry goods. Several manufacturers have launched bio-PP films for such applications. For example, Taghleef Industries introduced *Derprosa bioBLUE*, claiming it offers the same functionality, machinability, and quality as standard PP films. US-based Accredo Packaging announced in 2024 the first 100% bio-based stand-up pouch, in which both the film and the zipper are made from sugarcane-derived polyolefins. In this design, the film was made from bio-PE (Braskem), while the zipper component was likely a bio-based PE or PP.

5.6 Synthesis and production of poly(trimethylene terephthalate)

PTT is another bio-based linear aromatic polyester, obtained through the condensation of TPA or its dimethyl ester with 1,3-propanediol (PDO). PDO is derived via fermentation of plant sugars or by hydrogenation of biodiesel glycerol (Chanda et al., 2023). The use of bio-based PDO makes PTT a partially bio-based polymer (typically containing ~20%–30% renewable carbon content). However, since the TPA component in PTT is still fossil-derived, PTT is not 100% bio-based. Ongoing research efforts aim to produce bio-TPA or use furan-based monomers to further increase the renewable content. bio-PTT exhibits mechanical properties comparable to those of PET and poly(butylene terephthalate) (PBT) (Stoleru et al., 2021). Its excellent spinning processability has led to widespread use in fibers, carpets, textiles, and various sanitary and medical products. PTT can also serve as a construction material and as a film for packaging applications (Rydz et al., 2018).

Although bio-PTT is not yet widely used in packaging, it offers several advantages including good mechanical strength, chemical resistance, low moisture absorption, dimensional stability, processability, and recyclability. The production cost of bio-PTT is higher than that of bio-PET. The resulting PTT polymer is semicrystalline and can be melt-processed by extrusion, injection molding, fiber spinning, or film casting. PTT has a moderate T_g of around 45°C–60°C and T_m of ~220°C–230°C. Thus, PTT is less heat-resistant than PET, which limits its direct use in hot-fill packaging or high-temperature applications. However, it crystallizes faster than PET but slower than PBT, which can facilitate shorter molding cycle times and easier processing. PTT exhibits high strength and toughness, with a tensile strength of ~60 MPa (which is comparable to that of PET), but its Young's modulus is somewhat lower (1.6–3 GPa vs. 2–5 GPa for PET). Indeed, PTT is less stiff but more ductile than PET and its elongation at break can reach 200%–300% (Kobayashi et al., 2011). PTT fibers and films have excellent toughness, stretch recovery, and impact resistance, which are advantageous for packaging that requires durability and crack resistance. PTT is a hydrophobic and chemically resistant polyester with low moisture absorption and water repellency. The aromatic backbone of PTT also provides good barrier to gases and flavors. Although detailed permeability data for PTT are less frequently reported than for PET, it is described as having "excellent barrier properties for food and beverage packaging." For example, PTT bottles have been shown to retain carbonation and limit oxygen ingress similarly to PET bottles. Its hydrophobicity and chemical resistance make PTT compatible with both aqueous and fatty foods, and it resists many solvents and acids without stress cracking. PTT is naturally transparent and can be molded into clear and glossy packaging forms. Its ability to be biaxially oriented into films or bottles can further enhance gas-barrier properties. PTT is 100% recyclable via mechanical or chemical methods. Numerous studies indicated that moderate amounts of PTT mixed into PET recycling streams do not significantly degrade the properties of recycled PET. This compatibility suggests that dedicated recycling streams (or even comingled recycling with PET) could be feasible if PTT packaging becomes more widespread. PTT does not biodegrade under industrial composting or soil conditions because its aromatic ester bonds are hydrolytically stable and therefore, it is not considered a compostable/biodegradable bioplastic.

5.6.1 Commercialization and global market of poly(trimethylene terephthalate)

PTT was initially commercialized for textiles and carpets, but its application is now expanding into packaging films, sheets, and containers. The clarity and moldability of PTT make it suitable for bottles and rigid food containers. For example, PTT can be blow-molded into beverage bottles or formed into trays. It can also be extruded or biaxially oriented into films for snack packaging, laminates, or heat-sealable layers. DuPont (now Covation) has commercialized bio-PTT (with ~37% renewable content) in 2000 for use in packaging and high-performance engineering materials. DuPont's Sorona has been evaluated for packaging applications due to its partially bio-based composition. The bio-based process used to manufacture Sorona consumes ~30% less energy and produces 63% fewer greenhouse gas emissions compared with nylon 6 production. Sorona PTT fiber has been certified by the USDA BioPreferred Program for its renewable content, and some PTT resins carry OEKO-Tex or other environmental labels when used in consumer products. Industrial PTT is generally synthesized using titanium-based catalysts and avoids the antimony catalysts used in PET production. Several companies have also manufactured PTT bottles for water or juice. In Asia, Teijin has explored PTT bottles and found them compatible with PET/PBT in recycling trials. Emerging producers in China such as Covation, Zhangjiagang Glory Biomaterial, and Shenghong Group also market PTT resins for food packaging applications. Despite these initiatives, the current market for PTT in packaging is low compared with PET, primarily due to higher costs and the limited supply/availability of bio-PDO.

5.7 Synthesis and production of polyethylene furanoate

Polyethylene furanoate (PEF) is a bio-based polyester synthesized from renewable plant sugars, specifically 2,5-furandicarboxylic acid (FDCA) and EG. FDCA serves as the key building block for PEF and is typically obtained through the catalytic oxidation of 5-(hydroxymethyl)furfural (HMF), which itself is produced by the dehydration of sugars (Hwang et al., 2020). Several commercial routes have been developed for the production of FDCA (Fig. 5.6). DuPont and Archer Daniels Midland patented a one-step process for converting fructose to FDCA via HMF intermediates, while Novamont developed improved

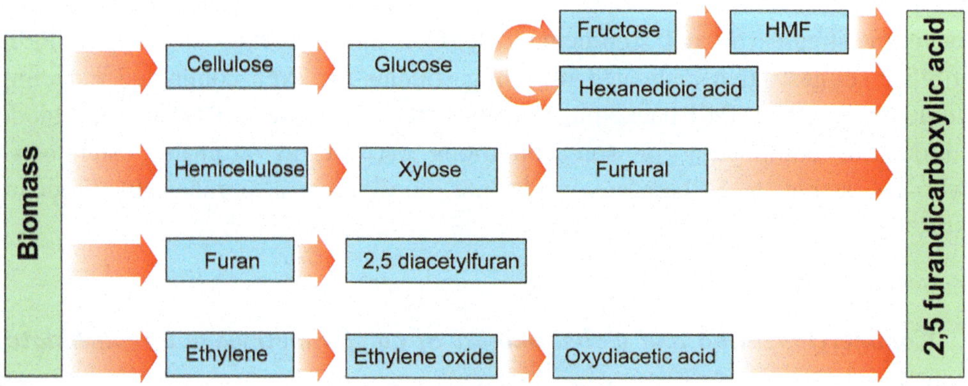

Figure 5.6
Pathways for the bio-based production of 2,5-furandicarboxylic acid (a key component for polyethylene furanoate production) from renewable biomass sources.

catalytic systems for HMF production and stabilization. Avantium (The Netherlands) pioneered an alternative route using methoxymethylfurfural (MMF) as a stable intermediate. In this route, fructose is first converted to MMF, which is then oxidized to FDCA. The first synthesis of PEF was reported in a 1946 patent (Drewitt & Lincoln) using melt polycondensation. In practice, the catalysts and reaction conditions for PEF polymerization are similar to those used for PET. Unlike bio-PE and bio-PET, PEF can be produced entirely from biomass. Its 100% renewable composition gives it a significant sustainability advantage and has driven its development as a potential replacement for PET.

PEF is increasingly recognized as a promising alternative to conventional plastics such as PET owing to its favorable performance characteristics and environmental credentials. PEF has excellent barrier properties against gases such as oxygen, carbon dioxide, and water vapor. PEF is far less permeable to gases like oxygen and carbon dioxide than PET. It has up to 10 times lower oxygen permeability (~2.9×10^{-9} cm^2/s for PEF vs. 3.2×10^{-8} for PET) and around 16–20 times lower carbon dioxide permeability compared to that of PET, making it highly effective for food and beverage packaging (Burgess et al., 2014, 2015). The origin of these superior barrier properties lies in the furanic structure of PEF. The furan rings in the PEF backbone restrict segmental mobility and create strong interchain interactions that impede gas diffusion. In fact, the asymmetric and polar furan rings hinder the ring-flipping motions that gases exploit to diffuse through PET. As a result, PEF can maintain barrier performance even under humid conditions. These features make PEF highly suitable for applications such as beverage bottles, films, and multilayer food packaging where gas barrier performance is critical (Jiang et al., 2020).

PEF exhibits a T_m of ~210°C–230°C and a T_g of ~88°C. The higher T_g gives PEF better resistance to deformation under moderate heat, which is a benefit for hot-filled or warm-storage applications. PEF also shows good thermal stability, with decomposition temperatures of ~350°C. It tends to crystallize more slowly and at higher temperatures than PET due to its rigid furan rings. Mechanically, PEF demonstrates high strength and stiffness with a Young's modulus of ~2.45 GPa and a tensile strength of ~35 MPa, which are higher than the mechanical performance of PET. PEF is also resistant to a wide range of chemicals, which enhances its durability and suitability for packaging applications. PEF is a transparent polymer and can yield high-clarity products similar to PET. PEF can be recycled via both mechanical recycling and chemical recycling using existing PET recycling infrastructure, thus supporting a circular economy. Indeed, PEF can be depolymerized to recover FDCA monomer, which can be purified and reused for new PEF synthesis. PEF is not biodegradable under normal conditions, but it shows partial degradation under composting conditions over several months and appears to be somewhat more biodegradable than PET. Enzymatic degradation studies have shown that PEF can depolymerize about 1.7 times faster than PET under specific conditions. The presence of furan ring and aliphatic ester bonds in PEF can make it more susceptible to enzymatic attacks (e.g., cutinases) compared with PET's aromatic ring. This suggests that PEF waste could biodegrade slightly more readily in the environment than PET, but it is still a durable plastic, and it is not designed for quick biodegradation.

5.7.1 Commercialization and global market of polyethylene furanoate

After years of research and development, PEF is now transitioning from laboratory and pilot scale to commercial reality. The most significant milestone to date has been achieved by Avantium, which is leading the commercialization of PEF. In 2024, Avantium completed construction of the world's first commercial FDCA production plant in Delfzijl (The Netherlands), with a capacity of 5000 tons per year (de Jong et al., 2022). Following this, Avantium began a phased start-up and produced the first commercial batches of FDCA in late 2024 and the first PEF resin in 2025. Avantium's production process converts plant-derived fructose into FDCA via the MMF route. In early 2025, Avantium aimed to accelerate PEF production and begin fulfilling supply agreements under the product brand name "releaf." In 2024, Avantium also obtained U.S. FDA food-contact approval for PEF, meaning that PEF bottles and films are officially cleared safe for use in food and drink packaging in the US market. Avantium has built broad partnerships and collaborations with major companies including Toyobo, Terphane, Refresco, Resilux, Sukano, Carlsberg Group, AmBev, LVMH, Monosuisse, Henkel, Origin Materials, and Kvadrat. This collaboration also extends to large consumer brands. Coca-Cola and Danone were early partners with Avantium in the "PEFerence" consortium (a European project) and intend to use 100% renewable PEF bottles for their products. In addition to Avantium, other companies are developing monomers for PEF production. Stora Enso (a Finnish Swedish forestry corporation) has announced its FuraCore process for producing FDCA from lignocellulosic sugars derived from woody biomass. Origin Materials (USA) is also developing a platform to convert wood residues into chloromethyl furfural and 5-hydroxymethylfurfural (HCMF), which can then be oxidized to FDCA.

The market outlook for PEF in packaging is highly promising if current scale-up efforts succeed. A recent report forecasts that the PEF market to grow from ~US$30 million in 2025 to ~US$406 million by 2030, representing a compound annual growth rate of ~68%. This rapid growth is expected to be driven by increasing demand for sustainable and recyclable packaging, stronger regulatory support for bioplastics, and an expanding range of applications across food, beverage, and consumer goods sectors. Within this growth, bottles are expected to be the fastest-growing application of PEF. Brand owners are eager to substitute PEF for PET to meet sustainability targets and new regulatory mandates while benefiting from the better performance of PEF. The potential to produce lighter-weight, recyclable bottles with extended shelf life is particularly attractive. Indeed, some beverage companies have already set internal targets to use PEF for selected product lines. Cost is the main hurdle for wider adoption of PEF. PEF currently costs ~2–3 times more than PET per kilogram at the 5000-ton production scale. However, analyses suggest that as global capacity increases to hundreds of thousands of tons, production costs will decrease sharply (Martínez Bello et al., 2025). Producers are also exploring co-production of high-value co-products and the use of cheaper second-generation feedstocks to improve process economics. The next few years will likely see announcements of larger FDCA plants as commercialization accelerates.

5.7.2 Applications of polyethylene furanoate in food packaging

PEF is emerging as a high-performance material for food and beverage packaging due to its superior gas-barrier properties and mechanical performance. PEF is currently used for the production of bottles and containers for water, juices, carbonated drinks, and alcoholic beverages. The superior barrier of PEF against carbon dioxide is particularly advantageous for carbonated beverages such as soft drinks and beer because a PEF bottle retains carbonation much longer than PET bottles of the same thickness (Martínez Bello et al., 2025). This can either extend shelf life or allow manufacturers to use thinner and lighter bottle walls without compromising beverage freshness. Oxygen-sensitive beverages such as juices or wine also benefit from the low oxygen permeability of PEF. To maximize its barrier advantages, multilayer bottle designs have been developed. For example, Avantium demonstrated bottles in which a thin inner layer of PEF is co-injected with an outer PET layer. The PEF layer acts as a gas barrier, while the PET layer provides rigidity and processability. Beyond bottles, PEF can also be processed into films for flexible packaging such as snack wrappers and pouch liners. PEF films have high tensile strength and heat stability, making them suitable for oriented film applications. Their gas-barrier performance could eliminate the need for metalized layers or allow significantly longer shelf life for dry products such as chips, coffee, cereals, and nuts, where both oxygen and moisture resistance are crucial. PEF is heat-sealable, and it can be used as an inner sealant layer in multilayer packaging systems. One challenge for this application is the relatively slow crystallization rate of PEF, which can affect processing speed and film orientation during manufacturing.

5.8 Conclusion

The development of biologically derived, non-biodegradable bioplastics is an important step toward reducing the environmental footprint of conventional plastics. Unlike synthetic fossil-based plastics, these bioplastics are produced from renewable carbon sources while maintaining chemical equivalence and full compatibility with existing manufacturing and recycling infrastructures. This dual advantage allows for their rapid integration into current supply chains without substantial investment in new production technologies. This chapter provided an overview of recent progress in bio-derived bioplastics such as bio-PE, bio-PET, PEF, bio-PP, and bio-PTT, along with their emerging applications in food packaging. The demand for bio-based plastics is huge particularly in the food packaging sector, where manufacturers and consumers increasingly prioritize materials that are high-performance and safe for the environment. Despite ongoing global economic challenges, most market reports consistently claim steady growth in the bioplastics sector. Given the limited availability of large-scale composting facilities for biodegradable plastics, those bio-based (non-biodegradable and degradable) bioplastics that can be seamlessly recycled within existing systems are expected to dominate the bioplastics market over the coming decade However, several challenges remain in realizing the full potential of bio-based, non-biodegradable polymers for sustainable food packaging. For some materials, such as bio-PP, and other emerging systems, issues related to feedstock cost and availability, production scalability, and performance consistency (including mechanical strength, barrier integrity, and processability) still persist. In addition, integration with existing recycling streams and the lack of clear standards and labeling frameworks can hinder market acceptance and circularity. Overcoming these barriers presents significant opportunities for innovation in polymer design, recycling technologies, and circular value chains, positioning bio-based plastics as key enablers of a more sustainable and resource-efficient packaging future.

References

AlNouss, A., McKay, G., Al-Ansari, T., 2020. Production of syngas via gasification using optimum blends of biomass. Journal of Cleaner Production 242, 118499. Available at https://doi.org/10.1016/j.jclepro.2019.118499.

Borrelle, S.B., Ringma, J., Lavender Law, K., Monnahan, C.C., Lebreton, L., McGivern, A., et al., 2020. Predicted growth in plastic waste exceeds efforts to mitigate plastic pollution. Science 369 (6509), 1515–1518. Available at https://doi.org/10.1126/SCIENCE.ABA3656.

Burgess, S.K., Karvan, O., Johnson, J.R., Kriegel, R.M., Koros, W.J., 2014. Oxygen sorption and transport in amorphous poly(ethylene furanoate). Polymer 55 (18), 4748–4756. Available at https://doi.org/10.1016/j.polymer.2014.07.041.

Burgess, S.K., Kriegel, R.M., Koros, W.J., 2015. Carbon dioxide sorption and transport in amorphous poly (ethylene furanoate). Macromolecules 48 (7), 2184–2193. Available at https://doi.org/10.1021/acs.macromol.5b00333.

Byun, Y., Kim, Y.T., 2014. Chapter 15 - Utilization of bioplastics for food packaging industry. In: Han, J.H. (Ed.), Innovations in Food Packaging, second ed. Academic Press, pp. 369–390. Available at 10.1016/b978-0-12-394601-0.00015-1.

Chanda, A., Adhikari, J., Ghosh, M., Saha, P., 2023. PTT-based green composites. In: Ajitha, A.R., Thomas, S. (Eds.), Poly Trimethylene Terephthalate: Based Blends, IPNs, Composites and Nanocomposites. Materials Horizons: From Nature to Nanomaterials. Springer, Singapore, pp. 167–185. Available at https://doi.org/10.1007/978-981-19-7303-1_9.

Chen, G.Q., Patel, M.K., 2012. Plastics derived from biological sources: present and future: a technical and environmental review. Chemical Reviews 112 (4), 2082–2099. Available at https://doi.org/10.1021/cr200162d.

de Jong, E., Visser, H.A., Dias, A.S., Harvey, C., Gruter, G.J.M., 2022. The road to bring FDCA and PEF to the market. Polymers 14 (5), 943. Available at https://doi.org/10.3390/polym14050943.

Dupaix, R.B., Boyce, M.C., 2005. Finite strain behavior of poly(ethylene terephthalate) (PET) and poly(ethylene terephthalate)-glycol (PETG). Polymer 46 (13), 4827–4838. Available at https://doi.org/10.1016/j.polymer.2005.03.083.

Ghasemlou, M., Adhikari, B., Naebe, M., Barrow, C.J., 2025. Bioengineering strategies to promote plastic circularity. Nature Reviews Bioengineering 3 (5), 353–354. Available at https://doi.org/10.1038/s44222-025-00287-5.

Ghasemlou, M., Barrow, C.J., Adhikari, B., 2024. The future of bioplastics in food packaging: an industrial perspective. Food Packaging and Shelf Life 43, 101279. Available at https://doi.org/10.1016/j.fpsl.2024.101279.

Hao, L.T., Ju, S., Hwang, D.K., Hwang, D.S., Ok, Y.S., Hwang, S.Y., et al., 2024. Optimizing bioplastics translation. Nature Reviews Bioengineering 2 (4), 289–304. Available at https://doi.org/10.1038/s44222-023-00142-5.

Hwang, K.R., Jeon, W., Lee, S.Y., Kim, M.S., Park, Y.K., 2020. Sustainable bioplastics: recent progress in the production of bio-building blocks for the bio-based next-generation polymer PEF. Chemical Engineering Journal 390, 124636. Available at https://doi.org/10.1016/j.cej.2020.124636.

Jariyasakoolroj, P., Leelaphiwat, P., Harnkarnsujarit, N., 2020. Advances in research and development of bioplastic for food packaging. Journal of the Science of Food and Agriculture 100 (14), 5032–5045. Available at https://doi.org/10.1002/jsfa.9497.

Ji, N., Zhang, T., Zheng, M., Wang, A., Wang, H., Wang, X., et al., 2008. Direct catalytic conversion of cellulose into ethylene glycol using nickel-promoted tungsten carbide catalysts. Angewandte Chemie International Edition 47 (44), 8510–8513. Available at https://doi.org/10.1002/anie.200803233.

Jiang, L., Gonzalez-Diaz, A., Ling-Chin, J., Malik, A., Roskilly, A.P., Smallbone, A.J., 2020. PEF plastic synthesized from industrial carbon dioxide and biowaste. Nature Sustainability 3 (9), 761–767. Available at https://doi.org/10.1038/s41893-020-0549-y.

Kobayashi, T., Wood, B.A., Blackman, G.S., Takemura, A., 2011. Morphology, dynamic mechanical, and electrical properties of bio-based poly(trimethylene terephthalate) blends. Part 1: Poly(trimethylene terephthalate)/poly(ether esteramide)/polyethylene glycol 400 bis(2-ethylhexanoate) blends. Journal of Applied Polymer Science 120 (6), 3519–3529. Available at https://doi.org/10.1002/app.33498.

Lange, J., Wyser, Y., 2003. Recent innovations in barrier technologies for plastic packaging—a review. Packaging Technology and Science 16 (4), 149–158. Available at https://doi.org/10.1002/pts.621.

Lee, H.J., Kim, S.J., Youk, J.H., Lee, K.H., 2025. Properties and application potential of mechanically and chemically recycled PET fibers for tire cord and airbag applications. Fibers and Polymers 26, 5393–5403. Available at https://doi.org/10.1007/s12221-025-01183-w.

Martínez Bello, L.I., Vázquez Cuevas, V.Y., González-Contreras, M., Gomez-Maldonado, D., Hernández-Guerrero, M., 2025. Bio-polyethylene furanoate production in Latin America as a response to the current needs for sustainable food packaging. Journal of Polymers and the Environment 33 (4), 1792–1813. Available at https://doi.org/10.1007/s10924-025-03530-7.

Nikles, D.E., Farahat, M.S., 2005. New motivation for the depolymerization products derived from poly(ethylene terephthalate) (PET) waste: a review. Macromolecular Materials and Engineering 290 (1), 13–30. Available at https://doi.org/10.1002/mame.200400186.

Niyogi, A., Sarkar, P., Bhattacharyya, S., Mukherjee, S., Pal, S., 2025. Bioplastics: harmonious coexistence of innovation and sustainability—a comprehensive review. Polymer Bulletin 82 (14), 8871–8953. Available at https://doi.org/10.1007/s00289-025-05871-8.

Pant, A., Sängerlaub, S., Müller, K., 2017. Gallic acid as an oxygen scavenger in bio-based multilayer packaging films. Materials 10 (5), 489. Available at https://doi.org/10.3390/ma10050489.

Rabnawaz, M., Wyman, I., Auras, R., Cheng, S., 2017. A roadmap towards green packaging: the current status and future outlook for polyesters in the packaging industry. Green Chemistry 19 (20), 4737–4753. Available at https://doi.org/10.1039/c7gc02521a.

Rahman, M.H., Bhoi, P.R., 2021. An overview of non-biodegradable bioplastics. Journal of Cleaner Production 294, 126218. Available at https://doi.org/10.1016/j.jclepro.2021.126218.

Ramachandra, T.V., Hebbale, D., 2020. Bioethanol from macroalgae: prospects and challenges. Renewable and Sustainable Energy Reviews 117, 109479. Available at https://doi.org/10.1016/j.rser.2019.109479.

Rydz, J., Musioł, M., Zawidlak-Węgrzyńska, B., Sikorska, W., 2018. Chapter 14 - Present and future of biodegradable polymers for food packaging applications. In: Grumezescu, A.M., Holban, A.M. (Eds.), Biopolymers for Food Design. Academic Press, pp. 431–467. Available at 10.1016/b978-0-12-811449-0.00014-1.

Sheldon, R.A., 2014. Green and sustainable manufacture of chemicals from biomass: state of the art. Green Chemistry 16 (3), 950–963. Available at https://doi.org/10.1039/c3gc41935e.

Shirvanimoghaddam, K., Ozen, I., Wu, P., Jafarzadeh, S., Yadav, R., Naebe, M., 2024. Surface functionality through periodical patterning on carbon nanotubes. Small Structures 5 (6), 2300539. Available at https://doi.org/10.1002/sstr.202300539.

Singh, A., Rangaiah, G.P., 2017. Review of technological advances in bioethanol recovery and dehydration. Industrial and Engineering Chemistry Research 56 (18), 5147–5163. Available at https://doi.org/10.1021/acs.iecr.7b00273.

Siracusa, V., Blanco, I., 2020. Bio-polyethylene (Bio-PE), bio-polypropylene (Bio-PP) and bio-poly(ethylene terephthalate) (Bio-PET): recent developments in bio-based polymers analogous to petroleum-derived ones for packaging and engineering applications. Polymers 12 (8), 1641. Available at https://doi.org/10.3390/polym12081641.

Soo, X.Y.D., Muiruri, J.K., Wu, W.-Y., Yeo, J.C.C., Wang, S., Tomczak, N., et al., 2024. Bio-polyethylene and polyethylene biocomposites: an alternative toward a sustainable future. Macromolecular Rapid Communications 45 (14), 2400064. Available at https://doi.org/10.1002/marc.202400064.

Spierling, S., Knüpffer, E., Behnsen, H., Mudersbach, M., Krieg, H., Springer, S., et al., 2018. Bio-based plastics - A review of environmental, social and economic impact assessments. Journal of Cleaner Production 185, 476–491. Available at https://doi.org/10.1016/j.jclepro.2018.03.014.

Stoleru, E., Irimia, A., Butnaru, E., 2021. Bio-based bioplastics in active food packaging. Bioplastics for Sustainable Development. Springer, Singapore, pp. 347–379. Available at https://doi.org/10.1007/978-981-16-1823-9_14.

Trucillo, P., Rizzo, M., Errico, D., Di Maio, E., 2025. Social, economic, and environmental impacts of bio-based versus fossil-derived polyethylene production. Advanced Sustainable Systems 9 (1), 2400392. Available at https://doi.org/10.1002/adsu.202400392.

van den Oever, M., Molenveld, K., 2017. Replacing fossil based plastic performance products by bio-based plastic products—technical feasibility. New Biotechnology 37 (Pt A), 48–59. Available at https://doi.org/10.1016/j.nbt.2016.07.007.

Wang, S., Muiruri, J.K., Soo, X.Y.D., Liu, S., Thitsartarn, W., Tan, B.H., et al., 2023. Bio-polypropylene and polypropylene-based biocomposites: solutions for a sustainable future. Chemistry, an Asian Journal 18 (2), e202200972. Available at https://doi.org/10.1002/asia.202200972.

Webb, H.K., Arnott, J., Crawford, R.J., Ivanova, E.P., 2012. Plastic degradation and its environmental implications with special reference to poly (ethylene terephthalate). Polymers 5 (1), 1–18.

CHAPTER 6

Fossil-based and biodegradable bioplastics for food packaging

Mehran Ghasemlou[1,2,3]

[1]Department of Chemical Engineering, The University of Melbourne, Parkville, VIC, Australia, [2]Center for Sustainable Bioproducts, Deakin University, Geelong Waurn Ponds Campus, Geelong, VIC, Australia, [3]School of Science, STEM College, RMIT University, Melbourne, VIC, Australia

6.1 Introduction

The modern world is facing a plastic pollution crisis (Ghasemlou et al., 2025; Pottinger et al., 2024). Since the mid-20th century, the massive production of plastic packaging has revolutionized the global food supply chain, allowing fresh produce to travel thousands of miles without spoiling (Ghasemlou et al., 2022; Jambeck et al., 2015). Fossil-based plastics, such as polyethylene (PE) and polyethylene terephthalate (PET), have been the undisputed backbone of the food packaging industry for over 50 years (Helmes et al., 2022). Beyond their low cost, the functionality of fossil-based plastics is difficult to replicate with other materials. They are lightweight, chemically inert, and have superior barrier properties against oxygen and moisture (Ghasemlou et al., 2024). However, the durability that makes these polymers ideal for preserving food renders them catastrophic for the environment. Once discarded, they persist in the environment indefinitely, accumulating in landfills or leaking into nature, where they pose a persistent threat to biodiversity. An estimated 11 million metric tons of plastic waste enter the oceans every year (Rosenboom et al., 2022). The vast majority of this debris originates from single-use packaging items that are used for minutes but engineered to last for centuries (Barone et al., 2025).

The response of the plastics industry has been to design materials that can safely return to nature. Within this emerging field, a debate is ongoing between two distinct families of materials (Fig. 6.1). On one hand are the bio-based plastics (e.g., polyhydroxyalkanoate (PHA) and polylactic acid (PLA)), derived from plants and bacteria (Meereboer et al., 2020). On the other hand is the often-overlooked but economically superior workhorse, which is fossil-based biodegradable plastics (e.g., polybutylene adipate terephthalate (PBAT) and poly (ε-caprolactone) (PCL)) (Itabana et al., 2024). While the former captures the public imagination with its "plant-based" marketing, it is the latter that offers the scalable, low-cost, and durable solution the world actually needs right now. Most consumers (and even

Bioplastics for Food Packaging. DOI: https://doi.org/10.1016/B978-0-443-41654-5.00007-X

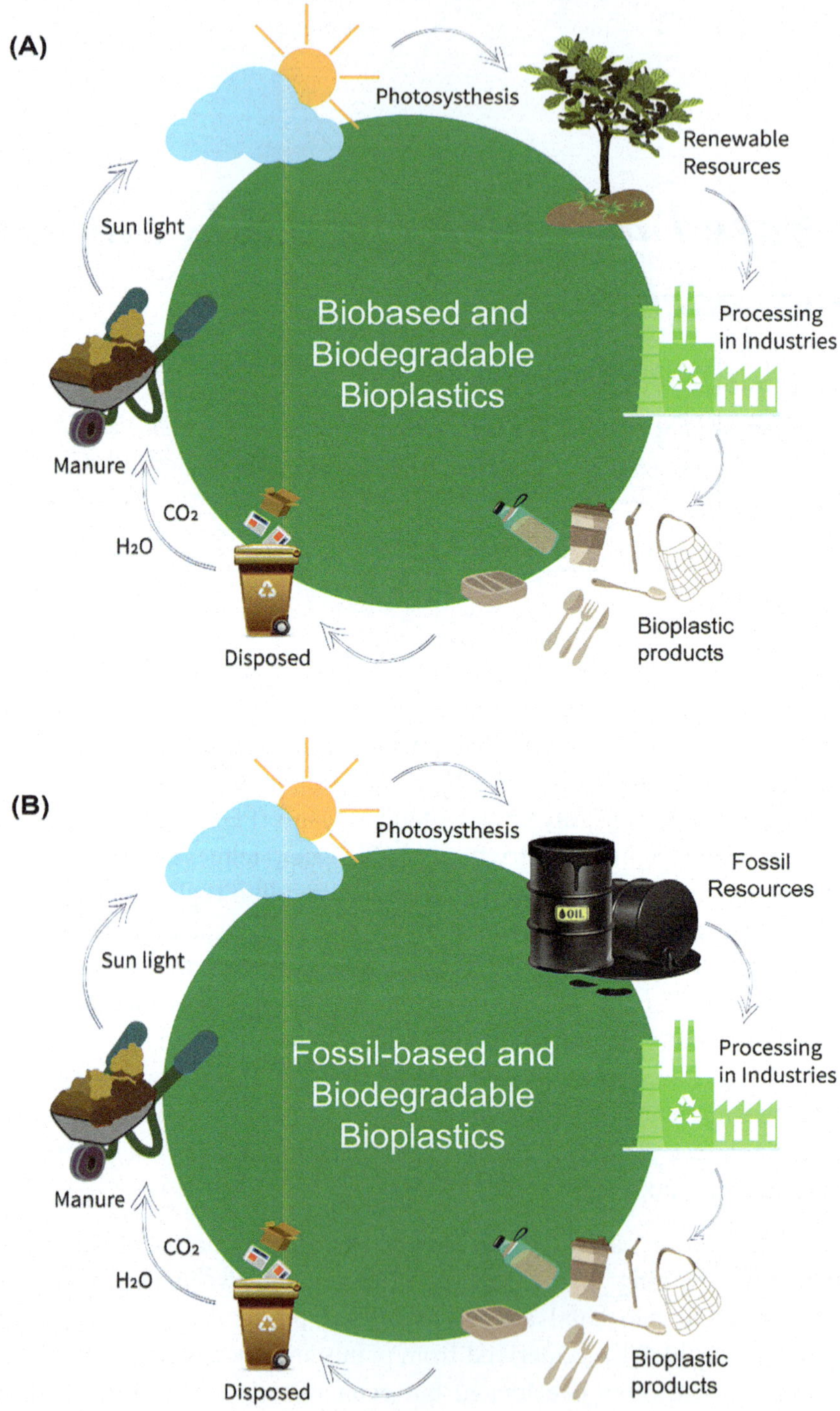

Figure 6.1

Conceptual life cycle comparison of biodegradable plastics derived from renewable resources (A) and those produced from petrochemical resources but engineered with hydrolyzable functional groups (B). The image is obtained under a license from Shutterstock.

policymakers) assume that if a plastic comes from oil (fossil-based), it must be non-biodegradable, and if it comes from a plant (bio-based), it must be biodegradable. This is factually incorrect. From a chemical engineering perspective, the environmental fate of a polymer (whether it persists for 500 years or degrades within six months) is determined not by the origin of its carbon atoms, but by its molecular architecture. Indeed, biodegradability is a function of chemical bonds, specifically the presence of hydrolyzable linkages (e.g., esters, amides, or urethanes) that microorganisms can enzymatically cleave (Hao et al., 2024). It is entirely possible to synthesize plastic from petroleum sources that contains these linkages and is fully biodegradable. Conversely, it is possible to synthesize a plastic from sugarcane that is chemically identical to standard PE and therefore non-biodegradable. While the transition toward a fully bio-based economy is the ultimate goal for decoupling industrial production from finite resources, the urgency of the plastic pollution crisis requires a more pragmatic and immediate intervention (Hahladakis et al., 2024). The argument for fossil-based biodegradable plastics as a faster solution to the global plastic waste challenge is built on three factors.

First, the global chemical industry is built on the processing of hydrocarbons, and fossil-based biodegradable plastics can be easily synthesized using these existing assets with relatively minor modifications compared to the construction of entirely new biorefineries (Walker and Rothman, 2020). For instance, the production of PBAT utilizes established monomers such as 1,4-butanediol, adipic acid, and terephthalic acid, whereas bio-based polymers such as PHA require the construction of new fermentation facilities and complex agricultural supply chains. Therefore, the plastics industry can scale production of fossil-based biodegradable alternatives rapidly and meet the high-volume demands of the global food packaging market much faster than bio-based alternatives could hope to achieve (de Castro et al., 2022). Second, cost is the single greatest barrier to the adoption of sustainable packaging. Fossil-based biodegradable plastics, while still more expensive than conventional PE, are significantly cheaper than their bio-based alternatives. This relative economic efficiency allows large companies to integrate biodegradability into their supply chains today, rather than waiting for an uncertain future where bio-based economy is achieved (Swetha et al., 2024). Thus, fossil-based biodegradable plastics act as an economic bridge and quicky enable the creation of a market for compostable goods. The "green premium" associated with bio-based plastics remains a significant barrier, largely due to the volatility of agricultural feedstocks and the complexity of fermentation processes. This economic feasibility is crucial for the immediate adoption of compostable packaging systems.

Third, fossil-based biodegradable polymers fill a critical functional gap left by leading bio-based alternatives. PLA, the most common bio-based plastic, exhibits inherent brittleness and poor extensibility, limiting its utility in flexible packaging applications such as cling films and bags (Guicherd et al., 2024). Fossil-based biodegradable plastics can fill this functional void. Indeed, PBAT is engineered to mimic the rheological and mechanical properties of PE. It possesses high ductility, high elongation at break, and excellent toughness. This allows it to be processed on existing blown-film extrusion lines used for conventional plastics without capital investment for new machinery (Roy et al., 2024).

This chapter provides a comprehensive review of the state-of-the-art regarding fossil-based biodegradable plastics. A detailed examination will be presented on the synthesis and chemistry of these polymers. This will be followed by an analysis of their physicochemical properties, such as mechanical, thermal, and barrier properties, and how these compare with both conventional petrochemical plastics and emerging bio-based ones. This chapter will also explore the commercialization landscape and global market trends and highlight production capacities, key industrial producers, and the economic drivers that shape the adoption of these materials. Their applications in food packaging will be critically reviewed, with an assessment of their ability to preserve shelf life and maintain food safety standards.

6.2 Global plastic crisis and the emergence of bioplastics

Since the mid-20th century, plastics have become indispensable to modern society due to their versatility. Plastics are cheap, lightweight, durable, transparent, and moldable and have found applications in everything from health care and construction to military supplies and agriculture (Vidal et al., 2024). Despite their benefits, the environmental cost of conventional plastics is staggering. Currently, 99% of commercially available plastics are derived from petrochemicals, relying heavily on finite fossil resources (Chen and Patel, 2012). The plastic life cycle is intrinsically linked to the climate crisis, and 99% of commercially available plastics are derived from petrochemicals, accounting for 6% of global oil consumption. This figure is expected to rise to 20% by 2050, with plastics and petrochemicals driving nearly half of the growth in oil demand (Kapustin and Grushevenko, 2023). In 2019 alone, plastics were responsible for 1.8 billion tons of CO_2 equivalent emissions, representing 3.4% of global emissions (Andreasi Bassi et al., 2021; Schade et al., 2024). Furthermore, waste management systems are failing to keep pace. Of the plastic waste generated, 79% accumulates in landfills or the environment, 12% is incinerated, and only 9% is recycled (Ghosh and Jones, 2021). This mismanagement leads to massive pollution in natural environments. To mitigate these threats, research has pivoted toward "bioplastics" an umbrella term covering a range of polymer chemistries. This definition encompasses two distinct concepts: the bio-based origin of raw materials and biodegradability at the end of life (Clark and Shaver, 2024). Beyond merely replacing fossil-based materials, bioplastics offer unique advantages in sectors such as food packaging, agriculture, and biomedical applications (Nanda et al., 2022). However, their adoption is currently limited to ~1% of the global market due to economic barriers, performance trade-offs, and widespread public confusion. This chapter details the current landscape of bioplastics and analyzes the scientific hurdles, environmental realities, and strategic pathways required to achieve a truly circular economy.

6.2.1 The opportunities of using bioplastics

The most fundamental opportunity presented by bioplastics is the decoupling of plastic production from finite fossil resources. Currently, 99% of commercially available plastics are derived from petrochemicals, and the industry accounts for 6% of global oil consumption. By shifting to renewable feedstocks such as maize, sugarcane, cellulose, and waste biomass, the industry can significantly reduce its carbon footprint (Mohanty et al., 2018; Stegmann et al., 2022). Bio-based plastics can be fully carbon-neutral or show a significantly lower carbon footprint than their conventional counterparts. Replacing approximately two-thirds of synthetic fossil-based plastics with bioplastics could reduce emissions by an estimated 241–316 million tons of CO_2 equivalent per year (Spierling et al., 2018). As the energy and transportation decarbonize, the plastics industry remains a major driver of oil demand. Thus, bioplastics offer a pathway for the polymer industry to align with the global energy transition (Borrelle et al., 2020). If the economic costs of extracting fossil resources rise due to carbon pricing or scarcity, bio-based polymers will likely reach a maximum point where they become the more viable alternative, akin to the shift seen in renewable energy (Chaudry et al., 2025). Bioplastics are not just material substitutions; instead, they enable new waste management strategies that are unattainable with conventional plastics. Indeed, certified compostable plastics can be processed alongside food waste in industrial composting facilities. This is particularly valuable for food-contaminated packaging, which is often rejected by recycling streams (Hobbs et al., 2021). By allowing packaging to be composted with the food scraps, bioplastics divert waste from landfills and return nutrients to the soil. Finally, while littering is never a solution, biodegradable plastics offer a "safety net" for accidental leakage into the environment (Keith et al., 2024). Materials such as PHA have shown high degradation rates in marine environments, which can significantly alleviate the environmental impact of accidental leakage (Degli Innocenti and Breton, 2020).

6.2.2 The challenges for using bioplastics

While bioplastics offer a promising pathway toward a circular economy and reduced greenhouse gas emissions, their widespread adoption is hindered by significant economic, technological, and infrastructural hurdles (Negrete-Bolagay and Guerrero, 2024).

First, cost is a primary obstacle. Bioplastics are currently more expensive to produce than established commodity plastics such as PE and polypropylene (PP). This price disparity is driven by the scale and efficiency of the long-established petrochemical industry and historically low crude oil prices (ten Boer et al., 2025). Furthermore, the conversion efficiency of biomass into polymer building blocks remains a bottleneck and current processes are often too expensive to compete with traditional refining (Coppola et al., 2021). However, scaling up production technologies could potentially render bioplastics cost-competitive within the next few years.

Second, replacing conventional plastics is difficult because bioplastics often face a trade-off between mechanical strength and biodegradability (Cywar and Beckham, 2022). For instance, PLA and PHA can be too brittle for certain applications, while amorphous PHA is often too soft. To improve durability, manufacturers frequently add fillers, stabilizers, or plasticizers. Unfortunately, these additives can be toxic, impede biodegradation, or result in the formation of microplastics.

Third, there is a critical disconnect between public perception and scientific reality regarding "biodegradability." A major challenge is that many biodegradable plastics do not degrade in all environments, and they typically require specific conditions found only in industrial composting facilities (Haider et al., 2019). For example, PLA degrades ~20 times faster than high-density PE on land, but it degrades at a similarly slow rate in marine environments (Bhagwat et al., 2020; Wang et al., 2021). Without access to specialized composting infrastructure, these materials often end up in landfills where their anaerobic degradation can release methane (a potent greenhouse gas). Fourth, the complexity of bioplastic terminology has created fertile ground for "greenwashing" (Zhu and Wang, 2020). Public understanding of bioplastics is often fueled by deceptive marketing. Advertisements frequently use images of corn or leave to imply that a product is environmentally safe. This leads consumers to falsely believe all bioplastics are biodegradable (Cao et al., 2025). Ironically, studies showed consumers prefer "degradable" labels over "bio-based" ones, even if the "degradable" product is actually an environmentally damaging oxo-degradable plastic. This confusion causes contamination in waste streams: compostable plastics contaminate mechanical recycling, while nondegradable plastics contaminate compost. Fifth, there is a persistent public concern that growing crops for bioplastics competes with food production. However, research indicates this is overstated and even if bioplastics completely replaced fossil-based plastics, they would require only roughly 1.5% of global agricultural land (Mattlar and Ekholm, 2025; Nogueira et al., 2024). Using inedible biomass and advanced agricultural technologies can minimize any potential conflict.

6.3 Polybutylene succinate

6.3.1 Physical and chemical characteristics

Bioplastics is a large section that consists of biodegradable and bio-based plastics. Biodegradable plastics may or may not be obtained from bioresources but can be degraded by microorganisms, whereas bio-based plastics may or may not be degraded by microorganisms (Lackner et al., 2023). Polybutylene succinate (PBS) is obtained from petroleum-based products but can be broken down using enzymes and microorganisms. PBS belongs to the group of aliphatic thermoplastic polyesters. PBS is one of the most interesting compostable polymers because of its unique combination of rigidity, ductility, toughness, and biodegradability (Barletta et al., 2022).

PBS is a biodegradable aliphatic polyester with properties that are comparable to PP. It is a white, semicrystalline thermoplastic polymer with a density of ~1.25 g/cm^3. Its degree of crystallinity varies depending on factors such as cooling rate and molecular weight. PBS has a melting temperature in the range of ~90°C–120°C, and a low glass transition temperature of –45°C to –10°C. The mechanical properties of PBS are generally comparable to those of polyolefins. It has good mechanical properties and excellent processability in textile filaments, injection molding, and extruded and blown products (Rajgond et al., 2024). In fact, PBS also has a wide temperature range, and it can be worked similarly to polyolefins in the range of 160°C–200°C. It has moderate tensile strength (~30–40 MPa), and elongation at break of up to 200%–300% (Barletta et al., 2022; Rafiqah et al., 2021). Moreover, its high flexibility guarantees its use in many applications involving film production. Although PBS film has high heat deflection and is tough, it yields an opaque or semitransparent bioplastic (Bumbudsanpharoke et al., 2022; Kim et al., 2023). PBS polymers are good components to blend with other bio-based polymers, such as PLA, PHAs, or thermoplastic starch (TPS).

The technical drawbacks of PBS can be improved by its copolymerization with adipic acid to produce poly(butylene succinate-co-adipate) (PBSA) (Jariyasakoolroj et al., 2020). The presence of the longer adipate blocks disrupts the crystallinity of the polymer. In terms of brittleness, PBS is more rigid than PBSA but slightly ductile (Ghasemlou et al., 2024). Both PBS and PBSA are categorized as aliphatic polyesters (no aromatic rings in the backbone), which is a key factor in their biodegradability. PBS and PBSA have relatively good biodegradability, and they can biodegrade into natural products (CO_2, water, and biomass) at end-of-life under appropriate conditions. This is because PBS and PBSA have chemically hydrolyzable backbones (unlike PE or PP) and no aromatic rings to hinder enzyme access (Aliotta et al., 2022). Thus, a wide range of naturally occurring enzymes (e.g., lipases, esterases, and cutinases) and microbes can eventually break them down. Therefore, PBS and PBSA can quickly degrade in environments where PLA or PBAT cannot substantially degrade. However, PBSA biodegrades faster than PBS due to its lower degree of crystallinity and faster hydrolysis (Platnieks et al., 2021; Zaborowska et al., 2023). This shows that the adipate units in PBSA increase biodegradability, likely by reducing crystallinity and providing more flexible chain segments.

PBS is not home compostable while PBSA is home compostable. Srikanth et al. (2022) performed enzymatic degradation of PBS and PBSA and reported that *Aspergillus niger*-derived lipase could degrade PBSA (Rafiqah et al., 2021). Furthermore, Maeda et al. (2005) reported that *Aspergillus oryzae*-derived cutinase degraded both PBS and PBSA. For this reason, PBSA is more attractive than PBS for many applications. Both polymers have similar heat resistance and mechanical properties to low-density PE. Similar to low-density PE, PBS is a soft and flexible semicrystalline polyester with excellent properties suitable for both blown and cast film extrusion (Nath et al., 2025). PBS/PBSA materials have been shown to meet industrial composting conditions when processed to a suitable thickness. For example, a

PBSA grade from PTT MCC Biochem (FD92PM) obtained home compostable certification at up to 502 μm thickness. bio-PBS and bio-PBSA from Xinjiang Blue Ridge in China are certified home compostable at thinner thicknesses (≈45–68 μm).

6.3.2 Synthesis and production

Research regarding the aliphatic polyester polymers has a long history. The synthesis of succinic acid-based polyesters can be traced back to 1863, when Portuguese chemist Agostinho Vicente Lourenço described a reaction between succinic acid and ethylene glycol, yielding a compound it was referred to as "succino-ethylenic acid." This early investigation laid the foundational understanding of polyester chemistry. In the early 20th century, further developments were carried out by Wallace H. Carothers of DuPont, who pioneered the synthesis of various aliphatic polyesters during his work on polymerization processes (Smith and Hounshell, 1985). Later, P.J. Floly successfully increased the molecular weight using an acid chloride as the raw material (Barrino et al., 2023; Kato et al., 2023). However, aliphatic polyesters were not commercially utilized for a long time due to their low heat stability in comparison with aromatic polyesters such as PET and PBT, and polyamides like nylon. A timeline of the development of PBS and PBSA research is presented in Fig. 6.2.

PBS is composed of repeating 1,4-butanediol and succinic acid, while PBSA contains an additional adipic acid (another diacid). Traditionally, all three monomers have been produced from petrochemical feedstocks. Succinic acid can be synthesized via routes such as the hydrogenation of maleic anhydride or the oxidation of butane. Adipic acid, used in PBSA, is typically produced by the nitric acid oxidation of cyclohexanol/cyclohexanone. Consequently, most commercially available PBS on the market is produced from fossil-based resources. With the emergence of succinic acid derived from bio-based renewable resources, PBS and PBSA can now be partially manufactured with bio-based content (Platnieks et al., 2021). Various substrates (e.g., corn glucose, glycerol, agricultural waste hydrolysates, etc.) can be fermented by strains of *Escherichia coli*, *Actinobacillus succinogenes*, and *Mannheimia succiniciproducens* to produce succinic acid with high yields. The polymerization process for PBS typically involves a two-step melt polycondensation method. Initially, 1,4-butanediol and succinic acid undergo esterification, followed by a polycondensation reaction conducted under reduced pressure (Fig. 6.3). This process (while effective) faces challenges such as the generation of by-products, including tetrahydrofuran and cyclic oligomers, which may result from thermal degradation (Henke et al., 2017). To mitigate such degradation pathways, process optimization focusing on catalyst selection, reaction time, and temperature control is essential. Innovations have also explored solid-state polymerization and solvent-assisted methods to improve molecular weight and mechanical performance (Guckert et al., 2022). Catalysts frequently used in the polymerization process are titanium-based compounds, stannous octoate, and enzymes such as Novozym 435. Adipic acid is more challenging to

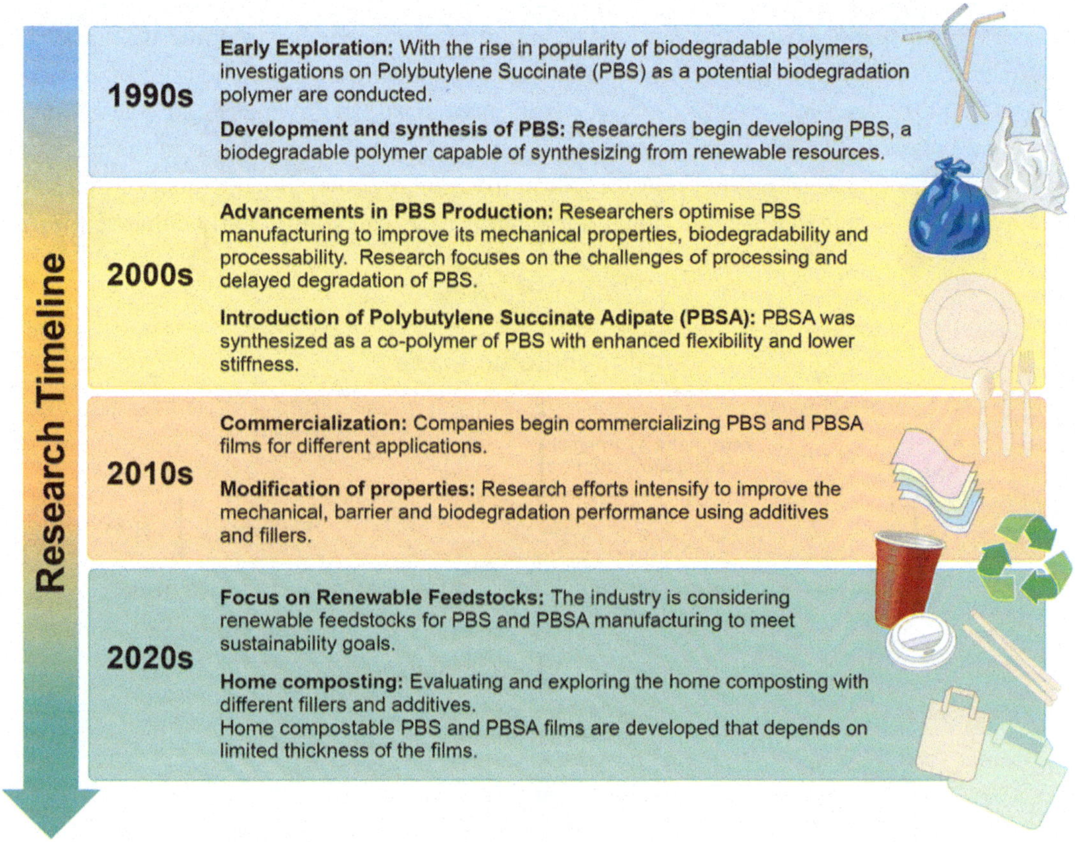

Figure 6.2

Key research milestones for synthesis, development, industrial adoption of polybutylene succinate and poly(butylene succinate-co-adipate).

Reprinted with permission from Nath, D., Misra, M., Al-Daoud, F., Mohanty, A.K., 2025. Studies on poly (butylene succinate) and poly (butylene succinate-co-adipate)-based biodegradable plastics for sustainable flexible packaging and agricultural applications: a comprehensive review. RSC Sustainability, 3, 1267–1302.

produce biologically, but research is ongoing to synthesize it from renewable resources. Although bio-adipic acid is not yet widely commercialized, life cycle studies indicated that fully bio-based adipic could greatly reduce environmental impacts (62%–78% lower greenhouse gas emissions than conventional) (Nag et al., 2020).

6.3.3 Commercialization and mass production

The global supply of PBS and PBSA is currently concentrated among a few major producers. PBS was first commercialized under the trade name Bionolle and remained on the market until 2016, when production was stopped due to insufficient demand. In 2003, Mitsubishi

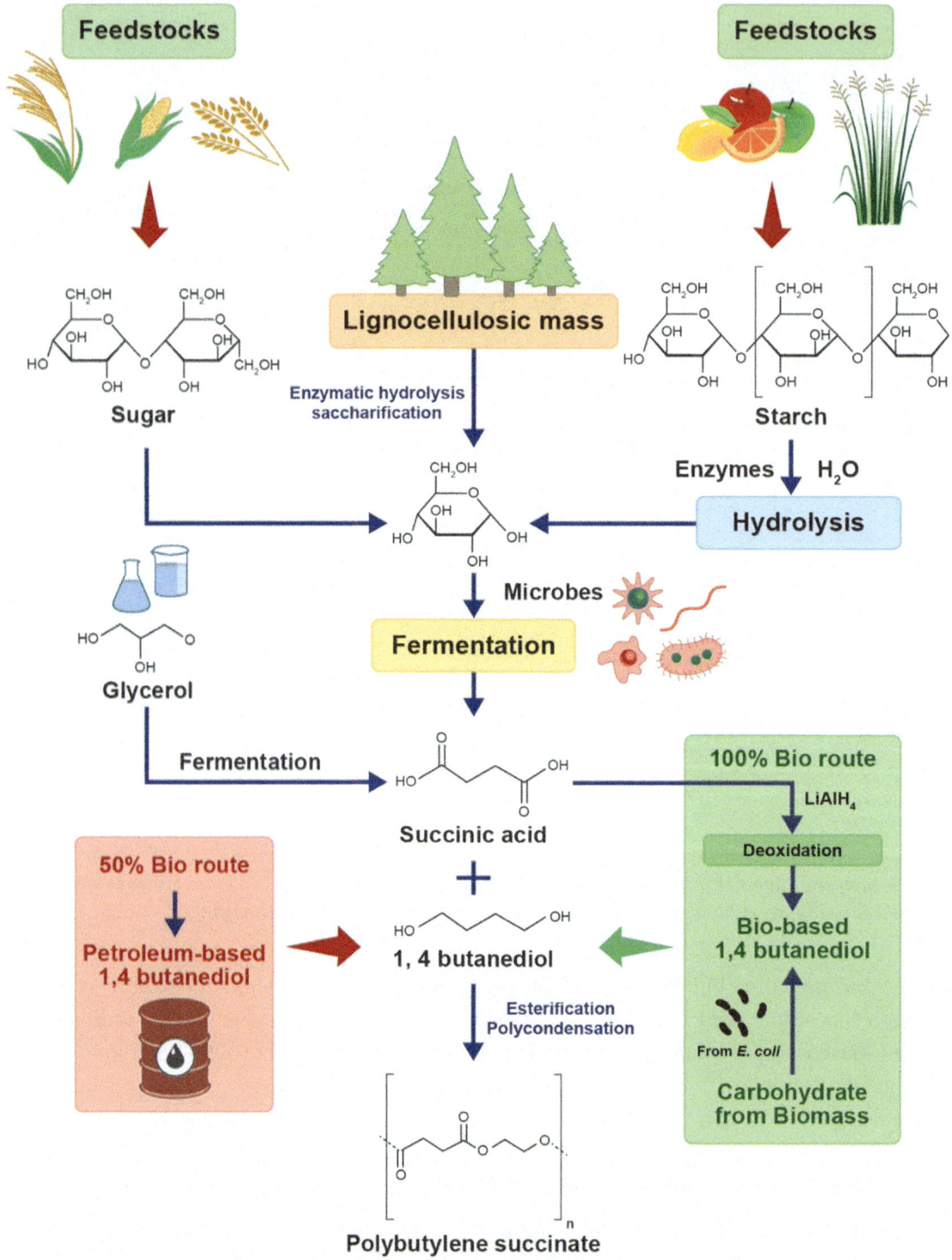
Feedstocks
Feedstocks
Lignocellulosic mass
Sugar
Enzymatic hydrolysis saccharification
Starch
Enzymes
H_2O
Hydrolysis
Microbes
Fermentation
Glycerol
Fermentation
Succinic acid
100% Bio route
$LiAlH_4$
Deoxidation
50% Bio route
Petroleum-based 1,4 butanediol
1, 4 butanediol
Bio-based 1,4 butanediol
From E. coli
Carbohydrate from Biomass
Esterification Polycondensation
Polybutylene succinate

Chemical successfully launched the commercial production of PBS under the brand name GS Pla, marking the first instance of high-molecular-weight PBS production achieved without the use of a chain extender. The "GS Pla" was manufactured for biodegradable applications in batch production using fossil-based succinic acid and 1,4-butanediol as the main raw materials.

In the following years, Mitsubishi Chemical intensified its efforts to transition away from fossil-based raw materials. Beginning around 2000, the company sought to develop chemical products with high functionalization while simultaneously reducing CO_2 emissions. This initiative led to significant research and development aimed at producing bio-based PBS, in which bio-succinic acid would replace its fossil-derived counterpart. However, at the onset of these efforts, bio-succinic acid derived from biomass was found to contain impurities inherent to the feedstock. These impurities posed significant challenges in producing PBS with high molecular weight and sufficient mechanical strength. Despite initial attempts, the resulting products showed poor designability, including undesirable color tones and the presence of foreign particles.

To overcome these challenges, Mitsubishi Chemical integrated biochemical conversion processes for biomass, chemical engineering techniques for purification, and polymer science for high-molecular-weight polymerization. This multidisciplinary approach enabled the production of high-quality bio-based PBS with the necessary mechanical properties and biodegradability control. This success led to the commercialization of BioPBS in partnership with Public Company Limited (PTT) in 2017. This collaboration resulted in the establishment of a full-scale commercial plant (20,000 tons/year) under the joint venture PTT MCC Biochem (Kato et al., 2023).

Today, BioPBS is produced using bio-succinic acid derived from renewable resources and fossil-derived 1,4-butanediol. As of 2025, the 1,4-butanediol monomer is derived from petroleum; however, PTT MCC Biochem plans to change to a bio-based raw material in the near future. BioPBS is available in two primary grades: FZ-type (PBS) and FD-type (PBSA). The FZ-type have a melting point analogous to PE (~115°C) but has lower transparency

Figure 6.3

Schematic representation of the synthesis of bio-based polybutylene succinate. Renewable feedstocks are converted to succinic acid via microbial fermentation, followed by catalytic conversion to 1,4-butanediol and subsequent polycondensation to produce bio-based polybutylene succinate. *Reprinted with permission from Nath, D., Misra, M., Al-Daoud, F., Mohanty, A.K., 2025. Studies on poly (butylene succinate) and poly (butylene succinate-co-adipate)-based biodegradable plastics for sustainable flexible packaging and agricultural applications: a comprehensive review. RSC Sustainability 3, 1267–1302.*

(Kato et al., 2023). However, the FD-type has a much lower melting point than that of PE (~88°C) but offers superior transparency.

In China, Xinjiang Blue Ridge Tunhe Polyester is another significant producer with a capacity of tens of thousands of tons for PBS and PBAT. However, commercial production of PBS/PBSA faces several challenges. A major issue is maintaining a consistent supply of bio-based feedstock at competitive prices. The bankruptcy of BioAmber (a prominent bio-succinic producer) in 2018 highlighted the economic difficulties of sustaining bio-monomer production. Another challenge is technical expertise because the polycondensation of succinic acid and 1,4-butanediol requires handling corrosive acids and achieving high-vacuum conditions for finishing. Furthermore, producers must obtain product certifications (e.g., compostability and food contact), which can delay the commercialization of new plants.

6.3.4 Food packaging applications

PBS and PBSA are generally regarded as safe polyesters and when in contact with food liquids, they do not release harmful levels of leachable substances (Uysal-Unalan et al., 2024; Vytejčková et al., 2017). The food packaging sector is indeed a key target for PBS and PBSA, owing to their favorable safety profile. For this reason, commercial products made from PBS/PBSA blends or composites are already on the market. One significant application area is biodegradable food service items. PBS and PBSA have been used (often in combination with other biopolymers or natural fillers) to produce compostable plates, cups, cutlery, and straws (Moliner et al., 2025). For example, disposable knives and forks made from a PBS/PBSA blend possess the necessary toughness and heat resistance to perform similarly to PP utensils, while remaining compostable after use (Genovese et al., 2016).

Another growing application is paper coating. PBS can act as a biodegradable coating on paperboard for cups or food trays, providing an effective barrier against moisture and oil. Rigid packaging (e.g., bottles or rigid trays) is less common for pure PBS due to its lower stiffness; however, composites, fiber-reinforced PBS or PLA/PBS blends showed potential for items such as yogurt cups or disposable food trays. Additionally, PBS films are suitable for fresh produce bags, bakery product wraps, and snack packaging, where their breathability helps prevent moisture condensation. It should be noted that the real-world use of PBS and PBSA in food packaging is still limited compared to PLA or starch-based plastics. However, as of 2025, compostable produce bags and shopping bags made of PBSA/PBAT blends have been introduced in selected retail outlets.

6.3.5 Current market

The global PBS/PBSA market is still relatively small but is rapidly growing. In 2023, global PBS resin production was estimated at ~120,000 tons, with a market value of around US$480 million, which is less than 0.1% of total plastics production. It is estimated that ~65% of PBS produced in 2023 was bio-based (i.e., derived partly from renewable monomers). Major end-use segments for PBS/PBSA are packaging (~40% of volume). The major end-use segments for PBS and PBSA are packaging (~40% of volume) and agriculture (~25%), followed by fibers/textiles and other applications. For example, roughly 48,000 tons were consumed in packaging applications in 2023, including compostable films and food service products, while another ~45,000 tons were used in agricultural mulch films. One of the primary barriers to scaling PBS production has been its high-cost relative to commodity plastics. PBS resin typically costs about 20%–30% more per kilogram than conventional polyolefins of similar grade. Reported prices for PBS resin range from US$2.50/kg to US$3.50/kg, which is significantly higher than PE or PP with average nearly US$1.00/kg–US$2.00/kg (Ghasemlou et al., 2024). This price difference stems from expensive monomers (particularly 1,4-butanediol and adipic acid), the smaller scale of production, and the limited number of suppliers. However, techno-economic analysis suggests that bio-based PBS could potentially reach a cost of ~US$1.37/kg at full commercial scale if using cheap feedstock such as sugar beet pulp.

6.4 Polybutylene adipate terephthalate

6.4.1 Physical and chemical characteristics

PBAT is a fossil-based synthetic polymer that is 100% biodegradable. It is used in various applications, such as packaging materials (trash bags, food containers, and cling films) and hygiene products (diaper back sheets and cotton swabs). PBAT is produced by polycondensation of three monomers: 1,4-butanediol as the diol, adipic acid (hexanedioic acid) as the aliphatic diacid, and terephthalic acid (1,4-benzenedicarboxylic acid) as the aromatic diacid (Ferreira et al., 2019). These monomers react to form a copolyester comprising butylene adipate (aliphatic) and butylene terephthalate (aromatic) repeating units. PBAT is a material of significant commercial interest due to its unique combination of flexibility (elongation at break: >700%), stiffness (Young's modulus: 75–150 MPa), and inherent biodegradability. The density of PBAT is approximately 1.22 g/cm³, which is higher than that of most polyolefins due to the presence of aromatic rings and ester linkages. PBAT is generally pale yellow or off-white in pellet form and can be readily colored with pigments or masterbatches (Itabana et al., 2024). PBAT is a semicrystalline random copolyester. Its crystallinity is generally limited (<30%) because the random sequence of adipate and terephthalate units disrupts highly regular packing. As a result, PBAT exhibits a broad melting range around 110°C–120°C. The T_g of PBAT is around −30°C, inherited from its flexible aliphatic segments.

PBAT is characterized by high ductility and toughness. It typically exhibits a low Young's modulus (20–35 MPa), moderate tensile strength (32–36 MPa), and a very high elongation at break (> 700%) (De Hoe et al., 2019). This flexibility makes PBAT highly suitable for film applications (e.g., bags, wraps), as it can undergo high draw ratios without tearing. It can be processed using standard methods such as blown-film extrusion, casting, or injection molding, generally requiring processing temperatures between 140°C and 170°C. Regarding barrier performance, pure PBAT provides a good barrier to liquids but only a moderate barrier to gases. The O_2, CO_2, and water vapor permeability coefficients of PBAT have been reported as 1.0 bar, 12.1 bar, and 365 g·mm/m per day per atm, respectively (Qin et al., 2021). Its water vapor barrier property is approximately 1/70 that of low-density PE and its O_2 and CO_2 barrier properties are far inferior to traditional barrier packaging materials, which are about 1/50 and 1/100 of PET, respectively (Li et al., 2018). Several studies have shown that PLA films have significantly lower O_2 and H_2O transmission than PBAT films of similar thickness. This is attributed to the more rigid and polar structure of PLA, which impedes gas diffusion, whereas the flexible chains of PBAT facilitate permeation.

Despite its outstanding toughness, PBAT lacks sufficient strength/rigidity for certain industrial applications (Kanwal et al., 2022). Therefore, PBAT is frequently blended with stronger, more brittle biodegradable polymers, such as PLA or starch, to increase its modulus. Importantly, starch-PBAT blend films can achieve improved gas-barrier properties due to the filler effect and the tortuous path created by starch granules (Gao et al., 2025).

A key attraction of PBAT is its ability to biodegrade under appropriate conditions. The degradation of PBAT under compost conditions was first reported by Witt et al. (1997), who reported that the copolymer degraded in a compost simulation test at 60°C. The biodegradability originates from its hydrolyzable ester bonds, which are readily attacked by microbial enzymes. Under industrial composting conditions (typically ~58°C, high humidity, and rich microbial consortium), PBAT can achieve near-complete biodegradation within a few months (Hasan et al., 2025; Šerá et al., 2020). Indeed, PBAT is certified to standards such as EN 13432 and ASTM D6400, which require >90% conversion to CO_2 within 180 days in compost.

6.4.2 Synthesis and production

Polyesters are generally synthesized via polycondensation from combinations of diols and dicarboxylic acids. Specifically, PBAT is synthesized through the polycondensation of 1,4-butanediol with both adipic (or its ester) and terephthalic acids (or its ester). The preparation of PBAT typically requires long reaction times, high vacuum, and temperatures usually exceeding 190°C (Díaz et al., 2014; Sousa et al., 2015). These rigorous conditions are necessary to favor condensation reactions and to efficiently remove small by-product molecules such as water or methanol (Fig. 6.4).

The synthesis of PBAT typically follows a two-stage polycondensation process. In the first stage, the diacids react with excess 1,4-butanediol to form low-molecular-weight oligomers

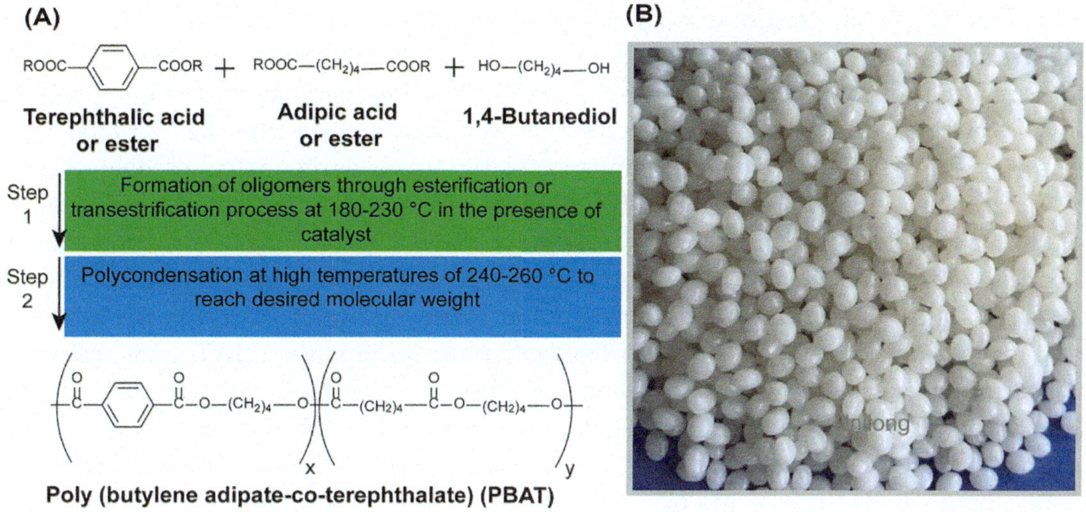

Figure 6.4

(A) Two-step polycondensation route for the production of polybutylene adipate terephthalate, involving esterification or transesterification of terephthalic acid (or ester), adipic acid (or ester), and 1,4-butanediol to form oligomers, followed by high-temperature polycondensation to achieve the target molecular weight. (B) Representative Polybutylene adipate terephthalate resin pellets obtained after polymerization and pelletization.

(prepolymers) via esterification (or transesterification). This is conducted at moderate temperatures (~200°C–220°C), where water or methanol is generated as a by-product and continuously removed by distillation. When the oligomeric intermediates are formed, the process enters the second stage using a high-vacuum, high-temperature polycondensation. Typically, the reaction mixture is raised to ~250°C–260°C under vacuum of <5 mbar to drive the condensation of oligomers into high-molecular-weight PBAT. Upon completion, the viscous molten polymer is discharged and cooled into strands or pellets. The polyesterification of PBAT is catalyzed by typical polycondensation catalysts. Organometallic compounds of titanium (e.g., tetrabutyl titanate), tin (e.g., dibutyltin dilaurate), antimony, or zinc are commonly used to accelerate ester exchange and condensation. Titanium-based catalysts are common in PBAT production due to their high activity (Lee et al., 2024).

The mechanical properties of PBAT are affected by monomer composition and molecular weight. Commercial PBAT typically contains 40–60 mol.% aromatic terephthalate units and 60–40 mol.% adipate units, but this can be tuned for desired properties. Higher aromatic content (more terephthalate) increases stiffness and thermal resistance, whereas higher aliphatic content (more adipate) increases flexibility and biodegradability. To further boost molecular weight or tailor rheology, manufacturers sometimes use chain extenders during or after polymerization (Nifant'ev et al., 2022). Chain extenders are reactive compounds (often di- or multifunctional epoxides, isocyanates, or anhydrides) that can react with terminal hydroxyl or carboxyl groups on the PBAT chains to link two chains together, effectively increasing molecular weight (Jian et al., 2020). For instance, multifunctional epoxide oligomers (like BASF's Joncryl ADR series) are well-known chain extenders for biodegradable polyesters. When added in small amounts (0.1%–0.5%) during extrusion, they react to form branching or bridging between PBAT chains. Additionally, nucleating agents can be used in the final polymerization process to improve PBAT crystallization behavior and avoid stickiness. The most suitable nucleating agents are generally inorganic compounds, such as talc, chalk, mica, or silicon oxides. Currently, the vast majority of PBAT on the market is fossil-based; however, significant effort has been made to produce PBAT from renewable feedstocks. Indeed, the adipic acid fraction of PBAT can be replaced with succinic acid derived from biomass, creating poly(butylene succinate-co-terephthalate) (PBST) (i.e., ~35% total bio-content) (Luo et al., 2010). However, the origin of the two remaining fractions is usually fossil-based. Partially bio-based PBST has better biodegradability than fully fossil-based PBAT. A fully bio-based PBAT containing bio-sourced 1,4-butanediol, bio-adipic acid, and bio-terephthalic acid remains an aspirational goal.

6.4.3 Commercialization and mass production

Since its introduction by BASF in the 1990s, PBAT has grown from a niche biodegradable polymer to a widely produced commodity bioplastic. BASF pioneered PBAT in 1998 as the first biodegradable aliphatic–aromatic copolyester (Siegenthaler et al., 2011). Commercial adoption of PBAT has accelerated intensely over the past decade, driven by global plastic ban policies, increased consumer awareness, and significant investment in large-scale production, most notably in China and Europe.

As of 2020, the global production capacity of PBAT reached 279,000 tons, and by 2021 this figure had expanded to 465,000 tons, with major contributions from BASF, Novamont, Kingfa, Xinjiang Blue Ridge, and other emerging producers (Sedita et al., 2025). Ecoflex is the trade name that BASF uses for its PBAT, and the compounded grade Ecovio is a PBAT/PLA blend. BASF has highlighted that all its PBAT grades meet global compostability standards, including home composting (for certain thicknesses) and even soil biodegradable mulch film standard EN 17033. Novamont incorporates PBAT in its Mater-Bi family of compostable bioplastics (often blended with starch). Novamont does not sell neat PBAT, but it is a major consumer/producer for internal use in products like compostable bags (Li, 2025). Kingfa (China) is another large manufacturer of PBAT and PBAT compounds with PLA or starch. According to its 2021 annual report, Kingfa has an annual PBAT production capacity of 180,000 tons of PBAT and it is significantly expanding its capacity to meet growing domestic and international demand (Jian et al., 2020).

6.4.4 Food packaging applications

PBAT has established itself in applications that require flexibility and compostability. PBAT and its composites are currently used in different areas, such as biomedical fields (Wei et al., 2016), food packaging (Dhali et al., 2023; Moustafa et al., 2017), and agricultural applications (da Silva et al., 2022). One of the earliest large-scale uses of PBAT was in compostable shopping bags and organic waste collection bags (Roy et al., 2024). These thin films (typically 10–30 μm thickness) leverage the high elongation and tear resistance of PBAT, allowing bags to carry weight without rupture. A common formulation involves blending PBAT with 30%–50% TPS plus a portion of PLA. The starch accelerates biodegradation and reduces cost. However, high starch content can reduce film strength and increase moisture sensitivity.

PBAT can be also used to produce stretchable cling films. PBAT film is transparent and can be made very thin (<10 μm) and stretchable, making it suitable for produce wrap (Pietrosanto et al., 2020). It is safe for food contact (compliant with EU and US Food and Drug Administration (FDA) regulations when properly formulated) and can directly contact food items. The elastic nature of PBAT allows a thin film to cling around food containers and fruits similarly to poly(vinyl chloride) (PVC) or PE cling wrap. Furthermore, PBAT can function as a seal layer or a heat-sealable inner layer in a multilayer structure. for example, BASF and Fabbri Group developed a cling film combining PBAT with a thin polyvinyl alcohol (PVOH) coating to improve barrier properties. The heat sealability of PBAT is excellent and it can seal at lower temperatures than PLA with a broader hot-tack range (Fig. 6.5).

Pure PBAT is generally too flexible to be used alone in rigid packaging such as cups and trays. However, it can be combined with PLA to manufacture goods such as compostable cutlery that is less brittle or used in foam trays in combination with starch (Aversa et al., 2022). The main role of PBAT in food packaging is to enable flexible compostable films and bags. While it cannot fully replace high-barrier plastics for long-shelf-life packaged foods, it is perfectly suited for short-life, postconsumer organic waste packaging such as produce bags, food scrap bin liners, yard waste bags, and food service items like cutlery pouches or compostable cling film for immediate consumption items.

6.4.5 Current market

The global market for PBAT has transitioned from a niche segment of the polymer industry into a rapidly expanding commodity market, driven by an urgent convergence of legislative mandates and shifting consumer preferences (Xu et al., 2023). PBAT has emerged as the dominant biodegradable copolyester, second only to TPS in production volume and frequently outpacing PLA in flexible packaging applications. As of 2023, the global PBAT market was valued at approximately US$1.4 billion, with production volumes exceeding 500,000 metric tons. Industry projections indicate an exponential growth trajectory, with a compound annual

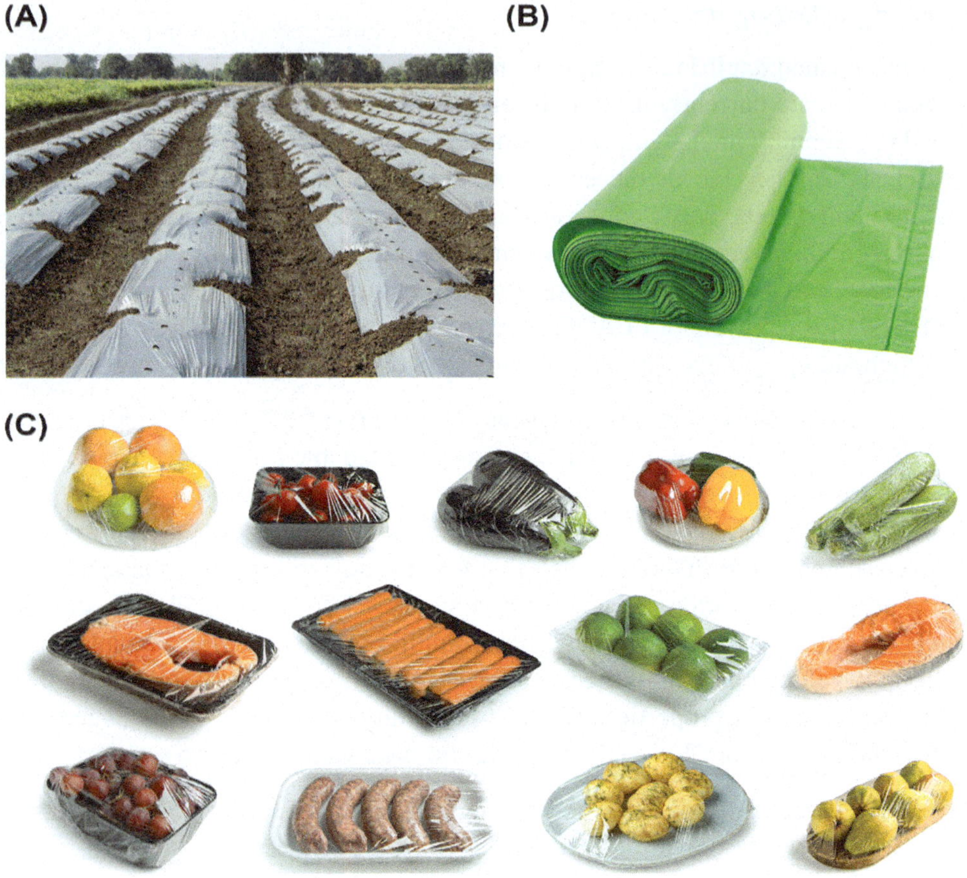

Figure 6.5

Representative applications of polybutylene adipate terephthalate. (A) Agricultural mulch films used for soil coverage and crop protection. Polybutylene adipate terephthalate films biodegrade after use. (B) Flexible compostable bags and films produced from polybutylene adipate terephthalate for waste collection and packaging applications. (C) Stretchable cling films and wraps for food packaging applications such as fresh produce, meat, and ready-to-eat foods.
The images are obtained under license from Shutterstock.com.

growth rate of 18.7% anticipated through 2030. If these trends hold, the market valuation is expected to surpass US$4.5 billion by the end of the decade.

Geographically, the center of gravity for PBAT production has shifted dramatically over the past decade. While the technology was pioneered in Europe by firms such as BASF and Novamont, the bulk of manufacturing capacity has migrated to the Asia-Pacific region, which now accounts for over 60% of global output. China has become the undisputed locus of this expansion. The Chinese government's aggressive regulatory framework effectively banned nondegradable bags in major cities, catalyzing a massive domestic investment in biodegradable capacity. Consequently, Chinese manufacturers such as Kingfa, Xinjiang Blue Ridge Tunhe, and Changchun Group have scaled operations rapidly, creating a supply glut that has begun to reshape global pricing dynamics. Conversely, the North American market remains fragmented, driven less by federal mandates and more by state-level legislation and corporate sustainability commitments.

The industrial adoption of PBAT is currently sustained by three primary drivers. First, regulatory pressure remains the single most effective catalyst. The European Union's Single-Use Plastics Directive and similar bans worldwide have mandated biodegradable alternatives for short-lived items such as shopping bags and cutlery. Second, the agricultural sector provides a critical volume baseline. Conventional PE mulch films often degrade into microplastics that permanently contaminate arable land. PBAT-based mulch films offer a functional solution to this "white pollution" crisis, as they can be thrown directly into the soil postharvest. Third, the surge in e-commerce has necessitated sustainable flexible packaging solutions, leading logistics giants to pilot PBAT-based courier bags as a substitute for low-density PE.

Despite this momentum, the market faces significant techno-economic headwinds. Cost is the primary barrier to ubiquitous adoption. PBAT resin prices typically range between US$2500 and US$3500 per metric ton, which is approximately 2–3 times the cost of conventional polyolefins. This "green premium" is exacerbated by the volatility of raw material costs, specifically 1,4-butanediol and adipic acid, which are currently derived from fossil fuel feedstocks. As a result, PBAT prices remain tethered to crude oil markets. To achieve long-term viability and decouple from fossil economics, the industry must transition toward bio-based monomers. While bio-based 1,4-butanediol derived from fermentation is technologically feasible, it has yet to achieve the economies of scale necessary to compete with petrochemical routes. Ultimately, the future success of the PBAT market will depend on narrowing this price gap and successfully integrating these materials into global composting infrastructures to ensure that the environmental promise of biodegradability is realized at the end of the product's life.

6.5 Poly(ε-caprolactone)

6.5.1 Physical and chemical characteristics

PCL is a semicrystalline, hydrophobic aliphatic polyester that occupies a significant niche in the bioplastics market. It has a very low T_g (nearly –60°C) and a moderate T_m of around 58°C–64°C. Therefore, PCL behaves as a tough rubbery elastomer at room temperature due to its low T_g (Sivalingam and Madras, 2003). The repeating unit consists of a hexanoate ester moiety, which confers substantial rotational freedom to the polymer backbone, resulting in high flexibility. Thermal degradation occurs at temperatures of >250°C, primarily driven by ester bond pyrolysis. However, these thermal parameters are strongly modulated by the polymer's molecular weight and crystalline fraction.

Mechanically, PCL is distinguished by its ductility rather than its strength. It exhibits lower stiffness and tensile strength (~10 MPa) compared to other bioplastics but has exceptional stretchability, with typical elongation at break reaching ~300%. While the density of PCL is modest (~1.10–1.15 g/cm³), its semicrystalline structure affords moderate resistance to oils and greases. However, PCL offers limited barrier performance against gases and moisture. It is relatively permeable, with reported water vapor transmission rates of ~177 g/m per day (at 25°C, 50% RH) and oxygen permeability of ~3.3×10^2 cm^3/m^{-2}·per day per bar (at 20°C). An interesting feature of PCL is its unusual miscibility with a broad spectrum of polymers, including PVC, styrene acrylonitrile (SAN), polycarbonate, nitrocellulose, and cellulose butyrate. This broad compatibility arises from favorable enthalpic interactions and the ability of PCL chains to interpenetrate amorphous matrices. This compatibility arises from favorable interactions and the ability of flexible PCL chains to interpenetrate amorphous matrices, allowing for the engineering of blends with tailored mechanical properties and degradation profiles. PCL is soluble in various chlorinated and aromatic solvents, such as dichloromethane, chloroform, and toluene, yet remains insoluble in water and short-chain alcohols. Due to its low melting temperature and rheological properties, PCL is an excellent candidate for extrusion, injection molding, and, increasingly, three-dimensional (3D) printing (Coudane et al., 2022).

PCL is biodegradable via the hydrolysis of its ester linkages. It biodegrades within several months to several years, depending on the molecular weight, the degree of crystallinity, and the environmental conditions. The polymer degrades by end-chain scission at higher temperatures, while it degrades by random chain scission at lower temperatures. For example, EPA testing showed PCL films lost 100% of their mass in <3 years in soil or compost.

Extracellular lipases and esterases from bacteria and fungi can cleave PCL chains into oligomers and monomers (e.g., 6-hydroxyhexanoic acid), which microbes further metabolize. While PCL can be enzymatically degraded in the environment, it cannot be degraded enzymatically in the human body. Although PCL is fossil-based, its degradation is slightly better than PLA under identical industrial composting conditions (Ghasemlou et al., 2024). However, at ambient temperatures or in cold marine environments, PCL degrades very slowly. For example, a pure PCL film achieved ~55% biochemical oxygen demand (BOD) in a marine test, whereas a modified composite with biotic additives reached ~96% BOD under the same test. In soils or compost at 20°C–60°C, complete biodegradation typically takes months or even years. With high biocompatibility, rubber-like characteristics, extendable biodegradability, and convenience to form and blend with other polymers, PCL is a good choice for biomedical use (Krumov et al., 2025). Overall, PCL offers good environmental performance if composted or biodegraded properly, but in open environments, it requires time and specific conditions to fully mineralize.

6.5.2 Synthesis and production

PCL was synthesized for the first time by Carothers and their co-workers in the early 1930s (Pawar et al., 2023). There are two main pathways to produce PCL: polycondensation of 6-hydroxyhexanoic acid and the ring-opening polymerization of ε-caprolactone.

Ring-opening polymerization of ε-caprolactone remains the most versatile and widely adopted route to produce high-molecular-weight PCL, largely due to excellent control. In this process, the monomer first coordinates to a metal center, after which nucleophilic attack by a metal–alkoxide initiator produces a growing chain end capable of successive insertions (Arbaoui and Redshaw, 2010). Lewis acids such as tin or aluminum halides can activate the monomer toward nucleophilic attack. This pathway minimizes unwanted side reactions and yields polymers with narrow molecular-weight distributions. The main drawbacks of this approach include the use of a metal-based catalytic system, which is usually toxic and not easily removed during purification. There are four general mechanisms for ring-opening polymerization of PCL: anionic, cationic, coordination insertion, and activated monomer (Labet and Thielemans, 2009).

The polycondensation of 6-hydroxyhexanoic acid is rarely used because it requires high temperature and longer times to reach only moderate molecular weights, often yielding PCL with a broader molecular-weight distribution. Recent research has found a few sustainable routes to produce PCL (Das et al., 2020). For example, glucose/fructose can be converted to cyclohexanone and then oxidized (via a Baeyer–Villiger monooxygenase) to ε-caprolactone, which is then polymerized into PCL. Enzymatic polymerization (e.g., with *Candida antarctica* lipase) can ring-open ε-caprolactone under mild conditions, but such methods remain at the laboratory scale due to high cost and low productivity (Duchiron et al., 2017).

6.5.3 Commercialization and mass production

PCL is a specialty polymer with a significantly smaller global production volume. However, demand is growing slowly, driven by the unique low-temperature rheology and biocompatibility of PCL. The global supply chain is consolidated among a few key chemical manufacturers who have the specialized capability to handle the ring-opening polymerization of ε-caprolactone safely and efficiently. Perstorp is the world's largest producer of ε-caprolactone and its derivatives. Under the trade name Capa, they offer a comprehensive portfolio of PCL grades ranging from low-molecular-weight polyols (used in coatings and elastomers) to high-molecular-weight thermoplastics (used in bioplastics and orthopedic applications). Perstorp has driven much of the innovation in PCL applications, particularly in home-compostable packaging blends. Daicel Corporation (Japan) is another major Asian producer. Daicel manufactures PCL under the brand names Placcel and Celgreen. They are particularly strong in the high-performance sector, supplying PCL for modification of other resins and for use in biodegradable films. While BASF is famous for their PBAT, they also produce PCL, primarily for use as a functional additive in their biodegradable compounds to improve processing or mechanical properties. Corbion also produces medical-grade PCL (under the Purrasorb line) specifically for high-precision biomedical applications like drug delivery and resorbable sutures. Other notable players involved in the distribution or specialty manufacturing of PCL include Gantrade Corporation (USA), Polysciences Inc. (USA), and emerging Chinese producers such as Haihang Industry and Shenzhen Esun Industrial, who are expanding capacity to meet the growing demand for biodegradable filaments in the 3D printing sector. The primary barrier to mass production of PCL remains the cost of the monomer, ε-caprolactone, which is derived from cyclohexanone (a petrochemical intermediate). The process requires peracetic acid oxidation, which is chemically hazardous and expensive to scale compared to the simple fermentation used for PLA. Consequently, PCL remains priced as a specialty material (often >US$5/kg–US$8/kg for industrial grades and significantly higher for medical grades), limiting its use in general food packaging to that of a high-performance additive rather than a bulk matrix.

6.5.4 Food packaging applications

PCL is approved by FDA for use in specific medical devices (e.g., sutures and drug delivery scaffolds) and is recognized as nontoxic and biocompatible. While its primary commercial success has been in the biomedical sector owing to its slow degradation rate and excellent physiological tolerance, these same safety characteristics make it a viable candidate for food-contact materials (Taghizadeh et al., 2024; Thakur et al., 2021). PCL is chemically stable, recyclable, and fully biodegradable under industrial composting conditions, aligning well with the requirements for sustainable packaging systems. Although rarely used as a standalone mainstream packaging resin due to high cost, PCL offers unique functional advantages when used in blends or composite structures. First, it acts as a superior toughening agent for brittle biopolymers. For example, when blended with PLA or TPS, PCL acts as a plasticizer and impact modifier. Even a moderate addition (e.g., 10%–20%) can significantly improve the elongation at break and tear resistance of the final film, making it suitable for flexible packaging applications like bags and pouches. Second, PCL serves as an excellent active packaging carrier. It exhibits high compatibility with various lipophilic bioactive compounds and is increasingly utilized as a carrier matrix for antimicrobial or antioxidant agents (e.g., essential oils and silver nanoparticles) (Wu et al., 2014). Its semipermeable nature allows for the controlled release of these agents, helping to extend the shelf life of perishable products such as meat and cheese. Third, its low melting point and favorable rheology make PCL easy to process using standard equipment, often serving as a processing aid to improve the melt stability of starch-based materials.

Despite its versatility, the widespread adoption of PCL in the food packaging sector is limited by three significant hurdles. The first is low thermal stability. With a melting point of ~60°C, PCL cannot be used for hot-fill applications, microwaveable containers, or products exposed to high ambient temperatures. It is strictly limited to cold-chain or room-temperature applications (e.g., refrigerated produce and frozen foods). The second is low barrier performance. PCL has relatively poor barrier properties against oxygen and water vapor compared to high-barrier bioplastics like polyhydroxybutyrate or conventional plastics like PET. Consequently, it offers limited protection for moisture-sensitive or oxidation-prone foods unless it is incorporated into a multilayer structure. The final hurdle is cost. The production cost of PCL remains significantly higher than that of commodity bioplastics like PLA, restricting its commercial use primarily to high-value specialty films or as a minor additive in blends rather than as a bulk structural material.

6.5.5 Current market

The global market for PCL is undergoing a structural transition from a niche specialty chemical to a strategic material in the advanced polymer economy. Unlike high-volume bioplastics such as PLA or TPS, which compete primarily on cost and compostability in single-use applications, PCL needs a market position defined by high-performance utility. Market analysis by Acumen Research and Consulting projects a vigorous expansion trajectory: the global PCL valuation is forecast to surge from a baseline of approximately US$600 million in 2024 to US$1.33 billion by 2032, corresponding to a robust compound annual growth rate of 10.7% (Jha et al., 2024). This double-digit growth rate underscores a shift in industrial consumption patterns, driven not by the substitution of commodity plastics, but by the material's indispensable role in high-value-added sectors where its unique rheological and biological properties justify a premium price point. Three distinct industrial pillars underpin this market expansion. First, the synthesis of high-performance thermoplastic polyurethanes remains the dominant volume driver. PCL polyols are utilized as "soft segments" in the polymer backbone to impart superior hydrolytic stability, low-temperature flexibility, and oil resistance. Second, the biomedical sector, while lower in total tonnage, represents the highest margin segment. PCL's established regulatory profile and predictable mechanism of hydrolytic bulk erosion have cemented its status as the scaffold material of choice for tissue engineering and long-acting drug delivery systems. Third, the market is increasingly buoyed by the nascent sector of advanced biodegradable composites. Driven by stringent environmental mandates in the EU and North America, converters are turning to PCL as a high-performance toughening agent for brittle biopolymers such as PLA, thereby enabling sustainable flexible packaging solutions that were previously technologically unfeasible.

6.6 Polyvinyl alcohol

6.6.1 Physical and chemical characteristics

PVOH (or PVA) is a synthetic, water-soluble polymer with the repeating unit [–CH_2–CHOH–]$_n$. PVOH is semicrystalline. It is biocompatible, biodegradable, nontoxic, and nonhazardous, making it safe to handle and environmentally friendly. Depending on the degree of hydrolysis of polyvinyl acetate, it consists of either 1,3-diol units or 1,2-diol units. The crystallinity of PVOH depends on the degree of hydrolysis and polymerization. Fully hydrolyzed grades have a crystallinity degree of ~50% after annealing, while partial hydrolysis yields less crystalline and more amorphous material. Fully hydrolyzed PVOH has a relatively high T_g of ~85°C and T_m of 200°C–220°C (Belay and Islam, 2023).

The water solubility of PVOH varies for different grades. Low-hydrolysis grades (≈85%–90%) dissolve in cold water, whereas high-hydrolysis grades (98%–99%) may require hot water or swelling before dissolving. PVOH is known for its excellent film-forming ability and can form strong intermolecular hydrogen bonds due to the presence of –OH groups in its backbone. Therefore, the mechanical properties of PVOH films are generally higher than those of other common biodegradable polymers. PVOH films have very low permeability to oxygen and carbon dioxide. However, PVOH is extremely hydrophilic and absorbs water and swells readily. Thus, its water vapor permeability is very high, and it offers essentially no moisture barrier (Rahman and Goswami, 2023; Saini et al., 2025).

Chemically, PVOH is stable and nontoxic. It is insoluble in most organic solvents but readily soluble in water (Lopes and Simoneau, 2014). However, dissolution in water does not necessarily equate to complete biodegradation. PVOH requires specific conditions for degradation, including the presence of specific microorganisms and enzymes. Indeed, PVOH is fully biodegradable within weeks under suitable conditions. A study examined the biodegradation and ecotoxicity of PVA in marine environments and found that PVOH has limited biodegradability under natural marine conditions. Another study confirmed that PVOH poses no significant aquatic toxicity. PVOH has a good affinity with other plastic materials. As such, it can be thermoformed with other thermoformable polymers, allowing segregation in a recycling process. Blending PVOH with biopolymers, such as starch, can significantly enhance its biodegradability and overall performance. Studies have shown that blends containing starch and PVOH had substantial potential for manufacturing biodegradable food packaging films with excellent properties. The material can replace other barrier polymers to enhance the overall biodegradability of composite packaging. PVOH can also be combined with paper and corrugated packages to provide an internal vapor barrier or to dissipate static electricity. Water-soluble labels made from PVOH can be attached as labels to other plastic bottles and simply washed off at the end of the bottle's life.

6.6.2 Synthesis and production

PVOH was discovered in 1924 by Dr. Hermann Staudinger (a Nobel Laureate) and was commercialized three decades later by the Japanese chemical manufacturer, Kuraray. Unlike most vinyl polymers, PVOH is not made by the direct polymerization of vinyl alcohol (which tautomerizes to acetaldehyde). Instead, PVOH is produced by the hydrolysis (saponification) of polyvinyl acetate. Industrially, vinyl acetate monomer is first polymerized by free-radical or suspension polymerization to yield polyvinyl acetate. Catalysts for this polymerization include Ziegler-type ($TiCl_4$) compounds or peroxides. The polyvinyl acetate is then dissolved in alcohol (e.g., methanol or ethanol) and treated with a strong base (NaOH or KOH) to hydrolyze the acetate groups into alcohol groups. In base-catalyzed hydrolysis, hydroxide ions attack the carbonyl of the acetate ester, forming a tetrahedral intermediate that collapses to release acetate and a polymer-bound alkoxide, which then protonates to give the alcohol. This produces PVOH and a sodium acetate by-product (Aslam et al., 2018).

By controlling temperature, base concentration, and reaction time, manufacturers can achieve various degrees of hydrolysis. Generally, there are two degrees of saponification or hydrolysis: "fully hydrolyzed" (≈98%–99% OH with only ~0.5%–1% residual acetate) and "partially hydrolyzed" (typically ~88% OH for food-grade PVA). In particular, fully hydrolyzed PVOH exhibits lower cold-water solubility when compared to partially hydrolyzed PVOH. Finally, the solvent and water content are removed (via stripping or filtration) to isolate PVOH resin (Aslam et al., 2018).

6.6.3 Commercialization and mass production

The PVOH market is competitive, involving many large chemical companies. Kuraray Co. Ltd. (Japan) is the world's largest PVOH producer. Other major global manufacturers are Sekisui Chemical Co. (Japan), DuPont (Chemours) (USA), Nippon Synthetic Chemical Industry Co. (Japan), Eastman Chemical Company (USA), Sinopec Sichuan Vinylon Works (China), and Anhui Wanwei Group (China). These companies control significant PVOH capacity and have global sales.

6.6.4 Food packaging applications

PVOH film was first used to package pesticides in unit-dose, water-soluble pouches to protect farmers from accidental chemical exposure. PVOH is approved as a coating agent (food additive) for food supplements. In the USA, the FDA allows PVOH as an edible film component for single-serve food pouches. The largest end-use of PVOH is food packaging. Within food packaging, PVOH is used as a packaging material itself (e.g., water-soluble pouches), as well as coatings for paper or fiber-based packaging to impart barrier. Other major uses include paper coatings (to improve grease/wet strength), textile sizing, and personal care/ adhesives (Rahman and Goswami, 2023).

The unique characteristics of PVOH enable novel packaging applications. The most prominent application in food and nonfood packaging is water-soluble films and pouches. For example, laundry detergent and dishwasher pods are packed using PVOH films to encapsulate concentrated detergents. The film dissolves entirely in the wash, delivering the dose and leaving no plastic residue. The single-use laundry and dishwasher detergent packets and tablets offer consumers convenience, safety and sustainability. Beyond the popular laundry and detergent packets, it is used in many household, medical, personal-care and industrial applications, including food packaging, textile yarns, paper products, and water treatment chemicals.

In addition, PVOH is used as a coating to improve barrier in compostable or paper-based packaging. For example, PVOH can be coated onto paperboard cartons or molded pulp as a moisture barrier. It also serves as a gas-barrier layer in multilayer films. Compared to other biodegradable films (PLA or cellulose), PVOH provides much better gas barrier. PLA or cellulose alone have moderate gas permeability, whereas a PVOH layer can reduce oxygen transmission by orders of magnitude, extending shelf life. This makes it valuable for gas-sensitive foods (e.g., chips, cereals) if humidity is controlled.

Depending on the grade and environmental humidity conditions, PVOH contains 3%–10% water. Therefore, it can be used to hold products with water content in the same range without dissolving or creating holes in the film packaging. Shelf life and storage conditions are taken into account as part of the film selection and compatibility testing process. For example, pool and spa water treatment chemicals are tested to simulate storage in an outdoor garage for up to 2 years.

PVOH has excellent biocompatibility, making it suitable for wound care and medical applications. For example, PVOH can be used to develop hydrogel wound dressings that provide a moist environment for wound healing (Jin, 2022). PVA hydrogels are biostable and have been studied for cartilage replacement, synthetic arteries, and wound dressings. PVOH can be also used in drug delivery systems such as controlled-release tablets, due to its water solubility and biocompatibility. For example, PVOH-based microspheres are used as embolic agents (blocking blood flow in tumors) and as carriers for injectable drugs.

6.6.5 Current market

PVOH is a high-volume polymer and is used in diverse markets such as textiles, paper, and adhesives. The global PVOH market has expanded steadily, driven by the demand for sustainable and water-soluble materials. The global PVOH market was estimated around US$1.0–1.1 billion in 2022 and is forecast to grow at ~6%– 7% annually, reaching ~US$1.6–1.7 billion by 2030.

China alone represents >50% of the global PVOH production capacity. Key regional markets include Japan, South Korea, India, Europe, and North America. Food packaging is the largest application of PVOH (~30% of market value). Textile sizing is another major segment (~30%–40% of volume). Other significant segments include paper and coatings, construction, and detergents/consumer goods.

6.7 Conclusion

Plastic waste is a major global environmental challenge. While conventional plastics offer superior performance, their recalcitrance causes them to persist in the environment long after disposal. This chapter shows that fossil-based biodegradable bioplastics, such as PBS and PBAT, provide a realistic and immediate pathway to mitigate this issue. These polymers contain hydrolysable ester linkages that enable biodegradation under composting conditions. Their compatibility with existing manufacturing infrastructure allows large-scale production without major changes to current industrial systems. These materials provide several practical advantages. They are compatible with current processing technologies, exhibit mechanical properties suitable for flexible packaging, and have already achieved industrial compostability certification. As a result, fossil-based biodegradable plastics are currently more scalable and economically accessible than many fully bio-based alternatives. However, their deployment should be viewed as a transitional strategy rather than a final solution. Looking ahead, future progress will require coordinated advances in material design, waste management infrastructure, and regulatory frameworks. Improving biodegradation under broader environmental conditions, increasing the bio-based content of existing polymers, and developing clearer end-of-life pathways will be critical. Rather than relying on a single material solution, sustainable food packaging will depend on integrated systems that combine material innovation with realistic implementation. In this context, fossil-based biodegradable bioplastics can play a meaningful role in reducing plastic pollution while the transition toward fully circular packaging systems continues to mature.

References

Aliotta, L., Seggiani, M., Lazzeri, A., Gigante, V., Cinelli, P., 2022. A brief review of poly (butylene succinate) (PBS) and its main copolymers: synthesis, blends, composites, biodegradability, and applications. Polymers 14 (4), 844. Available at https://doi.org/10.3390/polym14040844.

Andreasi Bassi, S., Tonini, D., Saveyn, H., Astrup, T.F., 2021. Environmental and socioeconomic impacts of poly (ethylene terephthalate)(PET) packaging management strategies in the EU. Environmental Science & Technology 56 (1), 501–511.

Arbaoui, A., Redshaw, C., 2010. Metal catalysts for ε-caprolactone polymerisation. Polymer Chemistry 1 (6), 801–826. Available at https://doi.org/10.1039/b9py00334g.

Aslam, M., Kalyar, M.A., Raza, Z.A., 2018. Polyvinyl alcohol: a review of research status and use of polyvinyl alcohol based nanocomposites. Polymer Engineering and Science 58 (12), 2119–2132. Available at https://doi.org/10.1002/pen.24855.

Aversa, C., Barletta, M., Cappiello, G., Gisario, A., 2022. Compatibilization strategies and analysis of morphological features of poly(butylene adipate-co-terephthalate) (PBAT)/poly(lactic acid) PLA blends: a state-of-art review. European Polymer Journal 173, 111304. Available at https://doi.org/10.1016/j.eurpolymj.2022.111304.

Barletta, M., Aversa, C., Ayyoob, M., Gisario, A., Hamad, K., Mehrpouya, M., et al., 2022. Poly (butylene succinate)(PBS): materials, processing, and industrial applications. Progress in Polymer Science 132, 101579.

Barone, A.S., Maragoni-Santos, C., Farias, P.Md, Cortat, C.M.G., Maniglia, B.C., Ongaratto, R.S., et al., 2025. Rethinking single-use plastics: Innovations, polices, consumer awareness and market shaping biodegradable solutions in the packaging industry. Trends in Food Science and Technology 158, 104906. Available at https://doi.org/10.1016/j.tifs.2025.104906.

Barrino, F., De La Rosa-Ramírez, H., Schiraldi, C., López-Martínez, J., Samper, M.D., 2023. Preparation and characterization of new bioplastics based on polybutylene succinate (PBS). Polymers 15 (5), 1212. Available at https://doi.org/10.3390/polym15051212.

Belay, M., Islam, M.M., 2023. Review on physicochemical modification of biodegradable plastic: focus on agar and polyvinyl alcohol (PVA). Advances in Materials Science and Engineering 2023, 1–11. Available at https://doi.org/10.1155/2023/4056020.

Bhagwat, G., Gray, K., Wilson, S.P., Muniyasamy, S., Vincent, S.G.T., Bush, R., et al., 2020. Benchmarking bioplastics: a natural step towards a sustainable future. Journal of Polymers and the Environment 28 (12), 3055–3075. Available at https://doi.org/10.1007/s10924-020-01830-8.

Borrelle, S.B., Ringma, J., Law, K.L., Monnahan, C.C., Lebreton, L., McGivern, A., et al., 2020. Predicted growth in plastic waste exceeds efforts to mitigate plastic pollution. Science 369 (6509), 1515–1518. Available at https://doi.org/10.1126/SCIENCE.ABA3656.

Bumbudsanpharoke, N., Wongphan, P., Promhuad, K., Leelaphiwat, P., Harnkarnsujarit, N., 2022. Morphology and permeability of bio-based poly(butylene adipate-co-terephthalate) (PBAT), poly(butylene succinate) (PBS) and linear low-density polyethylene (LLDPE) blend films control shelf-life of packaged bread. Food Control 132, 108541. Available at https://doi.org/10.1016/j.foodcont.2021.108541.

Cao, Y., Otsuka, A., Teah, H.Y., Kanematsu, Y., Kikuchi, Y., 2025. Consumers' knowledge, perceptions, and preferences regarding bioplastics: evidence from a survey in Japan. Sustainable Production and Consumption 55, 1–10. Available at https://doi.org/10.1016/j.spc.2025.02.003.

Chaudry, S., Cheng, K.Y., Speight, R., Kaksonen, A.H., 2025. Techno-economic framework for sustainable use of microalgae for industrial production – a bioplastics case study. Journal of Cleaner Production 536, 147158. Available at https://doi.org/10.1016/j.jclepro.2025.147158.

Chen, G.Q., Patel, M.K., 2012. Plastics derived from biological sources: present and future: a technical and environmental review. Chemical Reviews 112 (4), 2082–2099. Available at https://doi.org/10.1021/cr200162d.

Clark, R.A., Shaver, M.P., 2024. Depolymerization within a circular plastics system. Chemical Reviews 124 (5), 2617–2650. Available at https://doi.org/10.1021/acs.chemrev.3c00739.

Coppola, G., Gaudio, M.T., Lopresto, C.G., Calabro, V., Curcio, S., Chakraborty, S., 2021. Bioplastic from renewable biomass: a facile solution for a greener environment. Earth Systems and Environment 5 (2), 231–251. Available at https://doi.org/10.1007/s41748-021-00208-7.

Coudane, J., Nottelet, B., Mouton, J., Garric, X., Van Den Berghe, H., 2022. Poly (ε-caprolactone)-based graft copolymers: synthesis methods and applications in the biomedical field: a review. Molecules 27 (21), 7339.

Cywar, R.M., Beckham, G.T., 2022. Producing performance-advantaged bioplastics. Nature Chemistry 14 (9), 967–969. Available at https://doi.org/10.1038/s41557-022-01030-y.

da Silva, T.C.P., Fortes, A.Gd.S., de Abreu, I.R., de Carvalho, L.H., de Almeida, Y.M.B., Alves, T.S., et al., 2022. Development of biodegradable PLA/PBAT-based filaments for fertilizer release for agricultural applications. Materials 15 (19), 6764. Available at https://doi.org/10.3390/ma15196764.

Das, M., Mandal, B., Katiyar, V., 2020. Sustainable routes for synthesis of poly (ε-caprolactone): prospects in chemical industries. In: Katiyar, V., Kumar, A., Mulchandani, N. (Eds.), Advances in Sustainable Polymers:

Synthesis, Fabrication and Characterization. Materials Horizons: From Nature to Nanomaterials. Springer, Singapore, pp. 21–33. Available at https://doi.org/10.1007/978-981-15-1251-3_2.

de Castro, T.R., de Macedo, D.C., de Genaro Chiroli, D.M., da Silva, R.C., Tebcherani, S.M., 2022. The potential of cleaner fermentation processes for bioplastic production: a narrative review of polyhydroxyalkanoates (PHA) and polylactic acid (PLA). Journal of Polymers and the Environment 30 (3), 810–832. Available at https://doi.org/10.1007/s10924-021-02241-z.

De Hoe, G.X., Zumstein, M.T., Getzinger, G.J., Rüegsegger, I., Kohler, H.P.E., Maurer-Jones, M.A., et al., 2019. Photochemical transformation of poly(butylene adipate- co-terephthalate) and its effects on enzymatic hydrolyzability. Environmental Science and Technology 53 (5), 2472–2481. Available at https://doi.org/10.1021/acs.est.8b06458.

Degli Innocenti, F., Breton, T., 2020. Intrinsic biodegradability of plastics and ecological risk in the case of leakage. ACS Sustainable Chemistry & Engineering 8 (25), 9239–9249. Available at https://doi.org/10.1021/acssuschemeng.0c01230.

Dhali, K., Daver, F., Cass, P., Field, M.R., Adhikari, B., 2023. Development and characterisation of poly(butylene adipate-co-terephthalate):silane modified cellulose nanocrystals composite materials and films. Journal of Polymers and the Environment 31 (10), 4506–4521. Available at https://doi.org/10.1007/s10924-023-02896-w.

Díaz, A., Katsarava, R., Puiggalí, J., 2014. Synthesis, properties and applications of biodegradable polymers derived from diols and dicarboxylic acids: from polyesters to poly(ester amide)s. International Journal of Molecular Sciences 15 (5), 7064–7123. Available at https://doi.org/10.3390/ijms15057064.

Duchiron, S.W., Pollet, E., Givry, S., Avérous, L., 2017. Enzymatic synthesis of poly(ε-caprolactone-co-ε-thiocaprolactone). European Polymer Journal 87, 147–158. Available at https://doi.org/10.1016/j.eurpolymj.2016.12.024.

Ferreira, F.V., Cividanes, L.S., Gouveia, R.F., Lona, L.M.F., 2019. An overview on properties and applications of poly(butylene adipate-co-terephthalate)–PBAT based composites. Polymer Engineering and Science 59 (2), E7. Available at https://doi.org/10.1002/pen.24770.

Gao, S., McClements, D.J., Wang, B., Rashid, A., Geng, X., Sun, S., et al., 2025. Regulation of mechanical and barrier properties of starch/PBAT films from extrusion perspective: structural evolution, correlation insights, and predictive model development. ACS Sustainable Chemistry and Engineering 13 (43), 18776–18794. Available at https://doi.org/10.1021/acssuschemeng.5c07433.

Genovese, L., Lotti, N., Gazzano, M., Siracusa, V., Dalla Rosa, M., Munari, A., 2016. Novel biodegradable aliphatic copolyesters based on poly(butylene succinate) containing thioether-linkages for sustainable food packaging applications. Polymer Degradation and Stability 132, 191–201. Available at https://doi.org/10.1016/j.polymdegradstab.2016.02.022.

Ghasemlou, M., Barrow, C.J., Adhikari, B., 2024. The future of bioplastics in food packaging: an industrial perspective. Food Packaging and Shelf Life 43, 101279. Available at https://doi.org/10.1016/j.fpsl.2024.101279.

Ghasemlou, M., Adhikari, B., Naebe, M., Barrow, C.J., 2025. Bioengineering strategies to promote plastic circularity. Nature Reviews Bioengineering 3 (5), 353–354. Available at https://doi.org/10.1038/s44222-025-00287-5.

Ghasemlou, M., Daver, F., Murdoch, B.J., Ball, A.S., Ivanova, E.P., Adhikari, B., 2022. Biodegradation of novel bioplastics made of starch, polyhydroxyurethanes and cellulose nanocrystals in soil environment. Science of the Total Environment 815, 152684. Available at https://doi.org/10.1016/j.scitotenv.2021.152684.

Ghosh, K., Jones, B.H., 2021. Roadmap to biodegradable plastics-current state and research needs. ACS Sustainable Chemistry and Engineering 9 (18), 6170–6187. Available at https://doi.org/10.1021/acssuschemeng.1c00801.

Guckert, F.E., Sayer, C., de Oliveira, D., de Araújo, P.H.H., Oechsler, B.F., 2022. Synthesis of polybutylene succinate via lipase-catalyzed transesterification: enzyme stability, reuse and PBS properties in bulk polycondensations. European Polymer Journal 179, 111573.

Guicherd, M., Ben Khaled, M., Guéroult, M., Nomme, J., Dalibey, M., Grimaud, F., et al., 2024. An engineered enzyme embedded into PLA to make self-biodegradable plastic. Nature 631 (8022), 884–890. Available at https://doi.org/10.1038/s41586-024-07709-1.

Hahladakis, J.N., Iacovidou, E., Gerassimidou, S., 2024. Plastic waste in a circular economy. Environmental Materials and Waste: Circular Economy and Pollution Abatement. Elsevier, Qatar, pp. 99–134. Available at https://doi.org/10.1016/B978-0-443-22069-2.00003-6.

Haider, T.P., Völker, C., Kramm, J., Landfester, K., Wurm, F.R., 2019. Plastics of the Future? The impact of biodegradable polymers on the environment and on society. Angewandte Chemie - International Edition 58 (1), 50–62. Available at https://doi.org/10.1002/anie.201805766.

Hao, L.T., Ju, S., Hwang, D.K., Hwang, D.S., Ok, Y.S., Hwang, S.Y., et al., 2024. Optimizing bioplastics translation. Nature Reviews Bioengineering 2 (4), 289–304. Available at https://doi.org/10.1038/s44222-023-00142-5.

Hasan, Y., Abbassi, B., Mohanty, A.K., Misra, M., Bali, A., Tiessen, M., 2025. Life cycle assessment of poly (butylene adipate-co-terephthalate) (PBAT)-talc Ontario agri-film. Cleaner Environmental Systems 19, 100360. Available at https://doi.org/10.1016/j.cesys.2025.100360.

Helmes, R.J.K., Goglio, P., Salomoni, S., van Es, D.S., Vural Gursel, I., Aramyan, L., 2022. Environmental impacts of end-of-life options of biobased and fossil-based polyethylene terephthalate and high-density polyethylene packaging. Sustainability 14 (18), 11550. Available at https://doi.org/10.3390/su141811550.

Henke, L., Zarrinbakhsh, N., Endres, H.J., Misra, M., Mohanty, A.K., 2017. Biodegradable and bio-based green blends from carbon dioxide-derived bioplastic and poly(butylene succinate). Journal of Polymers and the Environment 25 (2), 499–509. Available at https://doi.org/10.1007/s10924-016-0828-x.

Hobbs, S.R., Harris, T.M., Barr, W.J., Landis, A.E., 2021. Life cycle assessment of bioplastics and food waste disposal methods. Sustainability (Switzerland) 13 (12), 6894. Available at https://doi.org/10.3390/su13126894 Available from.

Itabana, B.E., Mohanty, A.K., Dick, P., Sain, M., Bali, A., Tiessen, M., et al., 2024. Poly (butylene adipate-co-terephthalate) (PBAT)–based biocomposites: a comprehensive review. Macromolecular Materials and Engineering 309 (12), 2400179.

Jambeck, J.R., Geyer, R., Wilcox, C., Siegler, T.R., Perryman, M., Andrady, A., et al., 2015. Plastic waste inputs from land into the ocean. Science 347 (6223), 768–771. Available at https://doi.org/10.1126/science.1260352.

Jariyasakoolroj, P., Leelaphiwat, P., Harnkarnsujarit, N., 2020. Advances in research and development of bioplastic for food packaging. Journal of the Science of Food and Agriculture 100 (14), 5032–5045. Available at https://doi.org/10.1002/jsfa.9497.

Jha, S., Akula, B., Enyioma, H., Novak, M., Amin, V., Liang, H., 2024. Biodegradable biobased polymers: a review of the state of the art, challenges, and future directions. Polymers 16 (16), 2262. Available at https://doi.org/10.3390/polym16162262.

Jian, J., Xiangbin, Z., Xianbo, H., 2020. An overview on synthesis, properties and applications of poly(butylene-adipate-co-terephthalate)–PBAT. Advanced Industrial and Engineering Polymer Research 3 (1), 19–26. Available at https://doi.org/10.1016/j.aiepr.2020.01.001.

Jin, S.G., 2022. Production and application of biomaterials based on polyvinyl alcohol (PVA) as wound dressing. Chemistry, an Asian Journal 17 (21), e202200595. Available at https://doi.org/10.1002/asia.202200595.

Kanwal, A., Zhang, M., Sharaf, F., Li, C., 2022. Polymer pollution and its solutions with special emphasis on poly (butylene adipate terephthalate (PBAT)). Polymer Bulletin 79 (11), 9303–9330. Available at https://doi.org/10.1007/s00289-021-04065-2.

Kapustin, N.O., Grushevenko, D.A., 2023. Analysis of the "circular plastics economy" phenomena and its long-term implications for demand for petroleum market. Federation Environmental Science and Pollution Research 30 (36), 85889–85902. Available at https://doi.org/10.1007/s11356-023-28441-9.

Kato, S., Ueda, T., Aoshima, T., Kosaka, N., Nitta, S., 2023. BioPBS™(polybutylene succinate). Synthetic Biodegradable and Biobased Polymers: Industrial Aspects and Technical Products. Springer, pp. 269–304.

Keith, M., Koller, M., Lackner, M., 2024. Carbon recycling of high value bioplastics: a route to a zero-waste future. Polymers 16 (12), 1621. Available at https://doi.org/10.3390/polym16121621.

Kim, M.S., Chang, H., Zheng, L., Yan, Q., Pfleger, B.F., Klier, J., et al., 2023. A review of biodegradable plastics: chemistry, applications, properties, and future research needs. Chemical Reviews 123 (16), 9915–9939. Available at https://doi.org/10.1021/acs.chemrev.2c00876.

Krumov, N., Atanasova, N., Boyadzhieva, I., Petrov, K., Petrova, P., 2025. Biodegradation of poly (ε-caprolactone): microorganisms, enzymes, and mechanisms. International Journal of Molecular Sciences 26 (12), 5826.

Labet, M., Thielemans, W., 2009. Synthesis of polycaprolactone: a review. Chemical Society Reviews 38 (12), 3484–3504. Available at https://doi.org/10.1039/b820162p.

Lackner, M., Mukherjee, A., Koller, M., 2023. What are "bioplastics"? Defining renewability, biosynthesis, biodegradability, and biocompatibility. Polymers 15 (24), 4695. Available at https://doi.org/10.3390/polym15244695.

Lee, J., Park, C., Fai Tsang, Y., Andrew Lin, K.Y., 2024. Towards sustainable production of polybutylene adipate terephthalate: non-biological catalytic syntheses of biomass-derived constituents. ChemSusChem 17 (23), e202401070. Available at https://doi.org/10.1002/cssc.202401070.

Li, J., 2025. PBAT Plastics. Biodegradable PBAT Plastics and Composites. Springer, pp. 141–228.

Li, J., Lai, L., Wu, L., Severtson, S.J., Wang, W.J., 2018. Enhancement of water vapor barrier properties of biodegradable poly(butylene adipate- co-terephthalate) films with highly oriented organomontmorillonite. ACS Sustainable Chemistry and Engineering 6 (5), 6654–6662. Available at https://doi.org/10.1021/acssuschemeng.8b00430.

Lopes, J.F.A., Simoneau, C., 2014. Solubility of polyvinyl alcohol in ethanol. EFSA Supporting Publications 11 (9), 660E. Available at https://doi.org/10.2903/sp.efsa.2014.EN-660.

Luo, S., Li, F., Yu, J., Cao, A., 2010. Synthesis of poly(butylene succinate- co -butylene terephthalate) (PBST) copolyesters with high molecular weights via direct esterification and polycondensation. Journal of Applied Polymer Science 115 (4), 2203–2211. Available at https://doi.org/10.1002/app.31346.

Maeda, H., Yamagata, Y., Abe, K., Hasegawa, F., Machida, M., Ishioka, R., et al., 2005. Purification and characterization of a biodegradable plastic-degrading enzyme from Aspergillus oryzae. Applied Microbiology and Biotechnology 67 (6), 778–788. Available at https://doi.org/10.1007/s00253-004-1853-6.

Mattlar, T., Ekholm, T., 2025. The impact of bioplastics production on climate change mitigation, fossil fuels and land-use. Renewable and Sustainable Energy Reviews 212, 115234. Available at https://doi.org/10.1016/j.rser.2024.115234.

Meereboer, K.W., Misra, M., Mohanty, A.K., 2020. Review of recent advances in the biodegradability of polyhydroxyalkanoate (PHA) bioplastics and their composites. Green Chemistry 22 (17), 5519–5558. Available at https://doi.org/10.1039/d0gc01647k.

Mohanty, A.K., Vivekanandhan, S., Pin, J.M., Misra, M., 2018. Composites from renewable and sustainable resources: challenges and innovations. Science 362 (6414), 536–542. Available at https://doi.org/10.1126/science.aat9072.

Moliner, C., Drago, E., Lagazzo, A., Caputo, S., Pettinato, M., Finocchio, E., et al., 2025. Performance of PBS materials with degradable additives for food packaging. Journal of Food Engineering 404, 112769.

Moustafa, H., Guizani, C., Dupont, C., Martin, V., Jeguirim, M., Dufresne, A., 2017. Utilization of torrefied coffee grounds as reinforcing agent to produce high-quality biodegradable PBAT composites for food packaging applications. ACS Sustainable Chemistry & Engineering 5 (2), 1906–1916. Available at https://doi.org/10.1021/acssuschemeng.6b02633.

Nag, A., Matsumi, N., Kaneko, T., 2020. Bio-based aromatics: aminobenzoic acid derivatives for high-performance bioplastics. Sustainability & Green Polymer Chemistry: Vol. 2: Biocatalysis and Biobased Polymers. ACS Publications, pp. 99–121.

Nanda, S., Patra, B.R., Patel, R., Bakos, J., Dalai, A.K., 2022. Innovations in applications and prospects of bioplastics and biopolymers: a review. Environmental Chemistry Letters 20 (1), 379–395. Available at https://doi.org/10.1007/s10311-021-01334-4.

Nath, D., Misra, M., Al-Daoud, F., Mohanty, A.K., 2025. Studies on poly (butylene succinate) and poly (butylene succinate-co-adipate)-based biodegradable plastics for sustainable flexible packaging and agricultural applications: a comprehensive review. RSC Sustainability 3, 1267–1302.

Negrete-Bolagay, D., Guerrero, V.H., 2024. Opportunities and challenges in the application of bioplastics: perspectives from formulation, processing, and performance. Polymers 16 (18), 2561. Available at https://doi.org/10.3390/polym16182561.

Nifant'ev, I.E., Bagrov, V.V., Komarov, P.D., Ilyin, S.O., Ivchenko, P.V., 2022. The use of branching agents in the synthesis of PBAT. Federation Polymers 14 (9), 1720. Available at https://doi.org/10.3390/polym14091720.

Nogueira, G.P., Petrielli, G.P., Chagas, M.F., Sampaio, I.Ld.M., Martins, L.Zd.O., Junqueira, T.L., et al., 2024. Sustainability synergies and trade-offs considering circularity and land availability for bioplastics production in Brazil. Nature Communications 15 (1), 8836. Available at https://doi.org/10.1038/s41467-024-53201-9.

Pawar, R., Pathan, A., Nagaraj, S., Kapare, H., Giram, P., Wavhale, R., 2023. Polycaprolactone and its derivatives for drug delivery. Polymers for Advanced Technologies 34 (10), 3296–3316. Available at https://doi.org/10.1002/pat.6140.

Pietrosanto, A., Scarfato, P., Di Maio, L., Incarnato, L., 2020. Development of eco-sustainable PBAT-based blown films and performance analysis for food packaging applications. Materials 13 (23), 5395. Available at https://doi.org/10.3390/ma13235395.

Platnieks, O., Gaidukovs, S., Kumar Thakur, V., Barkane, A., Beluns, S., 2021. Bio-based poly (butylene succinate): recent progress, challenges and future opportunities. European Polymer Journal 161, 110855. Available at https://doi.org/10.1016/j.eurpolymj.2021.110855.

Pottinger, A.S., Geyer, R., Biyani, N., Martinez, C.C., Nathan, N., Morse, M.R., et al., 2024. Pathways to reduce global plastic waste mismanagement and greenhouse gas emissions by 2050. Science 386 (6726), 1168–1173. Available at https://doi.org/10.1126/science.adr3837.

Qin, P., Wu, L., Li, B., Li, N., Pan, X., Dai, J., 2021. Superior gas barrier properties of biodegradable PBST vs. PBAT copolyesters: a comparative study. Polymers 13 (19), 3449. Available at https://doi.org/10.3390/polym13193449.

Rafiqah, S.A., Khalina, A., Harmaen, A.S., Tawakkal, I.A., Zaman, K., Asim, M., et al., 2021. A review on properties and application of bio-based poly(butylene succinate). Polymers 13 (9), 1436. Available at https://doi.org/10.3390/polym13091436.

Rahman, L., Goswami, J., 2023. Poly(vinyl alcohol) as sustainable and eco-friendly packaging: a review. Journal of Packaging Technology and Research 7 (1), 1–10. Available at https://doi.org/10.1007/s41783-022-00146-3.

Rajgond, V., Mohite, A., More, N., More, A., 2024. Biodegradable polyester-polybutylene succinate (PBS): a review. Polymer Bulletin 81 (7), 5703–5752. Available at https://doi.org/10.1007/s00289-023-04998-w.

Rosenboom, J.G., Langer, R., Traverso, G., 2022. Bioplastics for a circular economy. Nature Reviews Materials 7 (2), 117–137. Available at https://doi.org/10.1038/s41578-021-00407-8.

Roy, S., Ghosh, T., Zhang, W., Rhim, J.-W., 2024. Recent progress in PBAT-based films and food packaging applications: a mini-review. Food Chemistry 437 (Pt 1), 137822. Available at https://doi.org/10.1016/j.foodchem.2023.137822.

Saini, T., Meena, J., Verma, V., Saini, S., Malik, R., 2025. Polyvinyl alcohol: recent advances and applications in sustainable materials. Polymer-Plastics Technology and Materials 64 (6), 794–825. Available at https://doi.org/10.1080/25740881.2024.2438046.

Schade, A., Melzer, M., Zimmermann, S., Schwarz, T., Stoewe, K., Kuhn, H., 2024. Plastic waste recycling—a chemical recycling perspective. ACS Sustainable Chemistry & Engineering 12 (33), 12270–12288. Available at https://doi.org/10.1021/acssuschemeng.4c02551.

Sedita, S.R., Di Maria, E., Mazzoni, L., Bekele, N.A., 2025. Decoding the biobased blueprint: key players and evolutionary trends in materials innovation. Polymers 17 (2), 177. Available at https://doi.org/10.3390/polym17020177.

Šerá, J., Kadlečková, M., Fayyazbakhsh, A., Kučabová, V., Koutný, M., 2020. Occurrence and analysis of thermophilic poly(butylene adipate-co-terephthalate)-degrading microorganisms in temperate zone soils.

International Journal of Molecular Sciences 21 (21), 7857. Available at https://doi.org/10.3390/ijms21217857.

Siegenthaler, K.O., Künkel, A., Skupin, G., Yamamoto, M., 2011. Ecoflex® and Ecovio®: Biodegradable, Performance-Enabling Plastics, vol. 245. Springer Science and Business Media, pp. 91–136. Available at https://doi.org/10.1007/12_2010_106.

Sivalingam, G., Madras, G., 2003. Thermal degradation of poly (ε-caprolactone). Polymer Degradation and Stability 80 (1), 11–16. Available at https://doi.org/10.1016/s0141-3910(02)00376-2.

Smith, J.K., Hounshell, D.A., 1985. Wallace H. Carothers and fundamental research at Du Pont. Science 229 (4712), 436–442. Available at https://doi.org/10.1126/science.229.4712.436.

Sousa, A.F., Vilela, C., Fonseca, A.C., Matos, M., Freire, C.S.R., Gruter, G.J.M., et al., 2015. Biobased polyesters and other polymers from 2,5-furandicarboxylic acid: a tribute to furan excellency. Polymer Chemistry 6 (33), 5961–5983. Available at https://doi.org/10.1039/c5py00686d.

Spierling, S., Knüpffer, E., Behnsen, H., Mudersbach, M., Krieg, H., Springer, S., et al., 2018. Bio-based plastics - a review of environmental, social and economic impact assessments. Journal of Cleaner Production 185, 476–491. Available at https://doi.org/10.1016/j.jclepro.2018.03.014.

Srikanth, M., Sandeep, T.S.R.S., Sucharitha, K., Godi, S., 2022. Biodegradation of plastic polymers by fungi: a brief review. Bioresources and Bioprocessing 9 (1), 42. Available at https://doi.org/10.1186/s40643-022-00532-4.

Stegmann, P., Daioglou, V., Londo, M., van Vuuren, D.P., Junginger, M., 2022. Plastic futures and their CO2 emissions. Nature 612 (7939), 272–276. Available at https://doi.org/10.1038/s41586-022-05422-5.

Swetha, T.A., Bora, A., Ananthy, V., Ponnuchamy, K., Muthusamy, G., Arun, A., 2024. A review of bioplastics as an alternative to petrochemical plastics: Its types, structure, characteristics, degradation, standards, and feedstocks. Polymers for Advanced Technologies 35 (6), e6482. Available at https://doi.org/10.1002/pat.6482.

Taghizadeh, F., Heidari, M., Mostafavi, S., Mortazavi, S.M., Haeri, A., 2024. A review of preparation methods and biomedical applications of poly(ε-caprolactone)-based novel formulations. Journal of Materials Science 59 (24), 10587–10622. Available at https://doi.org/10.1007/s10853-024-09774-3.

ten Boer, N., Soyhan, I., Abrishami, S., Parisi, D., Krooneman, J., Siebring, J., et al., 2025. Biodegradable bioplastics in healthcare: opportunities, challenges and sustainable recycling. RSC Sustainability 4 (2), 628–649. Available at https://doi.org/10.1039/d5su00557d.

Thakur, M., Majid, I., Hussain, S., Nanda, V., 2021. Poly(ε-caprolactone): a potential polymer for biodegradable food packaging applications. Packaging Technology and Science 34 (8), 449–461. Available at https://doi.org/10.1002/pts.2572.

Uysal-Unalan, I., Sogut, E., Realini, C.E., Cakmak, H., Oz, E., Espinosa, E., et al., 2024. Bioplastic packaging for fresh meat and fish: Current status and future direction on mitigating food and packaging waste. Trends in Food Science and Technology 152, 104660. Available at https://doi.org/10.1016/j.tifs.2024.104660.

Vidal, F., van der Marel, E.R., Kerr, R.W.F., McElroy, C., Schroeder, N., Mitchell, C., et al., 2024. Designing a circular carbon and plastics economy for a sustainable future. Nature 626 (7997), 45–57. Available at https://doi.org/10.1038/s41586-023-06939-z.

Vytejčková, S., Vápenka, L., Hradecký, J., Dobiáš, J., Hajšlová, J., Loriot, C., et al., 2017. Testing of polybutylene succinate based films for poultry meat packaging. Polymer Testing 60, 357–364. Available at https://doi.org/10.1016/j.polymertesting.2017.04.018.

Walker, S., Rothman, R., 2020. Life cycle assessment of bio-based and fossil-based plastic: a review. Journal of Cleaner Production 261, 121158. Available at https://doi.org/10.1016/j.jclepro.2020.121158.

Wang, G.-X., Huang, D., Ji, J.-H., Völker, C., Wurm, F.R., 2021. Seawater-degradable polymers—fighting the marine plastic pollution. Advanced Science 8 (1), 2001121. Available at https://doi.org/10.1002/advs.202001121.

Wei, D., Wang, H., Ziaee, Z., Chibante, F., Zheg, A., Xiao, H., 2016. Non-leaching antimicrobial biodegradable PBAT films through a facile and novel approach. Materials Science and Engineering: C 58, 986–991. Available at https://doi.org/10.1016/j.msec.2015.09.023.

Witt, U., Müller, R.J., Deckwer, W.D., 1997. Biodegradation behavior and material properties of aliphatic/aromatic polyesters of commercial importance. Journal of Environmental Polymer Degradation 5 (2), 81–89. Available at https://doi.org/10.1007/BF02763591.

Wu, Y., Qin, Y., Yuan, M., Li, L., Chen, H., Cao, J., et al., 2014. Characterization of an antimicrobial poly(lactic acid) film prepared with poly(ε -caprolactone) and thymol for active packaging. Polymers for Advanced Technologies 25 (9), 948–954. Available at https://doi.org/10.1002/pat.3332.

Xu, W., Ganesapillai, M., Abdullah, A.Z., 2023. A study on the synthesis, modification and current market status of PBAT. E3S Web of Conferences 385, 04007. Available at https://doi.org/10.1051/e3sconf/202338504007.

Zaborowska, M., Bernat, K., Pszczółkowski, B., Kulikowska, D., Wojnowska-Baryła, I., 2023. Assessment of biodegradability of cellulose and poly(butylene succinate)-based bioplastics under mesophilic and thermophilic anaerobic digestion with a view towards biorecycling. Waste Management 168, 413–422. Available at https://doi.org/10.1016/j.wasman.2023.06.022.

Zhu, J., Wang, C., 2020. Biodegradable plastics: Green hope or greenwashing? Marine Pollution Bulletin 161 (Pt B), 111774. Available at https://doi.org/10.1016/j.marpolbul.2020.111774.

CHAPTER 7

Bioplastics from agri-food wastes and by-products for food packaging applications

Sahar Akhavan-Mahdavi[1], Yasaman Esmaeili[2], Colin J. Barrow[3], Minoo Naebe[3,4], and Shima Jafarzadeh[2,3]

[1]Department of Food Materials and Process Design Engineering, Gorgan University of Agricultural Science and Natural Resources, Gorgan, Golestan, Iran, [2]School of Engineering, Deakin University, Geelong Waurn Ponds Campus, Geelong, VIC, Australia, [3]Center for Sustainable Bioproducts, Deakin University, Geelong Waurn Ponds Campus, Geelong, VIC, Australia, [4]Institute for Frontier Materials, Deakin University, Geelong Waurn Ponds Campus, Geelong, VIC, Australia

7.1 Introduction

The worsening plastic pollution problem has become one of the most critical environmental issues facing us today. Globally, plastic manufacturing tops out at nearly 390 million metric tons annually, with roughly 40% of that massive volume specifically dedicated to food packaging. This situation reveals a sharp paradox in modern convenience: while food packaging is vital for extending the shelf life of perishables and protecting consumer health, it simultaneously generates huge amounts of single-use waste that persist in the environment for hundreds of years (Ghasemlou et al., 2024). Each year, millions of tons of this waste bypass recycling systems and find their way into marine ecosystems and landfills, contributing significantly to ecological degradation and biodiversity loss (Rai and Choure, 2023; Castro-Criado et al., 2023). Fossil fuel-based conventional plastics still control the food packaging industry, mainly because they are cheap, tough, and highly adaptable. Yet, the fact that they do not break down naturally, combined with flawed waste management systems, makes their environmental toll huge. With food supply chains constantly growing and consumer demand soaring, finding truly transformative solutions is critical. This urgency is now driving researchers, policymakers, and industry leaders to find alternatives that can meet the practical needs of food packaging while upholding strong ecological responsibility (De Moura et al., 2017; Chan et al., 2021).

Bioplastics have emerged as a promising response to these challenges. Defined as materials that are either bio-based sourced from renewable biomass or biodegradable, these polymers offer a dual advantage (Visco et al., 2022). Bioplastics offer two major advantages. They decrease the use of fossil fuels and provide an end-of-life cycle that limits persistent pollution.

Bioplastics for Food Packaging. DOI: https://doi.org/10.1016/B978-0-443-41654-5.00003-2

In food applications, materials like polylactic acid (PLA), polyhydroxyalkanoates (PHAs), and starch composites perform as well as conventional plastics. They ensure freshness and safety by offering reliable protection against moisture, oxygen, and microbial contamination (Kumar et al., 2021). A sustainable approach now includes using agri-food waste and by-products as raw materials (feedstocks) for bioplastic manufacturing. The agri-food sector is responsible for generating massive amounts of residue, think sugarcane bagasse (SCB), spent coffee, cereal bran, and marine discards. These residues, which were traditionally seen as a burden on waste disposal, are now understood to be valuable resources, packed with cellulose, starch, proteins, and lipids. Repurposing this waste not only simplifies waste management but also cuts down on the consumption of new raw materials, which is central to building a circular economy (Mujtaba et al., 2023).

This chapter is dedicated to exploring the transformative potential of bioplastics derived from agri-food wastes and by-products for food packaging applications. It begins with a detailed characterization of these diverse feedstocks, examining their composition, availability, and suitability for bioplastic production. Subsequent sections will delve into the production technologies that convert these wastes into functional biopolymers, evaluate their mechanical, thermal, and barrier properties in the context of food packaging, and present case studies that demonstrate their practical applications. Finally, the chapter will address the current challenges to widespread adoption and propose future directions to fully harness the promise of this innovative, sustainable approach.

7.2 Overview of bioplastics

For the last few decades, concern over environmental threats, specifically resource depletion and worsening pollution, has accelerated the push for sustainable solutions. One of the most promising areas of innovation has been the development of bioplastics, which contribute significantly to reducing dependence on fossil fuels and lowering carbon dioxide emissions (Jayakumar et al., 2023). Bioplastics are broadly defined as materials that are either bio-based, biodegradable, or both. The European Committee for Standardization in its Technical Report 15,392 (2009) defines bio-based plastics as plastics derived from biomass, while biodegradable polymers are those whose properties deteriorate and completely degrade after exposure to these environmental conditions (Rashidi, 2022). Bio-based bioplastics are derived from renewable biomass sources such as agricultural residues, starch, and microbial fermentation, reducing dependence on fossil fuels. Meanwhile, biodegradable bioplastics can break down into natural elements under specific environmental conditions, mitigating long-term pollution (Krein et al., 2025; Skoczinski et al., 2023).

The global production capacities of bioplastics have been on a consistent rise, driven by technological advancements, consumer demand for sustainable products, and regulatory policies promoting eco-friendly materials. As illustrated in Fig. 7.1, global bioplastics production capacity is set to increase significantly from around 2.47 million tons in 2024 to

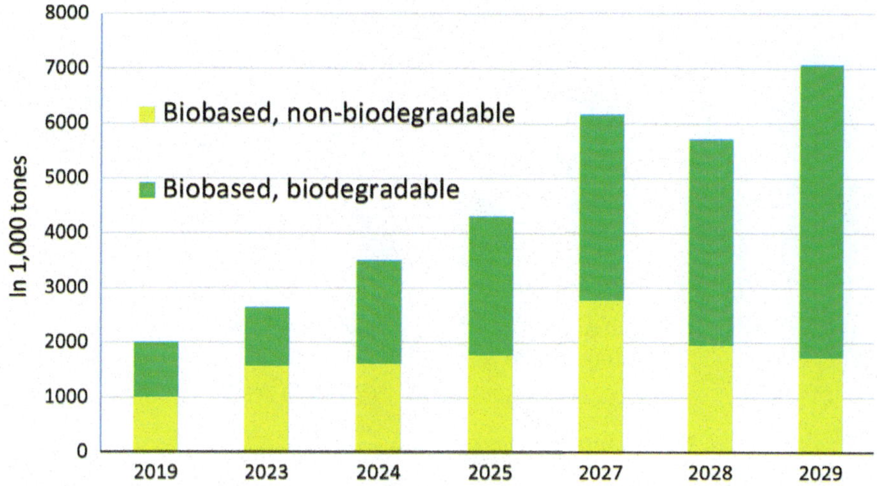

Figure 7.1
Global production capacities of bioplastics 2024–2029. Copyright European Bioplastics, nova-Institute (2021).

approximately 5.73 million tons in 2029 (Bioplastics Market Development Update 2024). This trend is largely attributed to the food packaging sector, which currently accounts for over 40% of bioplastic demand. Such widespread adoption is fueled by the need for sustainable packaging materials that offer comparable performance to fossil-based plastics while minimizing environmental impact (Paidari et al., 2024; Thomas et al., 2023).

Bioplastics can be classified based on their origin, production methods, and biodegradability (Fig. 7.2). The term bio-based refers to plastics derived from renewable biological sources, such as plants, while biodegradable describes plastics that can naturally break down in the environment, regardless of whether they are bio-based or fossil-based. Not all bio-based plastics are biodegradable, and some biodegradable plastics can still be sourced from fossil-based feedstocks (Boccalon and Gorrasi, 2022). Depending on their composition, bioplastics fall into three main categories: those that are both bio-based and biodegradable, those that are bio-based but non-biodegradable, and those that are fossil-based yet biodegradable. PLA, PHAs, and poly(butylene succinate) (PBS) are bioplastics derived from renewable sources that degrade under specific environmental conditions. In contrast, polyethylene (PE), polypropylene, polyethylene terephthalate, poly(trimethylene terephthalate), and total phenolic content (TPC)-Elastomer originate from renewable sources but have properties similar to conventional plastics, making them non-biodegradable. Additionally, fossil-based plastics like polybutylene adipate terephthalate (PBAT) and polycaprolactone (PCL) have been engineered to break down through microbial activity despite being petroleum derived (Cruz et al., 2022). The production methods used for bioplastics also influence their characteristics and applications. One approach is the modification of natural polymers, where naturally

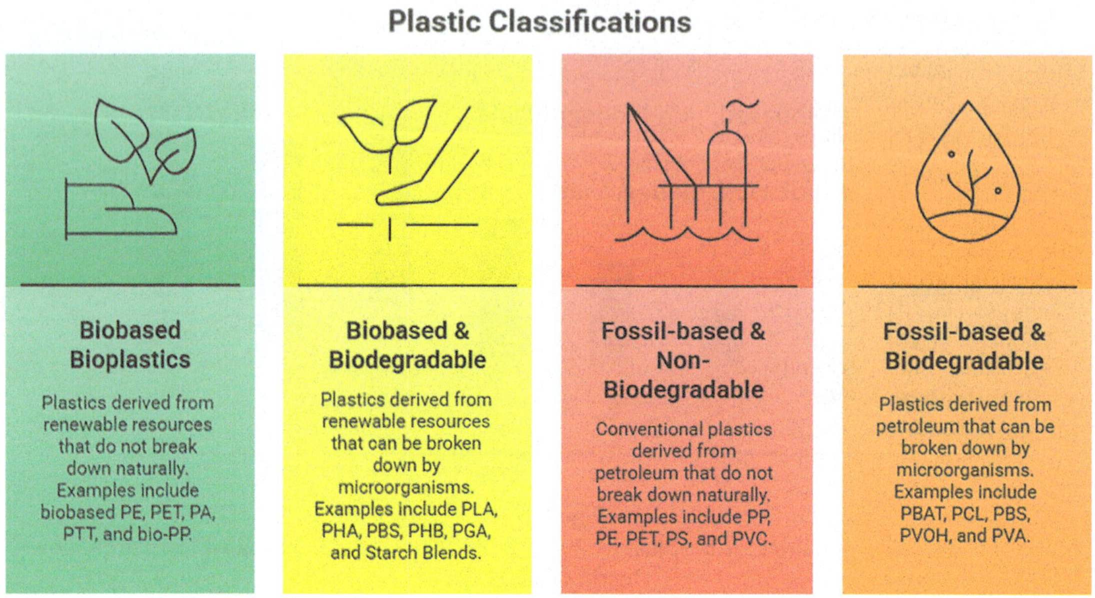

Figure 7.2
The different types of bioplastics.
European Bioplastics.

occurring biopolymers such as starch, cellulose, and lipids are extracted, separated, and chemically modified to meet performance requirements. Starch-based bioplastics, for example, are widely used in food packaging due to their biodegradability and nontoxicity, making them a sustainable alternative to conventional plastics. Another method involves bio-based monomers and conventional chemistry, where monomers are produced via fermentation or chemical processes and then polymerized to form plastics. A well-known example is PLA, which is produced by fermenting plant sugars, such as those derived from corn, into lactic acid before undergoing polymerization. PLA is commonly used in packaging, textiles, and biomedical applications because it combines high performance with environmental benefits (Siddiqui et al., 2024).

A third approach relies on fermentation-based polymers, in which microorganisms ferment agricultural waste or biomass to produce bioplastics. A key example is PHAs, which are derived from organic waste and are highly biodegradable. The capacity of PHAs to decompose in land and water environments makes them increasingly suitable for food packaging, medical products, and other applications where environmental sensitivity is paramount (Visco et al., 2022). The fermentation-based method is particularly promising because it makes use of agro-industrial waste, aligning with principles of waste valorization and circular economy. In addition to production methods, bioplastics are categorized by their biodegradability. Some bioplastics are derived directly from biomass and natural resources, such as cellulose-based and starch-based plastics, which naturally decompose without leaving harmful residues (Navasingh et al., 2023). Others, like microbially derived polymers, are produced through microbial fermentation of agricultural waste or organic matter, offering excellent biodegradability. PHAs are an example of such plastics, breaking down rapidly in both soil and marine environments. Another group consists of fermentation-derived biodegradable polymers, where biomass is fermented to produce monomers or oligomers that are then polymerized into bioplastics. PLA, which is derived from fermented sugars, falls into this category and decomposes efficiently under industrial composting conditions (Ali et al., 2023). Lastly, fossil-based biodegradable polymers represent a unique category in which plastics are synthesized from petroleum-derived feedstocks but are designed to degrade through microbial activity. PBAT, commonly used in compostable films and biodegradable bags, is an example of such materials. While not as sustainable as bio-based biodegradable plastics, fossil-based biodegradable alternatives offer environmental advantages over traditional nondegradable plastics (Narayanan et al., 2023).

A comparison between bioplastics and conventional plastics highlights their differences in key properties (Table 7.1). Bioplastics are derived from renewable sources such as corn

Table 7.1 Comparison of bioplastics and conventional plastics.

Property	Bioplastics	Conventional plastics	References
Raw materials	Renewable sources (e.g., corn starch, sugarcane, and agricultural residues)	Fossil fuels (petroleum-based)	Bishop et al. (2021)
Biodegradability	Some are biodegradable or compostable (e.g., PLA, PHAs) under specific conditions; others are not	Mostly non-biodegradable, persisting for centuries in the environment	Polman et al. (2021)
CO_2 emissions	Lower carbon footprint; can be carbon-neutral with renewable energy use	High carbon emissions due to fossil fuel extraction and processing	Chen et al. (2024)
Recyclability	Some types can be recycled, but infrastructure is limited and often requires separation from conventional plastics	Well-established recycling systems for some types, although contamination and low recycling rates remain challenges	Mneimneh et al. (2024)
Applications	Packaging, agriculture, medical devices, textiles, and consumer goods	Widespread use across all industries, including packaging, construction, and electronics	Muneer et al. (2021)
Durability	Generally lower durability than fossil-based plastics; performance varies by type	Highly durable and resistant, contributing to long-term environmental persistence	Ali et al. (2023)
Cost	Higher production costs due to renewable raw materials and less established production scales	Lower cost due to large-scale, mature production processes and abundant fossil fuel resources	Singh and Verma (2020)

PLA, polylactic acid; *PHAs*, polyhydroxyalkanoates.

starch, sugarcane, and agricultural residues, whereas conventional plastics are fossil fuel-based. In terms of biodegradability, certain bioplastics, such as PLA and PHAs, can degrade under specific conditions, while most conventional plastics persist in the environment for centuries. Another advantage of bioplastics is their lower carbon footprint, as they can be produced using renewable energy, making them a more sustainable option (Ohl et al., 2025). However, recyclability remains a challenge, since bioplastics require specialized recycling systems that are not as widespread as those for conventional plastics. Despite their environmental benefits, bioplastics generally have lower durability than traditional plastics, limiting their application in certain industries. Additionally, higher production costs due to renewable raw materials and less established production scales make bioplastics more expensive than fossil-based plastics. Conventional plastics, in contrast, are widely used across multiple industries due to their durability, cost-effectiveness, and established recycling systems, although issues such as contamination and low recycling rates persist (Ohl et al., 2025).

7.3 Bioplastics in food packaging

Bioplastics are a promising replacement for traditional food packaging materials. Their renewable sources, ability to biodegrade, and potential to reduce environmental damage make them a path toward greater sustainability. These materials are increasingly utilized across a range of food packaging applications, such as films, trays, containers, and coatings, driven by growing consumer demand for eco-friendly options and stricter government regulations on plastic waste. However, despite their potential, the widespread adoption of bioplastics is hindered by challenges related to their material properties, higher production costs, and limitations in recycling and composting infrastructure (Jabeen et al., 2015).

Bioplastics are an attractive option for a sustainable future in food packaging, offering both environmental and functional advantages. Unlike plastics derived from petroleum, bioplastics come from renewable plant sources (like corn starch, sugarcane, or microbial fermentation), greatly cutting down on the use of limited fossil fuels. Moreover, specific types, such as PLA and PHAs, can decompose in industrial composting settings, which helps to control plastic waste buildup (Table 7.2) (Rosenow et al., 2025). Additionally, bioplastics often emit fewer greenhouse gases during production compared to conventional plastics, particularly when derived from sustainably managed feedstock. From a safety perspective, many bioplastics are free of harmful additives like phthalates and bisphenol A, ensuring they are nontoxic and suitable for direct food contact, which enhances their appeal for packaging applications (Rahman et al., 2019; Kakadellis and Harris, 2020).

A variety of bioplastics are employed in food packaging, each bringing unique characteristics to the table. PLA, one of the most widely used bioplastics, is derived from fermented plant sugars and finds its place in food containers, cups, and films due to its transparency and lightweight nature; however, while it is compostable in industrial facilities, it does not break

Table 7.2 Bioplastics used in food packaging and their key properties.

Bioplastic type	Source materials	Key properties	Common applications	References
PLA (polylactic acid)	Corn starch and sugarcane	Transparent, compostable, and brittle	Cups, food trays, and films	Swetha et al. (2023)
PHAs (polyhydroxyalkanoates)	Microbial fermentation	Biodegradable in soil and marine environments	Food coatings, trays, and wraps	Plackett and Siró (2011)
Starch-based plastics	Starch (corn, potato, etc.)	Biodegradable and water-sensitive	Packaging films and disposable cutlery	Niranjana Prabhu and Prashantha (2018)
Bio-PE and Bio-PET	Sugarcane and biomass	Similar to conventional plastics and recyclable	Bottles and flexible packaging	McCurdy et al. (2022)
PBAT (polybutylene adipate terephthalate)	Fossil-based but biodegradable	Flexible and compostable	Compostable bags and films	Sciancalepore et al. (2022)
PCL (polycaprolactone)	Fossil-based	Biodegradable, flexible, and slow degradation rate	Biomedical packaging and compostable films	İlaslan (2024)
Cellulose-based plastics	Plant cellulose (wood and cotton)	Biodegradable and good barrier properties	Food wraps, coatings, and biodegradable films	Romão et al. (2022)
chitosan-based films	Chitosan (from crustacean shells and fungi)	Antimicrobial, biodegradable, and moisture-sensitive	Edible coatings and antimicrobial food wraps	Akhavan-Mahdavi et al. (2023)
Algae-based bioplastics	Algae biomass	Renewable, biodegradable, and high oxygen barrier	Food films, sachets, and edible packaging	Yap et al. (2023)

Bio-PE, bio-based polyethylene; *Bio-PET*, bio-based polyethylene terephthalate.

down in natural environments (Jariyasakoolroj et al., 2020). PHAs, produced through microbial fermentation, stand out for their biodegradability across diverse settings, including soil and marine ecosystems, making them ideal for food wraps, trays, and coatings. Starch-based plastics, crafted from modified starch, offer natural biodegradability and are commonly used in food trays, disposable cutlery, and flexible packaging. Meanwhile, bio-based polyethylene and bio-based polyethylene terephthalate mirror the chemical structure of their fossil-based counterparts but are made from renewable sources like sugarcane; these materials are recyclable within existing systems, although they lack biodegradability. Another option, PBAT, is a fossil-based yet biodegradable polymer often blended with PLA or starch-based plastics to boost flexibility and compostability, enhancing their practical use in packaging (Cruz et al., 2022) (Table 7.2). Infrastructure poses another hurdle, while some bioplastics are compostable, they require industrial composting facilities that maintain specific conditions, and such facilities are not yet widely accessible. Furthermore, when bioplastics are mixed with traditional plastics in recycling streams, they can cause contamination, complicating waste management efforts. Production costs also remain a barrier, as bioplastics are more expensive to manufacture than conventional plastics due to limited scale and the need for specialized processing techniques (Shlush and Davidovich-Pinhas, 2022).

7.4 Agri-food wastes and by-products as feedstocks for bioplastics

The production of bioplastics increasingly depends on sustainable and cost-effective feedstocks, with agri-food waste and by-products emerging as valuable resources. These materials, including crop residues, food processing by-products, and organic waste, are abundant, renewable, and often underutilized. Their integration into bioplastic production reduces reliance on virgin resources, lowers production costs, and minimizes the environmental footprint associated with plastic manufacturing. By leveraging these waste materials, industries can contribute to a circular bioeconomy, reduce greenhouse gas emissions, and address challenges related to plastic pollution and fossil fuel dependency.

A wide range of agri-food waste serves as potential feedstock for bioplastic synthesis, offering unique chemical compositions suited for different polymer types. Key sources include sugar industry by-products, cereal processing residues, spent coffee grounds (SCGs), agroforestry waste, and marine-derived biomass. These materials contain valuable biopolymers such as cellulose, starch, lignin, proteins, and lipids, which can be converted into biodegradable plastics through microbial fermentation, chemical modification, and biorefinery approaches. Table 7.3 provides a comparative overview of these feedstocks, detailing their chemical composition, suitability for biopolymer synthesis, estimated annual production, major producing regions, and associated challenges and opportunities as of 2025. The composition of these agri-food waste significantly influences their suitability for different types of biodegradable plastics. As illustrated in Fig. 7.3, the key chemical components cellulose,

Table 7.3 Comparative overview of agri-food waste feedstocks.

Feedstock type	Composition	Suitability for biopolymer synthesis	Global availability (2025)	Challenges	Opportunities	References
Sugar industry waste (e.g., molasses and bagasse)	Molasses: 50%–60% sugars, 5%–10% proteins; bagasse: 4%0–50% cellulose, 20%–30% hemicellulose, 20%–25% lignin	Fermentation of molasses for PHAs/PLA; hydrolysis of bagasse for cellulose-based bioplastics or PHAs	~300 million tons/year; major regions: Brazil, India, and Thailand	Variability in sugar content and pretreatment costs for bagasse	Abundant, low-cost, reduces CO_2 emissions with EU policy support (2024–2025)	Singh et al. (2022)
Cereal industry waste (e.g., bran, and straw)	Bran: 50%–60% starch, 10%–15% proteins, 5%–10% fibers; straw: 40%–50% cellulose, 20%–30% hemicellulose, 15%–20% lignin	Starch extraction from bran for starch-based bioplastics; fermentation of straw for PHAs	~500 million tons/year; major regions: China, the USA, and India	Inconsistent quality and high lignin content in straw	Widespread availability, support to circular economy, and growing demand for packaging	Deshwal et al. (2021)
Fruit and vegetable processing waste (e.g., peels and pomace)	40%–60% starch/cellulose, 5%–15% proteins, 5%–10% lipids (varies by fruit/vegetables)	Fermentation of sugars for PHAs; starch/cellulose for starch-based bioplastics or films	~200 million tons/year; major regions: South America, Europe, and Asia	Seasonal variability and high watercontent	Abundant, sustainable for packaging and alignment with food industry sustainability goals	Merino et al. (2022)
Dairy industry waste (e.g., whey)	Whey: 70%–80% water, 4%–5% lactose, 0.8%–1.5% proteins, 0.2%–0.5% lipids	Fermentation of lactose for PHAs; proteins/lipids for bio-based composites	~100 million tons/year; major regions: EU, North America, and India	High water content and processing complexity	Renewable, reduces dairy waste, and growing interest in 2025 for bioplastics	Zandona et al. (2021)
Meat and poultry industry waste (e.g., bones, feathers, and fat)	Bones: 60%–70% minerals, 20%–30% proteins; feathers: 90%–95% keratin; fat: 95%–99% lipids	Hydrolysis of proteins/keratin for chitosan-like bioplastics; lipids for bio-based polyesters	~50 million tons/year; major regions: the USA, Brazil, and China	Pathogen risks and complex processing	Sustainable waste reduction and potential for biodegradable films in packaging	Khodaei et al. (2021)
Oilseed industry waste (e.g., meal, hulls, and oilcake)	Meal: 40%–50% proteins, 20%–30% fibers, 5%–10% lipids; hulls: 60%–70% cellulose, 10%–20% lignin	Fermentation of proteins/lipids for PHAs; cellulose for bio-based composites	~150 million tons/year; Major regions: US, Argentina, Malaysia	High fiber/lignin content and cost of extraction	Abundant, support to bio-based plastics, and alignment with 2025 sustainability policies	Gupta et al. (2022)

(Continued)

Table 7.3 (Continued)

Feedstock type	Composition	Suitability for biopolymer synthesis	Global availability (2025)	Challenges	Opportunities	References
Winery and brewery waste (e.g., grape pomace and spent grains)	Grape pomace: 40%–50% cellulose, 20%–30% hemicellulose, 10%–15% proteins; spent grains: 50%–60% fibers, 15%–20% proteins	Fermentation of sugars for PHAs; cellulose/fibers for bio-based films or composites	~80 million tons/year; major regions: Europe, the USA, and Australia	Seasonal availability and processing costs	Renewable, reduces beverage industry waste, and growing use in packaging (2025)	Vannini et al. (2019)
Root and tuber crop waste (e.g., peels and stalks)	50%–70% starch, 20%–30% cellulose, 5%–10% proteins (varies by crop, e.g., potatoes and cassava)	Starch extraction for starch-based bioplastics; fermentation of sugars for PHAs	~150 million tons/year; major regions: Africa, Asia, and South America	High water content and storage issues	Abundant, sustainable, and support to food packaging demand	Aguilar-Rivera (2024)
Nut and seed processing waste (e.g., shells, husks, and meal)	Shells/husks: 60%–70% cellulose, 10%–20% lignin, 5%–10% oils; meal: 40%–50% proteins, 10%–15% fibers	Cellulose for bio-based composites; fermentation of oils/proteins for PHAs	~100 million tons/year; major regions: the USA, Turkey, and China	Hard shells require intensive processing	Renewable, reduces nut industry waste, and potential for durable bioplastics	Teixeira (2017)
Spent coffee bean waste	30%–40% cellulose, 10%–15% lipids, 10%–15% proteins, 5%–10% sugars	Fermentation of sugars/lipids for PHAs; cellulose for bio-based composites or films	~10 million tons/year; major regions: Brazil, Vietnam, and Colombia	Limited volume and processing complexity	Sustainable, urban waste reduction, and innovations in 2025 for packaging	Hernández-Varela and Medina (2023)
Agroforestry residue (e.g., leaves, stems, and bark)	40%–50% cellulose, 20%–30% hemicellulose, 20%–30% lignin	Hydrolysis/fermentation for PHAs/PLA; cellulose for bio-based films/composites	~1000 million tons/year; major regions: Southeast Asia, South America, and Africa	High lignin content and collection logistics	Renewable and support to rural economies, aligns with 2025 circular economy goals	Torres et al. (2019)
Marine waste (e.g., chitin from crustacean shells)	Chitin: 20%–30% polysaccharides, 30%–40% proteins, 10%–20% minerals; seaweed: 30%–50% polysaccharides (alginate, carrageenan)	Chitin deacetylation for chitosan-based bioplastics; seaweed fermentation for PHAs or biodegradable films	~40 million tons/year; major regions: Coastal Asia, Europe, and North America	Limited infrastructure and complex processing	Marine-friendly, reduces ocean pollution, and growing use in packaging	Swain et al. (2023)

PLA, polylactic acid; *PHAs*, polyhydroxyalkanoates.

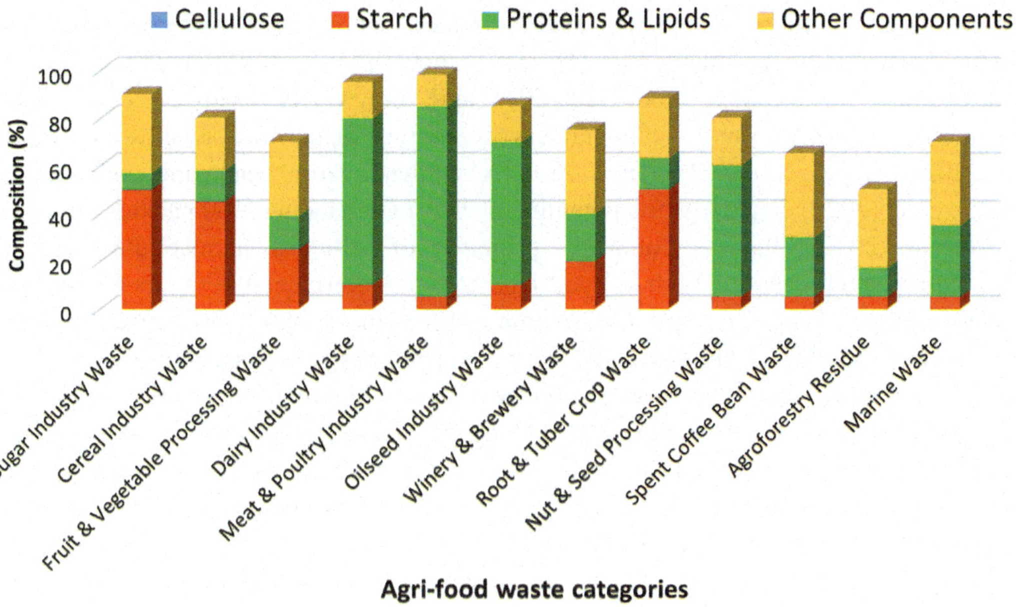

Figure 7.3
Composition of agri-food wastes for biopolymer synthesis.

starch, proteins and lipids, and other components vary across different waste categories, impacting their potential application in bioplastic production. For instance, cereal industry waste and root and tuber crop waste, which contain a high proportion of starch, are ideal feedstocks for starch-based biodegradable plastics. Similarly, sugar industry waste and fruit and vegetable processing waste, rich in fermentable sugars and cellulose, are particularly suited for PLA and PHA production via microbial fermentation. On the other hand, oilseed industry waste and dairy industry waste, with elevated protein and lipid content, can be explored for protein-based bioplastics or innovative polymer composites. Additionally, marine waste, primarily composed of chitin and proteins, serves as a valuable precursor for chitosan-based bioplastics (Brennan et al., 2023; Khandeparkar et al., 2024; Gupta et al., 2022; Carnaval et al., 2024).

7.4.1 Sugar industry waste

The sugar industry generates substantial waste, primarily SCB and molasses, which hold significant potential for bioplastic production. Bagasse, the fibrous residue left after juice extraction, and molasses, a sugar-rich syrup from refining, can be transformed into biopolymers for sustainable food packaging. This approach aligns with global efforts to reduce waste and develop eco-friendly alternatives to conventional plastics.

Molasses, with its high fermentable sugar content, serves as a valuable feedstock for microbial fermentation, producing biopolymers such as PHAs and lactic acid for PLA. These bioplastics exhibit excellent biodegradability and are widely used in food packaging applications, including films and containers. Recent studies, such as those by de Paula et al. (2021), highlight the efficiency of PHA synthesis from sugarcane molasses, while Narancic et al. (2020) emphasize the potential of PLA in enhancing food packaging properties (Singh et al., 2022). Bagasse, composed primarily of cellulose, hemicellulose (HC), and lignin, requires preprocessing such as hydrolysis to release fermentable sugars for biopolymer production. Research by Cruz et al. (2022) demonstrates the feasibility of converting bagasse-derived sugars into PHAs through fermentation. Additionally, bagasse serves as a source of cellulose for reinforcing bioplastic composites, improving their mechanical strength and barrier properties. Advanced applications include the production of nanocellulose (NC) from bagasse, which enhances bioplastic properties like tensile strength and water resistance, making it highly suitable for food packaging. Both PHAs and PLA derived from sugar industry waste are increasingly employed in sustainable food packaging solutions. PHA-based materials are ideal for single-use items such as cups and cutlery due to their compostability, whereas PLA is used for films, beverage bottles, and rigid containers. Moreover, incorporating bagasse-derived cellulose enhances the durability and barrier properties of bioplastics, extending product shelf life and ensuring food safety.

Molasses spentwash, a high-strength wastewater from the sugar industry, can be effectively utilized for bioplastic production. Khardenavis et al. (2009) investigated the utilization of molasses spentwash, a by-product of the sugar industry, to produce polyhydroxybutyrate (PHB) using waste-activated sludge. The study demonstrated that raw molasses spentwash with an optimized carbon-to-nitrogen (C:N) ratio resulted in 60% chemical oxygen demand (COD) removal and 31% PHB accumulation. The highest PHB yield was observed in deproteinized spentwash supplemented with nitrogen, indicating efficient carbon conversion. This process provides a sustainable approach to waste management by simultaneously treating molasses spentwash and producing biodegradable plastics (Khardenavis et al., 2009).

Jayasekara et al. (2022) evaluated the production of PLA pellets from corn and sugarcane molasses, comparing their environmental impacts using process simulation and life cycle analysis. The study found that sugarcane molasses-based PLA production consumed approximately 80% of the total life cycle net energy inputs compared to corn. However, it exhibited a net energy gain with a net energy ratio of 1.09 and a renewability indicator of 5.63, making it a more sustainable feedstock. Additionally, sugarcane molasses contributed only 25% of the climate change impact (1470 kg CO_2 eq/ton of PLA) compared to corn, highlighting its lower environmental footprint for PLA production (Jayasekara et al., 2022). Tripathi et al. (2019) explored the production ofPHAs using sugar refinery waste (cane molasses) as a substrate, with potential applications in food packaging. The study examined the effect of nutritional supplements, including amino acids (cysteine, isoleucine, and methionine) and vitamin (thiamin), on PHA production by *Alcaligenes sp. NCIM 5085* under submerged fermentation. Response surface methodology was used to optimize cultivation conditions, leading to a maximum PHA yield of 70.89%, which increased to 79.26% in a 7.5 L bioreactor with a productivity of 0.312 g/L/h. Fourier transform infrared spectroscopy confirmed the produced polymer as PHB. The extracted PHB exhibited high molecular weight, crystallinity (66%), and thermal stability (degradation at 302°C), making it a promising biodegradable material for food packaging applications (Tripathi et al., 2019). Ghaderi et al. (2014) developed an all-cellulose nanocomposite (ACNC) film using SCB cellulose nanofibers for food packaging applications. The nanofibers were obtained through disk grinding, reducing the fiber diameter from 14 μm to 39 nm. ACNC was produced via partial dissolution in N,N-dimethylacetamide/lithium chloride (DMAc/LiCl) solvent. The resulting film exhibited high tensile strength (140 MPa), compared to 101 MPa for nanofiber sheets and 8 MPa for untreated fiber sheets, although its thermal stability was slightly lower than that of pure cellulose nanofibers. Water vapor permeability (WVP) increased with longer dissolution times (e.g., from 10 min onward). Despite this, ACNC demonstrated excellent mechanical performance, full biodegradability, and suitability as a sustainable, high-barrier material for cellulose-based food packaging. Ghaderi et al. (2014) and Hamid et al. (2022) fabricated a biopolymer composite using starch and SCB pulp, aiming to replace traditional plastic packaging. Their findings demonstrate that this starch-bagasse bioplastic exhibits promising mechanical properties, enhanced by the addition of bagasse, while castor oil improves flexibility and elongation. The resulting bio-based film shows competitive performance compared to existing plastic films, underscoring its potential as a green material for future food packaging applications (Hamid et al., 2022). Azmin et al. (2020) combined cellulose from cocoa pod husks with bagasse fiber in ratios of 100:0, 75:25, 50:50, 25:75, and 0:100, finding that the 75:25 blend (75% cellulose, 25% fiber) offered the best properties, with low water absorption and WVP, ideal for packaging. As illustrated in "Composition of Agri-Food Wastes for Biopolymer Synthesis", bagasse's cellulose (35%–45%) and HC (25%–35%) enhance the bioplastic's performance, reducing moisture issues and supporting sustainability by valorizing waste (Azmin et al., 2020). Lu et al. (2020) developed a hydrogel from bagasse-derived NC, prepared via TEMPO-mediated oxidation and cross-linked with Zn^{2+}, to serve as a matrix for pH-responsive dyes (bromothymol blue and methyl red). This hydrogel functions as a freshness indicator, changing color from green to red to signal chicken breast spoilage, triggered by increased CO_2 levels and microbial growth (exceeding the acceptability limit by day three). Bagasse's high cellulose content (35%–45%) supports its transformation into NC, enhancing its utility in biodegradable packaging (Lu et al., 2020). Weligama Thuppahige et al. (2023) utilized powdered cassava bagasse, glycerol (30% and 35%), and water to create bioplastic films via hot pressing at temperatures of 60°C, 100°C, and 140°C under 0.28 t pressure, with further testing at 0.42 t and 0.84 t. The optimal conditions were identified as 35% glycerol, 100°C, and 0.42 t pressure, producing films with satisfactory morphology, good starch–glycerol compatibility, and slightly enhanced thermal stability, which highlights cassava bagasse's high cellulose (30%–40%) and residual starch content (Weligama Thuppahige et al., 2023). Silva Junior et al. (2024) emphasized the importance of mechanical properties and flexibility in packaging materials, as they influence processing, handling, storage, and distribution. Their study demonstrated that bacterial cellulose (BC) exhibited exceptional flexibility (>100 score), meaning it can withstand over 100 folding cycles without tearing. This superior flexibility is due to BC's nanofibrillar structure (fiber diameter ~83 nm), which allows molding into various shapes without losing structural integrity. In contrast, SCB showed lower flexibility (score of 26), likely due to its more rigid, fibrous nature (A). The 0.7 BC/0.3 SCB composite [70% (w/w) BC + 30% (w/w) SCB] (C,D) exhibited enhanced tensile strength (up to 46.22 MPa vs. 17.43 MPa for SCB alone) and flexibility, making it a promising candidate for sustainable, high-performance packaging (Fig. 7.4).

Despite its promise, the commercialization of sugar industry waste-derived bioplastics faces challenges. The primary barriers include the need for efficient hydrolysis techniques for bagasse and ensuring cost-effective fermentation processes for molasses. Enzymatic hydrolysis and genetic engineering of fermentation microbes could enhance bioplastic yields, although these

Figure 7.4

Appearance of bacterial cellulose (BC) and sugarcane bagasse (SCB) samples after undergoing flexibility tests. (A) SCB alone, (B) 50% (w/w) BC plus 50% (w/w) SCB, and (C,D) 70% BC plus 30% (w/w) SCB.

*da Silva Junior, C.J.G., de Medeiros, A.D.M., Cavalcanti, A.K.L.d.H., de Amorim, J.D.P., Durval, I.J.B., Cavalcanti, Y.d.F., Converti, A., Costa, A.F.d.S., Sarubbo, L.A., 2024. Towards sustainable packaging using microbial cellulose and sugarcane (*Saccharum officinarum L.*) bagasse. Materials 17 (15), 3732. https://doi.org/10.3390/ma17153732.*

approaches require further development before widespread adoption (Paixão et al., 2016). Nevertheless, the valorization of sugar industry waste contributes to the circular economy, reducing dependence on fossil-based plastics and lowering the environmental impact of food packaging. The exploration of NC from bagasse represents a key innovation, potentially expanding the applications of bioplastics in advanced packaging materials (Singh et al., 2022).

7.4.2 Cereal industry waste

Cereal production generates large quantities of waste, including bran, husks, and straw, which are often discarded or burned, contributing to environmental pollution. However, recent research highlights their potential as sustainable feedstocks for bioplastic production, particularly in food packaging. These waste materials are rich in cellulose, HC, and starch, making them valuable for conversion into biodegradable polymers (Rani et al., 2025). The conversion process generally involves hydrolysis to break down complex carbohydrates into fermentable sugars. These sugars can then be used in microbial fermentation to produce PHAs, a class of biodegradable plastics. Alternatively, cellulose and starch from cereal waste can be directly processed into bioplastics. For instance, wheat straw hydrolysate has been used to produce PHAs, while rice husks have been used to create cellulose nanocrystals, enhancing the mechanical properties of bioplastic composites (Bangar et al., 2023). A study by Chougan et al. (2022) evaluated enhancing PLA composites with wheat straw for potential packaging applications. The research applied an eco-friendly hybrid pretreatment (hot water and steam) and surface functionalization with attapulgite nanoclay (AT) and graphene nanoplatelets (G) to wheat straw, improving its compatibility with PLA. The optimized composites, 10 H+S-AT and 10 H+S-G, achieved tensile strengths of 28 and 27 MPa, respectively, nearly triple the 9.7 MPa of untreated samples, while the 10UN-G sample showed the lowest water absorption (1.6% after 24 hours) (Chougan et al., 2022). Table 7.4 highlights recent studies about cereal waste used in food packaging (Fig. 7.5).

Despite their potential, several challenges must be addressed. The variability in waste composition can affect bioplastic quality, while high lignin content in straw requires costly pretreatment. Additionally, cereal waste competes with other applications, such as animal feed and bioenergy, which may impact large-scale adoption. However, given the abundance and low cost of cereal waste, advancements in processing technologies could make its use in bioplastic production more feasible. Research efforts are focused on optimizing hydrolysis, improving fermentation efficiency, and enhancing bioplastic formulations to overcome these challenges (Baiju et al., 2024).

7.4.3 Fruit and vegetable processing waste

The utilization fruit and vegetable processing waste as a feedstock for bioplastic production marks a major step toward sustainable food packaging, aligning closely with the broader goal

Table 7.4 Studies on cereal waste in bioplastic production for food packaging.

Cereal waste source	Bioplastic type	Key findings	References
Wheat straw fibers (WSFs)	PHBV-based bioplastic composites	Studying plasticization of PHBV with biodegradable plasticizers (acetyl tributyl citrate, glycerol triacetate (GTA_, and polyethylene glycol). Found that WSF incorporation reduced flexibility due to poor fiber–matrix adhesion. Suggested surface modifications to improve compatibility.	Martino et al. (2015)
Wheat straw (WS)	PLA-based biocomposites	Developed PLA/wheat straw composites using solid-state shear milling (SSSM) for enhanced dispersion. Achieved improved crystallization, flexural modulus, and heat deflection temperature, making the material suitable for food packaging applications, especially for fresh produce.	Yang et al. (2018)
WS	Polystyrene (PS)-based biocomposites	Examined chemical pretreatments (NaOH, HCl, and H_2SO_4) to improve WS surface properties for PS composites. Found that alkali-treated WS had the best compatibility with PS, leading to enhanced mechanical, thermal, and barrier properties. Concluded its potential for green packaging applications.	Dixit and Yadav (2020)
WS	Nano-holocellulose and reduced graphene oxide (rGO) composites	Developed nano-holocellulose/rGO films for active packaging. Enhanced tensile strength (+639.8%), ultraviolet resistance, and water vapor barrier properties. Films showed electrical conductivity, with a 25-fold increase in conductivity for 10% rGO content. Demonstrated strong potential for biodegradable, intelligent food packaging.	Wu et al. (2022)
Cereal straw fiber	Pulp-molded packaging	Investigated feasibility of using cereal straw fiber for molded food trays as an alternative to EPS. Found improved tensile properties (0.47 MPa vs. 0.16 MPa for EPS) but lower flexural modulus. Adjustments in thickness allowed matching of performance parameters. Straw-based material exhibited good biodegradability (20% mass loss in soil after 4 weeks).	Curling et al. (2017)
WS	Polyethylene (PE)/ polypropylene (PP)/alkali-treated WS composites	Optimized PE/PP/alkali-treated WS film using Response surface methodology (1.8 g PE, 1.2 g PP, 1.192 g WS), achieving 45.018 MPa tensile strength, 120% elongation, and 51.740 $g/m^2 \cdot day$ water vapor transmission rates. Demonstrated suitability for green packaging applications.	Dixit and Yadav (2019)

Table 7.4 (Continued)

Cereal waste source	Bioplastic type	Key findings	References
WS	Lignocellulose-based biodegradable films	Developed biodegradable films using wheat straw lignocellulose with Zn^{2+} solubilization and Ca^{2+} cross-linking (200–800 mM). Achieved tensile strength of 6.61 MPa (2.5 times higher than that of PE films), transparency, and 90% biodegradation in soil within a month with 800 mM $CaCl_2$.	Ahmed et al. (2024)
WSFs	PHBV-based bioplastic composites	Investigated PHBV/WSF biocomposites for food contact; high migration in hydrophilic food simulants (e.g., water, ethanol) due to WSF leaching, low migration with fatty simulants (e.g., olive oil). Suitable for low/intermediate water activity foods (aw ≤ 0.90) like fruits, bread, and cheese.	Angellier-Coussy et al. (2020)
Rice husk fiber (RHF)	Starch-based antimicrobial biocomposite	Developed RHF-reinforced corn starch film with benzalkonium chloride (BKC); optimized at 20% RHF and 0.05% BKC, achieving 61% higher tensile strength, low water solubility, and antimicrobial properties. Extended strawberry shelf life by over 11 days.	Srivastava et al. (2023)
Rice husk	LDPE-based biocomposites	Developed LDPE/rice husk (RH) biocomposites (10–50 wt.% RH) via melt mixing; increased RH content reduced thermal stability, increased water vapor permeability, crystallinity, and biodegradability. Enhanced "Granny Smith" apple firmness and reduced respiration, with highest RH content films showing most degradation after 21 days composting.	Bajsic et al. (2021)

PLA, polylactic acid; *PHBV*, poly(3-hydroxybutyrate-co-3-hydroxyvalerate); *LDPE*, low-density polyethylene; *EPS*, extracellular polymeric substances.

of minimizing environmental impact through innovative waste management. This waste stream, encompassing peels, seeds, pomace, and pulp, is generated in vast quantities globally, approximately 2 billion tons annually according to the Food and Agriculture Organization, particularly in high-production regions like South America (e.g., Brazil for oranges), Europe (e.g., Italy for apples), and Asia (e.g., India for bananas). Traditionally relegated to low-value uses such as animal feed or composting, these organic residues are rich in valuable compounds like starch, cellulose, pectin, and fermentable sugars, making them ideal candidates for bioplastic synthesis. This approach not only mitigates waste disposal challenges but also diminishes dependence on fossil-based plastics, embodying circular economy principles (Chandarana and Chandra, 2021).

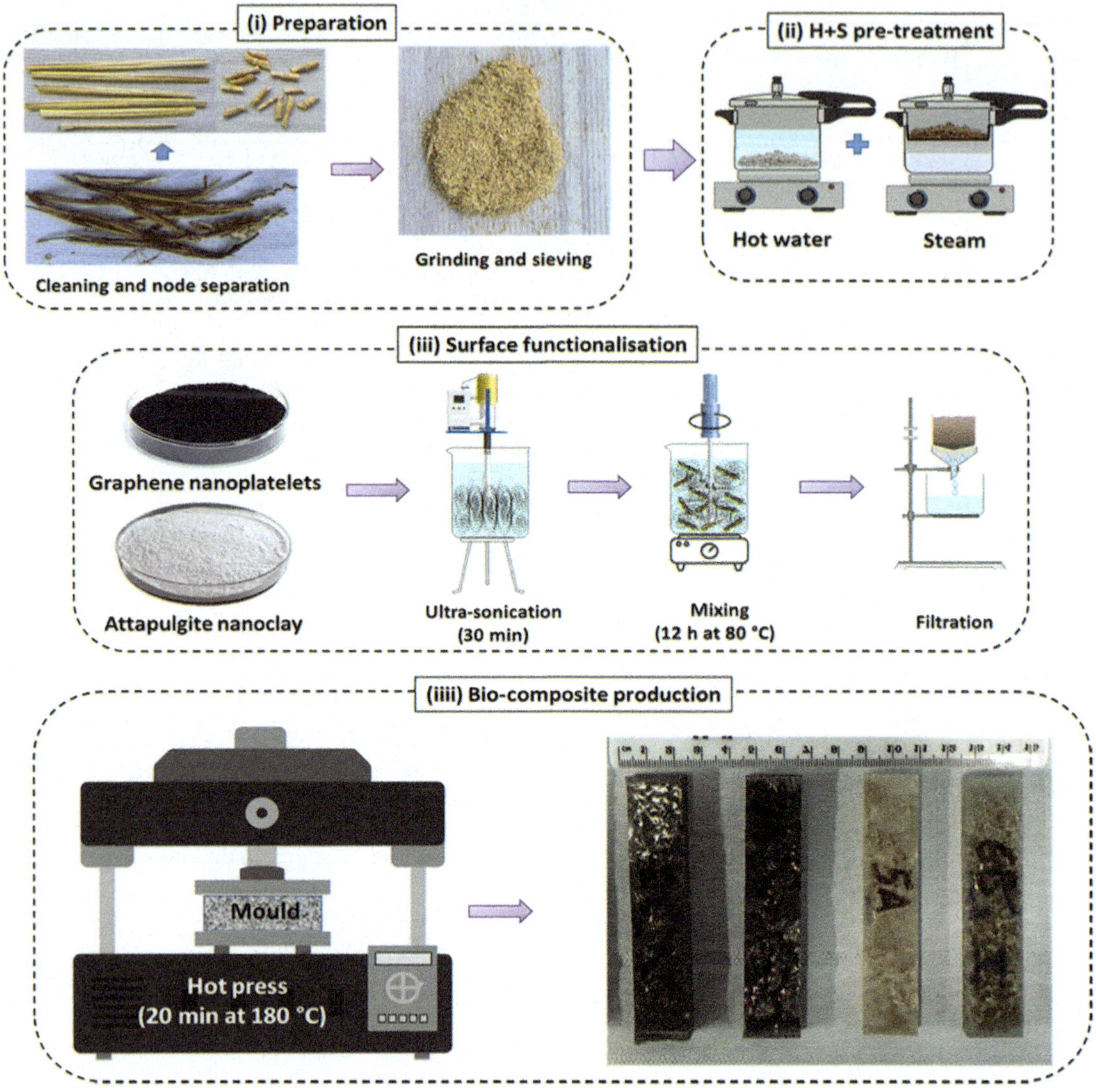

Figure 7.5

Schematic framework of different stages of wheat straw board preparation.
Chougan, M., Ghaffar, S.H., Mijowska, E., Kukułka, W., Sikora, P., 2022. High-performance polylactic acid compressed strawboard using pre-treated and functionalised wheat straw. Industrial Crops and Products. 184, 114996. https://doi.org/10.1016/j.indcrop.2022.114996.

The biochemical composition of fruit and vegetable processing waste underpins its potential for bioplastic production. For example, potato peels, containing 60%–70% starch, are well-suited for producing thermoplastic starch (TPS) bioplastics, which offer good tensile strength and water resistance for flexible packaging like bags and wrappers. Banana peels, with 10%–15% cellulose and lignin, can be processed into cellulose-based films or composites, enhancing mechanical properties for food wrapping applications. Orange peels, rich in pectin (20%–30%) and limonene, enable the creation of pectin-based bioplastics with excellent barrier properties, ideal for edible films or coatings that extend shelf life, an unexpected yet innovative application highlighted in recent studies (Sánchez-Orozco et al., 2024; Yaradoddi et al., 2022; Kharb and Saharan, 2025). Similarly, apple pomace, containing 10%–15% pectin and fermentable sugars, enables PHA production through fermentation and direct processing into edible films, expanding its applications in wrappers and trays. These compositions, as summarized in Table 7.5, highlight the versatility of fruit and vegetable waste as a resource for various bioplastic types (Loh et al., 2023).

Several processing methods transform these wastes into functional bioplastics. Starch extraction from potato peels, followed by plasticization with glycerol, yields TPS films, as demonstrated in studies showing their suitability for packaging dry goods. Cellulose extraction from banana peels produces robust films or reinforces composites, with research confirming enhanced mechanical strength for wrapping fresh produce. Pectin extraction from orange peels results in films with promising barrier properties, while fermentation of fruit and vegetable waste sugars using bacteria like *Cupriavidus necator* or *Bacillus subtilis* produces PHAs, achieving yields up to 40% for applications in cups and trays. Additionally, direct processing of apple pomace into edible films offers a simpler route, capitalizing on its natural polymer content. These bioplastics provide environmental benefits, such as biodegradability, with degradation occurring within weeks under composting conditions, and functional advantages, including adequate oxygen and moisture barriers, nontoxicity compliant with food safety regulations (e.g., EU Regulation 10/2011), and sustainability by reducing reliance on finite resources. Chandrasekar et al. (2023) investigated the transformation of banana peel waste into bioactive thermoplastic packaging films. The research fully recycled banana peels by extracting starch (88.55% amylopectin, thermally stable up to ~295°C), NC fibers, and bioactive compounds (e.g., β-sitosterol) with high antioxidants (74.43% 2,2-diphenyl-1-picrylhydrazyl (DPPH) inhibition) and antimicrobial properties (Fig. 7.6). These components were reconstituted into cellulose-reinforced starch-based bioactive thermoplastic films (BPT), with NC enhancing tensile strength sixfold and reducing oxygen permeability by 0.6-fold, while achieving ~98% ultraviolet (UV) blocking capacity. The bioactive BPT films extended bread shelf life by 10 days compared to PE controls, as assessed through moisture content, water activity, texture, color, acidity, and microbial counts, aligning with the cellulose content (10%–15%) (Fig. 7.6) (Chandrasekar et al., 2023). Other recent studies regarding fruit and vegetable waste are highlighted in Table 7.5.

7.4.4 Dairy industry waste

Dairy industry waste, primarily derived from the processing of cheese, yogurt, and other milk-based products, presents a significant environmental challenge due to its large volume and

Table 7.5 Studies on fruit and vegetable processing waste in bioplastic production for food packaging.

Waste type	Composition	Primary bioplastic	Production method	Food packaging application	References
Potato peels	60%–70% starch	Starch-based (TPS)	Starch extraction and plasticization	Bags and wrappers	Ebrahimian et al. (2022)
Banana peels	10%–15% cellulose and lignin	Cellulose composites	Cellulose extraction and processing	Films and coatings	Seth et al. (2025)
Orange peels	20%–30% pectin and limonene	Pectin-based films	Pectin extraction and processing	Edible films and coatings	Yaradoddi et al. (2022)
Apple pomace	10%–15% pectin and sugars	PHAs and edible films	Fermentation and direct processing	Wrappers and trays	Gustafsson et al. (2019)
Cassava pomace	High-residual starch and fiber	Starch-based reinforced films	Starch extraction and reinforcement with Fiber	Bags, films, and trays	Akmeemana et al. (2024)
Citrus peel	20%–30% pectin, cellulose, and sugars	Pectin-based films	Pectin extraction and processing	Edible films and coatings	Oluwasina and Awonyemi (2021)
Olive pomace	20%–30% cellulose, lignin, and sugars	Cellulose-based films and PHAs	Cellulose extraction and fermentation	Films and trays	Lammi et al. (2018)

PHAs, polyhydroxyalkanoates; *TPS*, thermoplastic starch.

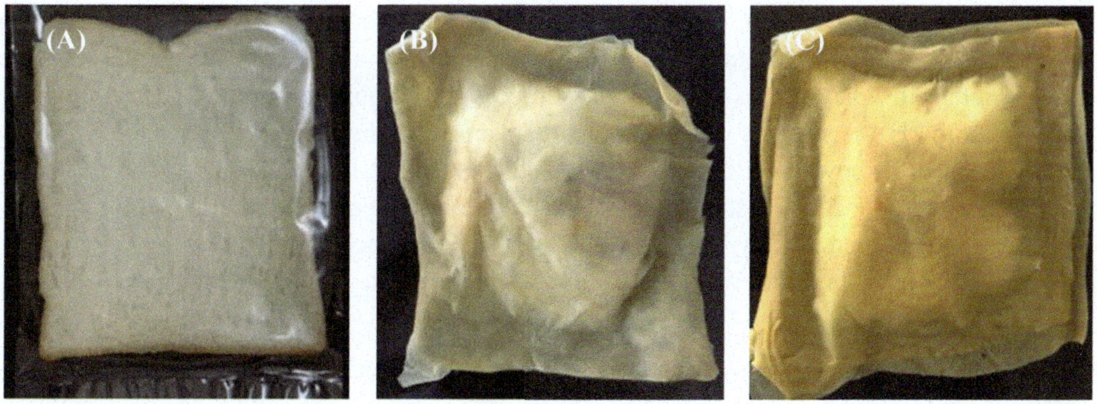

Figure 7.6

(A) Bread sample packed in polyethylene films. (B) Bread sample packed in bioactive thermoplasti film. (C) Bread sample packed in bioactive thermoplasti film.

The figure was reproduced from Chandrasekar, C.M., Krishnamachari, H., Farris, S., Romano, D., 2023. Development and characterization of starch-based bioactive thermoplastic packaging films derived from banana peels. Carbohydrate Polymer Technologies and Applications. 5, 100328. https://doi.org/10.1016/j.carpta.2023.100328, from with the permission of Elsevier.

disposal issues. Among these by-products, whey is the most abundant, consisting of approximately 4.5% lactose, 0.8% protein, 1.0% salts, and 0.1%–0.8% lactic acid. With global whey production estimated at around 200 million tons per year and nearly 47% being discarded, this waste contributes to environmental concerns such as oxygen depletion in water bodies. However, its rich composition makes it a promising raw material for bioplastic production, particularly in regions with high dairy output, such as Europe, North America, and India (Chalermthai et al., 2021).

Whey, a by-product of cheese and yogurt production, is rich in lactose and proteins, making it an ideal substrate for microbial fermentation. The utilization of whey for PHA and PHB synthesis not only provides a cost-effective alternative but also helps mitigate the environmental burden of whey disposal. However, several challenges remain, including whey pretreatment methods and the selection of efficient bacterial strains capable of converting whey into high-yield PHAs (Schmid and Müller, 2019). Recent advancements have proposed innovative solutions to overcome these hurdles, such as enzymatic hydrolysis to enhance lactose availability and genetic engineering of microbial strains to improve yield efficiency (Rana et al., 2023). The conversion of whey into bioplastics follows two primary pathways, including protein-based bioplastics and fermentation-derived PHAs. Whey proteins, particularly whey protein isolate (WPI) and whey protein concentrate (WPC), contain essential proteins like β-lactoglobulin and α-lactalbumin, which can be denatured and plasticized to create biodegradable films. These protein-based bioplastics, produced through reactive extrusion or copolymerization, exhibit excellent mechanical strength, elasticity, and oxygen barrier properties, making them suitable for food packaging applications. Alternatively, the lactose content in whey serves as a carbon source for microbial fermentation, enabling the production of PHAs, such as PHB. Microorganisms like *Escherichia coli* and *Ralstonia eutropha* convert whey into PHAs, which possess excellent thermal stability, water vapor resistance, and food contact safety approved by the U.S. Food and Drug Administration (FDA). PHAs are increasingly utilized in single-use food packaging, including trays, cups, and compostable films, offering a sustainable alternative to petroleum-based plastics (Chalermthai et al., 2019).

Recent research has aimed to enhance the properties and economic viability of these bioplastics. Advances in protein-based bioplastics focus on incorporating nanomaterials and antioxidants to improve mechanical properties, although high production costs and purification challenges remain barriers. For PHA-based bioplastics, efforts are being made to reduce costs by using low-cost carbon sources like whey and optimizing fermentation conditions. A recent study by Zolfaghari et al. (2022) proposed a novel biorefinery approach to valorizing cheese whey into eco-friendly food packaging materials. The research demonstrated that fungal chitosan extracted from *Mucor indicus* cultivated on raw cheese whey could be blended with starch to develop biodegradable films with enhanced oxygen barrier and water resistance properties. Additionally, the effluent from fungal cultivation was subjected to anaerobic digestion, leading to a significant reduction in COD by 89.6% and the production of biomethane as a valuable by-product (Zolfaghari et al., 2022). Chalermthai et al. (2019) investigated the production of whey protein–based polymers as biodegradable alternatives to polyethylene, utilizing WPI and WPC from dairy waste. The proteins were methacrylated and polymerized with poly(ethylene glycol) methyl ether methacrylate (PEGMA) using free-radical polymerization at two protein-to-PEGMA ratios (20:80 and 30:70). The results showed that polymers with a higher protein content (30:70) exhibited significantly better tensile strength (3.8 ± 0.2 MPa) compared to lower-protein-content samples (1.9 ± 0.4 MPa), demonstrating the influence of protein-to-copolymer balance on material properties. Thermal analysis revealed that WPI-based polymers were stable up to 277.8°C, while WPC-based samples were stable up to 273.0°C. The study highlights the potential of whey protein–based polymers as environmentally friendly plastics and supports the valorization of dairy industry by-products for sustainable material development (Chalermthai et al., 2019).

7.4.5 Meat and poultry industry waste

The transformation of meat and poultry industry waste into bioplastics for food packaging offers a sustainable solution to manage the substantial by-products generated during slaughter and processing, aligning with the chapter's emphasis on innovative waste valorization. Globally, meat production reached approximately 340 million tons in 2023, producing significant waste such as hides, bones, feathers, fat trimmings, blood, and offal, particularly in major producers like the USA, Brazil, and China. These wastes, rich in proteins (e.g., collagen, keratin) and lipids, are typically landfilled, contributing to environmental pollution, but their biochemical composition makes them promising feedstocks for bioplastic production, supporting circular economy goals (Khodaei et al., 2021).

The composition of meat and poultry waste underpins its bioplastic potential. Collagen, constituting 25%–30% of hides and also present in bones, is extracted via enzymatic or acid hydrolysis and cast into films with glycerol, yielding biodegradable coatings with good oxygen and moisture barriers for fresh produce packaging. Keratin, abundant in feathers (over 3 million tons annually), is processed through alkaline or enzymatic extraction into strong, flexible films suitable for trays and containers, although food safety evaluations are ongoing. Lipids from fat trimmings, comprising up to 15% of poultry carcasses, are hydrolyzed and fermented by microbes like *Cupriavidus necator* into PHAs, producing thermally stable, FDA-approved materials for cups, trays, and films. An intriguing aspect is the potential antibacterial properties inspired by protein-based bioplastics, which could enhance food safety applications (Shlush and Davidovich-Pinhas, 2022).

Recent studies on bioplastics derived from meat and poultry industry waste, as summarized in Table 7.6, highlight innovative approaches to sustainable food packaging. Dou et al. (2020) developed edible feather keratin (FK) films cross-linked with dialdehyde carboxymethyl cellulose (DCMC), achieving improved water resistance and UV-barrier properties, making them viable for edible coatings. Similarly, Song et al. (2014) advanced active packaging by incorporating clove oil into chicken feather protein/gelatin films, demonstrating significant antimicrobial activity and extended shelf life for smoked salmon, showcasing practical applications for poultry waste (Table 7.6). Fernández-d'Arlas (2019) reported tough keratin films from sheep wool with thermal cross-linking, offering UV-protective packaging potential with a toughness of 19 MJ/m^3 (Table 7.6). Collagen films are formed by casting extracted collagen, often enhanced with plasticizers, while keratin films are cast from dissolved feather extracts, offering biodegradability and mechanical strength. PHA production involves lipid hydrolysis, followed by microbial fermentation, yielding compostable packaging materials. Applications include collagen-based edible coatings, keratin-based trays, and PHA-based single-use items, reducing landfill waste. However, challenges persist, including high production costs, inconsistent waste quality, and regulatory compliance for food contact, with recent advances like nanomaterial integration and optimized fermentation aiming to address these issues. Table 7.6 summarizes key studies, highlighting the diversity and potential of meat and poultry waste in sustainable packaging.

7.4.6 Oilseed industry waste

The oilseed industry, driven by the high global demand for edible oils from seeds like soybean, sunflower, and palm, generates substantial waste streams including meal, hulls,

Table 7.6 Studies on meat and poultry industry waste in bioplastic production for food packaging.

Waste component	Composition	Primary bioplastic	Production method	Food packaging application	References
Collagen (hides)	~25%–30% collagen	Collagen/laponite nano-biocomposite films	Film casting with laponite addition	Films and coatings	Valencia et al. (2019)
Collagen (hides)	~25–30% collagen	Collagen/apatite composite films	In situ apatite synthesis and film casting	Films and coatings	Wang et al. (2016)
Keratin (feathers)	High keratin content	Keratin/cellulose bioplastic film	Sodium sulfide extraction and film synthesis with cellulose	Films and potential trays	Sharma et al. (2018)
Keratin (sheep wool)	High keratin content	Keratin-based bioplastic film	Alkaline oxidative extraction, thermal cross-linking with glycerol/formaldehyde	Films and UV-protective packaging	Fernández-d'Arlas (2019)
Keratin (feathers)	High keratin content	FK/DCMC cross-linked films	Glycerol plasticization, DCMC cross-linking, and casting	Edible films and coatings	Dou et al. (2020)
Keratin (feathers)	High keratin content	Keratin/PLA blend fibersPLA: polylactic acid	Sulphitolysis extraction and electrospinning with PLA	Potential films and fibrous mats	Isarankura Na Ayutthaya et al. (2015)
Keratin (feathers)	High keratin content	Keratin/OD films	Extraction, 1,8-octanediol plasticization, and formaldehyde cross-linking	Films and coatings	Liu et al. (2018)
Keratin (feathers)	High keratin content	Chicken feather protein/gelatin/clove oil film	Protein extraction, gelatin blending, and clove oil addition	Active packaging films	Song et al. (2014)
Keratin (feathers)	High keratin content	FK/DAS film	Glycerol plasticization, DAS cross-linking, and casting	Films and coatings	Dou et al. (2015)

PLA, polylactic acid; *UV*, ultraviolet; *DCMC*, dialdehyde carboxymethyl cellulose; *FK*, feather keratin; *DAS*, dialdehyde starch.

oilcake, and wastewater, posing significant environmental challenges. With oilseed production reaching approximately 600 million tons in 2023, vast amounts of residues accumulate in key regions such as the USA, Argentina, and Malaysia. Beyond solid wastes, the industry produces wastewater from processing operations like equipment washing, characterized by high COD, biological oxygen demand, organic content, minerals, and lipids (fat, oil, and grease) (Naseem et al., 2022). Untreated disposal of this nutrient-rich wastewater damages soil, water, and air, amplifying pollution concerns (Yadav and Ali, 2022). Traditionally underutilized as animal feed or discarded, these wastes, including soybean meal (40%–50% protein), sunflower meal (30%–40% protein, and 20%–30% fiber), and palm oil waste (high in cellulose and HC), are rich in proteins, fibers, and lipids, making them prime candidates for bioplastic production. This valorization not only mitigates environmental harm but also supports the recycling of waste into valuable products like food packaging, aligning with circular economy principles and offering economic and ecological benefits (Lomer et al., 2024).

The urgency to address oilseed waste is heightened by the environmental drawbacks of conventional oil-based plastics, which globally produce 350–400 million metric tons annually, with only 1% being bioplastics (Naseem et al., 2022). Widely used in packaging due to their low cost and barrier properties, these synthetic plastics are non-biodegradable, leading to waste accumulation and microplastic pollution. In contrast, biodegradable bioplastics from oilseed waste, such as PHAs from waste cooking oil (WCO) and plant oil mill effluents, provide a sustainable alternative. Approximately 4% of oils worldwide are used for bioplastic production annually, with WCO, generating 11 billion liters in the USA and 1 billion liters in the EU yearly, serving as a cost-effective carbon source for PHAs, yielding up to 85% PHAs by weight (Naseem et al., 2022). Proteins from soybean, sunflower, and rapeseed meals can be extracted and cast into films with glycerol, offering good barrier properties for coatings or wrappers. Lipids in wastewater, meal, and animal fats from slaughterhouses (e.g., lard, tallow) are fermented by microbes like *Cupriavidus necator* into PHAs, suitable for compostable cups and trays. Cellulose from palm oil empty fruit bunches enhances bioplastic strength when extracted and processed into reinforced films or composites, ideal for containers. Delgado et al. (2018) highlight rapeseed meal, a by-product of oil extraction with ~35% protein, as a viable source for injection-molded bioplastics, with mechanical properties improved by 50% at higher mold temperatures (120°C) and up to 200% when reinforced with 20% PCL, suggesting applications in rigid packaging like trays (Delgado et al., 2018).

Production methods capitalize on these compositions effectively. Protein films are formed via alkaline extraction and casting, PHAs through fermentation of hydrolyzed waste or unfiltered WCO, and cellulose-based bioplastics via chemical or enzymatic extraction, with rapeseed meal processed via injection molding (Delgado et al., 2018), each tailored to the waste's biochemical profile. Applications include edible coatings from protein films, PHA-based single-use items for fast-food packaging, cellulose-reinforced wrappers, and rapeseed-based biocomposites for trays, all compostable and reducing landfill burdens. Recent studies, as shown in Table 7.7,

Table 7.7 Studies on oilseed industry waste in bioplastic production for food packaging.

Waste type	Composition	Primary bioplastic	Production method	Food packaging application	References
Soybean meal	40%–50% protein and carbohydrates	PHAs and protein films	Fermentation and film casting	Cups, films, and coatings	Muhammad et al. (2017)
Sunflower meal	30%–40% protein, and 20%–30% fiber	PHAs and protein films	Fermentation and film casting	Trays and wrappers	Alves et al. (2024)
Palm oil waste	High cellulose and hemicellulose	Cellulose composites and PHAs	Cellulose extraction and fermentation	Reinforced films and containers	Siregar et al. (2022)
Waste cooking oil (WCO)	High lipids and fatty acids	PHAs	Fermentation	Films and trays	Omar et al. (2023)
Animal fats	High fatty acids (e.g., 32%–61% saturated fatty acid)	PHAs	Fermentation	Cups and containers	Shlush and Davidovich-Pinhas (2022)
Plant oil mill effluents	High COD, lipids, and organic content	PHAs	Fermentation	Films and coatings	Phang et al. (2022)

PHAs, polyhydroxyalkanoates; *COD*, chemical oxygen demand.

underscore this potential, with innovations like genetic engineering of microbial strains enhancing PHA yields from oily waste (Naseem et al., 2022). Challenges such as inconsistent waste quality, high processing costs, regulatory compliance, and scalability persist, but advancements in biorefinery approaches and cost optimization, including the use of rapeseed meal with PCL (Delgado et al., 2018), are making oilseed waste a vital resource for sustainable food packaging, reducing greenhouse gas emissions by up to 200% and fossil energy use by 95% (Naseem et al., 2022).

7.4.7 Winery and brewery waste

The winery and brewery industries, essential to global food and beverage production, generate significant organic waste streams that hold untapped potential for bioplastic development, especially in sustainable food packaging. Grape pomace, a by-product of winemaking comprising skins, seeds, and stems, and brewer's spent grain (BSG), the primary residue from brewing, are abundant in regions like Europe, the USA, and Australia. Grape pomace accounts for 20%–25% of processed grapes, while BSG constitutes approximately 85% of brewing waste, with over 6.4 million tons produced annually in the EU alone (Margioulas, 2023). Traditionally composted, landfilled, or used as low-value animal feed, these wastes are rich in cellulose, HC, proteins, and fermentable sugars, making them ideal feedstocks for bioplastics such as PHAs and bio-based composites. This valorization aligns with circular economy principles, reducing landfill burdens and the environmental impact of synthetic plastics, which contribute 350–400 million metric tons of waste globally each year (Nehra et al., 2024), By transforming winery and brewery waste into biodegradable packaging, these industries can address pollution concerns—such as soil and water contamination from high-COD winery wastewater (Naseem et al., 2022), while meeting the rising demand for eco-friendly alternatives.

The composition of these wastes underpins their suitability for bioplastic production. Grape pomace contains 25%–55% cellulose, 11%–50% HC, 15%–40% lignin, 5%–12% proteins, and unfermented sugars, along with pectins and polyphenols (Jafarzadeh et al., 2025; Ezati et al., 2022). This lignocellulosic profile supports PHA synthesis, while pectins enable edible films, and lipids from seeds enhance fermentation potential. Spent grains, on the other hand, offer 25%–30% cellulose, 20%–25% HC, 9%–17% lignin, and 20%–30% protein, with residual sugars providing a robust base for both PHA production and composite bioplastics (Qazanfarzadeh et al., 2024; Margioulas, 2023). An unexpected benefit of BSG is its inherent antioxidant properties, which can be amplified in fortified bioplastics, as demonstrated by Nehra et al. (2024). These compositional strengths position grape pomace and BSG as complementary resources to oilseed waste like soybean meal and palm oil residues, broadening the scope of agri-food waste valorization for food packaging applications (Nehra et al., 2024). Production processes for these bioplastics vary based on waste type and desired outcomes. For grape pomace, fermentation is a key method, where hydrolysis releases sugars that microbes like *Pseudomonas resinovorans* convert into medium-chain-length PHAs.

Follonier et al. (2014) demonstrated that Solaris grape pomace yielded 21.3 g/L of PHAs, highlighting its potential for biodegradable films and trays (Follonier et al., 2014). Nehra et al. (2024) developed edible bowls by fortifying BSG with finger millet and jaggery, achieving a threefold increase in hardness and a 25% boost in antioxidant activity, ideal for bakery packaging (Nehra et al., 2024). A more detailed exploration comes from Qazanfarzadeh et al. (2024), who employed a sequential biorefinery process to extract Pr-Lig, HC, and NC from BSG, fabricating two distinct composites: Pr-Lig/NC and HC/NC, with NC content varying from 3% to % wt. The HC-based films exhibited superior tensile strength (5.2 MPa), Young's modulus (45 MPa), WVP (0.83×10^{-10} g/m/s/Pa), and water solubility (79.75%), attributed to the branched structure of HC. In contrast, Pr-Lig films showcased higher UV-barrier properties, antioxidant activity (47.71% DPPH scavenging), and antibacterial effects against *E. coli* (21.03%) and *Staphylococcus aureus* (63.67%), owing to polyphenolic lignin. The addition of NC enhanced water vapor barriers (18.07% for HC and 41.30% for Pr-Lig), tensile strength (57.21% for HC and 61.90% for Pr-Lig), and Young's modulus (201.84% for HC and 170.13% for Pr-Lig), although it reduced Pr-Lig elongation by 45.83%, reflecting NC's crystalline reinforcement (Qazanfarzadeh et al., 2024).

These bioplastics find diverse applications in food packaging, enhancing sustainability across the supply chain. Grape pomace-derived PHA films and trays serve single-use needs, while pectin-based coatings extend the shelf life of fresh produce, leveraging their bioactivity (Jafarzadeh et al., 2025). Spent grains contribute PHA-based cups and films for disposable use, with composite trays and edible bowls catering to dry goods and bakery items (Nehra et al., 2024). The study by Qazanfarzadeh et al. (2024) further demonstrated that BSG-derived Pr-Lig (protein-lignin)/NC and HC/NC coatings extended blueberry shelf life by 12 days, delaying decay through improved freshness indicators like TPC and weight loss, with Pr-Lig/NC showing the most promise due to its bioactive properties. Visually, these composites differ markedly: Pr-Lig films are opaque and brownish due to lignin's chromophoric nature, while HC films are white and translucent (Fig. 7.7). Biodegradability is another key advantage, with Qazanfarzadeh et al. (2024) showing that Pr-Lig and HC control films degraded rapidly in soil within 15 days, though NC-reinforced composites resisted longer due to structural integrity, unlike polyethylene, which remained intact (Fig. 7.8).

7.4.8 Root and tuber crop waste

Root and tuber crops, including potatoes, cassava, sweet potatoes, and yams, are cornerstone agricultural products, particularly in regions such as Africa, Asia, and South America, where they serve as vital sources of carbohydrates. The processing of these crops, whether for food,

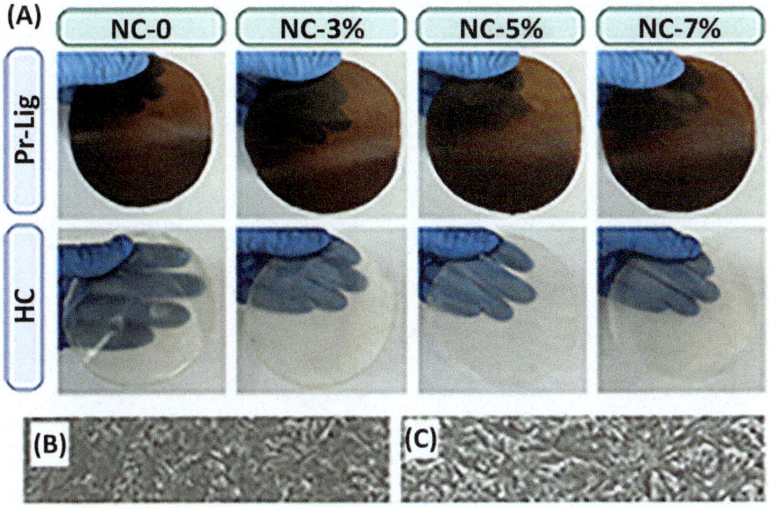

Figure 7.7
Photographs of protein-lignin and hemicellulose composites based on brewer's spent grain with/without nanocellulose.
The figure was reproduced from Qazanfarzadeh, Z., Masek, A., Chakraborty, S., Kumaravel, V., 2024. Development of brewer's spent grain-derived bio nanocomposites through a multiproduct biorefinery approach for food packaging. Industrial Crops and Products. 220, 119226. https://doi.org/10.1016/j.indcrop.2024.119226, with the permission of Elsevier.

starch extraction, or industrial uses, generates significant waste streams, such as peels, stalks, and residues like cassava bagasse. Rich in starch, cellulose, and fermentable sugars, these wastes present a valuable opportunity for bioplastic production, offering a sustainable alternative for food packaging (Teseme, 2020). Root and tuber crop waste provides a compelling feedstock for bioplastic development, complementing efforts with oilseed, winery, and brewery wastes in advancing sustainable food packaging. These crops, cultivated globally for their starchy roots and tubers, generate substantial residues during processing. For instance, potato processing yields approximately 15%–20% peel waste, while cassava processing produces millions of tons of bagasse annually (FAO Statistics). Predominantly found in developing regions, these wastes are often discarded or relegated to low-value uses like animal feed, yet their biochemical richness, high content of starch and lignocellulosic fibers, makes them ideal for bioplastics, reducing the environmental footprint of packaging and aligning with global sustainability goals (Sapuan et al., 2024).

The composition of root and tuber crop waste drives its potential for bioplastic synthesis. Potato peels, for example, contain 60%–70% starch, 10%–15% cellulose, and minor protein fractions (Sharma et al., 2018), offering a starch-heavy base for thermoplastic films. Cassava bagasse, the fibrous residue poststarch extraction, comprises 30%–40% cellulose, 20%–30% HC, and residual starch (Behera et al., 2022), suitable for fermentation into PHAs. Sweet

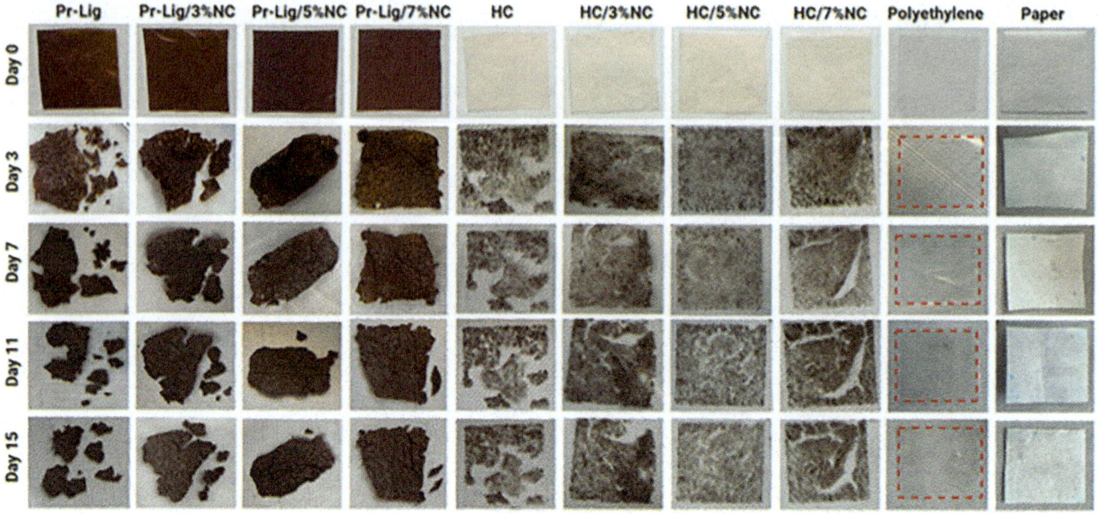

Figure 7.8
Photographs of composites, polyethylene, and paper derived from brewer's spent grain during 15 days of soil burial degradation.
The figure was reproduced from Qazanfarzadeh, Z., Masek, A., Chakraborty, S., Kumaravel, V., 2024. Development of brewer's spent grain-derived bio nanocomposites through a multiproduct biorefinery approach for food packaging. Industrial Crops and Products. 220, 119226. https://doi.org/10.1016/j.indcrop.2024.119226, with the permission of Elsevier.

potato peels, with 50%–60% starch and cellulose (Salah et al., 2025), and yam peels, rich in starch and fiber, further diversify the resource pool. An unexpected advantage lies in the presence of pectin in some wastes, such as potato peels, which can be extracted for edible films, enhancing shelf life and reducing waste (Maura et al., 2024). This compositional versatility parallels oilseed waste like soybean meal (40%–50% protein) and winery waste like grape pomace (25%–55% cellulose), yet root and tuber waste stand out for their high starch content, a direct precursor to bioplastic polymers (Jamal et al., 2025).

Production processes for these bioplastics split into two primary pathways starch-based and fermentation-based methods. For starch-based bioplastics, extraction from peels using water or enzymatic techniques yields high-purity starch, with potato peels. This starch is then plasticized with glycerol to form TPS, which is processed via reactive extrusion into films with robust tensile strength and water resistance, ideal for flexible packaging like wrappers (Sultan et al., 2024). Applications in food packaging highlight the practical utility of these bioplastics. Starch-based TPS from potato, sweet potato, and yam peels forms bags, wrappers, and trays for dry goods like snacks and cereals, offering good oxygen and moisture barriers. PHA-based bioplastics from cassava bagasse excel in broader uses, films, cups, and trays for fast-food packaging due to their FDA-approved status for food contact and superior barrier properties against water vapor and gases. Behera et al. (2022) developed yam starch-based bioplastic films reinforced with bentonite (0.5%, 1%, and 1.5% w/w) using glycerol and solvent casting, finding that the 1.5% bentonite film (B3) exhibited superior tensile strength, highest soil degradation (fully degraded in 55 days), lowest water absorption (degraded in 20–25 days), and reduced water vapor transmission due to bentonite's even distribution and silicate layers. Stable in salt and acidic conditions but not NaOH, these biodegradable films show promise as a sustainable food packaging alternative, with room for further enhancement (Fig. 7.9) (Behera et al., 2022).

7.4.9 Nut and seed processing waste

The processing of nuts and seeds, such as almonds, walnuts, sesame, and pumpkin, generates substantial waste streams, including shells, husks, and meal, that are rich in organic materials like cellulose, lignin, proteins, and lipids. These by-products, abundant in major producing regions like the USA, Turkey, and China, offer a promising avenue for bioplastic production, supporting sustainable food packaging. With global almond production at approximately 1.3 million tons in 2023 and sesame at 6 million tons annually (FAO Statistics), the sheer volume of waste underscores its potential to reduce reliance on virgin resources and mitigate plastic

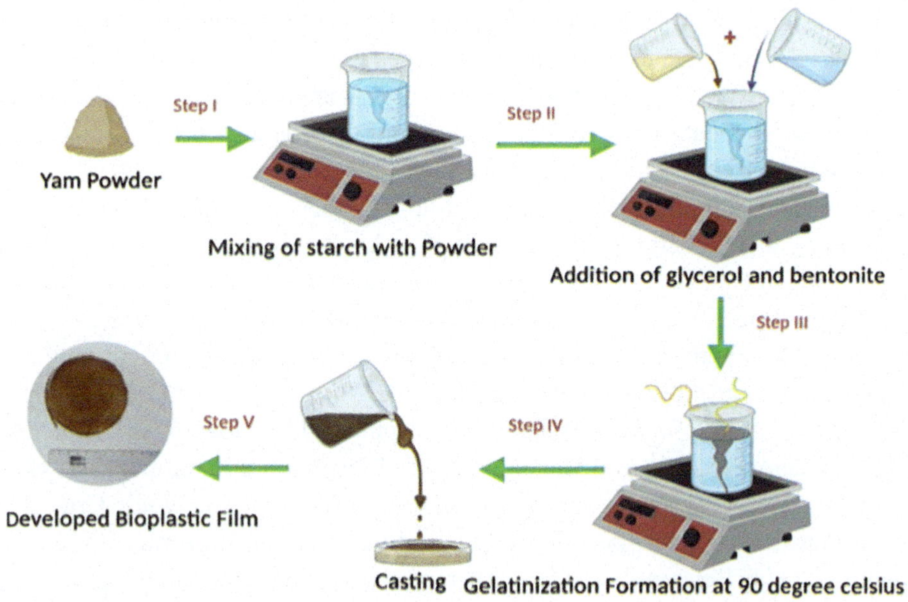

Figure 7.9
Schematic representation of the development of Yam starch bioplastic film using solvent casting method.
The figure was reproduced from Behera, L., Mohanta, M., Thirugnanam, A., 2022. Intensification of yam-starch based biodegradable bioplastic film with bentonite for food packaging application. Environmental Technology & Innovation. 25, 102180, with the permission of Elsevier.

pollution, aligning with circular economy goals (Senapati et al., 2024). Nut and seed processing waste provides a versatile feedstock for bioplastics, complementing efforts with oilseed, winery, brewery, and root/tuber waste in sustainable packaging innovation. In major nut-producing countries, shells from almonds and walnuts, husks from sesame and pumpkin seeds, and meals from oilseeds like soybean and sunflower are often underutilized, relegated to animal feed or fuel. Yet, their biochemical richness makes them ideal for bioplastics, offering an eco-friendly alternative to synthetic plastics, with significant potential in regions driving global nut and seed output (Valdés et al., 2022).

The composition of these wastes drives their bioplastic potential. Nutshells, such as almond (35% cellulose and 25% lignin) and walnut (20%–40% cellulose and 30%–50% lignin), are lignocellulosic powerhouses, ideal for reinforcing bioplastic films. Seed husks, like those from sesame (30% cellulose) and pumpkin, contain cellulose and polysaccharides, suitable for composites. Seed meals from oilseeds, rich in proteins (30%–50%) and lipids, supports PHA fermentation. An unexpected advantage emerges from walnut shells' phenolic compounds, which can enhance active packaging by inhibiting oxidation, adding functionality beyond structural use (Valdés et al., 2022).

Production processes leverage this diversity. Cellulose-based bioplastics from nut shells involve extraction via chemical or enzymatic methods, followed by film formation or reinforcement. For fermentation-based PHAs, seed meals are hydrolyzed to sugars, fermented by microbes like *Cupriavidus necator*, yielding thermally stable, FDA-approved bioplastics. Protein-based bioplastics from seed meals are processed into films or coatings, offering oxygen barriers and biodegradability. Recent studies highlight these approaches: McNeill et al. (2024) found that nutshells (walnut, almond, and peanut) as fillers in PLA and PBS increased stiffness but reduced flexibility due to agglomeration at higher loads, suggesting optimal use at lower percentages. Valdés et al. (2022) extracted antioxidant-rich compounds from almond shells, achieving high phenolic content for active films, enhancing shelf-life potential. Ali et al. (2019) reinforced starch films with walnut shell powder, improving tensile strength and water vapor barriers, indicating compatibility with starch matrices. Qian et al. (2025) developed walnut shell–carboxymethyl cellulose composites with exceptional UV-blocking and water resistance, broadening application scope. Leon-Bejarano et al. (2020) incorporated hazelnut skin extracts into starch films, reducing water solubility and boosting UV protection, while Cruz et al. (2024) combined walnut shells with pectin, yielding thicker, antioxidant-rich films with rapid biodegradability. Leon-Bejarano et al. (2020) reported that incorporating pecan nutshell and hazelnut skin extracts into the formulation enhanced water resistance by reducing solubility and increasing the contact angle, thereby improving hydrophobicity. Additionally, these extracts contributed to better UV-light-blocking properties and functioned as plasticizing agents. This suggests the potential of octenyl succinate starch combined with nut by-product extracts for developing active films suitable for food packaging and edible coatings, particularly for products requiring controlled humidity levels or UV protection (Fig. 7.10). Bianca (2022)

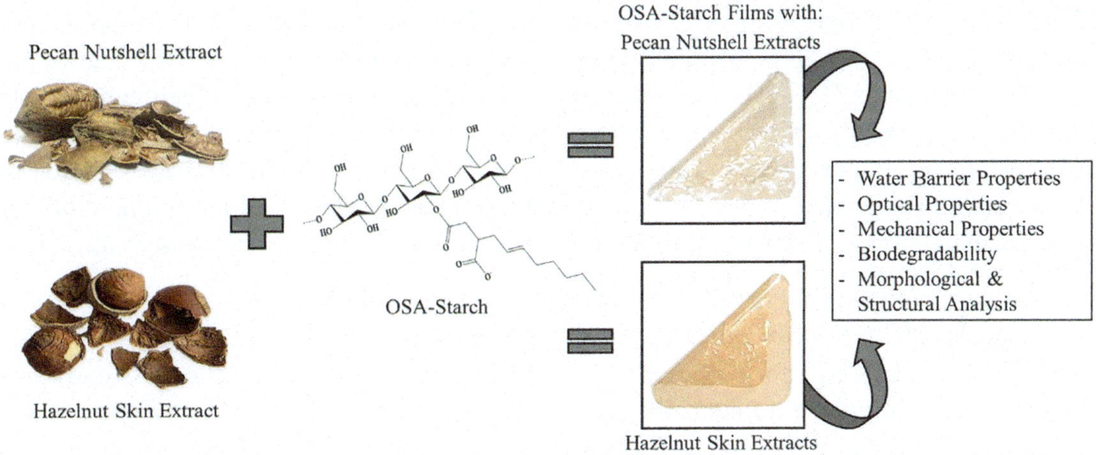

Figure 7.10
Development and properties of modified starch films with nut by-product extracts.
Leon-Bejarano, M., Durmus, Y., Ovando-Martínez, M., Simsek, S., 2020. Physical, barrier, mechanical, and biodegradability properties of modified starch films with nut by-products extracts. Foods. 9 (2), 226. https://doi.org/10.3390/foods9020226.

showed that nutshell fillers altered thermal and mechanical properties variably by particle size, with smaller grains enhancing modulus. These studies collectively demonstrate a progression from structural reinforcement to multifunctional films, with antioxidants adding value, although filler–matrix adhesion remains a challenge addressed by compatibilizers.

Applications for food packaging are broad. Cellulose-based bioplastics from shells form wrapping films and trays, offering mechanical strength for dry goods. PHA-based bioplastics from seed meals suit single-use cups and films, compostable postuse, while protein-based films from meal, enriched with antioxidants from walnut shells, serve as edible coatings for produce, extending shelf life. Challenges include waste variability, energy-intensive extraction, and scalability, with regulatory compliance per FDA standards critical. By March 7, 2025, innovations in biorefineries and compatibilizers are enhancing viability, positioning nut and seed waste as a key player in sustainable packaging, reducing environmental impact alongside other agri-wastes (Gupta et al., 2022).

7.4.10 Spent coffee bean waste

SCGs are the solid residues left after coffee brewing, a globally abundant by-product with over 6 million tons generated annually. Predominantly discarded in landfills or incinerated, SCGs are rich in organic compounds, making them a prime candidate for valorization (Oliveira et al., 2021). Their composition includes approximately 10% cellulose by dry weight, a structural carbohydrate suitable for bioplastic matrices. SCGs also contain about

15% lipids, rich in fatty acids like linoleic and palmitic acids, ideal for PHA production (Campos-Vega et al., 2015). Additionally, they comprise proteins (5%), HC, lignin, and phenolic compounds (1%–2% w/w GAE, primarily chlorogenic acids), which can enhance bioplastic functionality or serve as additives, aligning with circular economy principles by repurposing this abundant waste. The conversion of SCG into bioplastics leverages its diverse components through two primary methods include oil extraction for PHA production and cellulose extraction for bioplastic reinforcement (Oliveira et al., 2021; Nguyen et al., 2023).

SCG oil, extracted with solvents like n-hexane or supercritical CO_2, yields up to 15% oil rich in fatty acids, which *Cupriavidus necator* H16 ferments into PHAs like PHB, achieving high yields (YP/S = 0.82 g/g) for biodegradable plastics (Oliveira et al., 2021). Post extraction residues are hydrolyzed to fermentable sugars, further converted into PHAs by bacteria like *Bacillus megaterium* or *Burkholderia cepacia*, maximizing SCG use (Obruca et al., 2014). Meanwhile, SCGs' 10% cellulose content, extracted as nanofibers, reinforces bioplastics, enhancing tensile strength and water vapor resistance when combined with polymers like polyvinyl alcohol or TPS, with 10% SCG boosting elastic modulus in TPS blends (Nguyen et al., 2022). This dual approach enhances sustainability and bioplastic performance. SCG-derived bioplastics offer versatile applications in food packaging, leveraging both PHA and cellulose properties. PHA-based bioplastics, known for compostability and effective oxygen/moisture barriers, are ideal for single-use items like cups, trays, and films, suitable for wrapping fresh produce and dairy (Chen et al., 2022). Cellulose-based bioplastics, incorporated into forms like films or composites, enhance packaging for bread, fruits, and dried goods by improving mechanical strength and gas impermeability (Agüero et al., 2023). Nguyen et al. (2022) reported that TPS/SCG blends with 15%–20% SCG blocked 100% UV transmission at 320 nm and exhibited over 92% radical scavenging activity, suggesting active packaging potential due to SCG's antioxidants (Nguyen et al., 2022). Masssijaya et al. (2023) combined SCGs with PLA and TPS, noting enhanced thermal stability with 5% SCG, although mechanical properties declined at higher loadings, indicating optimal use at lower concentrations (Masssijaya et al., 2023).

7.4.11 Marine waste

Marine waste, encompassing discarded or underutilized marine organisms such as macroalgae (seaweed), microalgae, and fish industry by-products like scales, skins, and proteins, offers a promising resource for bioplastic production tailored to food packaging. With global plastic production nearing 390 million metric tons annually, approximately 40% of which is used for food packaging, the environmental impacts are severe, driving ocean pollution and ecosystem degradation (Florentino et al., 2022). Marine waste, rich in polysaccharides, proteins, and bioactive compounds, presents a renewable and abundant alternative that reduces reliance on fossil fuel–based plastics while aligning with sustainability goals and circular economy principles. This section explores how these materials are transformed into bioplastics, their properties, applications, and the challenges and future prospects of their use in food packaging (Agüero et al., 2023). The biochemical diversity of marine waste underpins its potential for bioplastic production. Seaweed, classified into green (Chlorophyta), red (Rhodophyta), and brown (Phaeophyta) types, provides distinct polysaccharides: alginate from brown seaweed, carrageenan from red seaweed, and cellulose from green seaweed. Microalgae, under nutrient-deficient conditions, produces PHAs internally, while fish waste provides an array of materials including muscle proteins, collagen, gelatin, and chitin from scales and shells (Fig. 7.11). Alginate and carrageenan excel at forming gels and films, cellulose can be processed into nanocrystals to boost strength, and PHAs offer biodegradable polymer options, collectively enabling a wide range of packaging solutions from edible coatings to robust films (Agüero et al., 2023).

Processing marine waste into bioplastics follows two primary pathways including direct extraction and fermentation. In direct extraction, polysaccharides like alginate and carrageenan are harvested from seaweed through washing, milling, drying, and chemical treatments, resulting in biodegradable films with effective oxygen and moisture barriers, some enhanced with antioxidants to extend shelf life (Debeaufort and Díaz-Moreno, 2021). Fish waste contributes significantly here, as showed in, with proteins, such as myofibrillar proteins from gilded catfish or Serra Spanish mackerel by-products and collagen from tilapia

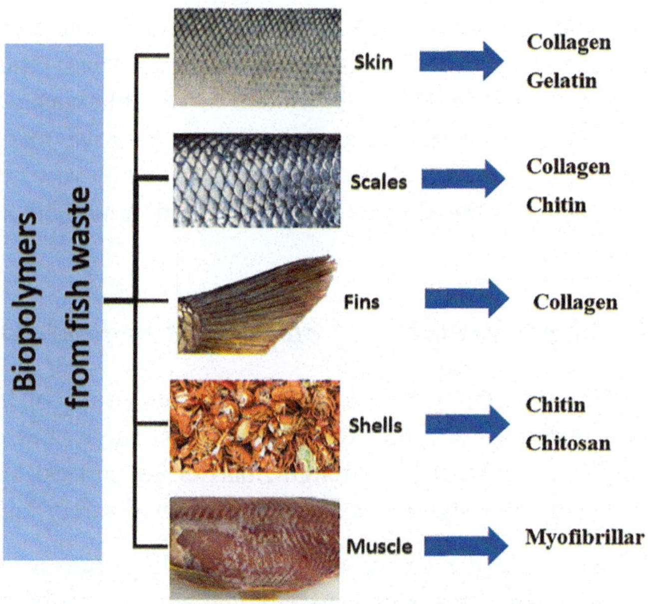

Figure 7.11
Biopolymers with potential food packaging application derived from fish industry waste.

skin, blended with starch or pectin and cast into films with optimized flexibility and strength, while fish scales paired with corn starch produce bioplastics via film casting (Arif et al., 2023). Alternatively, fermentation leverages algae sugars to feed microbes like *Cupriavidus necator*, producing PHAs from brown algae like *Laminaria japonica* or *Sargassum* sp., with microalgae like *Scenedesmus* sp. also yielding PHAs under nutrient stress, offering a scalable biopolymer source for packaging. These marine-derived bioplastics bring valuable properties to food packaging. Their biodegradability allows them to break down naturally within weeks under composting conditions, significantly reducing environmental pollution. Alginate, carrageenan, and fish protein films provide robust barriers against oxygen, moisture, and lipids, preserving food quality, while their nontoxic nature ensures compliance with food contact regulations like EU Regulation 10/2011. Sustainability is a key advantage, as utilizing marine waste minimizes dependence on finite land-based resources. Practical examples include alginate films extending the shelf life of sweet cherries and fish protein films offering low WVP, demonstrating their efficacy as alternatives to conventional plastics (Oliveira et al., 2022).

Applications of these bioplastics are advancing through both commercial and academic efforts. Companies like Notpla in the UK have pioneered seaweed-based edible wrappers and containers, scaling up by March 2025 to include broader food packaging solutions, while Kelpi develops compostable seaweed bioplastics and Evoware in Indonesia crafts edible seaweed sachets for food products (Yap et al., 2023). Academic research, such as Flinders University's seaweed coatings for fast-food packaging, complements these innovations, optimizing fish waste films for smart packaging that actively enhances food preservation through biopolymers like chitosan with antimicrobial properties (Hamed et al., 2022). However, challenges persist, including scalability due to variable waste composition, higher costs from energy-intensive processing, and the need to improve mechanical and barrier properties to rival traditional plastics. Table 7.8 summarizes key research on marine waste bioplastics, detailing waste types, bioplastic outputs, production methods, and food packaging applications.

7.5 Properties and performance of bioplastics in food packaging

The performance of bioplastics in food packaging hinges on a suite of functional properties that determines their suitability for applications like films, trays, and coatings. Table 7.9 defines these key properties, mechanical strength, barrier properties, biodegradability, and thermal stability, along with their significance in ensuring food safety, quality, and shelf life.

Mechanical strength, encompassing tensile strength and elongation at break, is crucial for ensuring bioplastics can endure handling and storage without failure. Research suggests alginate films from brown seaweed, with tensile strengths of 20–30 MPa and 10%–20% elongation, are suitable for lightweight films but fall short of polyethylene's 25–40 MPa and

Table 7.8 Recent studies about marine waste bioplastics for food packaging.

Waste type	Primary bioplastic	Production method	Food packaging application	References
Fish by-products (gilthead seabream)	Fish protein concentrate-gelatin films	Direct extraction and solvent casting	Biodegradable packaging films	Athanasopoulou et al. (2023)
Shrimp waste	Chitosan-rice straw composite films	Wet mixing and casting	Biodegradable food packaging films	Elhussieny et al. (2020)
Shrimp shells and cassava peel	Chitosan-starch bioplastic films	Direct processing and casting	Biodegradable food packaging	Dasumiati et al. (2019)
Marine actinomycetes	PHAs	Microbial fermentation of plastic waste	Biodegradable bioplastics for packaging	Oliveira et al. (2022)
Seaweed (brown)	Alginate-based films	Direct extraction and processing	Films for fruits and vegetables	Jayakody et al. (2022)
Seaweed (red)	Carrageenan-based films	Direct extraction and processing	Films for food packaging	Wang et al. (2024)
Oyster shells	LDPE-oyster shell blend films	Blending and thermal processing	Enhanced food preservation films	Qu et al. (2023)
Algae (microalgae)	PHAs	Fermentation using algae as carbon	Bioplastics for packaging	Arora et al. (2023)
Seaweed	PHAs	Fermentation using seaweed sugars	Bioplastics for packaging	Manikandan and Lens (2023)
Fish skin (tilapia)	Collagen-starch films	Direct extraction and film casting	Active food packaging films	Zhuang et al. (2019)
Fish scales	Corn starch-fish scale films	Direct extraction and film casting	Biodegradable packaging films	Surya et al. (2023)
Fish by-products	Protein-pectin films	Direct extraction and film casting	Sustainable food packaging	Florentino et al. (2022)
Lobster shells	Chitin-chitosan films	Extraction and film formation with TiO_2	Biodegradable packaging films	Raj et al. (2022)

PHAs, polyhydroxyalkanoates.

Table 7.9 Important properties and characteristics of bioplastics in food packaging.

Property	Definition	Significance in food packaging	References
Mechanical Strength	Ability to withstand tensile, compressive, or shear forces (e.g., tensile strength, elongation at break)	Ensures durability during handling, transport, and storage; prevents tearing or deformation	Zhao et al. (2019)
Barrier properties	Resistance to transmission of gases (oxygen) and water vapor	Maintains food freshness by preventing oxidation and moisture loss/gain	Arrieta et al. (2017)
Biodegradability	Rate and extent of decomposition by natural processes (e.g., composting)	Reduces environmental impact postuse; aligns with sustainability goals	Zhao et al. (2023)
Thermal stability	Resistance to degradation or softening at elevated temperatures	Enables use in hot-fill or high-temperature storage conditions	Chandra et al. (2024)
Antimicrobial properties	Ability to inhibit or kill microbial growth (e.g., bacteria, fungi)	Enhances food safety and extends shelf life by preventing spoilage	Abang et al. (2023)

200%–600% elongation, limiting their use in high-stress applications (Kajla et al., 2024). Chitin-chitosan films from lobster shells, enhanced with TiO_2, achieve tensile strengths of 3.34–3.96 MPa and 30%–50% elongation, making them ideal for flexible wraps, as demonstrated by Raj et al. (2022), who noted their suitability for packaging requiring flexibility. Fish protein films from tilapia skin, with around 15 MPa tensile strength and 10%–20% elongation, offer moderate strength for rigid trays, as shown by Zhuang et al. (2019), highlighting feedstock origin's influence on mechanical performance. Tan et al. (2022) investigated chitosan-reinforced starch-based bioplastics, achieving an optimized tensile strength of 5.19 MPa and elongation at break of 44.6%, demonstrating the importance of processing parameters such as starch concentration and glycerol loading (Tan et al., 2022). Iriani et al. (2019) reinforced bioplastics with NC from rice straw and palm empty fruit bunch, reporting tensile strength improvements of up to 23.3% and elongation increases of 24.76% compared to commercial bioplastics. Furthermore, Ratna et al. (2024) investigated the mechanical properties of bioplastic derived from tofu liquid waste, incorporating varying sorbitol concentrations. The results revealed that the bioplastic exhibited a thickness range of 0.049 to 0.365 mm, a WVP rate of 0.239–4.644 $g/m^2/h$, a tensile strength of 146.3–186.3 kgf/cm^2, and an elongation range of 8.52%–150.23%. The study determined that a 1% sorbitol concentration yielded optimal bioplastic properties, with a thickness of 0.110 mm, a WVP rate of 0.584 $/m^2/h$, a tensile strength of 146.3 kgf/cm^2, and an elongation of 97.04%. Further mechanical evaluations indicated that variations in sorbitol concentrations significantly influenced elongation values ($P < .01$), while tensile strength remained unaffected ($P > .05$). Additionally, Young's modulus analysis demonstrated that increased plasticizer concentrations reduced the rigidity of the bioplastic, as the starch network structure was modified (Fig. 7.12) (Ratna et al., 2024).

Barrier properties are critical for maintaining food freshness by preventing the transmission of oxygen and water vapor, thereby influencing the shelf life of packaged products. The permeability of gases through nonporous packaging materials occurs through adsorption at the primary interfaces, diffusion through the packaging material, and disruption at the irregular surfaces. This permeability is measured by three factors: transmission rate, permeance, and permeability (Jõgi and Bhat, 2020). Transmission rate refers to the weight or volume of a permeant, such as oxygen or moisture, passing through the packaging film per unit time and surface area under equilibrium conditions. Permeance is the ratio of the transmission rate to the difference in partial pressure of the permeant across the material, while permeability is the product of permeance and thickness (Mangaraj et al., 2015). These barrier properties are influenced by pressure, temperature, and relative humidity and are assessed under controlled equilibrium moisture conditions (Wang et al., 2018). Key parameters studied in packaging applications include oxygen, carbon dioxide, and water activity, as changes in these factors can alter the quality of packaged food, leading to color changes, pH fluctuations, weight loss, and microbial growth (Wang et al., 2018). Bioplastic food packaging derived from food waste exhibits a range of barrier properties that can match or even exceed those of conventional plastics in certain applications. Alginate films, for instance, demonstrate exceptional moisture control with WVP as low as 2.5–3 g·mm/m²·day·kPa, outperforming polyethylene in reducing moisture transmission (Mohammed et al., 2023). Oyster shell–LDPE blends enhance moisture resistance further by creating tortuous diffusion paths, while carrageenan films from red

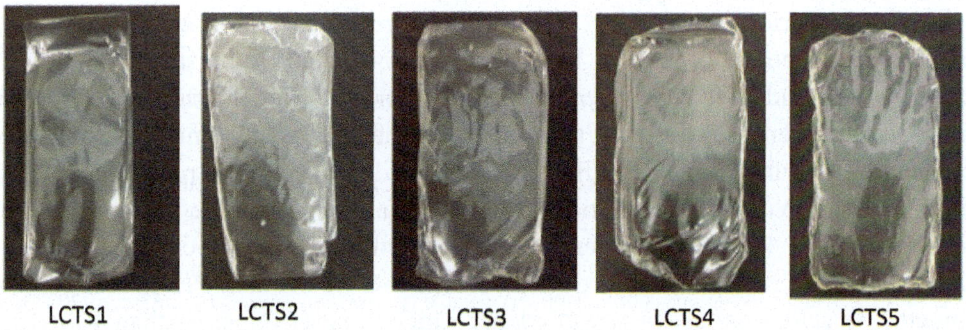

Figure 7.12

Bioplastic packaging is produced based on tofu liquit waste. LCTS1 (sorbitol 0%), LCTS2 (sorbitol 1%), LCTS3 (sorbitol 2%), LCTS4 (sorbitol 4%), and LCTS5 (sorbitol 8%).

The figure was reproduced from Ratna, R., Mutia, M., Darwin, D., Munawar, A.A., Fitriani, F., Handayani, L., 2024. Utilization of tofu liquid waste for the manufacture of bioplastic food packaging. Case Studies in Chemical and Environmental Engineering. 10, 100830. https://doi.org/10.1016/j.cscee.2024.100830, with the permission of Elsevier.

seaweed excel at lipid barrier properties, making them suitable for fatty food packaging (Nunes et al., 2024). Fish protein–pectin films have been shown to effectively reduce WVP, thereby extending shelf life, demonstrating how marine waste can be utilized as a sustainable alternative to conventional plastics (Arrieta et al., 2017; Florentino et al., 2022). The selection of packaging materials also depends on their structural and thermal properties. Polymer structural design plays a crucial role in mechanical properties and influences the processing methods used, such as sheet extrusion, thermoforming, injection molding, blow molding, and film formation. Thermal properties, including melting temperature and glass transition temperature, are particularly significant for food packaging materials that require low-temperature storage and are commonly assessed using differential scanning calorimetry (Krasnova et al., 2012). A key advantage of bioplastics over petrochemical-based plastics is their biodegradability and compostability. While all compostable biopolymers are biodegradable, not all biodegradable biopolymers are compostable. Compostability refers to the ability of a polymer to break down into compost under suitable conditions (Gutierrez et al., 2019). To optimize food preservation, packaging materials must maintain a balance between permeability to gases and moisture while ensuring mechanical and thermal stability, thereby extending the shelf life of packaged foods effectively (Birania et al., 2022).

Biodegradability refers to the natural decomposition of polymers into simpler compounds such as carbon dioxide, water, and humus. Various factors, including microbial activity, environmental conditions, and polymer composition, influence the degradation process. Bioplastics such as alginate and chitin-chitosan films degrade within weeks under composting conditions, while PHAs break down over months, contrasting sharply with the long persistence of conventional plastics like polyethylene. Biodegradation can occur under aerobic (soil and compost) and anaerobic (aquatic) conditions, with factors such as temperature and nutrient availability affecting the process (Birania et al., 2022). Testing methods for biodegradability include laboratory-controlled enzyme tests, simulation tests in soil and compost, and field tests in natural environments. Biodegradation rates in marine environments are generally higher than in soil due to greater microbial activity and environmental exposure. However, effective biodegradation requires proper waste management infrastructure to optimize conditions for decomposition. Evidence suggests that alginate films degrade within 4–6 weeks under composting conditions, chitin-chitosan films derived from lobster shells achieve complete degradation in 6–8 weeks, and PHAs break down over 6–12 months, in stark contrast to polyethylene, which persists for more than 100 years (Rai and Choure, 2023). This rapid breakdown, particularly in marine-derived bioplastics, aligns with sustainability goals, although it requires controlled disposal environments, highlighting the need for infrastructure development (Zhao et al., 2019). Thermal stability, the ability to resist degradation or softening at elevated temperatures, is vital for hot-fill or high-temperature storage. PHAs from marine actinomycetes withstand up to 250°C, suitable for heat-intensive applications, while alginate films hold to 200°C (Oliveira et al., 2022). PLA from sugar waste, however, softens near 60°C, limiting its range, and fish protein films offer moderate stability around 150°C, lagging behind polyethylene's 300°C threshold (Maluin, 2024). Feedstock origin thus influences thermal performance, with marine polysaccharides offering robust options for thermal challenges.

Antimicrobial properties, defined as the ability to inhibit microbial growth, play a crucial role in enhancing food safety and extending shelf life, an increasingly valued feature in bioplastics. Several studies highlight the potential of food waste-derived bioplastics in active food packaging applications. Fruit peel-derived bioplastics further expand this potential. Research by Kharb and Saharan (2024) highlights biodegradable films synthesized from orange, mango, banana, and sapodilla peels. These films exhibit significant antibacterial activity against *E. coli*, making them promising candidates for sustainable food packaging (Kharb and Saharan, 2024). The study also emphasizes the role of glycerol and sorbitol as plasticizers, affecting moisture content, solubility, and biodegradation rates, with mango peel-derived films showing up to 86.67% biodegradation (Kharb and Saharan, 2024). Biopolymers derived from tropical plant by-products also demonstrate antimicrobial and antioxidant properties. Jafarzadeh et al. (2020) review the potential of tropical plant extracts in bioactive food packaging, effectively reducing microbial spoilage and lipid oxidation. This highlights the role of bioactive compounds in enhancing food preservation while promoting environmentally friendly packaging solutions (Jafarzadeh et al., 2020). Another innovative approach utilizes cassava waste starch for developing antimicrobial bioplastics. Silveira et al. (2025) incorporated clove oil into cassava-based films, achieving inhibition zones averaging 11.87 mm against *S. aureus*, *Salmonella typhimurium*, and *Listeria monocytogenes* (Silveira et al., 2025).

Bioplastics derived from food waste present a compelling alternative to conventional plastics, offering a balance between functionality and sustainability. While traditional petroleum-based materials like LDPE excel in barrier properties and thermal stability, their environmental drawbacks, particularly non-biodegradability, remain a major concern. In contrast, bioplastics such as PLA and PHAs exhibit promising characteristics, with PHAs standing out for their exceptional biodegradability and thermal stability. However, challenges remain, including PLA's poor thermal resistance and the need for enhanced antimicrobial functionality in certain applications (Gaur et al., 2024).

7.6 Challenges and future directions

The adoption of bioplastics derived from agri-food wastes and by-products in food packaging encounters several challenges that must be addressed to realize their full potential. A key obstacle is the variability in feedstock composition. Agri-food wastes, such as SCB, cereal bran, fruit peels, and marine discards, originate from diverse sources and processes, leading to inconsistencies in their chemical makeup, starch, cellulose, proteins, and lipids, which can affect the uniformity and quality of the resulting bioplastics (Ghasemlou et al., 2024). This variability complicates the production of materials with consistent mechanical strength, barrier properties, and biodegradability, essential for food packaging applications. Another significant challenge is scalability. While laboratory-scale studies have demonstrated the feasibility of converting these wastes into bioplastics, transitioning to industrial-scale production is hindered by high costs, technological limitations, and the need for efficient supply chains to collect and process waste materials. For instance, pretreatment methods like enzymatic hydrolysis or microbial fermentation, required to standardize waste composition, remain energy-intensive and costly (Versino et al., 2023).

Regulatory compliance presents an additional challenge. Bioplastics designed for food contact must meet rigorous safety and quality requirements, such as EU Regulation 10/2011, which can be difficult to achieve with waste-derived materials due to potential contaminants or inconsistent properties. Moreover, consumer acceptance remains a significant obstacle. Even when safe and effective, packaging made from waste may encounter resistance from consumers familiar with traditional plastics, requiring targeted education and marketing initiatives to change perceptions. Finally, competition with alternative applications, such as animal feed or bioenergy production, may restrict the supply of agri-food wastes for bioplastics, especially for high-volume feedstocks like cereal straw or winery by-products (Jafarzadeh et al., 2022). Looking to the future, addressing these challenges requires targeted research and innovation. Developing standardized pretreatment techniques, such as advanced enzymatic processes or microbial consortia, could mitigate feedstock variability, ensuring consistent bioplastic quality. Advances in production technologies, including cost-effective biorefinery approaches and optimized fermentation methods, are critical to enhance scalability and reduce costs. For example, integrating bioplastic production with other waste valorization strategies, like bioenergy generation from winery wastewater or nutrient recovery from dairy could improve economic viability and sustainability. Collaborations across sectors between researchers, industry, and regulatory bodies will be vital to streamline approval processes and establish clear guidelines for waste-derived bioplastics. Exploring novel waste streams, such as residues from emerging food processing technologies or underutilized crops (e.g., tropical fruit waste), could expand feedstock options. Additionally, enhancing bioplastic properties through nanotechnology (e.g., NC reinforcement) or bioactive additives (e.g., antioxidants from coffee grounds) offer opportunities to meet diverse packaging needs, from antimicrobial coatings to high-strength trays. These directions promise to position agri-food waste-derived bioplastics as a cornerstone of sustainable food packaging (Babaee et al., 2022).

7.7 Conclusion

Bioplastics derived from agri-food wastes and by-products offer a transformative solution to the interconnected challenges of plastic pollution and waste management in the food packaging industry. By repurposing abundant renewable resources, such as SCB, cereal residues, fruit peels, dairy whey, and marine by-products into functional biopolymers, this strategy reduces dependence on fossil fuels and mitigates the environmental persistence of conventional plastics. The chapter demonstrates that these bioplastics, including PLA, PHAs, and starch-based composites, achieve performance metrics, mechanical strength, barrier properties, and biodegradability, comparable to traditional plastics, making them viable for applications like films, trays, and edible coatings. Case studies, such as PHA-based trays from winery waste and pectin films from orange peels, underscore their practical utility in extending shelf life and ensuring food safety. Beyond environmental benefits, this approach offers economic advantages, creating new markets for waste materials and reducing disposal costs, while aligning with circular economic principles. However, challenges remain, including feedstock variability, scalability, regulatory hurdles, and consumer perceptions, which temper immediate widespread adoption. Nevertheless, the field is dynamic, with ongoing research tackling these issues through innovative processing techniques, cross-sector partnerships, and the exploration of novel feedstocks. With continued investment in technology and collaboration, bioplastics from agri-food waste hold the potential to become a mainstream, sustainable solution, significantly advancing the future of food packaging and contributing to global efforts to combat plastic pollution.

References

Abang, S., Wong, F., Sarbatly, R., Sariau, J., Baini, R., Besar, N.A., 2023. Bioplastic classifications and innovations in antibacterial, antifungal, and antioxidant applications. Journal of Bioresources and Bioproducts 8 (4), 361–387. Available at https://doi.org/10.1016/j.jobab.2023.06.005.

Agüero, A., Lascano, D., Ivorra-Martinez, J., Gómez-Caturla, J., Arrieta, M.P., Balart, R., 2023. Use of bacterial cellulose obtained from kombucha fermentation in spent coffee grounds for active composites based on PLA and maleinized linseed oil. Industrial Crops and Products 202, 116971. Available at https://doi.org/10.1016/j.indcrop.2023.116971.

Aguilar-Rivera, N., 2024. Life cycle assessment of valorization of root and tuber crop wastes for bio-commodities and biofuels: cassava as a case study. In: Ray, R.C. (Ed.), Roots, Tubers, and Bulb Crop Wastes: Management by Biorefinery Approaches. Springer Nature, Singapore, pp. 333–350. Available at https://doi.org/10.1007/978-981-99-8266-0_15.

Ahmed, S., Janaswamy, S., Yadav, M.P., 2024. Biodegradable films from the lignocellulosic fibers of wheat straw biomass and the effect of calcium ions. International Journal of Biological Macromolecules 264 (Pt 1), 130601. Available at https://doi.org/10.1016/j.ijbiomac.2024.130601.

Akhavan-Mahdavi, S., Mirzazadeh, M., Alam, Z., Solaimanimehr, S., 2023. The effect of chitosan coating combined with cold plasma on the quality and safety of pistachio during storage. Food Science & Nutrition 11 (7), 4296–4307. Available at https://doi.org/10.1002/fsn3.3355.

Akmeemana, C., Somendrika, D., Wickramasinghe, I., Wijesekara, I., 2024. Cassava pomace-based biodegradable packaging materials: a review. Journal of Food Science and Technology 61 (6), 1013–1034. Available at https://doi.org/10.1007/s13197-023-05807-y.

Ali, A., Ali, S., Yu, L., Liu, H., Khalid, S., Hussain, A., et al., 2019. Preparation and characterization of starch-based composite films reinforced by apricot and walnut shells. Journal of Applied Polymer Science 136 (38), 47978. Available at https://doi.org/10.1002/app.47978.

Ali, W., Ali, H., Souissi, S., Zinck, P., 2023. Are bioplastics an ecofriendly alternative to fossil fuel plastics? Environmental Chemistry Letters 21 (4), 1991–2002. Available at https://doi.org/10.1007/s10311-023-01601-6.

Alves, L.T.D.O., Fronza, P., Gonçalves, I., da Silva, W.A., Oliveira, L.S., Franca, A.S., 2024. Development of polymeric films based on sunflower seed proteins and locust bean gum. Polymers 16 (13), 1905. Available at https://doi.org/10.3390/polym16131905.

Angellier-Coussy, H., Kemmer, D., Gontard, N., Peyron, S., 2020. Physical–chemical and structural stability of PHBV/wheat straw fibers based biocomposites under food contact conditions. Journal of Applied Polymer Science 137 (40), 49231. Available at https://doi.org/10.1002/app.49231.

Arif, A., Azeem, F., Rasul, I., Siddique, M.H., Zubair, M., Muneer, F., et al., 2023. Bioplastics from waste biomass of marine and poultry industries. Journal of Biosciences 48 (2), 11. Available at https://doi.org/10.1007/s12038-023-00332-8.

Arora, Y., Sharma, S., Sharma, V., 2023. Microalgae in bioplastic production: a comprehensive review. Arabian Journal for Science and Engineering 48 (6), 7225–7241.

Arrieta, M.P., Peponi, L., López, D., López, J., Kenny, J.M., 2017. 12 - An overview of nanoparticles role in the improvement of barrier properties of bioplastics for food packaging applications. In: Grumezescu, A.M. (Ed.), Food Packaging. Academic Press, pp. 391–424. Available at https://doi.org/10.1016/b978-0-12-804302-8.00012-1.

Athanasopoulou, E., Michailidi, A., Ladakis, D., Kalliampakou, K.I., Flemetakis, E., Koutinas, A., et al., 2023. Extraction of fish protein concentrates from discards and combined application with gelatin for the development of biodegradable food packaging. Sustainability 15 (15), 12062. Available at https://doi.org/10.3390/su151512062.

Azmin, S.N.H.M., Hayat, N.A.B.M., Nor, M.S.M., 2020. Development and characterization of food packaging bioplastic film from cocoa pod husk cellulose incorporated with sugarcane bagasse fibre. Journal of Bioresources and Bioproducts 5 (4), 248–255. Available at https://doi.org/10.1016/j.jobab.2020.10.003.

Babaee, M., Garavand, F., Rehman, A., Jafarazadeh, S., Amini, E., Cacciotti, I., 2022. Biodegradability, physical, mechanical and antimicrobial attributes of starch nanocomposites containing chitosan nanoparticles. International Journal of Biological Macromolecules 195, 49–58. Available at https://doi.org/10.1016/j.ijbiomac.2021.11.162.

Baiju, P., Mukkadan, M.J., Radhakrishnan, P., 2024. Husk and straw of cereals grains for sustainable food packaging. Agro-Waste Derived Biopolymers and Biocomposites: Innovations and Sustainability in Food Packaging. Wiley, pp. 153–185. Available at https://doi.org/10.1002/9781394175161.ch6.

Bajsic, E.G., Persic, A., Jemric, T., Buhin, J., Kucic Grgic, D., Zdraveva, E., et al., 2021. Preparation and characterization of polyethylene biocomposites reinforced by rice husk: application as potential packaging material. Chemistry (Weinheim an der Bergstrasse, Germany) 3 (4), 1344–1362. Available at https://doi.org/10.3390/chemistry3040096.

Bangar, S.P., Kajla, P., Ghosh, T., 2023. Valorization of wheat straw in food packaging: a source of cellulose. International Journal of Biological Macromolecules 227, 762–776.

Behera, L., Mohanta, M., Thirugnanam, A., 2022. Intensification of yam-starch based biodegradable bioplastic film with bentonite for food packaging application. Environmental Technology & Innovation 25, 102180.

Bianca, P., 2022. Bioplastics filled with nut shells particles: valorization of agro-wastes for the development of sustainable materials. [Laurea magistrale], Università di Bologna, Corso di Studio in Ingegneria chimica e di processo [LM-DM270]. https://amslaurea.unibo.it/id/eprint/25023.

Bioplastics Market Development Update 2024. https://www.european-bioplastics.org/bioplastics-market-development-update-2024/.

Birania, S., Kumar, S., Kumar, N., Attkan, A.K., Panghal, A., Rohilla, P., et al., 2022. Advances in development of biodegradable food packaging material from agricultural and agro-industry waste. Journal of Food Process Engineering 45 (1), e13930. Available at https://doi.org/10.1111/jfpe.13930.

Bishop, G., Styles, D., Lens, P.N., 2021. Environmental performance comparison of bioplastics and petrochemical plastics: a review of life cycle assessment (LCA) methodological decisions. Resources, Conservation and Recycling 168, 105451.

Boccalon, E., Gorrasi, G., 2022. Functional bioplastics from food residual: potentiality and safety issues. Comprehensive Reviews in Food Science and Food Safety 21 (4), 3177–3204. Available at https://doi.org/10.1111/1541-4337.12986.

Brennan, L., Francis, C., Jenkins, E.L., Schivinski, B., Jackson, M., Florence, E., et al., 2023. Consumer perceptions of food packaging in its role in fighting food waste. Sustainability 15 (3), 1917. Available at https://doi.org/10.3390/su15031917.

Campos-Vega, R., Loarca-Piña, G., Vergara-Castañeda, H.A., Oomah, B.D., 2015. Spent coffee grounds: A review on current research and future prospects. Trends in Food Science & Technology 45, 24–36. Available at https://doi.org/10.1016/j.tifs.2015.04.012.

Carnaval, L.D.S.C., Jaiswal, A.K., Jaiswal, S., 2024. Agro-food waste valorization for sustainable bio-based packaging. Journal of Composites Science 8 (2), 41.

Castro-Criado, D., Rivera-Flores, O., Abdullah, J.A.A., Castro-Osorto, E., Alonso-González, M., Ramos-Casco, L., et al., 2023. Valorization of Honduran agro-food waste to produce bioplastics. Polymers 15 (12), 2625. Available at https://doi.org/10.3390/polym15122625.

Chalermthai, B., Chan, W.Y., Bastidas-Oyanedel, J.R., Taher, H., Olsen, B.D., Schmidt, J.E., 2019. Preparation and characterization of whey protein-based polymers produced from residual dairy streams. Polymers 11 (4), 722. Available at https://doi.org/10.3390/polym11040722.

Chalermthai, B., Giwa, A., Schmidt, J.E., Taher, H., 2021. Life cycle assessment of bioplastic production from whey protein obtained from dairy residues. Bioresource Technology Reports 15, 100695. Available at https://doi.org/10.1016/j.biteb.2021.100695.

Chan, J.X., Wong, J.F., Hassan, A., Zakaria, Z., 2021. Bioplastics from agricultural waste. Biopolymers and Biocomposites from Agro-Waste for Packaging Applications. Elsevier, pp. 141–169.

Chandarana, J., Chandra, S., 2021. Production of bioplastics from banana peels. International Journal of Scientific Research & Engineering Trends 7 (1), 131–133.

Chandra, D.K., Kumar, A., Mahapatra, C., 2024. Ecofriendly bioplastics from biowaste: antimicrobial and functional enhancements for sustainable packaging. European Polymer Journal 113557.

Chandrasekar, C.M., Krishnamachari, H., Farris, S., Romano, D., 2023. Development and characterization of starch-based bioactive thermoplastic packaging films derived from banana peels. Carbohydrate Polymer Technologies and Applications 5, 100328. Available at https://doi.org/10.1016/j.carpta.2023.100328.

Chen, Y., Li, Q., Feng, C., Hu, Y., Liu, Y., Tian, J., 2022. Biocomposites based on spent coffee grounds and application in packaging. Interdisciplinary Research for Printing and Packaging. Springer, pp. 455–461.

Chen, G., Li, J., Sun, Y., Wang, Z., Leeke, G.A., Moretti, C., et al., 2024. Replacing traditional plastics with biodegradable plastics: impact on carbon emissions. Engineering 32, 152–162. Available at https://doi.org/10.1016/j.eng.2023.10.002.

Chougan, M., Ghaffar, S.H., Mijowska, E., Kukułka, W., Sikora, P., 2022. High-performance polylactic acid compressed strawboard using pre-treated and functionalised wheat straw. Industrial Crops and Products 184, 114996. Available at https://doi.org/10.1016/j.indcrop.2022.114996.

Cruz, R.M.S., Krauter, V., Krauter, S., Agriopoulou, S., Weinrich, R., Herbes, C., et al., 2022. Bioplastics for food packaging: environmental impact, trends and regulatory aspects. Foods 11 (19), 3087. Available at https://doi.org/10.3390/foods11193087.

Cruz, R.M.S., Rainer, B., Wagner, I., Krauter, V., Janalíková, M., Vicente, A.A., et al., 2024. Valorization of cork stoppers, coffee-grounds and walnut shells in the development and characterization of pectin-based composite films: physical, barrier, antioxidant, genotoxic, and biodegradation properties. Polymers 16 (8), 1053. Available at https://doi.org/10.3390/polym16081053.

Curling, S.F., Laflin, N., Davies, G.M., Ormondroyd, G.A., Elias, R.M., 2017. Feasibility of using straw in a strong, thin, pulp moulded packaging material. Industrial Crops and Products 97, 395–400. Available at https://doi.org/10.1016/j.indcrop.2016.12.042.

da Silva Junior, C.J.G., de Medeiros, A.D.M., Cavalcanti, A.K.L.H., de Amorim, J.D.P., Durval, I.J.B., Cavalcanti, Y.F., Converti, A., Costa, A.F.S., Sarubbo, L.A., 2024. Towards sustainable packaging using microbial cellulose and sugarcane (Saccharum officinarum L.) bagasse. Materials 17, 3732. Available at https://doi.org/10.3390/ma17153732.

Dasumiati, Saridewi, N., Malik, M., 2019. Food packaging development of bioplastic from basic waste of cassava peel (manihot uttilisima) and shrimp shell. IOP Conference Series: Materials Science and Engineering 1, 012053. IOP Publishing.

De Moura, I.G., De Sa, A.V., Abreu, A.S.L.M., Machado, A.V.A., 2017. Bioplastics from agro-wastes for food packaging applications. Food Packaging. Elsevier, pp. 223–263.

de Oliveira, P.Z., de Souza Vandenberghe, L.P., Soccol, C.R., 2024. Introduction About Second-and Third-Generation Bioplastics. Second and Third Generation Bioplastics. CRC Press, pp. 1–14.

de Paula, C.B.C., de Paula-Elias, F.C., Rodrigues, M.N., Coelho, L.F., de Oliveira, N.M.L., de Almeida, A.F., Contiero, J., 2021. Polyhydroxyalkanoate synthesis by burkholderia glumae into a sustainable sugarcane biorefinery concept. Frontiers in Bioengineering and Biotechnology 8. Available at https://doi.org/10.3389/fbioe.2020.631284.

Debeaufort, F., Díaz-Moreno, I., 2021. Active biopackaging produced from by-products and waste from food and marine industries. FEBS Open Bio 11 (4), 984–998. Available at https://doi.org/10.1002/2211-5463.13121.

Delgado, M., Felix, M., Bengoechea, C., 2018. Development of bioplastic materials: from rapeseed oil industry by products to added-value biodegradable biocomposite materials. Industrial Crops and Products 125, 401–407.

Deshwal, G.K., Alam, T., Panjagari, N.R., Bhardwaj, A., 2021. Utilization of cereal crop residues, cereal milling, sugarcane and dairy processing by-products for sustainable packaging solutions. Journal of Polymers and the Environment 29 (7), 2046–2061. Available at https://doi.org/10.1007/s10924-020-02034-w.

Dixit, S., Yadav, V.L., 2019. Optimization of polyethylene/polypropylene/alkali modified wheat straw composites for packaging application using RSM. Journal of Cleaner Production 240, 118228. Available at https://doi.org/10.1016/j.jclepro.2019.118228.

Dixit, S., Yadav, V.L., 2020. Comparative study of polystyrene/chemically modified wheat straw composite for green packaging application. Polymer Bulletin 77 (3), 1307–1326. Available at https://doi.org/10.1007/s00289-019-02804-0.

Dou, Y., Huang, X., Zhang, B., He, M., Yin, G., Cui, Y., 2015. Preparation and characterization of a dialdehyde starch crosslinked feather keratin film for food packaging application. RSC Advances 5 (34), 27168–27174. Available at https://doi.org/10.1039/C4RA15469J.

Dou, Y., Zhang, L., Zhang, B., He, M., Shi, W., Yang, S., et al., 2020. Preparation and characterization of edible dialdehyde carboxymethyl cellulose crosslinked feather keratin films for food packaging. Polymers 12 (1), 158. Available at https://doi.org/10.3390/polym12010158.

Ebrahimian, F., Denayer, J.F., Karimi, K., 2022. Potato peel waste biorefinery for the sustainable production of biofuels, bioplastics, and biosorbents. Bioresource Technology 360, 127609.

Elhussieny, A., Faisal, M., D'Angelo, G., Aboulkhair, N.T., Everitt, N.M., Fahim, I.S., 2020. Valorisation of shrimp and rice straw waste into food packaging applications. Ain Shams Engineering Journal 11 (4), 1219–1226. Available at https://doi.org/10.1016/j.asej.2020.01.008.

Ezati, P., Rhim, J.-W., Molaei, R., Priyadarshi, R., Han, S., 2022. Cellulose nanofiber-based coating film integrated with nitrogen-functionalized carbon dots for active packaging applications of fresh fruit. Postharvest Biology and Technology 186, 111845. Available at https://doi.org/10.1016/j.postharvbio.2022.111845.

Fernández-d'Arlas, B., 2019. Tough and functional cross-linked bioplastics from sheep wool keratin. Scientific Reports 9 (1), 14810. Available at https://doi.org/10.1038/s41598-019-51393-5.

Florentino, G.I.B., Lima, D.A.S., Santos, M.M.F., Ferreira, V.C.D.S., Grisi, C.V.B., Madruga, M.S., et al., 2022. Characterization of a new food packaging material based on fish by-product proteins and passion fruit pectin. Food Packaging and Shelf Life 33, 100920. Available at https://doi.org/10.1016/j.fpsl.2022.100920.

Follonier, S., Goyder, M.S., Silvestri, A.-C., Crelier, S., Kalman, F., Riesen, R., Zinn, M., 2014. Fruit pomace and waste frying oil as sustainable resources for the bioproduction of medium-chain-length polyhydroxyalkanoates. International Journal of Biological Macromolecules 71, 42–52. Available at https://doi.org/10.1016/j.ijbiomac.2014.05.061.

Gaur, V.K., Gaur, P., Telegin, A., Thakur, R.S., Sharma, P., Gupta, P., et al., 2024. Unlocking the potential of food waste chemistry for biodegradable plastics production: Recent advancements, perspectives, and life-cycle assessment. Trends in Food Science & Technology 156, 104836.

Ghaderi, M., Mousavi, M., Yousefi, H., Labbafi, M., 2014. All-cellulose nanocomposite film made from bagasse cellulose nanofibers for food packaging application. Carbohydrate Polymers 104 (1), 59–65. Available at https://doi.org/10.1016/j.carbpol.2014.01.013.

Ghasemlou, M., Barrow, C.J., Adhikari, B., 2024. The future of bioplastics in food packaging: an industrial perspective. Food Packaging and Shelf Life 43, 101279.

Gong, L., Passari, A.K., Yin, C., Kumar Thakur, V., Newbold, J., Clark, W., Jiang, Y., Kumar, S., Gupta, V.K., 2023. Sustainable utilization of fruit and vegetable waste bioresources for bioplastics production. Critical Reviews in Biotechnology 44, 236–254. Available at https://doi.org/10.1080/07388551.2022.2157241.

Gupta, P., Toksha, B., Rahaman, M., 2022. A review on biodegradable packaging films from vegetative and food waste. The Chemical Record 22 (7), e202100326.

Gustafsson, J., Landberg, M., Bátori, V., Åkesson, D., Taherzadeh, M.J., Zamani, A., 2019. Development of bio-based films and 3D objects from apple pomace. Polymers 11 (2), 289. Available at https://doi.org/10.3390/polym11020289.

Gutierrez, J.N., Royals, A.W., Jameel, H., Venditti, R.A., Pal, L., 2019. Evaluation of paper straws versus plastic straws: Development of a methodology for testing and understanding challenges for paper straws. BioResources 14 (4), 8345–8363. Available at https://doi.org/10.15376/biores.14.4.8345-8363.

Hamed, I., Jakobsen, A.N., Lerfall, J., 2022. Sustainable edible packaging systems based on active compounds from food processing byproducts: a review. Comprehensive Reviews in Food Science and Food Safety 21 (1), 198–226.

Hamid, L., Elhady, S., Abdelkareem, A., Fahim, I., 2022. Fabricating starch-based bioplastic reinforced with bagasse for food packaging. Circular Economy and Sustainability 2 (3), 1065–1076. Available at https://doi.org/10.1007/s43615-021-00139-5.

Hernández-Varela, J.D., Medina, D.I., 2023. Revalorization of coffee residues: advances in the development of eco-friendly biobased potential food packaging. Polymers 15 (13), 2823. Available at https://doi.org/10.3390/polym15132823.

İlaslan, K., 2024. Use of modified polycaprolactone polymer in food packaging applications: a review. Gıda ve Yem Bilimi Teknolojisi Dergisi 32, 13–26. Available at https://doi.org/10.56833/gidaveyem.1485689.

Iriani, E.S., Permana, A.W., Yuliani, S., Kailaku, S.I., Sulaiman, A.A., 2019. The effect of agricultural waste nanocellulose on the properties of bioplastic for fresh fruit packaging. IOP Conference Series: Earth and Environmental Science 309 (1), 012035. Available at https://doi.org/10.1088/1755-1315/309/1/012035.

Isarankura NaAyutthaya, S., Tanpichai, S., Wootthikanokkhan, J., 2015. Keratin extracted from chicken feather waste: extraction, preparation, and structural characterization of the keratin and keratin/biopolymer films and electrospuns. Journal of Polymers and the Environment 23 (4), 506–516.

Jabeen, N., Majid, I., Ahmad Nayik, G., Yildiz, F., 2015. Bioplastics and food packaging: a review. Cogent Food & Agriculture 1 (1), 1117749. Available at https://doi.org/10.1080/23311932.2015.1117749.

Jafarzadeh, S., Jafari, S.M., Salehabadi, A., Nafchi, A.M., Uthaya Kumar, U.S., Khalil, H.P.S.A., 2020. Biodegradable green packaging with antimicrobial functions based on the bioactive compounds from tropical plants and their by-products. Trends in Food Science and Technology 100, 262–277. Available at https://doi.org/10.1016/j.tifs.2020.04.017.

Jafarzadeh, S., Zargar, M., Forough, M., 2022. Renewable and recyclable polymeric materials for food packaging: a new open special issue in materials. Materials 15 (17), 5845.

Jafarzadeh, S., Qazanfarzadeh, Z., Parandi, E., Esmaeili, Y., Barrow, C.J., Timms, W., Naebe, M., 2025. Eco-friendly plastics from cereal-derived by-products and waste: A circular economy approach for sustainable packaging. Materials Today Chemistry 47, 102782. Available at https://doi.org/10.1016/j.mtchem.2025.102782.

Jamal, T., Sapuan, S., Siddiqui, V.U., Syamsir, A., 2025. Classification and properties of tuber-and root-based biocomposites. Plant Tuber and Root-Based Biocomposites. Elsevier, pp. 11–38.

Jariyasakoolroj, P., Leelaphiwat, P., Harnkarnsujarit, N., 2020. Advances in research and development of bioplastic for food packaging. Journal of the Science of Food and Agriculture 100 (14), 5032–5045.

Jayakody, M.M., Vanniarachchy, M.P.G., Wijesekara, I., 2022. Seaweed derived alginate, agar, and carrageenan based edible coatings and films for the food industry: a review. Journal of Food Measurement and Characterization 16 (2), 1195–1227.

Jayakumar, A., Radoor, S., Siengchin, S., Shin, G.H., Kim, J.T., 2023. Recent progress of bioplastics in their properties, standards, certifications and regulations: a review. Science of the Total Environment 878, 163156. Available at https://doi.org/10.1016/j.scitotenv.2023.163156.

Jayasekara, T., Wickrama Surendra, Y., Rathnayake, M., 2022. Polylactic acid pellets production from corn and sugarcane molasses: process simulation for scaled-up processing and comparative life cycle analysis. Journal of Polymers and the Environment 30 (11), 4590–4604.

Jõgi, K., Bhat, R., 2020. Valorization of food processing wastes and by-products for bioplastic production. Sustainable Chemistry and Pharmacy 18, 100326. Available at https://doi.org/10.1016/j.scp.2020.100326.

Kajla, P., Chaudhary, V., Dewan, A., Bangar, S.P., Ramniwas, S., Rustagi, S., et al., 2024. Seaweed-based biopolymers for food packaging: a sustainable approach for a cleaner tomorrow. International Journal of Biological Macromolecules 274 (Pt 1), 133166. Available at https://doi.org/10.1016/j.ijbiomac.2024.133166.

Kakadellis, S., Harris, Z.M., 2020. Don't scrap the waste: the need for broader system boundaries in bioplastic food packaging life-cycle assessment – a critical review. Journal of Cleaner Production 274, 122831. Available at https://doi.org/10.1016/j.jclepro.2020.122831.

Khandeparkar, A.S., Paul, R., Sridhar, A., Lakshmaiah, V.V., Nagella, P., 2024. Eco-friendly innovations in food packaging: a sustainable revolution. Sustainable Chemistry and Pharmacy 39, 101579. Available at https://doi.org/10.1016/j.scp.2024.101579.

Kharb, J., Saharan, R., 2024. Development of biodegradable and eco-friendly fruit peel-derived bioplastic film with antibacterial potential for food packaging application. Biomass Conversion and Biorefinery 15, 10839–10854.

Kharb, J., Saharan, R., 2025. Development of biodegradable and eco-friendly fruit peel-derived bioplastic film with antibacterial potential for food packaging application. Biomass Conversion and Biorefinery 15 (7), 10839–10854. Available at https://doi.org/10.1007/s13399-024-05834-5.

Khardenavis, A.A., Vaidya, A.N., Kumar, M.S., Chakrabarti, T., 2009. Utilization of molasses spentwash for production of bioplastics by waste activated sludge. Waste Management 29 (9), 2558–2565. Available at https://doi.org/10.1016/j.wasman.2009.04.008.

Khodaei, D., Álvarez, C., Mullen, A.M., 2021. Biodegradable packaging materials from animal processing co-products and wastes: an overview. Polymers 13 (15), 2561.

Krasnova, I., Dukalska, L., Seglina, D., Juhnevica, K., Sne, E., Karklina, D., 2012. Effect of passive modified atmosphere in different packaging materials on fresh-cut mixed fruit salad quality during storage. World Academy of Science, Engineering and Technology 6 (7), 908–916.

Krein, D.D.C., Rigueto, C.V.T., Rosseto, M., Gomes, K.S., Biduski, B., Costa, J.A.V., et al., 2025. Bioplastics: an alternative for circular economy in industrial processes. Environmental Hazards of Plastic Wastes: Bioremediation Approaches for Environmental Clean-up. Elsevier, Brazil, pp. 27–42. Available at https://doi.org/10.1016/B978-0-443-23599-3.00004-6.

Kumar, S.N., Jain, R., Anand, K., Ajay Kumar, H., 2021. Utilization of agro waste for the fabrication of bio composites and bio plastics—towards a sustainable green circular economy. In: Sandhu, K., Singh, S., Prakash, C., Subburaj, K., Ramakrishna, S. (Eds.), Sustainability for 3D Printing. Springer Tracts in Additive Manufacturing. Springer, Cham, pp. 131–147. Available at https://doi.org/10.1007/978-3-030-75235-4_7.

Lammi, S., Le Moigne, N., Djenane, D., Gontard, N., Angellier-Coussy, H., 2018. Dry fractionation of olive pomace for the development of food packaging biocomposites. Industrial Crops and Products 120, 250–261. Available at https://doi.org/10.1016/j.indcrop.2018.04.052.

Leon-Bejarano, M., Durmus, Y., Ovando-Martínez, M., Simsek, S., 2020. Physical, barrier, mechanical, and biodegradability properties of modified starch films with nut by-products extracts. Foods 9 (2), 226. Available at https://doi.org/10.3390/foods9020226.

Liu, S., Huang, K., Yu, H., Wu, F., 2018. Bioplastic based on 1,8-octanediol-plasticized feather keratin: a material for food packaging and biomedical applications. Journal of Applied Polymer Science 135 (30), 46516. Available at https://doi.org/10.1002/app.46516.

Loh, N.Y.L., Pang, H.Y., Tee, W.T., Hiew, B.Y.Z., Hanson, S., Chong, S., et al., 2023. Synthesis of renewable and cost-effective bioplastic from apple waste: physicochemical and biodegradability studies. Waste and Biomass Valorization 14 (10), 3235–3252. Available at https://doi.org/10.1007/s12649-022-01983-8.

Lomer, B.B., Khaledabad, M.A., Amiri, S., 2024. Valorization of oil extraction waste for food packaging. Agro-Wastes for Packaging Applications. CRC Press, pp. 219–242.

Lu, P., Yang, Y., Liu, R., Liu, X., Ma, J., Wu, M., et al., 2020. Preparation of sugarcane bagasse nanocellulose hydrogel as a colourimetric freshness indicator for intelligent food packaging. Carbohydrate Polymers 249, 116831. Available at https://doi.org/10.1016/j.carbpol.2020.116831.

Maluin, F.N., 2024. Enhancing chitosan nanofilm with agricultural waste fillers for sustainable and safe functional food packaging. ACS Agricultural Science and Technology 4 (11), 1136–1162. Available at https://doi.org/10.1021/acsagscitech.4c00398.

Mangaraj, S., Goswami, T., Panda, D., 2015. Modeling of gas transmission properties of polymeric films used for MA packaging of fruits. Journal of Food Science and Technology 52, 5456–5469.

Manikandan, N.A., Lens, P.N.L., 2023. Sustainable biorefining and bioprocessing of green seaweed (Ulva spp.) for the production of edible (ulvan) and non-edible (polyhydroxyalkanoate) biopolymeric films. Microbial Cell Factories 22 (1), 140. Available at https://doi.org/10.1186/s12934-023-02154-7.

Margioulas, C.A., 2023. Valorisation of brewer's spent grain for bacterial nanocellulose production and value added products for the development of biodegradable films as food packaging materials for salmon preservation. Master's thesis. https://doi.org/10.26216/heal.aua.2042.

Martino, L., Berthet, M.A., Angellier-Coussy, H., Gontard, N., 2015. Understanding external plasticization of melt extruded PHBV-wheat straw fibers biodegradable composites for food packaging. Journal of Applied Polymer Science 132 (10), 41611. Available at https://doi.org/10.1002/app.41611.

Masssijaya, S.Y., Lubis, M.A.R., Nissa, R.C., Nurhamiyah, Y., Nugroho, P., Antov, P., et al., 2023. Utilization of spent coffee grounds as a sustainable resource for the synthesis of bioplastic composites with polylactic acid, starch, and sucrose. Journal of Composites Science 7 (12), 512. Available at https://doi.org/10.3390/jcs7120512.

Maura, C.S., Athariq, M., Mulmeyda, R., 2024. Utilization of plantain peel (Musa sapientum) and sweet potato starch (Ipomea batatas) waste in combination with glycerol addition to produce biodegradable plastic. Chemistry and Materials 3 (2), 50–55.

McCurdy, C., Dixion, D., Archer, E., Dooher, T., Edwards, I., 2022. A comparison of the sealing, forming and moisture vapour transmission properties of polylactic acid (PLA), polyethene (PE) and polyethylene terephthalate (PET) coated boards for packaging applications. Journal of Packaging Technology and Research 6 (2), 91–100. Available at https://doi.org/10.1007/s41783-022-00131-w.

McNeill, D.C., Pal, A.K., Nath, D., Rodriguez-Uribe, A., Mohanty, A.K., Pilla, S., et al., 2024. Upcycling of ligno-cellulosic nutshells waste biomass in biodegradable plastic-based biocomposites uses - a comprehensive review. Composites Part C: Open Access 14, 100478. Available at https://doi.org/10.1016/j.jcomc.2024.100478.

Merino, D., Quilez-Molina, A.I., Perotto, G., Bassani, A., Spigno, G., Athanassiou, A., 2022. A second life for fruit and vegetable waste: a review on bioplastic films and coatings for potential food protection applications. Green Chemistry 24 (12), 4703–4727. Available at https://doi.org/10.1039/d1gc03904k.

Mneimneh, F., Haddad, N., Ramakrishna, S., 2024. Recycle and reuse to reduce plastic waste-a perspective study comparing petro-and bioplastics. Circular Economy and Sustainability 4 (3), 1983–2010.

Mohammed, A., Gaduan, A., Chaitram, P., Pooran, A., Lee, K.-Y., Ward, K., 2023. Sargassum inspired, optimized calcium alginate bioplastic composites for food packaging. Food Hydrocolloids 135, 108192. Available at https://doi.org/10.1016/j.foodhyd.2022.108192.

Muhammad, A., Rashidi, A.R., Roslan, A., Idris, S.A., 2017. Development of bio based plastic materials for packaging from soybeans waste. AIP Conference Proceedings 1885, 020230. Available at https://doi.org/10.1063/1.5002424 15517616.

Mujtaba, M., Fernandes Fraceto, L., Fazeli, M., Mukherjee, S., Savassa, S.M., Araujo de Medeiros, G., et al., 2023. Lignocellulosic biomass from agricultural waste to the circular economy: a review with focus on biofuels, biocomposites and bioplastics. Journal of Cleaner Production 402, 136815. Available at https://doi.org/10.1016/j.jclepro.2023.136815.

Muneer, F., Hussain, S., Sidra-tul-Muntaha, M.R., Nadeem, H., 2021. Plastics versus bioplastics. Degradation of Plastics 99, 193–237.

Narayanan, M., Lee, J., Barathi, S., Kandasamy, S., 2023. Marine bioresources are a boon for bioplastic production as an alternative to conventional plastics—a review. Biomass Conversion and Biorefinery 15, 25717–25733.

Naseem, A., Azeem, F., Siddique, M.H., Hussain, S., Rasul, I., Saleem, T., et al., 2022. Bioplastic materials from oils. Biodegradable Materials and their Applications. Wiley, Pakistan, pp. 715–733. Available at https://doi.org/10.1002/9781119905301.ch26.

Navasingh, R.J.H., Gurunathan, M.K., Nikolova, M.P., Królczyk, J.B., 2023. Sustainable bioplastics for food packaging produced from renewable natural sources. Polymers 15 (18), 3760. Available at https://doi.org/10.3390/polym15183760.

Nehra, A., Biswas, D., Roy, S., 2024. Fabrication of brewer's spent grain fortified bio-based edible bowls: a promising alternative to plastic containers. Biomass Conversion and Biorefinery 14 (14), 16425–16434.

Nguyen, V.H.T., MN, P., Lee, D.W., Ji, S., 2022. Spent coffee grounds: an intriguing biowaste reinforcement of thermoplastic starch with potential application in green packaging. Polymer Composites 43 (8), 5488–5499.

Nguyen, V.H.T., M.N, P., Lee, D.W., Lee, I.C., Song, J.I., 2023. Films derived from thermoplastic starch/alginate/spent coffee grounds for food packaging applications. Journal of Polymer Research 30 (6), 191. Available at https://doi.org/10.1007/s10965-023-03565-1.

Niranjana Prabhu, T., Prashantha, K., 2018. A review on present status and future challenges of starch based polymer films and their composites in food packaging applications. Polymer Composites 39 (7), 2499–2522. Available at https://doi.org/10.1002/pc.24236.

Nunes, A., Azevedo, G.Z., de Souza Dutra, F., dos Santos, B.R., Schneider, A.R., Oliveira, E.R., et al., 2024. Uses and applications of the red seaweed Kappaphycus alvarezii: a systematic review. Journal of Applied Phycology 36 (6), 3409–3450. Available at https://doi.org/10.1007/s10811-024-03270-6.

Obruca, S., Petrik, S., Benesova, P., Svoboda, Z., Eremka, L., Marova, I., 2014. Utilization of oil extracted from spent coffee grounds for sustainable production of polyhydroxyalkanoates. Applied Microbiology and Biotechnology 98, 5883–5890. Available at https://doi.org/10.1007/s00253-014-5653-3.

Ohl, U., Backhouse, M., Schlögl-Flierl, K., Kalch, A., Bilandzic, H., 2025. Bioplastics versus conventional plastics: an analysis from a sociological, ethical, and educational perspective. Ecological Ambivalence, Complexity, and Change: Perspectives from the Environmental Humanities. Taylor and Francis, Germany, pp. 193–209. Available at https://doi.org/10.4324/9781032627984-15.

Oliveira, J., Almeida, P.L., Sobral, R.G., Lourenço, N.D., Gaudêncio, S.P., 2022. Marine-derived actinomycetes: biodegradation of plastics and formation of pha bioplastics—a circular bioeconomy approach. Marine Drugs 20 (12), 760. Available at https://doi.org/10.3390/md20120760.

Oliveira, G., Passos, C.P., Ferreira, P., Coimbra, M.A., Gonçalves, I., 2021. Coffee by-products and their suitability for developing active food packaging materials. Foods 10 (3), 683. Available at https://doi.org/10.3390/foods10030683.

Oluwasina, O.O., Awonyemi, I.O., 2021. Citrus peel extract starch-based bioplastic: effect of extract concentration on packed fish and bioplastic properties. Journal of Polymers and the Environment 29 (6), 1706–1716. Available at https://doi.org/10.1007/s10924-020-01990-7.

Omar, A.A., Hanafi, M.H.M., Razak, N.H., Razak, N.A.A., Grasianto, G., 2023. Effect of epoxidized waste cooking oil plasticizer in improving the mechanical properties of polylactic acid (PLA). Chemical Engineering Transactions 106, 319–324. Available at https://doi.org/10.3303/CET23106054.

Paidari, S., Esmaeili, Y., Ibrahim, S.A., Vahedi, S., Al-Hilifi, S.A., Zamindar, N., 2024. Microbial Biotechnology in the Food Industry. pp. 75–102. Available at https://doi.org/10.1007/978-3-031-51417-3_4.

Paixão, S.M., Ladeira, S.A., Silva, T.P., Arez, B.F., Roseiro, J.C., Martins, M.L.L., et al., 2016. Sugarcane bagasse delignification with potassium hydroxide for enhanced enzymatic hydrolysis. RSC Advances 6 (2), 1042–1052. Available at https://doi.org/10.1039/c5ra14908h.

Phang, L.Y., Mohammadi, M., Jenol, M.A., Gozan, M., 2022. Bioplastic production from oil producing plants. Biorefinery of Oil Producing Plants for Value-Added Products. vol. 11. Wiley, Malaysia, pp. 543–562. Available at https://doi.org/10.1002/9783527830756.ch27.

Plackett, D., Siró, I., 2011. Polyhydroxyalkanoates (phas) for food packaging. Multifunctional and Nanoreinforced Polymers for Food Packaging. Woodhead Publishing Limited, Denmark, pp. 498–526. Available at https://doi.org/10.1533/9780857092786.4.498.

Polman, E.M.N., Gruter, G.J.M., Parsons, J.R., Tietema, A., 2021. Comparison of the aerobic biodegradation of biopolymers and the corresponding bioplastics: a review. Science of the Total Environment 753, 141953. Available at https://doi.org/10.1016/j.scitotenv.2020.141953.

Qazanfarzadeh, Z., Masek, A., Chakraborty, S., Kumaravel, V., 2024. Development of brewer's spent grain-derived bio nanocomposites through a multiproduct biorefinery approach for food packaging. Industrial Crops and Products 220, 119226. Available at https://doi.org/10.1016/j.indcrop.2024.119226.

Qian, Y., Qin, C., Zhang, J., Shi, B., Wei, Y., Wang, C., et al., 2025. Sustainable, biodegradable, and recyclable bioplastics derived from renewable carboxymethyl cellulose and waste walnut shell. International Journal of Biological Macromolecules 299, 140130. Available at https://doi.org/10.1016/j.ijbiomac.2025.140130.

Qu, C.L., Lin, S.M., Potiyaraj, P., Meng, L., Wu, C.S., Yuan, L., et al., 2023. Polymer packaging through the blending of biowaste oyster shell and low-density polyethylene: a sustainable approach for enhanced food preservation. Polymers 15 (19), 3977. Available at https://doi.org/10.3390/polym15193977.

Rahman, R., Sood, M., Gupta, N., Bandral, J.D., Hameed, F., Ashraf, S., 2019. Bioplastics for food packaging: a review. nternational Journal of Current Microbiology and Applied Sciences 8 (3), 2311–2321.

Rai, P.K., Choure, K., 2023. Agriculture waste to bioplastics: a perfect substitution of plastics. Value-Addition in Agri-Food Industry Waste through Enzyme Technology. Elsevier, India, pp. 299–314. Available at https://doi.org/10.1016/B978-0-323-89928-4.00011-0.

Raj, A.L.T., John, R.M., Deepakraj, S., Dinakarkumar, Y., 2022. Utilization and conversion of discarded lobster shells into valuable biopolymers - a pilot study. Journal of Sustainability and Environmental Management 1 (4), 376–382. Available at https://doi.org/10.3126/josem.v1i4.49998.

Rana, A., Kumar, V., Dhewa, T., Taneja, N.K., 2023. Bioplastic production using whey (polyhydroxyalkanoates and polyhydroxybutyrates). In: Poonia, A., Trajkovska Petkoska, A. (Eds.), Whey Valorization Innovations, Technological Advancements and Sustainable Exploitation. Springer, Singapore, pp. 103–113. Available at https://doi.org/10.1007/978-981-99-5459-9_6.

Rani, N., Kumar, A., Kaushik, R., Gehlot, R., Yadav, S., 2025. Valorization of wheat straw and its applications in green packaging. Agro-Wastes for Packaging Applications. CRC Press, pp. 14–41.

Rashidi, L., 2022. Standards and guidelines for testing biodegradability of bioplastic. In: Dutt Tripathi, A., Darani, K.K., Rai, D.C., Paul, V. (Eds.), Biodegradable Polymer-Based Food Packaging. Springer Nature, Singapore, pp. 297–325. Available at http://doi.org/10.1007/978-981-19-5743-7_15.

Ratna, R., Mutia, M., Darwin, D., Munawar, A.A., Fitriani, F., Handayani, L., 2024. Utilization of tofu liquid waste for the manufacture of bioplastic food packaging. Case Studies in Chemical and Environmental Engineering 10, 100830. Available at https://doi.org/10.1016/j.cscee.2024.100830.

Romão, S., Bettencourt, A., Ribeiro, I.A., 2022. Novel features of cellulose-based films as sustainable alternatives for food packaging. Polymers 14 (22), 4968.

Rosenow, P., Fernández-Ayuso, C., López-García, P., Minguez-Enkovaara, L.F., 2025. Design, new materials, and production challenges of bioplastics-based food packaging. Materials 18 (3), 673. Available at https://doi.org/10.3390/ma18030673.

Salah, R.A., Alia, Z., Gahtar, A., Belahcene, S., Khechekhouche, E.A., Amara, D., 2025. Influence of pH on bioplastic production from sweet potato starch (Ipomea batatas L.). Journal of Agriculture and Applied Biology 6 (1), 81–87.

Sánchez-Orozco, R., Timoteo-Cruz, B., García-Sánchez, J.J., Gomez-Espinosa, R.M., Bernal-Martínez, L.A., Torres-Blancas, T., 2024. Properties of eco-friendly orange peel-alginate-glycerol bioplastic films as potential food packaging applications. Journal of Macromolecular Science, Part A 61 (8), 528–540. Available at https://doi.org/10.1080/10601325.2024.2359412.

Sapuan, S.M., Abral, H., Jamal, T., Thakur, V.K., Nazrin, A., Sherwani, S.F.K., 2024. Plant Tuber and Root-Based Biocomposites Plant Tuber and Root-Based Biocomposites: Development, Characterization, and Applications. Elsevier, Malaysia, pp. 1–261. Available at https://doi.org/10.1016/C2022-0-02856-7.

Schmid, M., Müller, K., 2019. Chapter 11 - Whey protein-based packaging films and coatings. In: Deeth, H.C., Bansal, N. (Eds.), Whey Proteins. Academic Press, pp. 407–437.

Sciancalepore, C., Togliatti, E., Giubilini, A., Pugliese, D., Moroni, F., Messori, M., et al., 2022. Preparation and characterization of innovative poly(butylene adipate terephthalate)-based biocomposites for agri-food packaging application. Journal of Applied Polymer Science 139 (24), 52370. Available at https://doi.org/10.1002/app.52370.

Senapati, T., Dey, S., Ghosh, A.R., Samanta, P., 2024. Bioplastic as potential food packaging material. Encyclopedia of Green Materials. Springer Nature, India, pp. 339–346. Available at https://doi.org/10.1007/978-981-97-4618-7_89.

Seth, A.K., Vellaparambil Hariharan, S., Jaideva, R., 2025. Harnessing banana peel waste: development of starch-based bioplastics for green packaging. Starch-Stärke 77 (3), e202400178.

Sharma, S., Gupta, A., Kumar, A., Kee, C.G., Kamyab, H., Saufi, S.M., 2018. An efficient conversion of waste feather keratin into ecofriendly bioplastic film. Clean Technologies and Environmental Policy 20 (10), 2157–2167. Available at https://doi.org/10.1007/s10098-018-1498-2.

Shlush, E., Davidovich-Pinhas, M., 2022. Bioplastics for food packaging. Trends in Food Science & Technology 125, 66–80. Available at https://doi.org/10.1016/j.tifs.2022.04.026.

Siddiqui, S.A., Yang, X., Deshmukh, R.K., Gaikwad, K.K., Bahmid, N.A., Castro-Muñoz, R., 2024. Recent advances in reinforced bioplastics for food packaging – a critical review. International Journal of Biological Macromolecules 263, 130399. Available at https://doi.org/10.1016/j.ijbiomac.2024.130399.

Silveira, Y.D.O., Franca, A.S., Oliveira, L.S., 2025. Cassava waste starch as a source of bioplastics: development of a polymeric film with antimicrobial properties. Foods 14 (1), 113.

Singh, R.V., Sharma, P., Sambyal, K., 2022. Application of sugarcane bagasse in chemicals and food packaging industry: potential and challenges. Circular Economy and Sustainability 2 (4), 1479–1500.

Singh, P., Verma, R., 2020. Bioplastics: a green approach toward sustainable environment. Environmental Microbiology and Biotechnology. vol. 1: Biovalorization of Solid Wastes and Wastewater Treatment. Springer Singapore, India, pp. 35–53. Available at https://doi.org/10.1007/978-981-15-6021-7_3.

Siregar, E., Sinambela, Y., Anwar, K., 2022. Characterization of cellulose nano fiber from solid waste of palm oil palm empty fruits as basic materials of bioplastic for food packaging. In: Proceedings of the First Jakarta International Conference on Multidisciplinary Studies Towards Creative Industries, JICOMS 2022, 16 November 2022, Jakarta, Indonesia. European Alliance for Innovation, p. 60.

Skoczinski, P., Carus, M., Tweddle, G., Ruiz, P., Guzman, D.D., Ravenstijn, J., et al., 2023. Bio-based building blocks and polymers: global capacities, production and trends 2022–2027. Industrial Biotechnology 19 (4), 185–194. Available at https://doi.org/10.1089/ind.2023.29322.psk.

Song, N.-B., Lee, J.-H., Al Mijan, M., Song, K.B., 2014. Development of a chicken feather protein film containing clove oil and its application in smoked salmon packaging. LWT - Food Science and Technology 57 (2), 453–460. Available at https://doi.org/10.1016/j.lwt.2014.02.009.

Srivastava, V., Singh, S., Das, D., 2023. Rice husk fiber-reinforced starch antimicrobial biocomposite film for active food packaging. Journal of Cleaner Production 421, 138525.

Sultan, A., Sultan, H., Shahzad, W., Kareem, A., Liaqat, A., Ashraf, Z., et al., 2024. Comparative analysis of physical and mechanical properties of starch based bioplastic derived from the pulp and peel of potatoes. Journal of the Indian Chemical Society 101 (10), 101301. Available at https://doi.org/10.1016/j.jics.2024.101301.

Surya, P., Sundaramanickam, A., Nithin, A., Iswarya, P., 2023. Eco-friendly preparation and characterization of bioplastic films made from marine fish-scale wastes. Environmental Science and Pollution Research 30 (12), 34174–34187. Available at https://doi.org/10.1007/s11356-022-24429-z.

Swain, P.K., Das, P., Pal, V., 2023. Production of eco-friendly plastics and bioplastics from crustacean shells and their environmental applications. Novel Polymeric Materials for Environmental Applications. World Scientific, pp. 233–261.

Swetha, T.A., Bora, A., Mohanrasu, K., Balaji, P., Raja, R., Ponnuchamy, K., et al., 2023. A comprehensive review on polylactic acid (PLA) – synthesis, processing and application in food packaging. International Journal of Biological Macromolecules 234, 123715. Available at https://doi.org/10.1016/j.ijbiomac.2023.123715.

Tan, S.X., Ong, H.C., Andriyana, A., Lim, S., Pang, Y.L., Kusumo, F., et al., 2022. Characterization and parametric study on mechanical properties enhancement in biodegradable chitosan-reinforced starch-based bioplastic film. Polymers 14 (2), 278. Available at https://doi.org/10.3390/polym14020278.

Teixeira, A.J.A.M.F., 2017. Industrial waste as raw material-application of almond nuts by-products in a bio-composite. Master's thesis. Universidade do Porto, Portugal.

Teseme, W.B., 2020. Review on the manufacturing of biodegradable plastic packaging film from root and tuber starches. American Journal of Nano Research and Applications 8 (1), 1–8.

Thomas, A.P., Kasa, V.P., Dubey, B.K., Sen, R., Sarmah, A.K., 2023. Synthesis and commercialization of bioplastics: organic waste as a sustainable feedstock. Science of the Total Environment 904, 167243. Available at https://doi.org/10.1016/j.scitotenv.2023.167243.

Torres, F., Rodriguez, S., Saavedra, A., 2019. Green composite materials from biopolymers reinforced with agroforestry waste. Journal of Polymers and the Environment 27, 2651–2673.

Tripathi, A.D., Raj Joshi, T., Kumar Srivastava, S., Darani, K.K., Khade, S., Srivastava, J., 2019. Effect of nutritional supplements on bio-plastics (PHB) production utilizing sugar refinery waste with potential application in food packaging. Preparative Biochemistry and Biotechnology 49 (6), 567–577. Available at https://doi.org/10.1080/10826068.2019.1591982.

Valdés, A., Garrigós, M.C., Jiménez, A., 2022. Extraction and characterization of antioxidant compounds in almond (Prunus amygdalus) shell residues for food packaging applications. Membranes 12 (8), 806.

Valencia, G.A., Luciano, C.G., Lourenço, R.V., Bittante, A.M.Q.B., do Amaral Sobral, P.J., 2019. Morphological and physical properties of nano-biocomposite films based on collagen loaded with laponite® Food Packaging and Shelf Life 19, 24–30. Available at https://doi.org/10.1016/j.fpsl.2018.11.013.

Vannini, M., Marchese, P., Sisti, L., Celli, A., Ferri, M., Monari, S., et al., 2019. From wine pomace and potato wastes to novel PHA-based bio-composites: examples of sustainable routes for full valorisation of the agro-wastes. In: Proceedings of the 7th International Conference on Sustainable Solid Waste Management, 2019, National Technical University of Athens, School of Chemical Engineering, pp. 44–45.

Versino, F., Ortega, F., Monroy, Y., Rivero, S., López, O.V., García, M.A., 2023. Sustainable and bio-based food packaging: a review on past and current design innovations. Foods 12 (5), 1057. Available at https://doi.org/10.3390/foods12051057.

Visco, A., Scolaro, C., Facchin, M., Brahimi, S., Belhamdi, H., Gatto, V., et al., 2022. Agri-food wastes for bioplastics: european prospective on possible applications in their second life for a circular economy. Polymers 14 (13), 2752. Available at https://doi.org/10.3390/polym14132752.

Wang, J., Gardner, D.J., Stark, N.M., Bousfield, D.W., Tajvidi, M., Cai, Z., 2018. Moisture and oxygen barrier properties of cellulose nanomaterial-based films. ACS Sustainable Chemistry and Engineering 6 (1), 49–70. Available at https://doi.org/10.1021/acssuschemeng.7b03523.

Wang, X., Guo, C., Guo, H., 2024. Progress of carrageenan-based films and coatings for food packaging applications. Packaging Technology and Science 37 (6), 533–550.

Wang, W., Liu, Y., Liu, A., Zhao, Y., Chen, X., 2016. Effect of in situ apatite on performance of collagen fiber film for food packaging applications. Journal of Applied Polymer Science 133 (44), 44154. Available at https://doi.org/10.1002/app.44154.

Weligama Thuppahige, V.T., Moghaddam, L., Welsh, Z.G., Karim, A., 2023. Investigation of morphological, chemical, and thermal properties of biodegradable food packaging films synthesised by direct utilisation of cassava (Monihot esculanta) bagasse. Polymers 15 (3), 767. Available at https://doi.org/10.3390/polym15030767.

Wu, H., Xu, C., Li, J., Guan, Y., Mao, J., Kong, F., et al., 2022. Development and properties of wheat straw nano-holocellulose and reduced graphene oxide composite films for active packaging materials. Industrial Crops and Products 189, 115816. Available at https://doi.org/10.1016/j.indcrop.2022.115816.

Yadav, V., Ali, U., 2022. By-products of oilseeds industry (defatted cake waste): biodegradable alternatives for plastics. Edible Food Packaging: Applications, Innovations and Sustainability. Springer Nature, India, pp. 247–262. Available at https://doi.org/10.1007/978-981-16-2383-7_13.

Yang, S., Bai, S., Wang, Q., 2018. Sustainable packaging biocomposites from polylactic acid and wheat straw: enhanced physical performance by solid state shear milling process. Composites Science and Technology 158, 34–42.

Yap, X.Y., Gew, L.T., Khalid, M., Yow, Y.Y., 2023. Algae-based bioplastic for packaging: a decade of development and challenges (2010–2020). Journal of Polymers and the Environment 31 (3), 833–851. Available at https://doi.org/10.1007/s10924-022-02620-0.

Yaradoddi, J.S., Banapurmath, N.R., Ganachari, S.V., Soudagar, M.E.M., Sajjan, A.M., Kamat, S., et al., 2022. Bio-based material from fruit waste of orange peel for industrial applications. Journal of Materials Research and Technology 17, 3186–3197. Available at https://doi.org/10.1016/j.jmrt.2021.09.016.

Zandona, E., Blažić, M., Režek Jambrak, A., 2021. Whey utilization: sustainable uses and environmental approach. Food Technology and Biotechnology 59 (2), 147–161.

Zhao, X., Ji, K., Kurt, K., Cornish, K., Vodovotz, Y., 2019. Optimal mechanical properties of biodegradable natural rubber-toughened PHBV bioplastics intended for food packaging applications. Food Packaging and Shelf Life 21, 100348. Available at https://doi.org/10.1016/j.fpsl.2019.100348.

Zhao, X., Wang, Y., Chen, X., Yu, X., Li, W., Zhang, S., et al., 2023. Sustainable bioplastics derived from renewable natural resources for food packaging. Matter 6 (1), 97–127. Available at https://doi.org/10.1016/j.matt.2022.11.006.

Zhuang, Y., Ruan, S., Yao, H., Sun, Y., 2019. Physical properties of composite films from tilapia skin collagen with Pachyrhizus starch and Rambutan peel phenolics. Marine Drugs 17 (12), 662. Available at https://doi.org/10.3390/md17120662.

Zolfaghari, S., Hashemi, S.S., Karimi, K., Sadeghi, M., 2022. Valorization of cheese whey to eco-friendly food packaging and biomethane via a biorefinery. Journal of Cleaner Production 366, 132870. Available at https://doi.org/10.1016/j.jclepro.2022.132870.

CHAPTER 8

Processing and manufacture of bioplastics

Ahmadreza Abedinia[1,2], Shaghayegh Khajehali Chaleshtori[3,4], Ali Nikmanesh[5], and Simin Hedayatnia[6]

[1]Department of Food Science and Technology, Da.C., Islamic Azad University, Damghan, Iran, [2]Postgraduate School, Universitas Brawijaya, Malang, East Java, Indonesia, [3]Department of Food Science and Technology, Faculty of Agriculture, Ferdowsi University of Mashhad, Mashhad, Iran, [4]Member of Iran National Standards Organization, Tehran, Iran, [5]Department of Food and Drug, Semnan University of Medical Sciences, Semnan, Iran, [6]Department of Food Biotechnology, Faculty of Food Science and Technology, Universiti Putra Malaysia, Selangor, Malaysia

8.1 Introduction

Since the 1940s, traditional materials like paper, glass, and metals have been replaced by conventional plastics, which are made from petroleum-based polymers and are inexpensive, versatile, and long-lasting (Thakur et al., 2018). Plastics are now essential in many industries, and the packaging sector, which mainly serves the food industry, accounts for almost 40% of total production. However, there are now major environmental and health concerns, such as greenhouse gas emissions and toxic by-products, due to their persistence in the environment, slow rates of degradation, and harmful emissions during incineration (Shlush and Davidovich-Pinhas, 2022). Recycling has been marketed as a solution, but its efficacy is limited by biological residue contamination and the process's high energy requirements (Kale et al., 2007).

Bioplastics are referred to the polymeric materials that are either (i) bio-based (derived entirely or in part from renewable biological resources), (ii) biodegradable (able to be broken down by biological processes under specific conditions), or (iii) both. By lowering life-cycle carbon intensity and dependence on fossil feedstocks, the production of such polymers from food- and lignocellulosic-derived feedstocks (such as starch, corn, sugarcane, and other biomass fractions) promotes a transition toward circular material flows (Chinthapalli et al., 2019; Roy Chong et al., 2022). Bioplastics can be broadly categorized based on their sources and production processes, which dictate their structural makeup, functionality, and uses. Generally speaking, they fall into three main categories: (i) bioplastics that are made directly from natural biomass, (ii) bioplastics that are made through microbial fermentation, and (iii) bioplastics that are chemically synthesized from bio-based monomers (Huang et al., 2021). This emphasizes that not all bioplastics are biodegradable; bio-based does not mean biodegradable. For instance, poly(butylene adipate terephthalate) (PBAT) is a biodegradable

Bioplastics for Food Packaging. DOI: https://doi.org/10.1016/B978-0-443-41654-5.00010-X

non-bio-based plastic. Low-molecular-weight polyolefins can also lead to products usable by microbes via photodegradation and thermal degradation, essential for significant biodegradation, which requires an average molecular weight below 5000 Da. A reduction in carbonyl groups during degradation indicates microbial growth, while the final degradation products of these polyolefins are water and carbon dioxide. Additionally, prodegradant additives, often metal salts, can be included to enhance degradation rates (Nandakumar et al., 2021). Fig. 8.1 illustrates global plastic consumption by sector and a schematic of bioplasctic used in each sectors and the origin of bioplastics.

A practical system divides plastics into bio-based and fossil-based groups, yet bio-based materials do not always break down naturally because bio-polyethylene (PE) and bio-based polyethylene terephthalate share identical chemical structures with their petrochemical counterparts. The total global bioplastics production reached 2.36 million tons during 2021 with degradable and nondegradable items producing 1.55 million tons and 0.86 million tons, respectively (Ali et al., 2022). Among biodegradable, bio-based resins, polylactic acid (PLA), polyhydroxyalkanoates (PHA), and thermoplastic starch (TPS) constituted a large share of the market (~40% in 2020), reflecting their commercial maturity and suitability for many applications (Van Roijen and Miller, 2022). Typical end-uses for these materials span food packaging, medical implants, and components for construction and building envelopes (Sidek et al., 2019).

The life cycle assessment (LCA) serves as the standard evaluation method to measure environmental and economic results throughout the entire lifespan of polymers, which enables fair comparisons between bioplastics and petrochemical plastics (Andreasi Bassi et al., 2021). The LCA demonstrates that bio-based plastics reduce climate change effects and fossil fuel usage but certain research shows they create higher eutrophication and toxicity levels, which demand careful handling of these trade-offs (Chen et al., 2016). The LCAs need to include total production through use and end-of-life stages for valid benchmarking instead of partial system boundaries (Rosenboom et al., 2022). Proper policy frameworks, which support the use of alternative feedstocks, require proof of net sustainability advantages before implementation because this policy approach demonstrates its effectiveness through the European strategy regarding recycling of plastics (Foschi and Bonoli, 2019). The circular economy model functions by promoting restorative design along with renewable energy systems and knowledge-driven manufacturing to produce environmental and social and economic benefits for regional stakeholders such as governments and investors and workers and consumers (Spierling et al., 2018). Circular practices do not always benefit the environment because recycling pathways that concentrate toxic chemicals and fail to assess comprehensive sustainability measures lead to decreased environmental advantages, thus industry leaders and policymakers need to analyze sustainability compromises before expanding circular interventions (Blum et al., 2020).

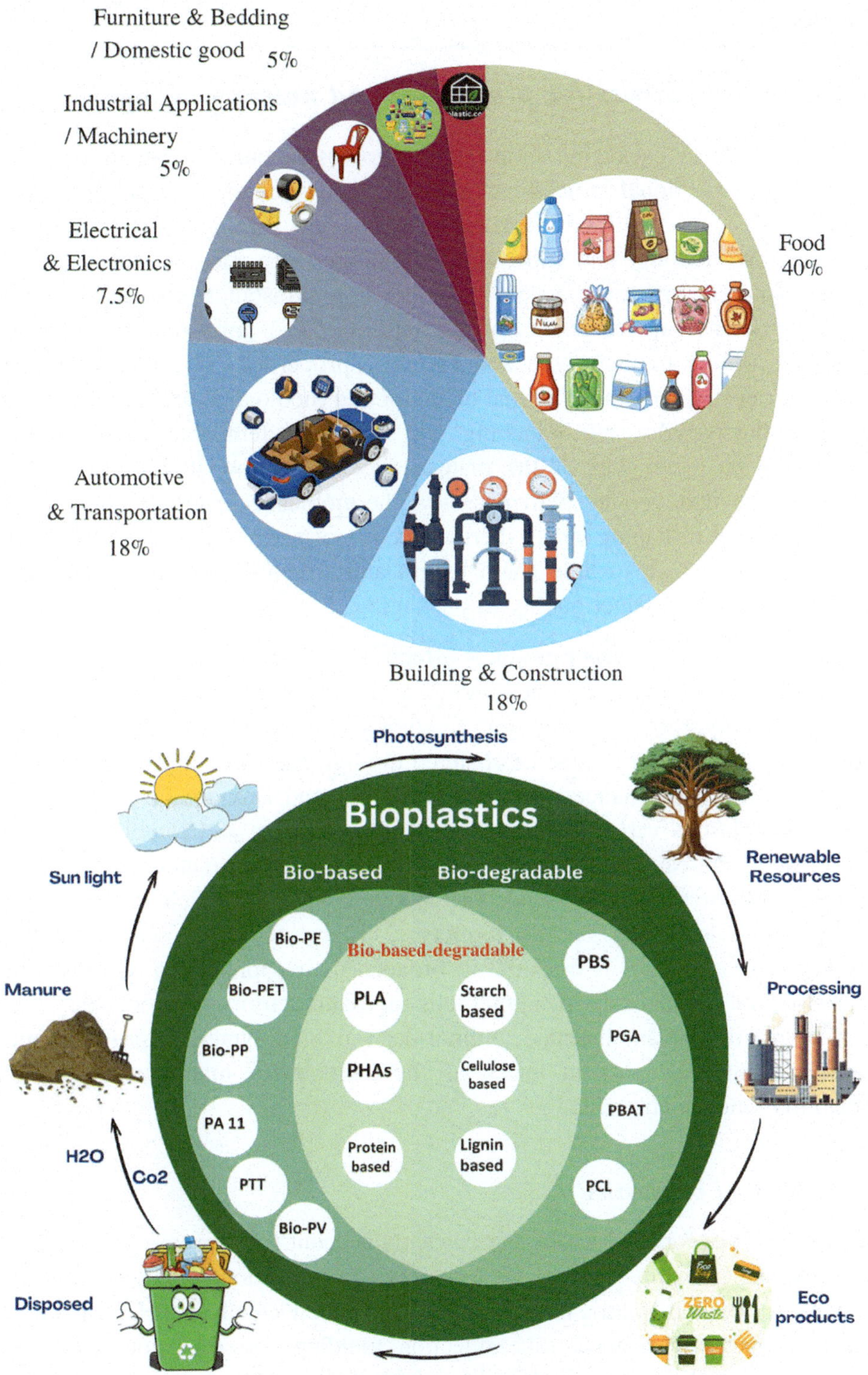

Figure 8.1

An outline of the bioplastics utilized in each area and their origins, as well as the worldwide plastic consumption by sector.

8.2 Types of bioplastics for processing and manufacturing

The production of food packaging bioplastics involves manufacturing different bioplastic materials through traditional processing methods and emerging techniques. Traditional manufacturing methods for food packaging bioplastics include extrusion, compression molding, injection molding, and blow molding along with modern approaches such as casting, electrospinning, microwave-assisted processing, ultrasonication, high-pressure homogenization (HPH), and supercritical carbon dioxide ($scCO_2$) impregnation. Plastics receive classification according to their type and essential properties, which include mechanical strength and barrier resistance and thermal behavior and rheological properties that stem from polymer molecular weight and branching and transition temperatures and density and crystallinity. The characteristics of the material determine how it will be processed and what properties the final product will possess. Thermoplastic materials allow for multiple cycles of melting and reshaping which thermosets do not permit. The chemical properties of packaging materials, which include chain length and structure determine gas permeability rates that directly affect product quality in packaging.

Packaging systems need to protect products effectively while keeping their design adaptable for different uses. Processing techniques change according to polymer characteristics because polyethylene and polypropylene thermoplastics possess strength and light weight properties. The properties of these materials can be changed through chemical or physical modifications, which improve their performance according to Shlush and Davidovich-Pinhas (2022). The processing of bioplastics varies with their specific properties, and established methods for conventional polymers can also apply. However, processing parameters require adjustments based on the bioplastic material used. Certain methods may lack efficiency, sustainability, or economic viability depending on the bioplastic. Therefore, additional research into the processability of biopolymers is essential to broaden their applications (Costa et al., 2023). Table 8.1 lists some of the most typical methods for manufacturing of bioplastics. Each of the sources of bioplastics—plant biomass, animal-derived biopolymers, algae, and microorganisms—offers distinct benefits in terms of the method for manufacturing, availability, biodegradability, and functionality.

8.2.1 Plant-based bioplastics

Plant-based bioplastics for food packaging currently account for approximately 40% of plastic production and are primarily used in food packaging applications (Shlush and Davidovich-Pinhas, 2022). The main bioplastic materials consist of starch, cellulose, PHAs, and PLA, which traditional manufacturing methods like injection molding and extrusion and compression molding can process (Srinivasan and Kulshreshtha, 2020). The development of bioplastics focuses on improving natural biodegradable polymer performance through physical, chemical, and enzymatic treatments, synthesizing new biodegradable polymers, and scaling up production

Table 8.1 Potential for processing of common commercial biopolymers.

	Injection molding	Extrusion	Extrusion blow molding	Cast film extrusion	Blow molding	Fiber spinning	Thermoforming
Starch	×	×	×	×			
Cellulose	×	×			×		
PHB	×	×	×	×	×		×
PHB-PHV	×	×	×	×	×	×	×
PLA	×	×		×	×	×	×
PBS	×	×					
PCL	×	×	×		×	×	×
PBST	×	×		×			×
PBAT		×	×	×			
PTMAT		×	×	×		×	
PVA	×	×		×		×	×
PP, PE + additives	×	×	×	×	×	×	×
Starch + PVA	×	×		×	×	×	
Starch + cellulose acetate	×	×	×		×		×

PHB, poly(3-hydroxybutyrate); *PHV*, poly(hydroxyl valerate); *PBS*, poly(butylenes succinate); *PCL*, poly(ε-caprolactone); *PBST*, poly(butylene succinate terephthalate); *PBAT*, poly(butylene adipate terephthalate); *PTMAT*, poly(tetramethylene adipate terephthalate); *PVA*, poly(vinyl alcohol); *PE*; polyethylene.
Clarinval, A., 2002. Classification and comparison of thermal and mechanical properties of commercialized polymers. International Congress & Trade Show, The Industrial Applications of Bioplastics, (n.d.).

processes (Shlush and Davidovich-Pinhas, 2022). The materials support active packaging systems, which use components to protect products from oxidation and harmful microorganisms (Srinivasan and Kulshreshtha, 2020). Due to its abundance, affordability, and biodegradability, starch—which can be derived from a variety of crops, including corn, cassava, rice, barley, potatoes, and others—has found extensive use in food packaging, disposable cutlery, and films (Ismail et al., 2016; Jiang et al., 2020; Yang et al., 2023). Likewise, the most prevalent biopolymer in plants, fungi, and green algae, cellulose, can be utilized in its natural state or chemically altered to create films, fibers, and pharmaceutical packaging (Di Bartolo et al., 2021; Yadav et al., 2022). Fig. 8.2A shows the plant-based biopolymers used in food packaging applications (Fig. 8.2) (García-Guzmán et al., 2022).

8.2.2 Animal-derived bioplastics

The food packaging industry processes animal-based bioplastics through leftover materials, which emerge from meat processing operations and slaughter facilities and dairy processing plants (Thakur et al., 2023). The primary biopolymers obtained from this process include

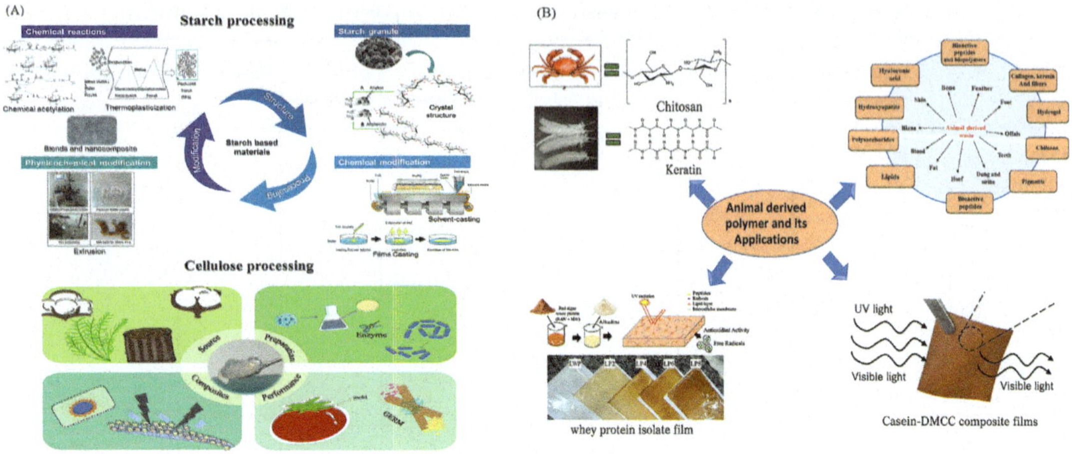

Figure 8.2
(A) Plant based–derived biopolymers for food packaging. (B) Animal-derived biopolymers for food packaging applications with permission.
(A) García-Guzmán, L., Cabrera-Barjas, G., Soria-Hernández, C.G., Castaño, J., Guadarrama-Lezama, A.Y., Rodríguez Llamazares, S., 2022. Progress in starch-based materials for food packaging applications. Polysaccharides 3 (1), 136–177. https://doi.org/10.3390/polysaccharides3010007. (B) Rahman, S., Gogoi, J., Dubey, S., Chowdhury, D., 2024. Animal derived biopolymers for food packaging applications: a review. International Journal of Biological Macromolecules 255, 128197. https://doi.org/10.1016/j.ijbiomac.2023.128197.

chitosan and gelatin and collagen and keratin and myofibrillar proteins and casein and whey protein and hyaluronic acid and silk fibroin (Rahman and Gogoi, 2024). The production process transforms waste materials including skin and bones and feathers and blood and scales and offal which would otherwise be discarded into these materials (Khodaei et al., 2021). The processing methods aim to improve mechanical durability and heat resistance and barrier capabilities by adding nanofillers such as montmorillonites and essential oils and plant extracts and additional polymers. They mainly include protein- and collagen/gelatin-based polymers. Biopolymers with promising qualities for films, coatings, and packaging have been fubricated from proteins derived from sources like whey, and even fish and poultry by-products (Abedinia et al., 2017, 2026; Bourbon et al., 2019; Jones et al., 2013; Sothornvit et al., 2003; Abedinia et al., 2020). Rich in collagen, gelatin, and chitin, fish waste has been used to produce bioplastics, which helps with waste valorization and environmental sustainability (Claverie et al., 2020). Fig. 8.2B illustrates the animal-based biopolymers applied for food packaging applications.

8.2.3 Microbial-derived bioplastics

This group constitute a major category, with PHAs and polyhydroxybutyrate (PHB) as notable examples. Under nutrient-limited conditions, bacteria like *Ralstoniaeutropha*,

Pseudomonas, and *Bacillusmegaterium* naturally synthesize these polyesters using sugars, lipids, or organic wastes as feedstocks (Bhuwal et al., 2014; Rajan et al., 2019; Vasudevan and Natarajan, 2025). Because of their biocompatibility and biodegradability, PHAs and PHB have found use in packaging, biomedical devices, and controlled drug delivery systems (Michalak et al., 2013; Mostafa et al., 2020). PLA, which is produced by fermenting sugars with genetically modified strains of *Escherichia coli* and *Pseudomonasputida*, is another significant microbial product. PLA's processability, safety, and advantageous thermal properties make it a popular choice for biomedical, textile, and compostable packaging applications (Ahsan et al., 2023; Lee et al., 2021).

Bacterial bioplastic manufacturing consists of several integrated processes: choice of bacterial strain, molecular manipulation (Upadhyaya et al., 2014), substrate conditioning (Cesário et al., 2014), fermentation, and downstream recovery. Choice of bacterial strain relies on indigenous enzymatic capability and stability in industrial uses, whereas genetic manipulations refine polymer-synthesis pathways (Jose et al., 2022). The utilized feedstock—from purified sugars to pretreated lignocellulosic hydrolysates or waste lipids—impacts carbon availability, cost of production, and resulting polymer properties. Fermentation conditions like pH, temperature, dissolved oxygen levels, and nutrient starvation may impact the yield, molecular weight, and monomer content of the final polymer considerably. Downstream processing targets biomass separation and polymer extraction using processes like centrifugation, solvent or enzymatic extraction, membrane filtration, and chromatography polishing to achieve a purified product. Optimal process design, monitoring at all times, and scalability practices are all part of bioplastic synthesis success (de Souza and Gupta, 2024). Following the production of the biopolymer, a variety of various processing methods, including solvent casting, injection molding, melt extrusion, and three-dimensional (3D) printing, can be used to form the polymer into shape.

8.8.2.3.1 Polylactic acid

PLA is being investigated as a potential packaging material, since it can be processed using a variety of methods, including thermoforming, blow molding, injection molding, extrusion over cast film, and 3D printing (Bogaert and Coszach, 2000; Drumright et al., 2000; Jamshidian et al., 2010; Rasal et al., 2010; Subramaniam et al., 2019). Technical issues with PLA include brittleness, limited gas barrier qualities, and low temperature stability (around 60°C). Many of these restrictions have been removed by recent developments, such as the use of nanofillers and stereocomplex PLA with improved temperature resistance over 230°C (Ujcic et al., 2022). Fig. 8.3A illustrates the common process for PLA production.

8.8.2.3.2 Polyhydroxyalkanoates

PHAs exist as biodegradable biopolymers that bacteria produce through fermentation to create sustainable manufacturing solutions (Mehrpouya and Vahabi, 2021). The production of PHAs

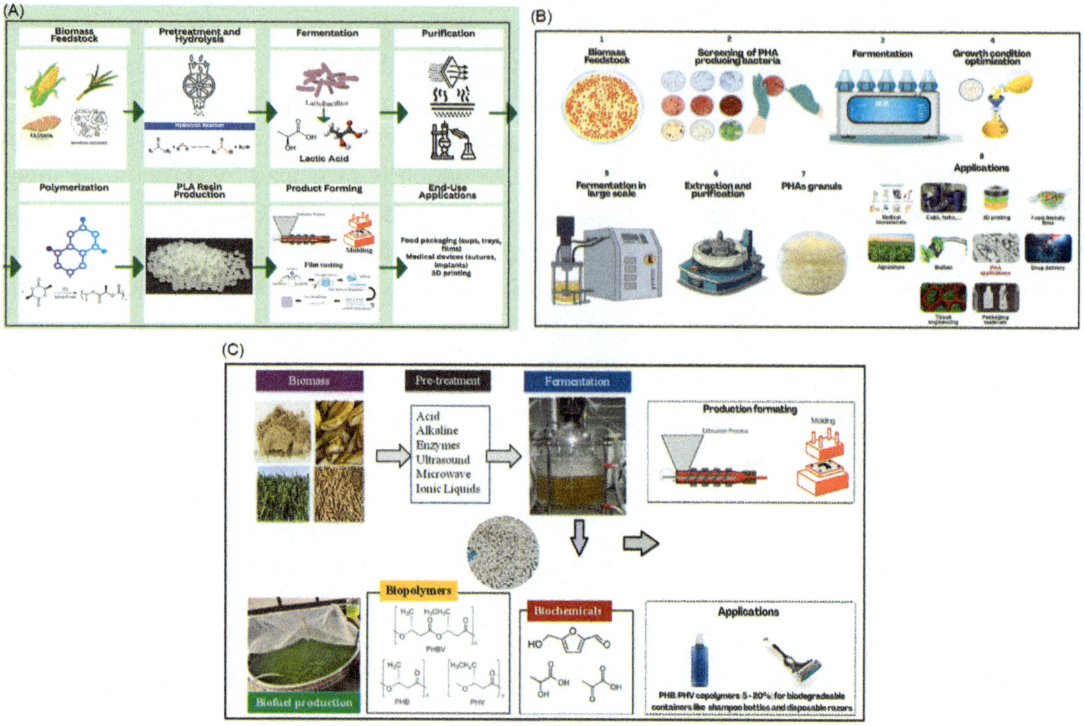

Figure 8.3

(A) Common process for bioplastics manufacturing and applications (polylactic acid). (B) Common process for bioplastics manufacturing and applications (polyhydroxyalkanoates). (C) Common process for bioplastics manufacturing and applications (polyhydroxybutyrate).
Mehrpouya, M., Vahabi, H., Barletta, M., Laheurte, P., Langlois, V., 2021. Additive manufacturing of polyhydroxyalkanoates (PHAs) biopolymers: materials, printing techniques, and applications. Materials Science and Engineering: C 127, 112216. https://doi.org/10.1016/j.msec.2021.112216.

requires both traditional manufacturing processes and modern fabrication methods. The manufacturing of medical devices depends on conventional production techniques, which include solvent casting and melt spinning and electrospinning and additive manufacturing (AM) systems according to Puppi et al. (2019). The manufacturing industry shows great interest in AM because it provides automated production with the capability to create customized products. Selective laser sintering and fused deposition modeling and computer-aided wet spinning have all been used to process various PHA types through different AM techniques (Puppi and Chiellini, 2020). These methods allow for the creation of sophisticated medical equipment and bone replacement devices, which would normally require expensive manufacturing costs (Mehrpouya et al., 2021). The commercialization process demands essential attention to three main areas, which include upstream processing parameters and downstream extraction methods and technoeconomic assessments and process scale-up from laboratory to pilot scale (Kumar et al., 2021). Fig. 8.3B shows a summary of the production, manufacturing, and application processes of PHAs.

8.8.2.3.3 Polyhydroxybutyrate

PHB emerges as a microbial bioprocessed polyester that belongs to the PHA family and demonstrates biocompatible and nontoxic qualities. Specialized bacteria produce PHB, which serves as a biodegradable substitute for petrochemical polymers such as high-density polyethylene because of its beneficial mechanical characteristics. The market expansion of this product faces obstacles because of its expensive manufacturing process and low output rates and difficult production methods and challenging postproduction treatment (de Mello et al., 2023). Fig. 8.3C shows a summary of its production, manufacturing, and application processes. The bacteria species cyanobacteria stands out for producing this compound from glycerol (GLY) and agro-industrial waste materials. Its production becomes cost-effective when appropriate feedstock materials are selected. The process through photobioreactors needs cyanobacteria to use carbon dioxide and light energy, which requires subsequent biomass collection and PHB extraction procedures. PHB serves as a base material for manufacturing agricultural products and medical supplies and packaging materials (Sirohi et al., 2020).

8.2.4 Algal-based bioplastics

Algal-based bioplastics use microalgae and macroalgae as renewable feedstocks for production according to studies by Anli Dino and Kishore (2024). The processing methods involve three main techniques that include PHA production from microalgae biomass under nutrient restrictions and polymer reinforcement through algae fillers and bioplastic manufacturing from bioactive compounds extracted from microalgae (Roy Chong et al., 2022). The manufacturing process includes solvent casting together with injection molding and extrusion methods (Santana et al., 2024). Key algal polysaccharides including galactans (agar and carrageenan), alginates, and ulvan offer excellent physicochemical properties with improved barrier, mechanical, antioxidant, and antimicrobial characteristics (Thiviya et al., 2023). Seaweed-derived alginate from brown seaweed functions as a gel-forming substance that scientists have examined for multiple industrial applications (Santana et al., 2024). The worldwide production of algal bioplastics occurs through 81 entities that operate under 55 patents from 2011 to now to show expanding business opportunities despite market obstacles (Mogany et al., 2024). They represent an emerging field, exploiting the high carbon-fixing capacity and rapid growth of microalgae. Rich in proteins and carbohydrates, microalgal biomass can be used as a feedstock for protein-based polymers, PHA, PHB, and PLA. These biomasses help reduce greenhouse gas emissions and offer a substitute for food (Amiri et al., 2025; Rahman and Miller, 2017; Yew et al., 2020). Fig. 8.4 shows a summary of its production, manufacturing, and applications.

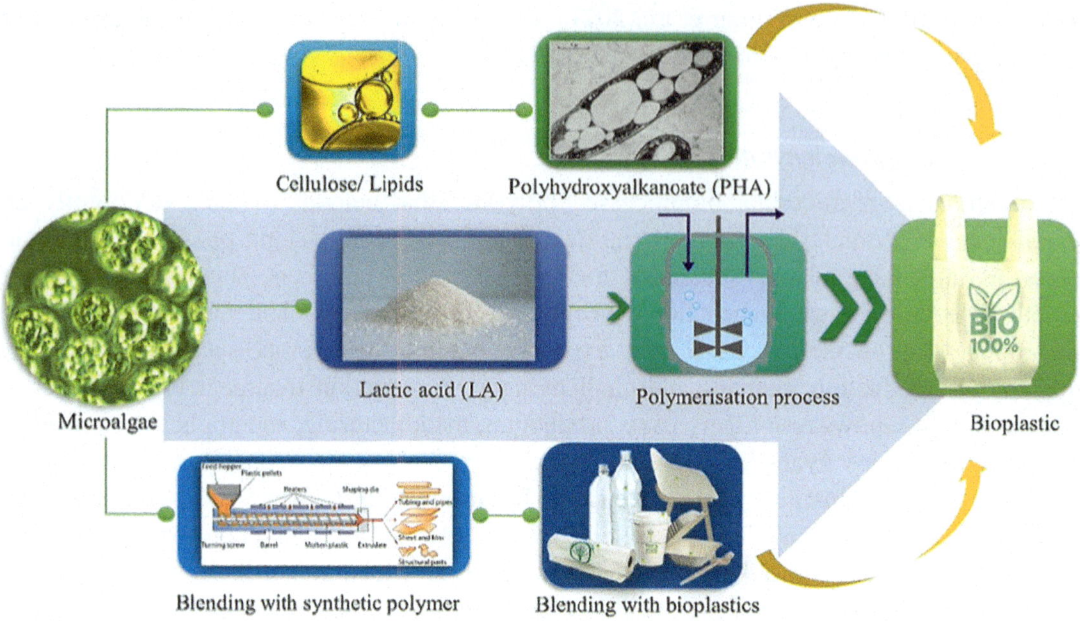

Figure 8.4
Microalgae-based bioplastics.
Roy Chong, J.W., Tan, X., Khoo, K.S., Ng, H.S., Jonglertjunya, W., Yew, G.Y. et al., 2022. Microalgae-based bioplastics: future solution towards mitigation of plastic wastes. Environmental Research 206, 112620. https://doi.org/10.1016/j.envres.2021.112620, http://www.elsevier.com/inca/publications/store/6/2/2/8/2/1/index.htt.

8.3 Techniques and methods for processing and manufacturing of bioplastics

8.3.1 Extrusion process

The food packaging and distribution industries are increasingly adopting bioplastics for packaging applications. Among manufacturing techniques, extrusion is particularly promising for enhancing the economic viability of these bio-based materials. It enables high-volume production and reduces manufacturing costs compared to other methods, thereby accelerating their entry into the market. Thus, extrusion is the most interesting method for achieving scale-up and moving on to industrial production of these bio-based polymers. Extrusion could enable bio-based materials to be developed more profitably, making them more competitive and making it easier to bring them to market.

Extrusion is commonly defined as a continuous manufacturing process wherein bioplastic pellets or granules undergo a phase transition into a molten state or dough-like material and are subsequently propelled through a restriction or die of specific geometry to produce films, sheets, or profiles (Alam et al., 2014; Bouvier and Campanella, 2014). Extrusion technology is a multivariate process that involves transport, compression, melting, mixing, cooking, texturization, and shaping raw material components into unique structures. This process relies on a combination of thermal energy and mechanical shear, generated by rotating screws within the extruder barrel, to induce melting and homogenization of the polymer material and widely employed method for producing sustainable packaging materials, utilizing thermal energy and mechanical shear to melt and homogenize biopolymers, which are then forced through a die to create desired shapes (Amesho et al., 2025). Extrusion facilitates the direct processing of pelletized or granulated recycled plastic, circumventing the necessity for filament formation, thereby reducing preprocessing costs and energy consumption (Ni et al., 2024). Extrusion-based printing has been commercialized by several companies, hence a wide selection of cheap 3D printers are available for researchers (Subramaniam et al., 2019). Maintaining meticulous control over critical process parameters, including barrel temperatures, screw speed, and melt pressure, is paramount for ensuring consistent product quality and dimensional accuracy in the extrusion process; deviations in these parameters can induce undesirable variations in the final product's mechanical properties and appearance (Jiang et al., 2012). Furthermore, the integration of sophisticated process analytical technology and control strategies is essential, particularly when dealing with formulations that exhibit narrow processing windows, necessitating a thorough understanding of material behavior to ensure process robustness and consistency (Laske et al., 2017). The melting zone within twin-screw extruders (TSEs) is of paramount significance, functioning as the primary site for homogenization and the development of targeted material attributes through efficient blending; consequently, the effectiveness of melting and mixing processes within this zone exerts a direct influence on the ultimate properties of the material (Bartilla et al., 1986).

8.3.2 Principal of extrusion process

Extrusion-based systems employ three principal deposition platforms: fused filament fabrication, big area AM, and direct write; each platform possesses distinct printing parameters that influence the thermophysical and rheological properties of the material being processed (Duty et al., 2018). An essential prerequisite for achieving extruded lines with well-defined cross sections is the pseudoplastic behavior of the material (Shang et al., 2019). The utilization of bioplastics in extrusion processes necessitates careful consideration of their unique thermal and rheological characteristics, along with considerations of potential degradation mechanisms. The optimization of extrusion parameters, including temperature profiles, screw design, and die geometry, is crucial for achieving desired product attributes while minimizing material degradation and energy consumption.

8.3.3 Equipment and technologies for bioplastic extrusion

An extrusion line typically comprises a hopper for material feeding, an extruder barrel housing one or two screws for melting and mixing, a shaping die, and downstream cooling and cutting units (Kumar et al., 2021). The screw is the key component of an extruder, and it has been divided into three main functional/geometrical zones (i.e., solids conveying or feed, melting or compression, and metering or pumping), which are generally based on the primary operations of an extruder (Abeykoon, 2016). Extrusion processing technology often uses two distinct equipment designs: The single-screw extruder and the TSE (co-rotating and counter-rotating) (Table 8.2). Single-screw extrusion (Fig. 8.5A) (Delvar et al., 2025) is most common for bioplastics due to its efficient melting and pumping capabilities and generally costs less and is ideal for processing simple materials and producing cost-effective products, while TSEs are more expensive and better suited for handling complex materials that require a higher degree of mixing and greater process flexibility (Bouvier and Campanella, 2014). However, co-rotating TSEs (Fig. 8.5B) have demonstrated superior economic performance in some cases, producing products typically made with single-screw extruders (Yacu, 2020). Starch/poly(vinyl alcohol)/montmorillonite nanocomposites (Navarchian et al., 2015) and nanocomposite based on thermoplasticized starch and CNFs are prepared in a TSE for food packaging film application (Fourati et al., 2020).

Both equipment designs can be compared to conventional pumps in their performance. Single-screw extruders operate as a frictional drag flow pump in the solid conveying zone, a viscous drag flow in the melt-conveying zone. This process relies heavily on the friction between the material and the barrel wall, allowing the material to advance through the extruder. Single-screw extruders require design features that ensure material does not slip. These features include grooves in the barrel, interrupted flights (spaces between the flights on the screw), or restriction mechanism to product flow (e.g., throttle rings, kneading disks, steam locks, or shear locks) (Yacu, 2020). In TSEs, the dominant mixing mechanism is elongational flow. This occurs as the material is forced through the narrow gap—typically on the order of a few microns—between the intermeshing screw elements, and this action generates intensive shear and elongation stresses that are effectively transferred to embedded fibers (Fourati et al., 2020). Despite the differences, the operating principle remains similar for both types of extruders: raw materials are added from the feeder and mixed with water that

Table 8.2 Comparison of single-screw vs. twin-screw extruders for bio-based film production.

Feature	Single-screw extruder	Twin-screw extruder
Mixing efficiency	Low to moderate	Very high
Shear control	Limited	Precise
Feed flexibility	Poor	Excellent
Self-wiping action	No	Yes
Process flexibility	Low	Very high
Cost	Lower	Higher

is injected into the extruder barrel, where the mixture is subjected to compression, mixing, shearing, kneading by the rotating screw(s), while also being exposed to a set temperature from the extruder barrel (Guy, 2001). The injected water acts as both a solvent and a plasticizer and liquefied the material into a viscoelastic mass (Luo et al., 2023). At the end of the barrel, the material is forced through a die, which is a channel with a specific cross-sectional shape and a relatively smaller profile area than the barrel. Cooling systems—such as air knives or water baths—solidify the extruded product quickly to preserve shape and surface quality. Automation and process monitoring have improved quality consistency and throughput (Mariano et al., 2017). Co-extrusion technology enables production of multilayer films with varied barrier and mechanical properties, combining different bioplastics in one product (Siracusa, 2012).

8.3.4 Applications and key factors influencing their properties

Food packaging is the process of enclosing food products in a protective material to protect them from environmental elements including light exposure, oxidation, moisture, microbial or chemical decay, and convenience from production to consumption. More specifically, food packaging regulates the optimum temperature and atmospheric conditions, so helping to lower food waste and also protect from chemical and microbial contamination (Schudel et al., 2023). After its use, the packaging is often discarded and recycled, incinerated, or disposed of in landfills. In the worst case, it may end up in the environment. Recent widespread use of petroleum-based plastic packaging in food packaging has resulted in a major global waste disposal crisis, inadequate management of waste products, pollution, significant challenges due to the non-biodegradable and non,renewable nature of the plastic and the risk of hazardous chemicals seeping into food and the environment (Kurt et al., 2017; Kumar et al., 2024). Regarding the negative impact of plastic pollution on environment and the depletion of fossil resources, there are several methods for eliminating conventional plastic wastes; which one of the best sustainable solution for food packaging is eco-friendly and biodegradable alternatives produced from renewable biological to replace conventional plastic (Kumar et al., 2024). The European Bioplastics Organization defines bioplastics as either plastics that are biodegradable and/or compostable or as plastics based on renewable resources (bio-based). Biologically based plastics are derived from biological materials, that is, those originating in plants (cellulose, lignin, hemicellulose, and pectin), bacteria, algae, among others. Recently, the attention in the packaging industry regarding the use of bioplastics has been shifting from compostable/biodegradable materials toward bio-based materials (Asgher et al., 2020). If bioplastics are biodegradable, they undergo decomposition into CO_2, CH_4, H_2O, inorganic compounds, or biomass, while their degradation capacity depends on the environment required by bacteria or microorganisms that give rise to enzymatic or chemical reactions produced and the origin of the material and its chemical structure (Boey et al., 2022). Extruded bioplastic films and sheets are extensively used for flexible packaging, food wraps, bags, trays, lids, and containers (Haider et al., 2019). Biodegradable films are typically categorized as lipid-, polysaccharides-, or protein-based, depending on their polymer structure (Hedayatnia et al., 2019). When synthesizing novel bioplastics, it is important to address key factors, including cost-effectiveness, thermal and mechanical properties, barrier performance,

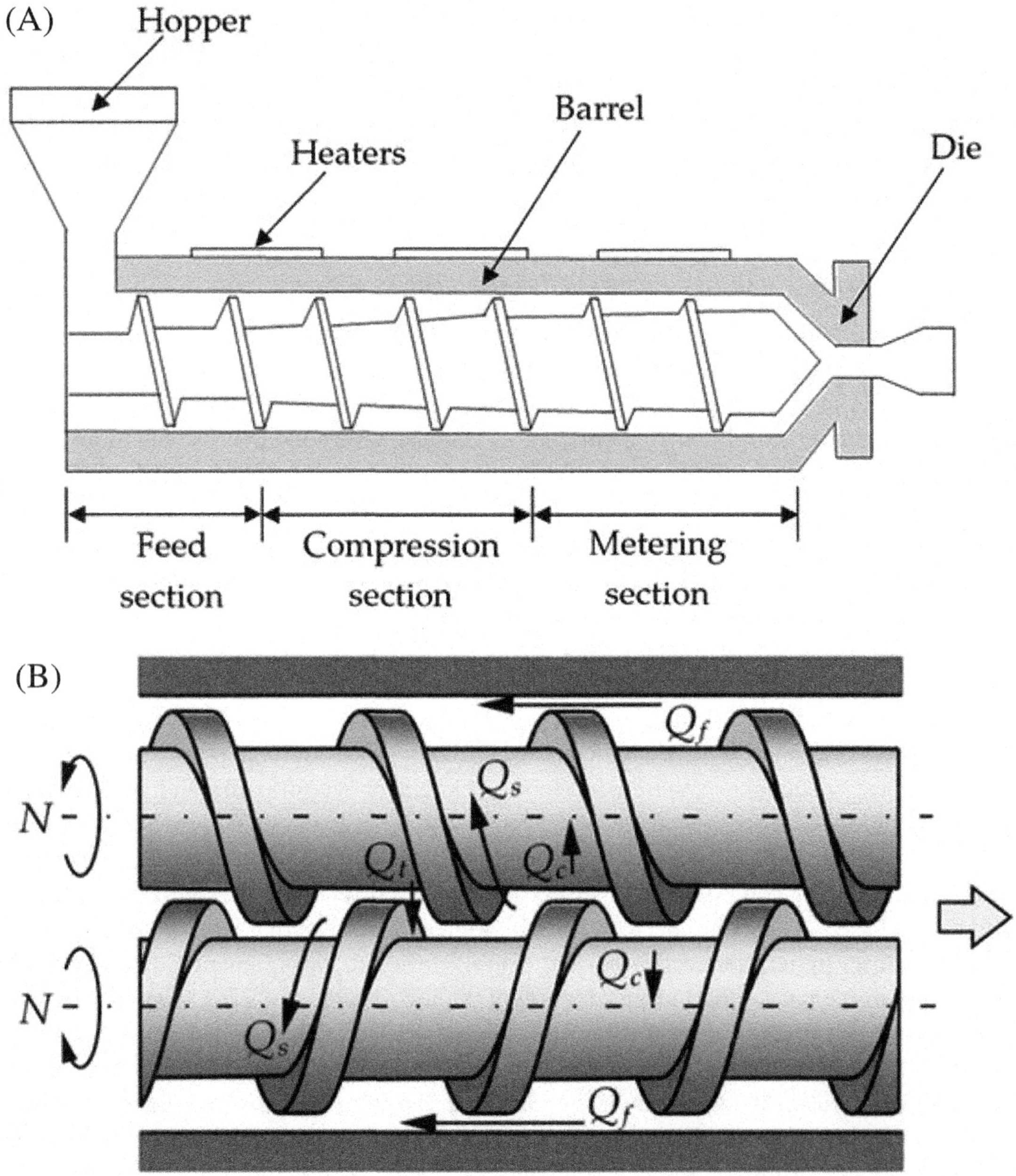

upscaling hurdles, and market demand (Attallah et al., 2021). Table 8.3 illustrates some papers that have used bioplastic extrusion to produce films and their properties.

The extrusion process determines the mechanical properties and thermal characteristics and barrier capabilities of bioplastics through its specific processing settings. The speed of the screw and temperature levels determine how polymer chains move and how fast they break down, which affects their molecular alignment and results in specific values for tensile strength and modulus and elongation at break (Mysiukiewicz et al., 2020; Pelissari et al., 2011). The starch/chitosan systems generate their best mechanical performance through moderate conditions which consist of 130°C and 35 rpm. The process of increasing screw speed and thermal energy levels leads to chain breakage, which reduces molecular weight (Kosmalska et al., 2022).

The mechanical and barrier properties depend on the concentration levels of plasticizers and surfactants within the formulation. The addition of plasticizer or acetyl tributyl citrate in PLA–TPS blends leads to better flexibility and elongation performance but results in decreased tensile modulus because it weakens the bonding between molecules (Calambás Pulgarin and Caicedo, 2024; Gálvez et al., 2020). The addition of PLA to starch-based matrices creates enhanced water resistance and reduced water vapor permeability because of PLA's crystalline structure and low polarity (Shirai et al., 2013). The thermal properties of the material show glass transition temperatures between 59°C and 65°C and melting temperatures from 120°C to 179°C because of the structural changes made during processing. The mechanical strength decreases when extrusion temperatures rise and material undergoes multiple recycling cycles because these conditions decrease molecular weight and lower T_g values (Mysiukiewicz et al., 2020; Shojaeiarani et al., 2019).

Figure 8.5

(A) Modern extrusion technology for bioplastics integrates temperature and pressure control, screw design optimization, and advanced dies to accommodate the unique thermal and rheological properties of biopolymers like polylactic acid (PLA), poly(butylenes succinate) (PBS), and starch blends. Schematic of components in single-screw extruder. (B) Modern extrusion technology for bioplastics integrates temperature and pressure control, screw design optimization, and advanced dies to accommodate the unique thermal and rheological properties of biopolymers like PLA, PBS, and starch blends. Types of twin-screw extruders, flow in counter-rotating extruder, Q_f is leakage flow (screw flight and barrel), Q_c is calendaring flow (screw flight and opposite screw), Q_t is tetrahedron region, and Q_s is leakage flow (between the flanks of the screw).
Delvar, E., Oliveira, I., Brito, M.S.C.A., Silva, C.G., Santamaria-Echart, A., Barreiro, M.-F., Santos, R.J., 2025. Literature review on single and twin-screw extruders design for polymerization using CFD simulation. Fluids 10 (1), 9. https://doi.org/10.3390/fluids10010009.

Table 8.3 Effects of extrusion parameters on extruded bioplastic films properties.

Bioplastic material system	Extrusion type	Primary parameters investigated	Optimal conditions	Thermal properties	Mechanical properties	Barrier and transmission properties	References
Starch/chitosan/ oregano oil	Single-screw	Temperature (120°C–140°C) and screw speed (25–45 rpm)	130°C/ 35 rpm	No mention found	Higher screw speed increased blow-up ratio; decreased opacity, tensile strength (TS), elongation; low temperature decreased strength, and modulus	Higher speed increased water vapor permeability (WVP); low temperature decreased WVP; less hydrophilic at optimal	Pelissari et al. (2011)
Thermoplastic starch/ polylactic acid (PLA) blends (Bio-Flex FX 1130), plasticizers, and surfactants	Twin-screw, co-extrusion	Temperature (115°C–145°C), screw speed (15–80 rpm), melt flow index (1–5 g/10 min), plasticizer/ surfactant content	Melt flow index 1–5 g/10 min, TPS50-PLA50-T5	Glass transition temperature 64°C; melting temperature 120°C (thermoplastic starch), 160°C PLA; degradation temperature increased by 20°C	TS 4.276 MPa, elongation 2.63%; plasticizer/surfactant increased miscibility	water vapor transmission rates (WVTR): 9.0×10^{-4}/ 7.5×10^{-4}; TPS75-PLA25-T5-A10: 2 orders difference between sides; contact angle 103.7°–30.3°	Calambás Pulgarin and Caicedo (2024)
PLA (Ingeo 2500HP, 3100HP, 3260HP, 4032D)	Twin-screw	Temperature (up to 260°C), screw speed, melt flow index, grade	High melt flow index, low viscosity, high speed	Glass transition temperature 59°C–62°C; melting temperature 172°C–179°C; higher temperature decreased glass transition temperature; thermogravimetric analysis: lower degradation temperature for 4032D	High temperature/ low speed increased degradation, decreased molecular weight, and color change	No mention found	Mysiukiewicz et al. (2020)

PLA (Ingeo 2003-D)/ acetyl tributyl citrate blends	Twin-screw	Temperature profile (up to 160°C), screw speed (60/ 150 rpm), and acetyl tributyl citrate (10–30%)	20% Acetyl tributyl citrate and 60 rpm	Glass transition temperature decreased with acetyl tributyl citrate/speed; maximum degradation temperature 362°C–370°C; lower speed increased stability	Acetyl tributyl citrate decreased modulus/ strength, increased elongation; higher speed decreased modulus; best performance at 20% acetyl tributyl citrate/ 60 rpm	Contact angle measured; no water vapor transmission rate/oxygen transmission rate	Gálvez et al. (2020)
Thermoplastic starch/ PLA blends, glycerol, citric acid	Single-screw	Polylactic acid/ glycerol/citric acid content, temperature, and screw speed	Response surface methodology-optimized	Differential scanning calorimetry used; no values	Optimum for mechanical properties similar to petroleum plastics	No mention found	García-Cruz et al. (2020)
Thermoplastic starch/ polybutylene adipate terephthalate/PLA blends	Blown film	Extrusion process (2 types), PLA content	20% PLA	No mention found	PLA increased modulus/strength, decreased elongation/ blow-up ratio; process had little effect	PLA decreased WVP due to hydrophobicity	Shirai et al. (2013)
PLA, polyhydroxyalkanoate, polyhydroxybutyrate, green polyethylene, bio polyamide 11, polybutylene adipate terephthalate, polybutylene succinate, polybutylene succinate-co-terephthalate, and polyethylene glycol	Sheet extrusion	Melt temperature, calander temperature, draw ratio, and rotation speed	No mention found	Thermogravimetric analysis and differential scanning calorimetry used; no values	Process parameters affect processability and film quality	Surface free energy measured	Svidler et al. (2014)
PLA (LX175)	Twin-screw	Temperature, screw configuration (2/4 kneading zones), speed (200–600 rpm)	210°C, 200 rpm, and 2 kneading zones	Glass transition temperature 59.7°C–61.7°C; melting temperature 152.7°C–156.3°C; higher temperature	Higher temperature/ speed increased melt flow rate, decreased molecular weight; optimal minimizes degradation	No mention found	Kosmalska et al. (2022)

(*Continued*)

Table 8.3 (Continued)

Bioplastic material system	Extrusion type	Primary parameters investigated	Optimal conditions	Thermal properties	Mechanical properties	Barrier and transmission properties	References
				increased degradation			
Low-density polyethylene, thermoplastic starch—potato, thermoplastic starch—corn, PLA (BioComp®BF 7210)	Blown film	Temperature (150°C–160°C), screw speed, and melt flow rate	Individually selected	Glass transition temperature PLA 61–65°C; melting temperature PLA 147°C; stable to 250°C	PLA highest melt flow rate, thermoplastic starch—potato lowest; higher speed increased mass flow	No mention found	Sikora et al. (2020)
Solanyl, bioflex, polylactic acid, and polyhydroxybutyrate-co-valerate	Twin-screw, compression molding	Recycling cycles, temperature, and pressure	No mention found	PLA melting temperature 150°C–152°C; polyhydroxybutyrate-co-valerate degrades 251°C–306°C; glass transition temperature decreased with recycling	More cycles decreased molecular weight, modulus, impact strength (especially polylactic acid/ polyhydroxybutyrate-co-valerate)	No mention found	Shojaeiarani et al. (2019)

The water vapor transmission rate exists as the sole barrier property that scientists have measured in specific systems, yet scientists still lack data about UV and gas transmission rates (Calambás Pulgarin and Caicedo, 2024; Pelissari et al., 2011). The current research shows that scientists have not conducted complete LCA together with polymer migration studies for food-contact materials, which creates two main research gaps for sustainability and safety evaluations (Svidler et al., 2014). The study results demonstrate that bioplastic production requires exact extrusion parameter control together with blend composition management to achieve the desired balance between mechanical strength and processability and barrier performance. Future studies should integrate multiscale characterization (mechanical–thermal–LCA–migration) to establish standardized processing–property correlations for industrial optimization.

8.3.5 Challenges in extrusion of bioplastics

The significant challenges exist, particularly in the extraction and purification processes of their constituent monomers and polymers (Patreau et al., 2023) compared to conventional plastics. Biodegradable plastics face three key technical challenges during extrusion (Mysiukiewicz et al., 2020). First, thermal and mechanical degradation—manifested as chain scission, cross-linking, and a reduction in molecular weight—is frequently observed. Starch-based polymers show preferential cleavage of larger molecules, while PLA is sensitive to high temperatures and low screw speeds, which reduce viscosity and impair flow (Liu et al., 2010). Second, process parameter optimization is critical. Studies report that precise control of temperature (with recommended ranges around 180°C–220°C for several formulations), screw speed, and residence time is needed to minimize degradation and maintain product quality (Mysiukiewicz et al., 2020). Third, material formulation and equipment adaptation play substantial roles (Dimonie et al., 2016). For example, while plasticizers can improve flexibility, they may also increase crystallinity and alter weldability, and suboptimal processing conditions can lead to issues such as screw clogging (Ljungberg et al., 2003). Equipment modifications, including tailored screw designs and pressure management strategies, are also necessary to address these challenges during both conventional and industrial-scale extrusion (Abayev and Shved, 2025).

Some main technical, economic, and regulatory challenges facing adoption of bioplastics are summarized in Table 8.4.

Starch-based polymers break down their long chains, leading to more uniform molecular weight distributions (Liu et al., 2010). PLA materials degrade quickly when exposed to high temperatures and extended exposure because they have poor thermal resistance and hydrolyze easily (Mysiukiewicz et al., 2020). Starch/PVOH blends show different degradation patterns based on their material composition because starch-rich blends undergo chain scission, while PVOH-rich blends experience cross-linking (Dimonie et al., 2016). Protein-based plastics face two main problems: excessive cross-linking and high softening points that restrict flowability and require plasticizers to improve their processability (Verbeek and Van Den, 2010).

Table 8.4 Key challenges in extruding bio-based plastics.

Polymer type	Extrusion method	Primary challenge focus	Key findings	References
Starch (amylopectin/amylose, water/glycerol plasticized)	No mention found	Degradation mechanisms during extrusion	Preferential cleavage of larger molecules; narrowing of size distribution; water/glycerol influence degradation	Liu et al. (2010)
Polylactic acid (PLA), plasticized with triacetine, and tributyl citrate	Film extrusion (type not specified)	Brittleness, weldability, crystallinity changes	Plasticizers improve flexibility; storage increases crystallinity, complicating welding; welding alters crystallinity	Ljungberg et al. (2003)
PLA, four grades, and no additives	Twin-screw, conventional extrusion	Degradation under processing parameters	High temperature/low screw speed increase degradation; high melt flow index (MFI) grades less susceptible; viscosity is a sensitive indicator	Mysiukiewicz et al. (2020)
Starch/poly(vinyl alcohol) blends	No mention found (MFI method)	Process sensitivity and degradation pathways	Corn starch blends: chain scission; PVOH blends: cross-linking; each blend needs tailored conditions	Dimonie et al. (2016)
PLA, polyhydroxyalkanoates, poly (butylene adipate terephthalate), with additives	Cascade disk-gear, extrusion blow molding	Industrial-scale processing, thermal/ultraviolet sensitivity	Precise temperature control and stabilizers needed; industrial extrusion feasible with minimal equipment changes	Abayev and Shved (2025)
PLA (PLA, Ingeo 4043D)	Single-screw, conventional extrusion	Process parameter effects and screw clogging	Low temperature causes screw clogging/high resistance; higher temperature/speed improve flow; PLA processable on standard machines	Fabijański (2023)
Biomaterials, recycled materials, and biocomposites	Material extrusion/additive manufacturing	Feedstock quality, sustainability, and energy/emissions	Recycled/biomaterial feedstock leads to inferior product quality; energy/emissions can be managed by process/material selection	Patti (2024)
PLA, thermoplastic starch—potato, thermoplastic starch—corn, low-density polyethylene (LDPE, with fillers/additives)	Single-screw and blown film extrusion	Flow properties, thermal stability, and processability	Biodegradable plastics have poor flow, narrow processing windows, high temperature/moisture sensitivity; higher error bars/mass flow rates than LDPE	Sikora et al. (2020)
Biodegradable polyesters and blends/composites	Reactive extrusion	Interfacial adhesion, process efficiency	Reactive extrusion enables cost-effective production, improved adhesion in blends/composites; technical challenges not detailed	Raquez et al. (2008)
Protein-based plastics (with plasticizers)	No mention found	Cross-linking, softening temperature, processability	Plasticizers needed to overcome high softening temperature/cross-linking; processability highly dependent on protein/conditions	Verbeek and Van Den (2010)

The process parameter optimization plays a vital role in achieving the lowest possible degradation while maintaining consistent product quality. Research shows that PLA and PBAT processing requires temperatures between 180°C and 220°C together with specific screw speeds and residence times to achieve optimal flow while avoiding molecular degradation (Abayev and Shved, 2025; Fabijański, 2023). Mysiukiewicz et al. (2020) demonstrated that using lower processing temperatures together with higher screw speeds results in shorter residence times, which decrease thermal degradation. Sikora et al. (2020) demonstrated that precise moisture and temperature control is necessary for starch/PVOH- and PLA-based blends to stop screw blockages while keeping product consistency.

Material formulation and equipment adaptation further influence extrusion performance. Plasticizers like triacetin and tributyl citrate improve flexibility, but they also raise crystallinity levels, which makes welding and storage stability more difficult (Sikora et al., 2020). The industrial-scale extrusion process gains enhanced thermal and ultraviolet (UV) resistance through stabilizers and antioxidants (Abayev and Shved, 2025). Equipment-related adaptations—including optimized screw design, die geometry, and pressure management—are often required to counteract high melt viscosity, flow instability, and pulsation effects observed in biopolymer melts (Fabijański, 2023). Although conventional extruders can process certain biodegradable polymers like PLA, tailored configurations are often necessary to ensure smooth processing and uniform product quality at industrial scales. The extrusion process for biodegradable plastics needs material science knowledge together with process engineering expertise and equipment optimization capabilities. Each polymer system exhibits unique sensitivities to heat, moisture, and shear, requiring customized solutions to balance biodegradability with mechanical and barrier performance. The field has shown promising advancements through reactive extrusion and nanocomposite integration and real-time process monitoring (Raquez et al., 2008), but no single solution has proven to be universally applicable. Future developments in adaptive screw design, predictive control systems, and formulation engineering will be pivotal to achieving reliable, scalable, and sustainable extrusion of biodegradable plastics across diverse applications.

8.3.6 Innovations in extrusion technology

Food packaging extrusion innovations include PLA-based composites with antioxidants and antimicrobials, nanotechnology-enhanced multilayer films, and physical surface treatments using UV and cold plasma to improve barrier and functional properties. Han Lyn et al. (2024) explain in their review paper how PLA and PLA-based composites develop active packaging properties through their integration with antioxidants and antimicrobials and natural pigments. Research findings indicate that extrusion methods achieve faster drying times and improved flexibility while producing films with uniform thickness and minimal structural defects compared to traditional solvent-based manufacturing techniques.Anukiruthika et al. (2020)

explain how multilayer films use nanotechnology and surface treatments (UV and cold plasma) to improve their barrier performance and mechanical strength and functional capabilities. The research findings demonstrate that these modern solutions tackle environmental sustainability issues by using their recycling capabilities and biodegradable properties. The research shows that developing advanced materials through extrusion process optimization leads to sustainable food packaging solutions, which become commercially viable. Other researches demonstrates that the technology of extrusion has advanced through bioactive nanomaterial coatings and nanotechnology applications, which enhance material strength by 64.3%. Hot-melt extrusion serves as a technique to modify chemicals through its process. The industry has developed new mechanical systems, which include oscillating die extrusion and 3D printing integration for expanding industrial possibilities.

The main focus of coating and surface engineering developments concentrates on creating tools that last longer while reducing friction and achieving biocompatibility. The production of printable filaments for medical and dental purposes became possible through the use of in-line bioactive coatings, which include nanohydroxyapatite and nanofluorapatite (Strähle et al., 2024). Physical and plasma-enhanced vapor deposition methods extended die lifespan and enhanced output quality during metal and composite extrusion processes (Leśniak et al., 2024), although numerical evidence stays inadequate. Nanotechnology integration has produced remarkable mechanical and functional improvements. Polyamide 12 with binary nanoinclusions showed improved tensile strength by 45.5% and modulus strength by 30.7% and impact strength by 64.3%, according to Vidakis et al. (2022). The extrusion process with nanotechnology has introduced self-healing and antibacterial and stimuli-responsive capabilities to polymers, which now serve biomedical and structural purposes (Jiang et al., 2020). Hot-melt extrusion technology has brought about a complete transformation in pharmaceutical production through its advancement of polymer and chemical materials. The addition of polyvinylpyrrolidone, polycaprolactone, and PLA polymers resulted in better drug solubility and bioavailability and processability and made possible sustained-release implants and amorphous drug dispersions (Mohsen et al., 2024; Patil et al., 2016). The development of friction management systems together with computational modeling continues to represent a major obstacle for engineers.

The combination of extrusion with 3D printing technologies including fused deposition modeling and pressure-assisted microsyringe printing and extrusion-based AM allows for the creation of customized pharmaceutical products and precision medical devices at the point of need (Azad et al., 2020; Strähle et al., 2024). The combination of these materials allows for better design flexibility together with lower costs and adjustable material properties. The new extrusion systems operate with intelligence and energy efficiency, and they perform specific tasks according to their design. The research needs to continue its work on achieving coating uniformity while addressing scalability in future projects. Overall, extrusion technology continues to evolve rapidly to meet the dual demands of environmental responsibility and high performance in food packaging. A large amount of studies has investigated different strategies to improve the properties of bioplastics (Table 8.5).

Table 8.5 Characteristics of some new innovative extruded bioplastics.

Study focus	Innovation type	Application	Key technical approach	Study
Advances in extruded polylactic acid–based composites for food packaging	Extrusion technology; incorporation of functional additives (antioxidants, antimicrobials, and pigments)	Active and intelligent food packaging; focus on industrial scalability and sustainability	Review article	Han Lyn et al. (2024)
Advances in multilayer packaging preparation techniques and emerging food applications	Nanotechnology; physical modification (ultraviolet and cold plasma surface treatments); multilayer film design	Multilayer food packaging films; focus on enhanced barrier/mechanical/functional properties and sustainability challenges	Review article	Anukiruthika et al. (2020)
Material design strategies for extrusion three-dimensional (3D) printing of polymers	Material innovation	Engineering, art, education, medicine, and aerospace	Enhanced 3D printability of polymers via material design for advanced properties (mechanical, conductance, self-healing, and stimuli responsiveness)	Jiang et al. (2020)
Hot-melt extrusion in pharmaceutical formulation	Process modification and equipment design	Pharmaceutical	Hot-melt extrusion for drug delivery, twin-screw extruders, quality by design, process analytical technology, and continuous manufacturing	Patil et al. (2016)
Polymers for extrusion-based 3D printing in pharmaceuticals	Additive manufacturing approach and process modification	Pharmaceutical	Fused deposition modeling and pressure-assisted microsyringe 3D printing, polymer selection, rheology, and personalized dosage forms	Azad et al. (2020)
Polyamide 12 nanocomposites with binary inclusions for medical 3D printing	Material innovation	Medical	Polyamide 12/titanium nitride/copper or copper (I) oxide nanocomposites, thermomechanical extrusion, and mechanical/antibacterial enhancement	Vidakis et al. (2022)
Coating machine for filaments with bioactive nanomaterials	Equipment design	Medical and dental	Additive-manufactured coating machine, in-line filament coating with nanohydroxyapatite/nanofluorapatite and extrusion 3D printing	Strähle et al. (2024)

(*Continued*)

Table 8.5 (Continued)

Study focus	Innovation type	Application	Key technical approach	Study
Advances in hot-melt extrusion techniques and technologies	Material innovation, process modification, equipment design, additive manufacturing approach	Pharmaceutical	Hot-melt extrusion with new polymers, single/double-screw extruders, spheronization, and 3D printing integration	Mohsen et al. (2024)
Hot-melt extrusion in pharmaceutical dosage form design	Process modification	Pharmaceutical	Hot-melt extrusion for amorphous drug solids, sustained release, films, implants, and integration with 3D printing/nanotechnology	Salave et al. (2022)

8.4 Injection molding

Injection molding of bioplastics for food packaging transforms heated, bio-based polymers into products such as containers, films, capsules, and cups. Studies report that PLA is the most common material—often blended with polybutylene succinate (PBS), PBAT, PHAs, starch, or agricultural fillers—to enhance performance (Aversa et al., 2020). Researchers describe process adjustments that include setting injection speeds (typically 36–39 mm/s), cycle times (95–102 s), and optimum melt blending temperatures; these are refined through techniques such as reactive compounding, compatibilization, and the addition of nanofillers or plasticizers (Opran et al., 2018). For example, a study using a poly(butylene succinate-co-adipate) (PBSA)/almond shell powder composite reports a 125% boost in flexural strength, moduli increases of 368%–385%, a 24% higher heat deflection temperature, and a 65% drop in viscosity (Root et al., 2023). Reported challenges include thermal degradation, flow issues with lignocellulosic fillers, and brittleness, with solutions involving chemical modification and precise parameter control. Food packaging applications span coffee capsules, rigid packaging, active or barrier films, and more, demonstrating the method's versatility when processing bioplastics under controlled conditions (Aversa et al., 2020) Fig. 8.6 shows a schematic of simple injection molding machine for introduction (Svečko et al., 2013).

8.4.1 Bioplastic materials suitable for injection molding

The food packaging industry utilizes PLA and its combination materials as the primary biodegradable polymers for injection molding because these materials derive from renewable

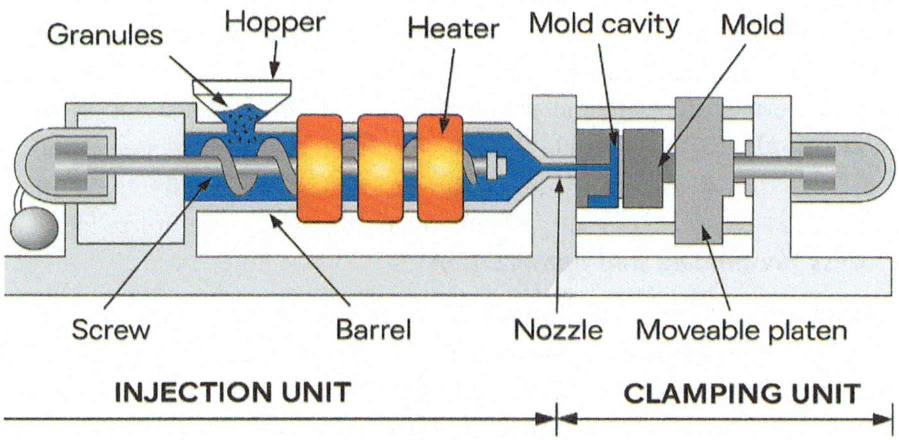

Figure 8.6

Schematic of a simple injection molding machine.

Svečko, R., Kusić, D., Kek, T., Sarjaš, A., Hančič, A., Grum, J., 2013. Acoustic emission detection of macro-cracks on engraving tool steel inserts during the injection molding cycle using PZT sensors. Sensors 13 (5), 6365–6379. https://doi.org/10.3390/s130506365.

sources and demonstrate excellent stiffness and can be processed at moderate temperatures (Aversa et al., 2020; Boey et al., 2021). The material properties of pure PLA prevent its application in rigid packaging and high-temperature uses because it exhibits both brittleness and insufficient heat resistance. The combination of PLA with PBS, PBAT, and PHA results in improved ductility and thermal stability and toughness (Opran et al., 2018; Root et al., 2023). PHAs consist of PHB and poly(3-hydroxybutyrate-co-3-hydroxyvalerate) (PHBV), which provide both biodegradability and barrier protection yet their processing becomes difficult because they begin to degrade thermally near their melting point. Blending PHAs with PLA or TPS and the addition of acetyl tributyl citrate or isosorbide diester plasticizers enhances flexibility and decreases brittleness during molding (Boey et al., 2021; Imam et al., 2008).

The availability together with low expenses makes TPS and cellulose-derived composites an attractive option for various applications. The materials possess hydrophilic characteristics, but their unstable thermal properties prevent them from being used directly in injection molding processes. The solution involves adding compatibilizers and coupling agents to enhance the bond between hydrophilic fillers and hydrophobic matrices (Pacheco et al., 2023). The researchers managed to use almond shell powder together with wheat straw and coffee husks as agricultural residues to create biocomposites through PLA or PBSA matrices, resulting in materials with improved flexural strength by 125% and higher heat deflection temperature (Root et al., 2023). The molding process benefits from nanofillers, which include talc and cellulose nanofibers and layered silicates to improve the rigidity and barrier properties of bioplastics (Jamshidian et al., 2010). The formulations enhance the dimensional stability of molded products, which include coffee capsules and trays and rigid containers while enabling bioplastics to be used in active and barrier packaging applications. The food packaging industry requires PLA, PBS, PBAT, PHAs, and TPS-based materials, which undergo blending, compatibilization, and nanofiller incorporation for injection molding because these materials offer optimal processability and performance and biodegradability (Jariyasakoolroj et al., 2020; Shi et al., 2025).

8.4.2 Process parameters and optimization

The manufacturing parameters for bioplastic injection molding determine the ultimate characteristics and durability and operational success of food packaging products. The standard process parameters for bioplastic molding need to achieve injection speeds between 36 and 39 mm/s and screw rotation speeds near 80 rpm and cycle durations from 95 to 102 seconds and melt temperatures close to 180°C (Opran et al., 2018). The correct mold temperature and moisture level must stay at proper levels because even small amounts of hydrolytic degradation cause polymers like PLA and PBSA to lose molecular weight and mechanical strength (Aversa et al, 2020).

Manufacturers achieve their objectives through combined material content optimization and manufacturing process monitoring techniques. Reactive compounding and solid-state extrusion and compatibilization techniques enhance thermal resistance and interfacial bonding strength between polymer phases (Pacheco et al., 2023; Root et al., 2023). The addition of nanofillers and agricultural by-products like almond shell powder improves stiffness and thermal stability and processing efficiency and leads to a 65% reduction in melt viscosity (Root et al., 2023). Plasticizers including acetyl tributyl citrate and isosorbide diester together with coupling agents decrease brittleness and improve flow characteristics during molding operations (Boey et al., 2021). From an operational perspective, optimization targets a balance between flowability and mechanical performance. Excessive shear or residence time may induce chain scission and thermal degradation, while insufficient pressure or cooling time can cause incomplete filling and dimensional instability. The manufacturing process needs exact control of temperature and pressure and cooling phases, which make up 50%–80% of the total cycle time to achieve structural integrity and reproducibility (Pacheco et al., 2023). The combined optimization efforts result in high-quality bioplastic food packaging production, which includes coffee capsules and rigid containers and active and barrier films that deliver mechanical performance and biodegradation and barrier properties similar to traditional plastics (Boey et al., 2021; Shi et al., 2025).

8.4.3 Challenges in injection molding of bioplastics

Injection molding of bioplastics encounters several technical and material-related challenges that limit performance and scale. The main problem involves thermal degradation which affects PLA and PBSA polymers because they experience molecular weight loss and mechanical strength deterioration when exposed to high melt temperatures in the presence of moisture and oxygen (Aversa et al, 2020; Root et al., 2023). Lignocellulosic fillers create flow difficulties and filler clustering, resulting in reduced uniformity and development of defects in molded parts (Pacheco et al., 2023). Unmodified biopolymers display brittleness and inadequate barrier properties, making them unsuitable for many applications (Boey et al., 2021). The commercial adoption of these technologies faces obstacles because of expensive raw materials and limited manufacturing time and insufficient production facilities (Imam et al., 2008; Root et al., 2023). Environmental factors including moisture and high temperature levels influence both the manufacturing process and the final product stability. The reported solutions consist of chemical modification and exact parameter control during processing and composite formulation for achieving balanced structural and functional characteristics.

8.4.4 Case studies: food packaging applications

Injection molding technology serves various functions when producing different types of food packaging materials. The study by Aversa et al. (2020) demonstrates a method for coffee capsules that applies reactive compounding extrusion together with process optimization to enhance both thermal and mechanical properties of PLA-based bioplastics. The study by Chalfant (2007) demonstrates a process for embedding RFID inlays into polypropylene containers through in-mold labeling, which enables smart features for packaging. Tsagdi et al. (2021) demonstrate a technique that applies food-grade foaming agents to polypropylene pellets for producing caps and bowls with a 14% void fraction and 20% lower thermal conductivity; laboratory- and industrial-scale testing confirmed these results. Through the use of simulation with a nested Taguchi design, Iliopoulou and Vosniakos (2024) identify melt temperature as the most vital factor for optimizing injection molding parameters of polypropylene containers. The case studies demonstrate how new packaging solutions emerge through material development and process optimization and smart technology integration, which suits particular packaging needs (Aversa et al., 2020).

In another study conducted by Plasencia et al. (2025), the creation of injectable zein-based biopolymers brings forth an important progression for environmentally friendly food packaging materials. This research proved that standard industrial injection molding methods can generate thermoplastic zein components. The authors achieved their main breakthrough by combining plasticizers GLY with urea (UREA) to establish a thermoplastic matrix system that operates through both extrusion and injection molding processes. The results demonstrated how plasticizer combinations need to be optimized for packaging applications to achieve the best mechanical performance. The ZEI-15GLY-10UREA formulation reached its highest tensile strength value at 21.3 MPa. The ZEI-20GLY-5UREA formulation showed better material flexibility and toughness through its 9.1% elongation at break, which resulted from its lower urea content. The optimized plasticizer combination created a more ductile structure, which the FESEM micrographs verified. Processing methods enable zein-based materials to become reality because researchers have mainly focused on film production with zein-based materials. The production of mechanical testing specimens became feasible through exact process parameter control, which involved extrusion and injection temperatures at 135°C. The testing of all formulations showed complete degradation within 10 weeks under composting conditions, which made them suitable sustainable alternatives to plastic materials. Table 8.6 summarizes some characteristics of case studies.

8.5 Casting method

The casting method stands as the most common laboratory-scale procedure that produces bioplastic films for food packaging applications. The process involves solvent casting and solution casting and casting evaporation methods that require polymers like PLA and starch and proteins and cellulose derivatives and chitosan to dissolve in solvents before being spread onto flat surfaces for drying into uniform films (Drago et al., 2023; Swaroop and Shukla, 2018; YousefniaPasha et al., 2021). Fig. 8.7 illustrates the casting method steps.

Table 8.6 Characteristics of some food packaging applications produced by injection molding.

Study focus	Application type	Material system	Key innovation	References
Enhancing performance of polylactic acid (PLA)-based bioplastics for food packaging via process and material optimization	Coffee capsules	PLA, polybutylene succinate (PBS), talc, compatibilizers, and processing additives	Reactive compounding extrusion; process optimization to prevent thermal degradation	Aversa et al. (2020)
Embedding radio-frequency identification (RFID) inlays into injection molded food containers using in-mold labeling (IML)	Injection molded food containers	Polypropylene (PP) matrix with polyethylene terephthalate (PETE), IML film, RFID inlays, and adhesives	Integration of RFID via IML technology	Chalfant (2007)
Development of PP foams for food packaging using food-grade foaming agents	Caps and bowls	PP, sodium bicarbonate, citric acid, stearic acid, polyethylene oxide (PEO), lipid esters, and talc	Coating PP pellets with food-grade foaming agents; process optimization for foaming	Tsagdi et al. (2021)
Simulation-based optimization of injection molding parameters for PP containers	Containers	PP	Nested Taguchi design of experiments for process parameter optimization	Iliopoulou and Vosniakos (2024)
Development and characterization of plasticized thermoplastic zein	Food packaging	Zein plasticized with glycerol (GLY) and urea (UREA)	Using injection molding (instead of common film casting) to process zein-based biopolymers, enabled by an optimized GLY/UREA plasticizer system	Plasencia et al. (2025)

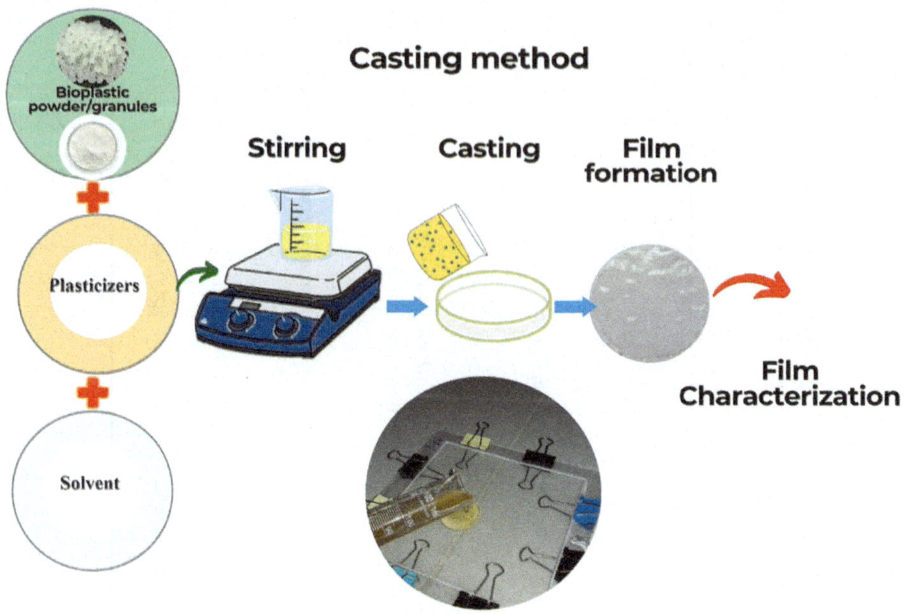

Figure 8.7
Steps of film formation as packaging materials by solvent casting method.

The film properties depend on polymer concentration together with solvent type and evaporation speed and drying temperature and humidity levels and film thickness (Sanyang et al., 2016; Abedinia et al., 2018; Loghmani et al., 2026). The films that were made from chloroform and dried at room temperature before controlled heating at 45°C produced films with enhanced uniformity and mechanical properties (Swaroop and Shukla, 2018). The ethanol–acetic acid solvent systems produced compatible zein/chitosan blends, which showed best performance at a 1:2 ratio (Drago et al., 2023).

8.5.1 Parameters, challenges, applications, and characteristics

The casting method determines how the films perform in terms of mechanical properties and barrier resistance and thermal characteristics. The studies showed that tensile strength values ranged from 10.8 to 26.6 MPa and elastic moduli reached between 56.2 and 169.2 GPa and plasticizer addition resulted in 81.1% elongation at break (Rivera et al., 2024; YousefniaPasha et al., 2021). The addition of tartaric acid plasticizers enhanced material flexibility, yet it caused a decrease in stiffness properties and MgO nanoparticles used as fillers produced a 29% increase in tensile strength at specific loading levels (Swaroop and Shukla, 2018). The barrier properties showed different results across formulations because cinnamic acid and lignin nanoparticles improved oxygen barrier performance according to Ordoñez et al. (2022) and Martín-Sampedro et al. (2025) but plasticizers together with MgO nanoparticles raised water vapor permeability according to YousefniaPasha et al. (2021). The lowest reported water vapor permeability was 4.9×10^{-10} g/m/s/Pa, which indicates that optimizing casting formulations can produce films with excellent barrier properties. The thermal properties of polymers undergo changes because casting affects how their molecules organize and their crystalline structure. The addition of cinnamic acid led to lower glass transition temperatures but simultaneously improved thermal stability according to Ordoñez et al. (2022), yet PEG–tartaric acid plasticization showed better thermal stability according to YousefniaPasha et al. (2021). Nanoparticles demonstrated decreased stability when exposed to catalytic degradation effects according to Swaroop and Shukla (2018). The casting method provides basic control, but it cannot be scaled up effectively and struggles with solvent recovery, which makes it ideal for laboratory-scale biofilm production and small batches but not suitable for industrial continuous production (Ahari et al., 2022). The method functions as an essential tool for optimizing formulations and creating nanocomposites and functional films, which serve as active and smart food packaging applications. The mechanical strength and barrier properties and thermal characteristics of bioplastic films depend on casting parameters that include solvent composition and polymer—additive ratio and drying conditions (Drago et al., 2023; Ordoñez et al., 2022; Swaroop and Shukla, 2018; YousefniaPasha et al., 2021). Table 8.7 summarizes some studies that have produced films by solvent casting method and their characterizations.

8.6 Electrospinning

Food packaging functions as a vital component of the food supply chain because it establishes a protective barrier that keeps food separated from environmental factors. Traditional passive packaging, which functions as a storage container, proves inadequate because food safety problems stem from contamination that occurs during harvesting as well as processing and transportation and storage. New packaging systems are under development to solve these problems. In this area, the active packaging materials fabricated using natural polymers is increasing in recent years. Active systems function to release bioactive compounds and remove unwanted substances, which stop biological and chemical changes in the food matrix. Electrospinning stands as one of the available techniques for this process. The process of electrospinning and electrospray allows for the encapsulation of bioactive compounds into fibers and particles, which serve as active packaging materials (Charles et al., 2021). The primary distinction between electrospray and electrospinning continues to depend on the level of molecular bonding between molecules. Scientists face difficulties when trying to measure molecular cohesion directly, although they can determine it through analyzing solution viscosity and concentration, according to Chakraborty et al. (2009). The droplet size decreases because of electrostatic potential, which leads to better atomization at fixed flow rates according to the study by Hao et al. (2013). Electrospray and electrospinning follow the same physics; however, if the level of molecular cohesion is less than the critical droplet value, particles are formed instead of fibers (Charles et al., 2021). Fig. 8.8A–D shows the schematic diagram of the electrospray and electrospinning operations.

Table 8.7 Characterizations of produced films by solvent casting method.

Biopolymers	Barrier	Process effects	Comparative	Mechanical	Solvent system	Drying conditions	Thickness control	Optimal conditions identified	References
Polylactic acid (PLA) with cinnamic acid	No mention found	Cast films more plastic and stretchable than thermally processed films; cinnamic acid reduced stiffness	Casting vs thermal processing	No mention found	No mention found	No mention found	No mention found	No mention found	Ordoñez et al. (2022)
Poly(acrylic acid)/chitosan/ linseed mucilage	Not applicable	Higher linseed mucilage content increased elastic modulus and TS, decreased elongation	Different linseed mucilage/ chitosan/poly (acrylic acid) ratios	TS: 10.8-16.4 MPa, elastic modulus: 56.2–169.2 GPa	No mention found	No mention found	No mention found	No mention found	Rivera et al. (2024)
PLA with magnesium oxide nanoparticles	Oxygen permeability decreased 25% (2 wt% magnesium oxide) and water vapor permeability increased 20%	Best properties at 2 wt% magnesium oxide due to uniform dispersion	Different magnesium oxide concentrations (1-4 wt%)	TS increased 29% at 2 wt% magnesium oxide	Chloroform	Room tempera-ture 2–3 days, then 45°C for 24 h	35-45 μm using film applicator	2 wt% magnesium oxide for optimal properties	Swaroop and Shukla (2018)
Polysaccharides and plant proteins	No mention found	Plasticizing effect observed with all plasticizers	Different plasticizers (glycerol, sorbitol, and triethanolamine)	No mention found	No mention found	No mention found	No mention found	Careful plasticizer selection required	Jost and Stramm (2015)

Cellulose nanofibers with lignin nanoparticles	No mention found	Properties unchanged or slightly improved up to 5%–10% lignin nanoparticles	Casting vs filtration plus drying	No mention found	No mention found	No mention found	No mention found	Casting evaporation preferred over filtration	Martín-Sampedro et al. (2025)
PLA with plasticizers	Lowest water vapor permeability: 4.9×10^{-10} g/m/s/Pa	Plasticizers increased elongation and decreased TS	Different plasticizer concentrations 10–30 wt%	Highest TS: 26.6 MPa (10 wt % tartaric acid), highest elongation: 81.1% (30 wt% tartaric acid)	No mention found	No mention found	No mention found	10 wt%tartaric acid for best overall performance	YousefniaP-asha et al. (2021)
Zein and chitosan blend	Two orders of magnitude decrease in blend compared to pure zein	Blend improved properties compared to pure polymers	Different blend ratios, pure vs blend films	TS exceeded 3.5 MPa minimum	Ethanol (80%–90% by volume), acetic acid (10% by volume)	60°C for 2 hours (zein), 60°C for 1 day then desiccator (chitosan/ blend)	Digital micrometer mea-surement	1:2 zein:chitosan ratio optimal	Drago et al. (2023)
Soy protein, wheat gluten, and corn zein	No mention found	Protein films generally lack mechanical strength	Review of various approaches	No mention found	Suitable solvents with plasticizers	Subsequent drying	No mention found	No mention found	Zhang and Mittal (2010)
Sugar palm starch	Not applicable	Not applicable	Different plasticizer types and concentrations	Not applicable	8% by weight aqueous dispersion	40°C for 24 h	Glass petri dishes	Sorbitol plasticizer performed best	Sanyang et al. (2016)
Various bio-nanocomposites	No mention found	Improvements possible through material combinations	Review of casting vs other techniques	No mention found	No mention found	No mention found	No mention found	Optimization needed for all techniques	Ahari et al. (2022)

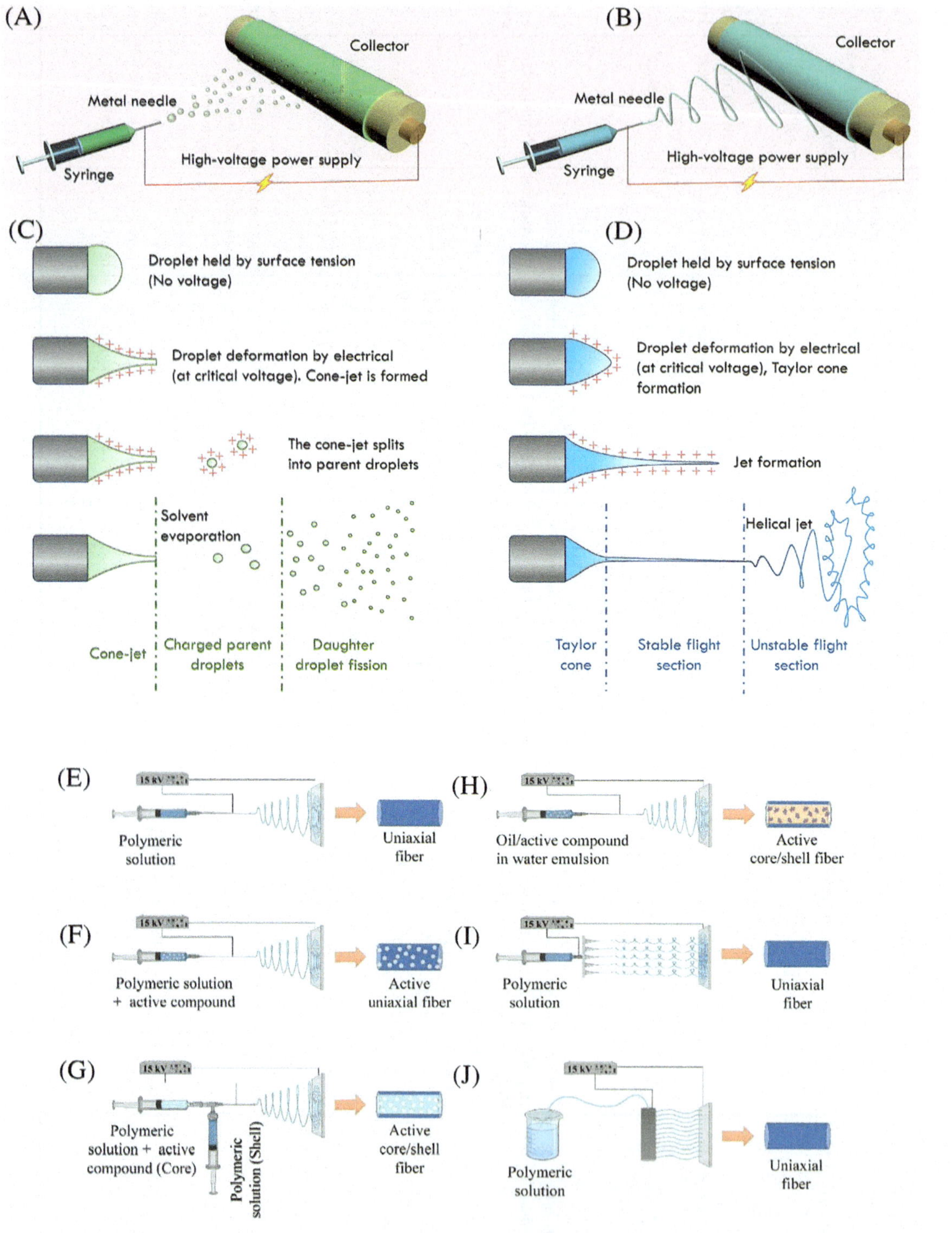

(A)
Collector
Metal needle
Syringe
High-voltage power supply
(B)
Collector
Metal needle
Syringe
High-voltage power supply
(C)
Droplet held by surface tension
(No voltage)
Droplet deformation by electrical
(at critical voltage). Cone-jet is formed
The cone-jet splits
into parent droplets
Solvent
evaporation
Cone-jet
Charged parent
droplets
Daughter
droplet fission
(D)
Droplet held by surface tension
(No voltage)
Droplet deformation by electrical
(at critical voltage), Taylor cone
formation
Jet formation
Helical jet
Taylor
cone
Stable flight
section
Unstable flight
section
(E)
15 kV
Polymeric
solution
Uniaxial
fiber
(F)
15 kV
Polymeric solution
+ active compound
Active
uniaxial fiber
(G)
15 kV
Polymeric
solution + active
compound (Core)
Polymeric
solution (Shell)
Active
core/shell
fiber
(H)
15 kV
Oil/active compound
in water emulsion
Active
core/shell fiber
(I)
15 kV
Polymeric
solution
Uniaxial
fiber
(J)
15 kV
Polymeric
solution
Uniaxial
fiber

8.6.1 Types of electrospinning

The electrospinning method for biodegradable materials serves to create standard and specialized materials that function as packaging solutions for food products. The method of operation depends on the type of spinning solution and needle, which determines fiber construction and material characteristics. The most basic system of uniaxial electrospinning (Fig. 8.8E) operates through a single needle that produces fibers from a uniform polymeric solution. The method enables basic and efficient active compound storage inside fibers, producing nanofibers that release compounds gradually (Fig. 8.8F). Reforms have reduced active component compatibility with polymers and environmental factor sensitivity, causing quick substance release in core–shell fiber structure through coaxial electrospinning using two polymer solutions inside concentric needles to produce fibers (Fig. 8.8G and H). The complex structure contains a polymeric core layer that holds the active compound and a polymeric shell layer that serves as a protective barrier for the active compound. The production of biodegradable mats through electrospinning systems based on emulsion and multijet (Fig. 8.8I) and needleless methods proves less effective than those produced by uniaxial and coaxial electrospinning techniques. The direct one-nozzle spinning of stable emulsions creates emulsion electrospinning, but industrial usage remains limited because of polymer selection challenges and low production rates and difficult polymeric emulsion processing. The spinning solution undergoes active agitation, while high voltage is applied during needleless electrospinning (Fig. 8.8J), which represents a new alternative method (Patiño Vidal et al., 2024).

8.6.2 Challenges, applications, and characteristics

Electrospinning functions through high-voltage electric fields between 5 and 25 kV, which operate on polymer solutions at flow rates from 0.2 to 5 mL/hour and maintain a fixed distance between the tip and collector of 15–30 cm to create customizable nanofibrous mats

Figure 8.8

Mechanisms and experimental sets for electrospray and electrospinning operations: (A and B) Diagrammatic representation of the experimental E-spinning and E-spraying sets. A syringe, an injection pump, a metal needle, a DC-mode high-voltage power source, and a collector are all included in each one. (C and D) The E-spinning and E-spraying system is shown schematically. Types of electrospinning processing: (E) uniaxial, (F) uniaxial to encapsulate active compounds, (G) coaxial, (H) emulsion, (I) multiaxial, and (J) needleless.

(C,D) Adapted with permission from Si, Y., Shi, S., Hu, J., 2024. Electrospinning and electrospraying synergism: twins-tech collaboration across dimensions. Matter 7 (4), 1373–1405. https://doi.org/10.1016/j.matt.2024.01.009. (E–J) Adapted from Patiño Vidal, C., Muñoz-Shugulí, C., Guivier, M., Puglia, D., Luzi, F., Rojas, A. et al., 2024. PLA- and PHA-biopolyester-based electrospun materials: development, legislation, and food packaging applications. Molecules 29 (22), 5452. https://doi.org/10.3390/molecules29225452.

(Marcoaldi et al., 2023; Yang et al., 2025). Process parameters such as voltage, solution flow rate, polymer concentration, ambient temperature, and humidity play critical roles in determining fiber morphology, porosity, and mechanical strength. The best fiber production occurs at 15–25 kV voltage and 30°C with 30% relative humidity because these conditions create both uniform fibers and minimal bead formation and better alignment of fibers (Yang et al., 2025). The material system selection determines how well the process works and what type of film properties will result in the end product. The mechanical properties of chitosan-based systems improve when they dissolve in acidic solvents such as acetic or formic or lactic acids and mix with polyvinyl alcohol (PVA or PVOH) or polyethylene oxide or PLA to reach tensile strength between 5 and 8 MPa and elongation from 13% to 17% and modulus reaching 648 MPa when polymer ratios and solvent parameters are optimized (Yang et al., 2025). The electrospinning process under controlled conditions produces multilayer films of PHBV with copolyamide 1010/1014 and SiO_2 nanoparticles and both high ductility and superhydrophobic properties (contact angle ≈ 146°) (Marcoaldi et al., 2023) (Table 8.8).

Parameter optimization techniques help to improve the functional performance of systems. The reduction in fiber diameter occurs when voltage increases, which leads to better barrier performance, and stable jet formation at moderate flow rates produces uniform mats (Charles et al., 2021). Environmental control stands as an essential factor because high humidity levels produce bead defects, which lead to weak fiber strength, yet moderate temperatures between 25°C and 35°C help solvents evaporate properly to create uniform deposits. The electrospun chitosan-casein systems show their best performance through their water vapor permeability rates of 3.6 g·mm/kPa/day/m^2 and oxygen transmission rates that approach zero (Yang et al., 2025). Thermally, electrospinning induces structural modifications that enhance stability; chitosan-based blends show an increase in decomposition temperature from 225°C to 310°C, while PHBV/copolyamide structures maintain a melting point near 167°C with 18% crystallinity (Marcoaldi et al., 2023). The nanofiber mats that have been created function as packaging materials that deliver active and intelligent capabilities because they have large surface areas and controlled permeability, and they can carry natural bioactive substances (Amorim et al., 2024; Si et al., 2024).

8.7 Assisted processing

8.7.1 Ultrasonication-assisted processing

Ultrasonication has become an emerging auxiliary technique that manufacturers now use to produce bioplastics for food packaging materials. The process functions as a strong pretreatment and in-process modification method, which improves component dispersion and creates molecular interactions and customized film microstructure (Abral et al., 2020; Jiménez-Rosado et al., 2020).

Table 8.8 Characteristics of electrospin bioplastics and their effective electrospinning parameters.

Study focus	Biopolymer type	Parameters investigated	Properties measured	References
Review of chitosan-based electrospun nanofibers for food packaging	Chitosan (CS), blends with polyvinyl alcohol (PVA), polyethylene oxide (PEO), polycaprolactone (PCL), polylactic acid (PLA), polyethylene glycol (PEG), gelatin, starch, alginate, and zein	Voltage, flow rate, solvent, and environmental conditions	Mechanical, barrier, thermal, fiber morphology, surface, and crystallinity	Yang et al. (2025)
Review of PLA- and polyhydroxyalkanoate (PHA)-biopolyester-based electrospun materials for food packaging	PLA and PHA	Techniques (uniaxial, coaxial, emulsion, multiaxial, and needleless)	Material properties, biodegradability, and antioxidant/antimicrobial activity	Patiño Vidal et al. (2024)
Review of polysaccharide-based electrospun fibers with natural colorants	Chitosan, starch, cellulose, and alginate	General process description	Barrier, mechanical, antioxidant, antimicrobial, and colorant incorporation	Amorim et al. (2024)
Review of electrohydrodynamic processing of natural polymers for active packaging	Natural polymers	General discussion	Structural, functional, barrier, and mechanical (qualitative)	Charles et al. (2021)
Review of electrospun biopolymer materials for antimicrobial packaging of produce	Proteins and polysaccharides	General discussion	Antimicrobial, mechanical, barrier, and thermal (qualitative)	Si et al. (2024)
Experimental study on multilayered films (poly(3-hydroxybutyrate-co-3-hydroxyvalerate) (PHBV), copolyamide 1010/1014, silicon dioxide (SiO_2))	PHBV, copolyamide 1010/1014, and SiO_2	Voltage, flow rate, distance, concentration, environmental conditions, and collector type/speed	Mechanical, barrier, thermal, fiber morphology, surface, and crystallinity	Marcoaldi et al. (2023)
Experimental study on multilayer starch-based films with biopolyester coatings	Thermoplastic corn starch, PCL, PLA, and polyhydroxybutyrate (PHB)	No mention found	Barrier (oxygen and water vapor) and morphology (scanning electron microscopy (SEM))	Fabra et al. (2016)
Review of electrospun PLA nanofibers for food packaging	PLA	General discussion	Barrier, mechanical, and thermal (qualitative)	Wu et al. (2022)
Review of electrospun biopolymer nanofibers	Edible biopolymers	General process discussion	Mechanical, surface, and functional (qualitative)	Kriegel et al. (2008)
Review of electrospun plant protein nanofibers for food packaging	Plant proteins	General discussion	Fabrication, challenges, and potential applications (qualitative)	Aghababaei et al. (2024)

The research studies have operated within 20–53 kHz frequency range while using 40 W/L to 360 W power levels for 2.5 to 60 minutes through probe systems and homogenizers and bath systems. Research focused on biopolymers including PVA and starch and pullulan and soy and oat proteins and waste-derived composites, which consist of potato peel and microalgae and sweet lime pomace (Sabathini et al., 2018). The following results demonstrate increases in tensile strength up to 29% and elongation at break up to 2.5 times and water vapor permeability reductions of approximately 11% (Abral et al., 2020; Kang et al., 2023). The testing showed that protein-based and polysaccharide-based matrices achieved better thermal stability and produced more uniform materials with enhanced cross-linking (Sabathini et al., 2018). Nevertheless, the technique remains supplementary rather than foundational in bioplastic manufacturing. These materials function as additives that help in the process of mixing and dispersion and structural changes, but they do not serve as independent forming materials. The process of sonication at excessive levels leads to polymer chain scission, which results in reduced mechanical integrity, as demonstrated by Liu et al. (2019), which makes it essential to adjust power and frequency and duration settings for each biopolymer system. Ultrasonication functions as an essential process intensification method, which enables scientists to modify bioplastic film characteristics for food packaging materials. The method functions as a secondary enhancement tool because it contains sensitive processing parameters, which restricts its ability to function as a main production method.

8.7.2 Microwave-assisted processing

Microwave-assisted processing stands as a new manufacturing approach that scientists have not yet fully studied for bioplastic production that serves food packaging needs. The technology of microwave processing delivers fast heating throughout the entire volume of materials in an energy-efficient way, which produces molecular changes that strengthen biopolymer matrix structures and their operational characteristics (Tekiner, 2024; Wang et al., 2016; Nilsen-Nygaard et al., 2021). The technology faces challenges because scientists need to handle microwave power distribution and sample uniformity, while no standardized equipment exists for industrial polymer processing systems (Nilsen-Nygaard et al., 2021). The research findings show that short microwave exposures at 350–360 W power levels with 45 seconds to 2 minutes duration and 60°C–85°C temperature range lead to improved material strength (5%–25% increase), decreased water absorption (up to 44.7%), and better thermal properties including glass transition temperature and latent heat (Tekiner, 2024; Wang et al., 2016). The best results appear with protein- and starch-based matrices, but PLA shows restricted improvement because its dipolar movement and chain relaxation capabilities stay limited when exposed to microwave fields (Nilsen-Nygaard et al., 2021). The microwave-assisted processing method shows potential as an emerging technology for bioplastic production, yet it remains a specialized approach. The method encounters two major obstacles because it lacks standardized industrial protocols and reproducibility standards, but it provides excellent potential for fast and eco-friendly biodegradable polymer modifications that can be customized. The research should proceed by integrating this technology with extrusion or casting methods for industrial production and building exact power and temperature control systems while conducting thorough environmental and economic assessments to confirm sustainable food packaging production.

8.7.3 High-pressure homogenization

HPH operates as a nonthermal method that physically alters bioplastics through clean processes to enhance their performance. The method operates as a nonthermal physical modification technique to boost bioplastic performance, yet it remains seldom used in food packaging production when compared to standard manufacturing methods including extrusion and casting and solvent evaporation (Ahmed et al., 2022; Cui et al., 2021). The method operates as a secondary or auxiliary process to adjust bioplastic film structures and conduct postprocessing operations, which need equipment capable of reaching 600 MPa, but it does not function as a primary film production technique (Cheng et al., 2022). The research shows that pressure values between 200 and 600 MPa enhance tensile strength and elongation and barrier performance of PLA and PPI films, and these effects appear most strongly in PLA/ZnO nanocomposites and PPI-GLY films (Cheng et al., 2022; Cui et al., 2021). The material shows better performance because of pressure-induced changes in its microstructure and increased crystallinity and stronger hydrogen bonds and polymer chain alignment (hydrogen bonding and polymer chain alignment) under pressure. The research shows that water-filled PLA loses its rigidity when exposed to 450–600 MPa pressure, which indicates that process optimization remains difficult and depends heavily on material characteristics (Ahmed et al., 2022). The research lacks sufficient data about how materials maintain thermal stability during aging and how they scale up for extended periods of time. The green solvent-free technique of HPH shows promise for modifying bioplastic packaging materials, although it remains primarily used to modify films. The research faces obstacles because HPH equipment costs remain high and experimental procedures become complex, while its application to biopolymer systems requires new exploration. The investigation of process–structure–property relationships together with industrial practicality stands as the essential direction for future studies.

8.7.4 Supercritical carbon dioxide impregnation

scCO_2 impregnation stands as a new environmentally safe method that operates without heat to embed bioactive elements including antimicrobial and antioxidant agents into bioplastic materials used for food packaging (De Souza et al., 2014; Lucic Skoric et al., 2022; Rivera et al., 2024). The manufacturing process of bioplastics mainly depends on extrusion and casting methods, but scCO_2 impregnation operates as an additional processing step that functions as a secondary technique. The process functions as a postproduction step for film or sheet materials, which allows engineers to adjust both surface characteristics and internal attributes of preformed bioplastics.

The process functions at moderate to high pressures between 30–250 MPa and temperatures from 35 to 130°C to use CO_2 solvation properties and diffusivity for active molecule impregnation without needing harmful solvents (Lucic Skoric et al., 2022). Studies have shown that $scCO_2$ impregnation enhances barrier performance by decreasing water vapor permeability and boosts thermal stability through higher melting and glass transition temperatures (De Souza et al., 2014). According to Rivera et al. (2024), the polymer blends demonstrate improved active compound distribution and higher porosity along with better foam structures according to morphological analysis. The retrieved studies did not perform any quantitative evaluation of mechanical properties, which creates an essential knowledge gap about how this method influences structural strength and flexibility. The limited number of studies exists because supercritical processing involves complex technical requirements and expensive equipment, which makes it harder to scale up compared to traditional extrusion or casting methods. The method functions best to add bioactive substances to materials at their surface level instead of working with entire materials, which explains its limited yet important contribution to bioplastic technology development. The $scCO_2$ impregnation method functions as a specialized technique that serves to enhance standard manufacturing processes. The technology allows scientists to create active smart food packaging materials that control substance release and maintain polymer biodegradability and chemical composition. Research should focus on process scaling and mechanical impact measurement and industrial integration of continuous bioplastic production for commercial viability (De Souza et al., 2014; Lucic Skoric et al., 2022; Rivera et al., 2024).

8.8 Conclusions

Bioplastic processing and manufacturing operations serve as the essential connection, which transforms laboratory discoveries into commercial production for sustainable food packaging solutions. The manufacturing industry relies on extrusion and injection molding as two primary scalable and cost-efficient methods for production. The polymer industry now can use ultrasonication and microwave-assisted treatment and $scCO_2$ impregnation as assisted processes to improve material properties. The material properties of bioplastics depend on their composition and the way they are processed and the equipment used during manufacturing. The combination of material composition and processing conditions and equipment settings produces significant effects on bioplastic structural properties and thermal behavior and barrier performance according to this research. The integration of monitoring systems with reactive extrusion and nanocomposite materials allows for precise control of biopolymer matrices to create improved mechanical performance and biodegradability. A complete strategy that merges material science with process optimization and LCA becomes necessary for achieving environmentally friendly bioplastic production because processing inefficiencies can damage sustainability advantages. The food packaging industry embraces circular economy principles through extrusion-based manufacturing and bioengineered bioplastics to create sustainable packaging solutions that combine ecological responsibility with superior performance.

References

Abayev, A., Shved, M., 2025. Innovative polymers for a sustainable future: extrusion of bio- and photo-degradable polymer films. Technical Sciences and Technologies 2 (40), 158–163. Available at https://doi.org/10.25140/2411-5363-2025-2(40)-158-163.

Abedinia, A., Ariffin, F., Huda, N., Nafchi, A.M., 2017. Extraction and characterization of gelatin from the feet of Pekin duck (Anas platyrhynchos domestica) as affected by acid, alkaline, and enzyme pretreatment. International Journal of Biological Macromolecules 98, 586–594. Available at https://doi.org/10.1016/j.ijbiomac.2017.01.139.

Abedinia, A., Ariffin, F., Huda, N., Mohammadi Nafchi, A., 2018. Preparation and characterization of a novel biocomposite based on duck feet gelatin as alternative to bovine gelatin. International Journal of Biological Macromolecules 109, 855–862. Available at https://doi.org/10.1016/j.ijbiomac.2017.11.051.

Abedinia, A., Mohammadi Nafchi, A., Sharifi, M., Ghalambor, P., Oladzadabbasabadi, N., Ariffin, F., Huda, N., 2020. Poultry gelatin: Characteristics, developments, challenges, and future outlooks as a sustainable alternative for mammalian gelatin. Trends in Food Science & Technology 104, 14–26. Available at https://doi.org/10.1016/j.tifs.2020.08.001.

Abedinia, A., Serri, A., Elahi, M., Gómez, A.M.A., Nurdiani, R., Huda, N., 2026. Poultry collagen: structural and functional properties, biomaterial potential, and the molecular mechanisms and biological interactions of its hydrolysates in wound healing and tissue regeneration—a review. International Journal of Biological Macromolecules 27, 151693. Available at https://doi.org/10.1016/j.ijbiomac.2026.151693.

Abeykoon, C., 2016. Single screw extrusion control: a comprehensive review and directions for improvements. Control Engineering Practice 51, 69–80. Available at https://doi.org/10.1016/j.conengprac.2016.03.008.

Abral, H., Atmajaya, A., Mahardika, M., Hafizulhaq, F., Kadriadi, Handayani, D., Sapuan, S.M., et al., 2020. Effect of ultrasonication duration of polyvinyl alcohol (PVA) gel on characterizations of PVA film. Journal of Materials Research and Technology 9 (2), 2477–2486. Available at https://doi.org/10.1016/j.jmrt.2019.12.078.

Aghababaei, F., McClements, D.J., Martinez, M.M., Hadidi, M., 2024. Electrospun plant protein-based nanofibers in food packaging. Food Chemistry 432, 137236. Available at https://doi.org/10.1016/j.foodchem.2023.137236.

Ahari, H., Golestan, L., Anvar, S.A.A., Cacciotti, I., Garavand, F., Rezaei, A., et al., 2022. Bio-nanocomposites as food packaging materials: the main production techniques and analytical parameters. Advances in Colloid and Interface Science 310, 102806. Available at https://doi.org/10.1016/j.cis.2022.102806.

Ahmed, J., Mulla, M.Z., Vohra, A., 2022. High-pressure treatment of water-filled co-extruded polylactide films: effect on microstructure, barrier, thermal, and rheological properties. Journal of Food Science 87 (4), 1754–1766. Available at https://doi.org/10.1111/1750-3841.16096.

Ahsan, W.A., Hussain, A., Lin, C., Nguyen, M.K., 2023. Biodegradation of different types of bioplastics through composting—a recent trend in green recycling. Catalysts 13 (2), 294. Available at https://doi.org/10.3390/catal13020294.

Alam, A.K.M.M., Mina, M.F., Beg, M.D.H., Mamun, A.A., Bledzki, A.K., Shubhra, Q.T.H., 2014. Thermo-mechanical and morphological properties of short natural fiber reinforced poly (lactic acid) biocomposite: effect of fiber treatment. Fibers and Polymers 15 (6), 1303–1309. Available at https://doi.org/10.1007/s12221-014-1303-8.

Ali, S.S., Elsamahy, T., Abdelkarim, E.A., Al-Tohamy, R., Kornaros, M., Ruiz, H.A., et al., 2022. Biowastes for biodegradable bioplastics production and end-of-life scenarios in circular bioeconomy and biorefinery concept. Bioresource Technology 363, 127869. Available at https://doi.org/10.1016/j.biortech.2022.127869.

Amesho, K.T.T., Edoun, E.I., Shangdiar, S., Shopati, A.K., Kadhila, T., Sithole, N.T., et al., 2025. Advanced technologies for the production of bioplastics. Microplastic Pollution: Occurrence, Health Risk and Challenges. CRC Press, Namibia, pp. 286–302. Available at https://doi.org/10.1201/9781032706573-16.

Amiri, M., Hassani, B., Babapour, H., Nikmanesh, A., Hosseini, S.E., Asadi, G., et al., 2025. Optimization of enzyme hydrolysis to improve functional and structural properties of microalgae protein extract. Journal of Food Science 90 (4), e70129. Available at https://doi.org/10.1111/1750-3841.70129.

Amorim, L.F.A., Mouro, C., Gouveia, I.C., 2024. Electrospun fiber materials based on polysaccharides and natural colorants for food packaging applications. Cellulose 31 (10), 6043–6069. Available at https://doi.org/10.1007/s10570-024-05956-z.

Andreasi Bassi, S., Boldrin, A., Frenna, G., Astrup, T.F., 2021. An environmental and economic assessment of bioplastic from urban biowaste. The example of polyhydroxyalkanoate. Bioresource Technology 327, 124813. Available at https://doi.org/10.1016/j.biortech.2021.124813.

Anli Dino, A., Kishore, G., 2024. Microalgae: an emerging source of bioplastics production. Discover Environment 2 (1), 10. Available at https://doi.org/10.1007/s44274-024-00038-0.

Anukiruthika, T., Sethupathy, P., Wilson, A., Kashampur, K., Moses, J.A., Anandharamakrishnan, C., 2020. Multilayer packaging: advances in preparation techniques and emerging food applications. Comprehensive Reviews in Food Science and Food Safety 19 (3), 1156–1186. Available at https://doi.org/10.1111/1541-4337.12556.

Asgher, M., Qamar, S.A., Bilal, M., Iqbal, H.M.N., 2020. Bio-based active food packaging materials: sustainable alternative to conventional petrochemical-based packaging materials. Food Research International 137, 109625. Available at https://doi.org/10.1016/j.foodres.2020.109625.

Attallah, O.A., Mojicevic, M., Garcia, E.L., Azeem, M., Chen, Y., Asmawi, S., et al., 2021. Macro and micro routes to high performance bioplastics: bioplastic biodegradability and mechanical and barrier properties. Polymers 13 (13), 2155. Available at https://doi.org/10.3390/polym13132155.

Aversa, C., Puopolo, M., Barletta, M., 2020. Tailor-made bioplastics for environmentally friendly food packaging: a methodological approach to a challenging problem. In: Hashmi, S., Choudhury, I.A. (Eds.), Encyclopedia of Renewable and Sustainable Materials. Elsevier, pp. 605–616. Available at https://doi.org/10.1016/B978-0-12-803581-8.10799-4.

Azad, M.A., Olawuni, D., Kimbell, G., Badruddoza, A.Z.M., Hossain, M.S., Sultana, T., 2020. Polymers for extrusion-based 3D printing of pharmaceuticals: a holistic materials–process perspective. Pharmaceutics 12 (2), 124. Available at https://doi.org/10.3390/pharmaceutics12020124.

Bartilla, T., Kirch, D., Nordmeier, J., Proemper, E., Strauch, T., 1986. Physical and chemical changes during the extrusion process. Advances in Polymer Technology: Journal of the Polymer Processing Institute 6 (3), 339–387. Available at https://doi.org/10.1002/adv.1986.060060308.

Bhuwal, A.K., Singh, G., Aggarwal, N.K., Goyal, V., Yadav, A., 2014. Poly-β-hydroxybutyrate production and management of cardboard industry effluent by new bacillus sp. NA10. Bioresources and Bioprocessing 1 (1), 9. Available at https://doi.org/10.1186/s40643-014-0009-5.

Blum, N.U., Haupt, M., Bening, C.R., 2020. Why "circular" doesn't always mean "sustainable. Resources, Conservation and Recycling 162, 105042. Available at https://doi.org/10.1016/j.resconrec.2020.105042.

Boey, J.Y., Lee, C.K., Tay, G.S., 2022. Factors affecting mechanical properties of reinforced bioplastics: a review. Polymers 14 (18), 3737. Available at https://doi.org/10.3390/polym14183737.

Boey, J.Y., Mohamad, L., Khok, Y.S., Tay, G.S., Baidurah, S., 2021. A review of the applications and biodegradation of polyhydroxyalkanoates and poly(lactic acid) and its composites. Polymers 13 (10), 1544. Available at https://doi.org/10.3390/polym13101544.

Bogaert, J.C., Coszach, P., 2000. Poly (lactic acids): a potential solution to plastic waste dilemma. Macromolecular symposia 153 (1), 287–303.

Bourbon, A.I., Pereira, R.N., Pastrana, L.M., Vicente, A.A., Cerqueira, M.A., 2019. Protein-based nanostructures for food applications. Gels 5 (1), 9. Available at https://doi.org/10.3390/gels5010009.

Bouvier, J.M., Campanella, O.H., 2014. Extrusion Processing Technology: Food and Non-Food Biomaterials. Wiley, pp. 1–518. Available at https://doi.org/10.1002/9781118541685.

Calambás Pulgarin, H.L., Caicedo, C., 2024. Barrier, mechanical, thermal, and rheological properties of plasticized biopolymeric films manufactured by co-extrusion. Processes 12 (3), 524. Available at https://doi.org/10.3390/pr12030524.

Cesário, M.T., Raposo, R.S., Almeida, M.C.M.Dd, van Keulen, F., Ferreira, B.S., Telo, J.P., et al., 2014. Production of poly(3-hydroxybutyrate-co-4-hydroxybutyrate) by Burkholderia sacchari using wheat straw hydrolysates and gamma-butyrolactone. International Journal of Biological Macromolecules 71, 59–67. Available at https://doi.org/10.1016/j.ijbiomac.2014.04.054.

Chakraborty, S., Liao, I.C., Adler, A., Leong, K.W., 2009. Electrohydrodynamics: a facile technique to fabricate drug delivery systems. Advanced Drug Delivery Reviews 61 (12), 1043–1054. Available at https://doi.org/10.1016/j.addr.2009.07.013.

Chalfant, A.M., 2007. Embedding radio frequency identification inlays into injection molded food containers using in mold labeling technology. Dissertation. University of Florida.

Charles, A.P.R., Jin, T.Z., Mu, R., Wu, Y., 2021. Electrohydrodynamic processing of natural polymers for active food packaging: a comprehensive review. Comprehensive Reviews in Food Science and Food Safety 20 (6), 6027–6056. Available at https://doi.org/10.1111/1541-4337.12827.

Chen, L., Pelton, R.E.O., Smith, T.M., 2016. Comparative life cycle assessment of fossil and bio-based polyethylene terephthalate (PET) bottles. Journal of Cleaner Production 137, 667–676. Available at https://doi.org/10.1016/j.jclepro.2016.07.094.

Cheng, J., Li, Z., Wang, J., Zhu, Z., Yi, J., Chen, B., et al., 2022. Structural characteristics of pea protein isolate (PPI) modified by high-pressure homogenization and its relation to the packaging properties of PPI edible film. Food Chemistry 388, 132974. Available at https://doi.org/10.1016/j.foodchem.2022.132974.

Chinthapalli, R., Skoczinski, P., Carus, M., Baltus, W., de Guzman, D., Käb, H., et al., 2019. Biobased building blocks and polymers—global capacities, production and trends, 2018–2023. Industrial Biotechnology 15 (4), 237–241. Available at https://doi.org/10.1089/ind.2019.29179.rch.

Claverie, M., McReynolds, C., Petitpas, A., Thomas, M., Fernandes, S.C.M., 2020. Marine-derived polymeric materials and biomimetics: an overview. Polymers 12 (5), 1002. Available at https://doi.org/10.3390/polym12051002.

Costa, A., Encarnação, T., Tavares, R., Todo Bom, T., Mateus, A., 2023. Bioplastics: innovation for green transition. Polymers 15 (3), 517. Available at https://doi.org/10.3390/polym15030517.

Cui, R., Fan, C., Dong, X., Fang, K., Li, L., Qin, Y., 2021. Effect of ultrahigh-pressure treatment on the functional properties of poly(lactic acid)/ ZnO nanocomposite food packaging film. Journal of the Science of Food and Agriculture 101 (12), 4925–4933. Available at https://doi.org/10.1002/jsfa.11136.

De Souza, A.C., Dias, A.M.A., Sousa, H.C., Tadini, C.C., 2014. Impregnation of cinnamaldehyde into cassava starch biocomposite films using supercritical fluid technology for the development of food active packaging. Carbohydrate Polymers 102 (1), 830–837. Available at https://doi.org/10.1016/j.carbpol.2013.10.082.

Delvar, E., Oliveira, I., Brito, M.S.C.A., Silva, C.G., Santamaria-Echart, A., Barreiro, M.-F., Santos, R.J., 2025. Literature review on single and twin-screw extruders design for polymerization using CFD simulation. Fluids 10 (1), 9. Available at https://doi.org/10.3390/fluids10010009.

Di Bartolo, A., Infurna, G., Dintcheva, N.T., 2021. A review of bioplastics and their adoption in the circular economy. Polymers 13 (8), 1229. Available at https://doi.org/10.3390/polym13081229.

Dimonie, D., Petrache, M., Damian, C., Anton, L., Musat, M., Dima, Ş.O., et al., 2016. New evidences on the process sensitivity of some renewable blends based on starch considering their melt rheological properties. International Journal of Polymer Science 2016 (1), 7873120. Available at https://doi.org/10.1155/2016/7873120.

Drago, E., Campardelli, R., Lagazzo, A., Firpo, G., Perego, P., 2023. Improvement of natural polymeric films properties by blend formulation for sustainable active food packaging. Polymers 15 (9), 2231. Available at https://doi.org/10.3390/polym15092231.

Drumright, R.E., Gruber, P.R., Henton, D.E., 2000. Polylactic acid technology. Advanced Materials 12 (23), 1841–1846. Available at https://doi.org/10.1002/1521-4095(200012) 12:23<1841::AID-ADMA1841>3.3.CO;2-5.

Duty, C., Ajinjeru, C., Kishore, V., Compton, B., Hmeidat, N., Chen, X., et al., 2018. What makes a material printable? A viscoelastic model for extrusion-based 3D printing of polymers. Journal of Manufacturing Processes 35, 526–537. Available at https://doi.org/10.1016/j.jmapro.2018.08.008.

Fabijański, M., 2023. Study of the single-screw extrusion process using polylactide. Polymers 15 (19), 3878. Available at https://doi.org/10.3390/polym15193878.

Fabra, M.J., López-Rubio, A., Cabedo, L., Lagaron, J.M., 2016. Tailoring barrier properties of thermoplastic corn starch-based films (TPCS) by means of a multilayer design. Journal of Colloid and Interface Science 483, 84–92. Available at https://doi.org/10.1016/j.jcis.2016.08.021.

Foschi, E., Bonoli, A., 2019. The Commitment of Packaging Industry in the Framework of the European Strategy for Plastics in a Circular Economy. Administrative Sciences 9 (1), 18. Available at https://doi.org/10.3390/admsci9010018.

Fourati, Y., Magnin, A., Putaux, J.-L., Boufi, S., 2020. One-step processing of plasticized starch/cellulose nanofibrils nanocomposites via twin-screw extrusion of starch and cellulose fibers. Carbohydrate Polymers 229, 115554. Available at https://doi.org/10.1016/j.carbpol.2019.115554.

Gálvez, J., Correa Aguirre, J., Hidalgo Salazar, M., Vera Mondragón, B., Wagner, E., Caicedo, C., 2020. Effect of extrusion screw speed and plasticizer proportions on the rheological, thermal, mechanical, morphological and superficial properties of PLA. Polymers 12 (9), 2111. Available at https://doi.org/10.3390/polym12092111.

García-Cruz, H.I., Jaime-Fonseca, M.R., Von-Borries-Medrano, E., Vieyra, H., 2020. Extrusion parameters to produce a PLA-starch derived thermoplastic polymer. Revista Mexicana de Ingeniería Química 19 (s1), 395–412. Available at https://doi.org/10.24275/rmiq/poly1529.

García-Guzmán, L., Cabrera-Barjas, G., Soria-Hernández, C.G., Castaño, J., Guadarrama-Lezama, A.Y., Rodríguez Llamazares, S., 2022. Progress in starch-based materials for food packaging applications. Polysaccharides 3 (1), 136–177. Available at https://doi.org/10.3390/polysaccharides3010007.

Guy, R., 2001. Extrusion Cooking: Technologies and Applications, Vol. 61. Woodhead Publishing.

Haider, T.P., Völker, C., Kramm, J., Landfester, K., Wurm, F.R., 2019. Plastics of the future? The impact of biodegradable polymers on the environment and on society. Angewandte Chemie International Edition 58 (1), 50–62. Available at https://doi.org/10.1002/anie.201805766.

Han Lyn, F., Ismail-Fitry, M.R., Noranizan, M.A., Tan, T.B., Nur Hanani, Z.A., 2024. Recent advances in extruded polylactic acid-based composites for food packaging: a review. International Journal of Biological Macromolecules 266 (Pt 2), 131340. Available at https://doi.org/10.1016/j.ijbiomac.2024.131340.

Hao, S., Wang, Y., Wang, B., Deng, J., Liu, X., Liu, J., 2013. Rapid preparation of pH-sensitive polymeric nanoparticle with high loading capacity using electrospray for oral drug delivery. Materials Science and Engineering: C 33 (8), 4562–4567. Available at https://doi.org/10.1016/j.msec.2013.07.009.

Hedayatnia, S., Tan, C.P., Joanne Kam, W.L., Tan, T.B., Mirhosseini, H., 2019. Modification of physicochemical and mechanical properties of a new bio-based gelatin composite films through composition adjustment and instantizing process. LWT 116, 108575. Available at https://doi.org/10.1016/j.lwt.2019.108575.

Huang, S., Xue, Y., Yu, B., Wang, L., Zhou, C., Ma, Y., 2021. A review of the recent developments in the bioproduction of polylactic acid and its precursors optically pure lactic acids. Molecules 26 (21), 6446. Available at https://doi.org/10.3390/molecules26216446.

Iliopoulou, V., Vosniakos, G.C., 2024. Injection molding control parameter assessment by nested Taguchi design of simulation experiments: a case study. Materials Research Proceedings 41, 2757–2766. Available at https://doi.org/10.21741/9781644903131-302.

Imam, S., Glenn, G., Chiou, B.-S., Shey, J., Narayan, R., Orts, W., 2008. Types, production and assessment of biobased food packaging materials. Environmentally Compatible Food Packaging: Woodhead Publishing Series in Food Science, Technology and Nutrition. Elsevier, pp. 29–62. Available at https://doi.org/10.1533/9781845694784.1.29.

Ismail, N.A., Mohd Tahir, S., Yahya, N., Abdul Wahid, M.F., Khairuddin, N.E., Hashim, I., et al., 2016. Synthesis and characterization of biodegradable starch-based bioplastics. Materials Science Forum 846, 673–678. Available at https://doi.org/10.4028/www.scientific.net/MSF.846.673.

Jamshidian, M., Tehrany, E.A., Imran, M., Jacquot, M., Desobry, S., 2010. Poly-lactic acid: production, applications, nanocomposites, and release studies. Comprehensive Reviews in Food Science and Food Safety 9 (5), 552–571. Available at https://doi.org/10.1111/j.1541-4337.2010.00126.x.

Jariyasakoolroj, P., Leelaphiwat, P., Harnkarnsujarit, N., 2020. Advances in research and development of bioplastic for food packaging. Journal of the Science of Food and Agriculture 100 (14), 5032–5045. Available at https://doi.org/10.1002/jsfa.9497.

Jiang, Z., Diggle, B., Tan, M.L., Viktorova, J., Bennett, C.W., Connal, L.A., 2020. Extrusion 3D printing of polymeric materials with advanced properties. Advanced Science 7 (17), 2001379. Available at https://doi.org/10.1002/advs.202001379.

Jiang, Z., Yang, Y., Mo, S., Yao, K., Gao, F., 2012. Polymer extrusion: From control system design to product quality. Industrial & Engineering Chemistry Research 51, 14759–14770. Available at https://doi.org/10.1021/ie301036c.

Jiménez-Rosado, M., Bouroudian, E., Perez-Puyana, V., Guerrero, A., Romero, A., 2020. Evaluation of different strengthening methods in the mechanical and functional properties of soy protein-based bioplastics. Journal of Cleaner Production 262, 121517. Available at https://doi.org/10.1016/j.jclepro.2020.121517.

Jones, A., Zeller, M.A., Sharma, S., 2013. Thermal, mechanical, and moisture absorption properties of egg white protein bioplastics with natural rubber and glycerol. Progress in Biomaterials 2 (1), 12. Available at https://doi.org/10.1186/2194-0517-2-12.

Jose, A.A., Hazeena, S.H., Lakshmi, N.M., B, A.K., Madhavan, A., Sirohi, R., et al., 2022. Bacterial biopolymers: from production to applications in biomedicine. Sustainable Chemistry and Pharmacy 25, 100582. Available at https://doi.org/10.1016/j.scp.2021.100582.

Jost, V., Stramm, C., 2015. Influence of plasticizers on the mechanical and barrier properties of cast biopolymer films. Journal of Applied Polymer Science 133. Available at https://doi.org/10.1002/app.42513.

Kale, G., Kijchavengkul, T., Auras, R., Rubino, M., Selke, S.E., Singh, S.P., 2007. Compostability of bioplastic packaging materials: an overview. Macromolecular Bioscience 7 (3), 255–277. Available at https://doi.org/10.1002/mabi.200600168.

Kang, L., Liang, Q., Rashid, A., Qayum, A., Chi, Z., Ren, X., et al., 2023. Ultrasound-assisted development and characterization of novel polyphenol-loaded pullulan/trehalose composite films for fruit preservation. Ultrasonics Sonochemistry 92, 106242. Available at https://doi.org/10.1016/j.ultsonch.2022.106242.

Khodaei, D., Álvarez, C., Mullen, A.M., 2021. Biodegradable packaging materials from animal processing co-products and wastes: an overview. Polymers 13 (15), 2561. Available at https://doi.org/10.3390/polym13152561.

Kosmalska, D., Janczak, K., Raszkowska-Kaczor, A., Stasiek, A., Ligor, T., 2022. Polylactide as a substitute for conventional polymers—biopolymer processing under varying extrusion conditions. Environments 9 (5), 57. Available at https://doi.org/10.3390/environments9050057.

Kriegel, C., Arrechi, A., Kit, K., McClements, D.J., Weiss, J., 2008. Fabrication, functionalization, and application of electrospun biopolymer nanofibers. Critical Reviews in Food Science and Nutrition 48 (8), 775–797. Available at https://doi.org/10.1080/10408390802241325.

Kumar, L.R., Yadav, B., Kaur, R., Yellapu, S.K., Pokhrel, S., Pandey, A., Tyagi, R.D., 2021. Chapter 19 - Process engineering and commercialization of polyhydroxyalkanoates. In: Binod, P., Raveendran, S., Pandey, A. (Eds.), Biomass, Biofuels, Biochemicals. Elsevier, pp. 517–549. Available at https://doi.org/10.1016/B978-0-12-821888-4.00005-8.

Kumar, V., Lal, S., Kumar, S., Kumar, P., Kumar, H., 2024. Mechanical, thermal and biodegradation studies of highly degradable composite films of pectin/guar gum/ polyvinylpyrrolidone with rice crop residues: a future vision of packaging and mulch films. Indian Chemical Engineer 66, 621–635. Available at https://doi.org/10.1080/00194506.2024.2377572.

Kurt, A., Toker, O.S., Tornuk, F., 2017. Effect of xanthan and locust bean gum synergistic interaction on characteristics of biodegradable edible film. International Journal of Biological Macromolecules 102, 1035–1044. Available at https://doi.org/10.1016/j.ijbiomac.2017.04.081.

Laske, S., Paudel, A., Scheibelhofer, O., Sacher, S., Hoermann, T., Khinast, J., et al., 2017. A review of PAT strategies in secondary solid oral dosage manufacturing of small molecules. Journal of Pharmaceutical Sciences 106 (3), 667–712. Available at https://doi.org/10.1016/j.xphs.2016.11.011.

Lee, K., Jeon, Y., Kim, D., Kwon, G., Kim, U.-J., Hong, C., et al., 2021. Double-crosslinked cellulose nanofiber based bioplastic films for practical applications. Carbohydrate Polymers 260, 117817. Available at https://doi.org/10.1016/j.carbpol.2021.117817.

Leśniak, D., Zasadziński, J., Libura, W., Gronostajski, Z., Śliwa, R., Leszczyńska-Madej, B., et al., 2024. Latest advances in extrusion processes of light metals. Archives of Civil and Mechanical Engineering 24 (3), 184. Available at https://doi.org/10.1007/s43452-024-00988-5.

Liu, W.C., Halley, P.J., Gilbert, R.G., 2010. Mechanism of degradation of starch, a highly branched polymer, during extrusion. Macromolecules 43 (6), 2855–2864. Available at https://doi.org/10.1021/ma100067x.

Liu, Y., Wang, S., Lan, W., Qin, W., 2019. Development of ultrasound treated polyvinyl alcohol/tea polyphenol composite films and their physicochemical properties. Ultrasonics Sonochemistry 51, 386–394. Available at https://doi.org/10.1016/j.ultsonch.2018.07.043.

Ljungberg, N., Andersson, T., Wesslén, B., 2003. Film extrusion and film weldability of poly(lactic acid) plasticized with triacetine and tributyl citrate. Journal of Applied Polymer Science 88 (14), 3239–3247. Available at https://doi.org/10.1002/app.12106.

Loghmani, R., Leila, N., Jafari, S.M., Abedinia, A., 2026. Development of active composite film based on chitosan-polyvinyl alcohol (PVA) blend containing basil essential oil and carvacrol nanoemulsions for enhanced packaging and shelf-life extension of toast bread. Elsevier, LWT 248, 119360. Available at https://doi.org/10.1016/j.lwt.2026.119360.

Lucic Skoric, M., Milovanovic, S., Zizovic, I., Ortega-Toro, R., Santagata, G., Malinconico, M., et al., 2022. Supercritical CO2 impregnation of thymol in thermoplastic starch-based blends: chemico-physical properties and release kinetics. Polymers 14 (20), 4360. Available at https://doi.org/10.3390/polym14204360.

Luo, Y., Sun, W., Bao, M., Zhu, X., Ning, C., Zhang, W., et al., 2023. Process fundamentals and quality investigation in extrusion 3D printing of shear thinning materials: extrusion process based on Nishihara model. The International Journal of Advanced Manufacturing Technology 124 (1–2), 245–264. Available at https://doi.org/10.1007/s00170-022-10506-7.

Marcoaldi, C., Pardo-Figuerez, M., Prieto, C., Arnal, C., Torres-Giner, S., Cabedo, L., et al., 2023. Electrospun multilayered films based on poly(3-hydroxybutyrate-co-3-hydroxyvalerate), copolyamide 1010/1014, and electrosprayed nanostructured silica. Nanomaterials 13 (6), 972. Available at https://doi.org/10.3390/nano13060972.

Mariano, M., Pilate, F., De Oliveira, F.B., Khelifa, F., Dubois, P., Raquez, J.M., et al., 2017. Preparation of cellulose nanocrystal-reinforced poly(lactic acid) nanocomposites through noncovalent modification with PLLA-based surfactants. ACS Omega 2 (6), 2678–2688. Available at https://doi.org/10.1021/acsomega.7b00387.

Martín-Sampedro, R., Valdehita, A., Navas, J.M., Ibarra, D., Eugenio, M.E., 2025. The influence of the film formation process on the properties of nanocellulose-lignin nanoparticle films: casting versus filtration. International Journal of Biological Macromolecules 322, 119360. Available at https://doi.org/10.1016/j.ijbiomac.2025.146907.

Mehrpouya, M., Vahabi, H., Barletta, M., Laheurte, P., Langlois, V., 2021. Additive manufacturing of polyhydroxyalkanoates (PHAs) biopolymers: materials, printing techniques, and applications. Materials Science and Engineering C 127, 112216. Available at https://doi.org/10.1016/j.msec.2021.112216.

de Mello, A.F.M., Vandenberghe, L.Pd.S., Machado, C.M.B., Valladares-Diestra, K.K., de Carvalho, J.C., Soccol, C.R., 2023. Polyhydroxybutyrate production by Cupriavidus necator in a corn biorefinery concept. Bioresource Technology 370, 128537. Available at https://doi.org/10.1016/j.biortech.2022.128537.

Michalak, M., Marek, A.A., Zawadiak, J., Kawalec, M., Kurcok, P., 2013. Synthesis of PHB-based carrier for drug delivery systems with pH-controlled release. European Polymer Journal 49 (12), 4149–4156. Available at https://doi.org/10.1016/j.eurpolymj.2013.09.021.

Mogany, T., Bhola, V., Bux, F., 2024. Algal-based bioplastics: global trends in applied research, technologies, and commercialization. Environmental Science and Pollution Research 31 (26), 38022–38044. Available at https://doi.org/10.1007/s11356-024-33644-9.

Mohsen, M.M.A., Patil, A.B., Alkanad, M., Patil, D., 2024. Beyond the horizon: recent advances in hot melt extrusion techniques and technologies. International Journal of Applied Pharmaceutics 16 (5), 12–21. Available at https://doi.org/10.22159/ijap.2024v16i5.51425.

Mostafa, Y.S., Alrumman, S.A., Alamri, S.A., Otaif, K.A., Mostafa, M.S., Alfaify, A.M., 2020. Bioplastic (poly-3-hydroxybutyrate) production by the marine bacterium Pseudodonghicola xiamenensis through date syrup valorization and structural assessment of the biopolymer. Scientific Reports 10 (1), 8815. Available at https://doi.org/10.1038/s41598-020-65858-5.

Mysiukiewicz, O., Barczewski, M., Skórczewska, K., Matykiewicz, D., 2020. Correlation between processing parameters and degradation of different polylactide grades during twin-screw extrusion. Polymers 12 (6), 1333. Available at https://doi.org/10.3390/polym12061333.

Nandakumar, A., Chuah, J.-A., Sudesh, K., 2021. Bioplastics: A boon or bane? Renewable and Sustainable Energy Reviews 147, 111237. Available at https://doi.org/10.1016/j.rser.2021.111237.

Navarchian, A.H., Jalalian, M., Pirooz, M., 2015. Characterization of starch/poly(vinyl alcohol)/clay nanocomposite films prepared in twin-screw extruder for food packaging application. Journal of Plastic Film and Sheeting 31 (3), 309–336. Available at https://doi.org/10.1177/8756087914568904.

Ni, Y., Li, X., Wang, T., Meyer, P., Ramos, J., Simmons, K., 2024. Toward recycling polyolefin wastes via single-step solid-state friction extrusion. Polymer Engineering & Science 64 (7), 3379–3387. Available at https://doi.org/10.1002/pen.26776.

Nilsen-Nygaard, J, Fernández, E.N., Radusin, T., Rotabakk, B.T., Sarfraz, J., Sharmin, N., Sivertsvik, M., Sone, I., Pettersen, M.K., 2021. Current status of biobased and biodegradable food packaging materials: Impact on food quality and effect of innovative processing technologies. Comprehensive Reviews in Food Science and Food Safety 20, 1333–1380. Available at https://doi.org/10.1111/1541-4337.12715.

Opran, C.G., Grosu, E., Enachescu, M., 2018. Research regarding the mechanical properties of some biodegradable polymeric composites for food packaging products. Materiale Plastice 55 (4), 498–501.

Ordoñez, R., Atarés, L., Chiralt, A., 2022. Properties of PLA films with cinnamic acid: effect of the processing method. Food and Bioproducts Processing 133, 25–33. Available at https://doi.org/10.1016/j.fbp.2022.02.002.

Pacheco, A., Evangelista-Osorio, A., Muchaypiña-Flores, K.G., Marzano-Barreda, L.A., Paredes-Concepción, P., Palacin-Baldeón, H., et al., 2023. Polymeric materials obtained by extrusion and injection molding from lignocellulosic agroindustrial biomass. Polymers 15 (20), 4046. Available at https://doi.org/10.3390/polym15204046.

Patil, H., Tiwari, R.V., Repka, M.A., 2016. Hot-melt extrusion: from theory to application in pharmaceutical formulation. AAPS PharmSciTech 17 (1), 20–42. Available at https://doi.org/10.1208/s12249-015-0360-7.

Patiño Vidal, C., Muñoz-Shugulí, C., Guivier, M., Puglia, D., Luzi, F., Rojas, A., et al., 2024. PLA- and PHA-biopolyester-based electrospun materials: development, legislation, and food packaging applications. Molecules 29 (22), 5452. Available at https://doi.org/10.3390/molecules29225452.

Patreau, V., Bernard, S., Leroux, J., Bellemare, M., Morissette, J., 2023. Consumer interest and willingness to pay for in-bulk products with reusable packaging options. Frontiers in Sustainability 4, 1228917. Available at https://doi.org/10.3389/frsus.2023.1228917.

Patti, A., 2024. Challenges to improve extrusion-based additive manufacturing process of thermoplastics toward sustainable development. Macromolecular Rapid Communications 45 (17), 2400249. Available at https://doi.org/10.1002/marc.202400249.

Pelissari, F.M., Yamashita, F., Grossmann, M.V.E., 2011. Extrusion parameters related to starch/chitosan active films properties. International Journal of Food Science and Technology 46 (4), 702–710. Available at https://doi.org/10.1111/j.1365-2621.2010.02533.x.

Plasencia, L., Arrieta, M.P., Lazaro-Hdez, C., Gomez-Caturla, J., Quiles-Carrillo, L., 2025. Innovative development and characterization of thermoplastic zein-based biopolymers for food packaging via injection molding. Food Hydrocolloids 163, 111134. Available at https://doi.org/10.1016/j.foodhyd.2025.111134.

Puppi, D., Chiellini, F., 2020. Additive manufacturing of PHA. The Handbook of Polyhydroxyalkanoates. CRC Press, pp. 119–136. Available at https://doi.org/10.1201/9781003087663.

Puppi, D., Pecorini, G., Chiellini, F., 2019. Biomedical processing of polyhydroxyalkanoates. Bioengineering 6 (4), 108. Available at https://doi.org/10.3390/bioengineering6040108.

Rahman, A., Miller, C.D., 2017. Chapter 6 - Microalgae as a source of bioplastics. In: Rastogi, R.P., Madamwar, D., Pandey, A. (Eds.), Algal Green Chemistry. Elsevier, pp. 121–138. Available at https://doi.org/10.1016/B978-0-444-63784-0.00006-0.

Rahman, S., Gogoi, J., Dubey, S., Chowdhury, D., 2024. Animal derived biopolymers for food packaging applications: a review. International Journal of Biological Macromolecules 255, 128197. Available at https://doi.org/10.1016/j.ijbiomac.2023.128197.

Rajan, K.P., Thomas, S.P., Gopanna, A., Chavali, M., 2019. Polyhydroxybutyrate (PHB): a standout biopolymer for environmental sustainability. Handbook of Ecomaterials. Springer, pp. 1–23. Available at https://doi.org/10.1007/978-3-319-48281-1_92-1.

Raquez, J.M., Narayan, R., Dubois, P., 2008. Recent advances in reactive extrusion processing of biodegradable polymer-based compositions. Macromolecular Materials and Engineering 293 (6), 447–470. Available at https://doi.org/10.1002/mame.200700395.

Rasal, R.M., Janorkar, A.V., Hirt, D.E., 2010. Poly(lactic acid) modifications. Progress in Polymer Science 35 (3), 338–356. Available at https://doi.org/10.1016/j.progpolymsci.2009.12.003.

Rivera, P., Torres, A., Romero, J., Alarcón, Á., Martínez, S., Arrieta, M.P., et al., 2024. Effect of operational variables on supercritical foaming of caffeic acid-loaded poly(lactic acid)/poly(butylene adipate-co-terephthalate) blends for the development of sustainable materials. Polymers 16 (7), 948. Available at https://doi.org/10.3390/polym16070948.

Root, K.P., Pal, A.K., Pesaranhajiabbas, E., Mohanty, A.K., Misra, M., 2023. Injection moulded composites from high biomass filled biodegradable plastic: properties and performance evaluation for single-use applications. Composites Part C: Open Access 11, 100358. Available at https://doi.org/10.1016/j.jcomc.2023.100358.

Rosenboom, J.G., Langer, R., Traverso, G., 2022. Bioplastics for a circular economy. Nature Reviews Materials 7 (2), 117–137. Available at https://doi.org/10.1038/s41578-021-00407-8.

Roy Chong, J.W., Tan, X., Khoo, K.S., Ng, H.S., Jonglertjunya, W., Yew, G.Y., et al., 2022. Microalgae-based bioplastics: future solution towards mitigation of plastic wastes. Environmental Research 206, 112620. Available at https://doi.org/10.1016/j.envres.2021.112620.

Sabathini, H.A., Windiani, L., Dianursanti, M., Gozan, E., Kusrini, F.H., Juwono, A., et al., 2018. Setiawan, mechanical physicial properties of chlorella-pva based bioplastic with ultrasonic homogenizer. E3S Web of Conferences 67, 03046. Available at https://doi.org/10.1051/e3sconf/20186703046.

Salave, S., Prayag, K., Rana, D., Amate, P., Pardhe, R., Jadhav, A., Jindal, A.B., Benival, D., 2022. Recent progress in hot melt extrusion technology in pharmaceutical dosage form design. Recent Advances in Drug Delivery and Formulation 16, 170–191. Available at https://doi.org/10.2174/2667387816666220819124605.

Santana, I., Felix, M., Bengoechea, C., 2024. Seaweed as basis of eco-sustainable plastic materials: focus on alginate. Polymers 16 (12), 1662. Available at https://doi.org/10.3390/polym16121662.

Sanyang, M.L., Sapuan, S.M., Jawaid, M., Ishak, M.R., Sahari, J., 2016. Effect of plasticizer type and concentration on physical properties of biodegradable films based on sugar palm (arenga pinnata) starch for food packaging. Journal of Food Science and Technology 53 (1), 326–336. Available at https://doi.org/10.1007/s13197-015-2009-7.

Schudel, S., Shoji, K., Shrivastava, C., Onwude, D., Defraeye, T., 2023. Solution roadmap to reduce food loss along your postharvest supply chain from farm to retail. Food Packaging and Shelf Life 36, 101057. Available at https://doi.org/10.1016/j.fpsl.2023.101057.

Shang, J., Li, X., Wang, Z., Wang, R., Zhu, H., 2019. Rheological and printable behavior of resin – class materials for 3D printing applications. Rapid Prototyping Journal 25 (5), 801–808. Available at https://doi.org/10.1108/rpj-11-2016-0183.

Shi, X., Cui, L., Xu, C., Wu, S., 2025. Next-generation bioplastics for food packaging: sustainable materials and applications. Materials 18 (12), 2919. Available at https://doi.org/10.3390/ma18122919.

Shirai, M.A., Olivato, J.B., Garcia, P.S., Müller, C.M.O., Grossmann, M.V.E., Yamashita, F., 2013. Thermoplastic starch/polyester films: effects of extrusion process and poly (lactic acid) addition. Materials Science and Engineering C 33 (7), 4112–4117. Available at https://doi.org/10.1016/j.msec.2013.05.054.

Shlush, E., Davidovich-Pinhas, M., 2022. Bioplastics for food packaging. Trends in Food Science & Technology 125, 66–80. Available at https://doi.org/10.1016/j.tifs.2022.04.026.

Shojaeiarani, J., Bajwa, D.S., Rehovsky, C., Bajwa, S.G., Vahidi, G., 2019. Deterioration in the physico-mechanical and thermal properties of biopolymers due to reprocessing. Polymers 11 (1), 58. Available at https://doi.org/10.3390/polym11010058.

Si, Y., Shi, S., Hu, J., 2024. Electrospinning and electrospraying synergism: twins-tech collaboration across dimensions. Matter 7 (4), 1373–1405. Available at https://doi.org/10.1016/j.matt.2024.01.009.

Sidek, I.S., Draman, S.F.S., Abdullah, S.R.S., Anuar, N., 2019. Current development on bioplastics and its future prospects: an introductory review. i TECH MAG 1, 3–8. Available at https://doi.org/10.26480/itechmag.01.2019.03.08.

Sikora, J., Majewski, Ł., Puszka, A., 2020. Modern biodegradable plastics—processing and properties: Part I. Materials 13 (8), 1986. Available at https://doi.org/10.3390/ma13081986.

Siracusa, V., 2012. Food packaging permeability behaviour: a report. International Journal of Polymer Science 2012, 1–11. Available at https://doi.org/10.1155/2012/302029.

Sirohi, R., Prakash Pandey, J., Kumar Gaur, V., Gnansounou, E., Sindhu, R., 2020. Critical overview of biomass feedstocks as sustainable substrates for the production of polyhydroxybutyrate (PHB). Bioresource Technology 311, 123536. Available at https://doi.org/10.1016/j.biortech.2020.123536.

Sothornvit, R., Olsen, C.W., McHugh, T.H., Krochta, J.M., 2003. Formation conditions, water-vapor permeability, and solubility of compression-molded whey protein films. Journal of Food Science 68 (6), 1985–1999. Available at https://doi.org/10.1111/j.1365-2621.2003.tb07006.x.

de Souza, F.M., Gupta, R.K., 2024. Bacteria for bioplastics: progress, applications, and challenges. ACS Omega 9 (8), 8666–8686. Available at https://doi.org/10.1021/acsomega.3c07372.

Spierling, S., Knüpffer, E., Behnsen, H., Mudersbach, M., Krieg, H., Springer, S., et al., 2018. Bio-based plastics - a review of environmental, social and economic impact assessments. Journal of Cleaner Production 185, 476–491. Available at https://doi.org/10.1016/j.jclepro.2018.03.014.

Srinivasan, B., Kulshreshtha, G., 2020. Recent developments in food-based bioplastics production. Bioprocess Engineering for Bioremediation: Valorization and Management Techniques. Springer, Cham, pp. 107–127. Available at https://doi.org/10.1007/698_2020_578.

Strähle, U.T., Pütz, N., Hannig, M., 2024. A coating machine for coating filaments with bioactive nanomaterials for extrusion 3D printing. Heliyon 10 (12), e33223. Available at https://doi.org/10.1016/j.heliyon.2024.e33223.

Subramaniam, S.R., Samykano, M., Selvamani, S.K., Ngui, W.K., Kadirgama, K., Sudhakar, K., et al., 2019. 3D printing: Overview of PLA progress. AIP Conference Proceedings 2059, 020015. Available at https://doi.org/10.1063/1.5085958.

Svečko, R., Kusić, D., Kek, T., Sarjaš, A., Hančič, A., Grum, J., 2013. Acoustic emission detection of macro-cracks on engraving tool steel inserts during the injection molding cycle using PZT sensors. Sensors 13 (5), 6365–6379. Available at https://doi.org/10.3390/s130506365.

Svidler, R., Hartmann, T., Buschbeck, S., Rinberg, R., Kroll, L., 2014. Extruded films of bio-based plastics for packaging applications. In: Shaping the future by engineering: Proceedings of the 58th IWK, Ilmenau Scientific Colloquium, Technische Universität Ilmenau.

Swaroop, C., Shukla, M., 2018. Nano-magnesium oxide reinforced polylactic acid biofilms for food packaging applications. International Journal of Biological Macromolecules 113, 729–736. Available at https://doi.org/10.1016/j.ijbiomac.2018.02.156.

Tekiner, İ.H., 2024. Effects of different processing techniques and sol-gel coating on physical properties of starch and clay-based biocomposite material. Turkish Journal of Agriculture - Food Science and Technology 12 (6), 966–977. Available at https://doi.org/10.24925/turjaf.v12i6.966-977.6632.

Thakur, R., Santhosh, R., Kumar, Y., Suryavanshi, V.R., Singhi, H., Madhubabu, D., et al., 2023. Characteristics and application of animal byproduct-based films and coatings in the packaging of food products. Trends in Food Science & Technology 140, 104143. Available at https://doi.org/10.1016/j.tifs.2023.104143.

Thakur, S., Chaudhary, J., Sharma, B., Verma, A., Tamulevicius, S., Thakur, V.K., 2018. Sustainability of bioplastics: opportunities and challenges. Current Opinion in Green and Sustainable Chemistry 13, 68–75. Available at https://doi.org/10.1016/j.cogsc.2018.04.013.

Thiviya, P., Gamage, A., Liyanapathiranage, A., Makehelwala, M., Dassanayake, R.S., Manamperi, A., et al., 2023. Algal polysaccharides: structure, preparation and applications in food packaging. Food Chemistry 405, 134903. Available at https://doi.org/10.1016/j.foodchem.2022.134903.

Tsagdi, A., Drossos, I., Georgiou, D., Exarhopoulos, S., Karasiotas, G., Kallitsis, J.K., et al., 2021. Injection molded PP foams using food ingredients for food packaging applications. Polymers 13 (2), 1–18. Available at https://doi.org/10.3390/polym13020288.

Ujcic, A., Fortelny, I., Krejcikova, S., Pavlova, E., Hodan, J., Slouf, M., 2022. Effects of thermal treatment and nucleating agents on crystallinity, toughness, and stiffness of PLA/PCL blends. Express Polymer Letters 16 (3), 221–233. Available at https://doi.org/10.3144/expresspolymlett.2022.18.

Upadhyaya, B.P., DeVeaux, L.C., Christopher, L.P., 2014. Metabolic engineering as a tool for enhanced lactic acid production. Trends in Biotechnology 32 (12), 637–644. Available at https://doi.org/10.1016/j.tibtech.2014.10.005.

Van Roijen, E.C., Miller, S.A., 2022. A review of bioplastics at end-of-life: linking experimental biodegradation studies and life cycle impact assessments. Resources, Conservation and Recycling 181, 106236. Available at https://doi.org/10.1016/j.resconrec.2022.106236.

Vasudevan, M., Natarajan, N., 2025. Towards achieving sustainable bioplastics production and nutrient recovery from wastewater—a comprehensive overview on polyhydroxybutyrate. Biomass Conversion and Biorefinery 15 (2), 1729–1748. Available at https://doi.org/10.1007/s13399-022-02399-z.

Verbeek, C.J.R., Van Den, L.E., 2010. Berg, Extrusion processing and properties of protein-based thermoplastics. Macromolecular Materials and Engineering 295 (1), 10–21. Available at https://doi.org/10.1002/mame.200900167.

Vidakis, N., Petousis, M., Mountakis, N., Korlos, A., Papadakis, V., Moutsopoulou, A., 2022. Trilateral multi-functional polyamide 12 nanocomposites with binary inclusions for medical grade material extrusion 3d printing: the effect of titanium nitride in mechanical reinforcement and copper/cuprous oxide as antibacterial agents. Journal of Functional Biomaterials 13 (3), 115. Available at https://doi.org/10.3390/jfb13030115.

Wang, N., Gao, Y.Z., Wang, P., Yang, S., Xie, T.M., Xiao, Z.G., 2016. Effect of microwave modification on mechanical properties and structural characteristics of soy protein isolate and zein blended film. Czech Journal of Food Sciences 34 (2), 180–188. Available at https://doi.org/10.17221/442/2015-CJFS.

Wu, J.H., Hu, T.G., Wang, H., Zong, M.H., Wu, H., Wen, P., 2022. Electrospinning of PLA nanofibers: recent advances and its potential application for food packaging. Journal of Agricultural and Food Chemistry 70 (27), 8207–8221. Available at https://doi.org/10.1021/acs.jafc.2c02611.

Yacu, W., 2020. Extruder sew, barrel, and die assembly: general design principles and operation. In: Ganjyal, G.M. (Ed.), Extrusion Cooking. Woodhead Publishing, pp. 73–117. Available at https://doi.org/10.1016/B978-0-12-815360-4.00003-1 [Chapter 3].

Yadav, D., Srivastava, A., Yadav, A., Mehla, B., Srivastava, M., 2022. Development and sustainability of bioplastics: a review. Asian Journal of Green Chemistry 6 (2), 112–128. Available at https://doi.org/10.22034/ajgc.2022.2.2.

Yang, J., Xu, S., Wang, W., Ran, X., Ching, Y.C., Sui, X., et al., 2023. Preparation and characterization of bioplastics from silylated cassava starch and epoxidized soybean oils. Carbohydrate Polymers 300, 120253. Available at https://doi.org/10.1016/j.carbpol.2022.120253.

Yang, J., Wang, H., Lou, L., Meng, Z., 2025. A review of chitosan-based electrospun nanofibers for food packaging: From fabrication to function and modeling insights. Nanomaterials 15, 1274. Available at https://doi.org/10.3390/nano15161274.

Yew, G.Y., Khoo, K.S., Chia, W.Y., Ho, Y.C., Law, C.L., Leong, H.Y., et al., 2020. A novel lipids recovery strategy for biofuels generation on microalgae Chlorella cultivation with waste molasses. Journal of Water Process Engineering 38, 101665. Available at https://doi.org/10.1016/j.jwpe.2020.101665.

YousefniaPasha, H., Mohtasebi, S.S., Tabatabaeekoloor, R., Taherimehr, M., Javadi, A., Soltani Firouz, M., 2021. Preparation and characterization of the plasticized polylactic acid films produced by the solvent-casting method for food packaging applications. Journal of Food Processing and Preservation 45 (12), e16089. Available at https://doi.org/10.1111/jfpp.16089.

Zhang, H., Mittal, G., 2010. Biodegradable protein-based films from plant resources: A review. Environmental Progress & Sustainable Energy 29, 203–220. Available at https://doi.org/10.1002/ep.10463.

CHAPTER 9

Applications of bioplastics for food packaging

Nazila Oladzadabbasabadi[1,2], Mehran Ghasemlou[1,3,4], Elena P. Ivanova[1], Benu Adhikari[1,5], and Shima Jafarzadeh[2,3]

[1]School of Science, STEM College, RMIT University, Melbourne, VIC, Australia, [2]School of Engineering, Deakin University, Geelong Waurn Ponds Campus, Geelong, VIC, Australia, [3]Center for Sustainable Bioproducts, Geelong Waurn Ponds Campus, Geelong, VIC, Australia, [4]Department of Chemical Engineering, The University of Melbourne, Parkville, VIC, Australia, [5]Centre for Advanced Materials and Industrial Chemistry (CAMIC), RMIT University, Melbourne, VIC, Australia

9.1 Introduction

The increasing environmental impact of traditional plastic waste has led to global efforts to create more sustainable packaging, especially for food products (Suwanamornlert et al., 2020). Petroleum-based plastics, long valued for their durability, low cost, and excellent barrier properties, now face critical scrutiny due to their non-biodegradability and dependence on finite fossil resources (Shlush and Davidovich-Pinhas, 2022). Globally, over 400 million tons of plastic are produced annually, with food packaging alone accounting for approximately one-third of this volume. Alarmingly, more than 85% of plastic packaging ends up in landfills or escapes into terrestrial and marine ecosystems, where it persists for decades or even centuries without decomposing. This underscores the urgency of transitioning toward materials that can reduce environmental persistence and ecological harm without compromising packaging functionality.

Within this context, bioplastics, a broad class of materials that are bio-based, biodegradable, or both, have emerged as viable alternatives for food packaging applications (Gurunathan et al., 2025; Shlush and Davidovich-Pinhas, 2022). Derived from renewable sources such as starch, cellulose, sugarcane, and microbial fermentation products, bioplastics like polylactic acid (PLA), polyhydroxyalkanoates (PHAs), polybutylene succinate (PBS), and poly(butylene adipate-co-terephthalate) (PBAT) can offer end-of-life options that include industrial composting and, in some cases, biodegradation under natural conditions (Correa et al., 2017; Gürler et al., 2023; Hubackova et al., 2013; Rajgond et al., 2024). The appeal of these materials lies in their reduced carbon footprint as well as their potential to support circular economy principles, whereby materials are designed with environmental impact minimization

Bioplastics for Food Packaging. DOI: https://doi.org/10.1016/B978-0-443-41654-5.00002-0

across their full life cycle. This concept is illustrated in Fig. 9.1, which presents a circular economy model highlighting the material flow from resource extraction to manufacturing, consumption, recycling, and waste management. It emphasizes how bioplastics, when integrated into such systems, can reduce linear material loss and promote closed-loop sustainability.

Modern bioplastic systems are no longer confined to passive containment; they are increasingly engineered for active functionality, enabling them to contribute to food preservation (Freire et al., 2024; Zhou et al., 2023). Through advancements in polymer blending, nanocomposites, and bioactive incorporation, researchers have enhanced the mechanical strength, barrier properties, thermal stability, and processability of bioplastic films (Ahmadi et al., 2024; Taechutrakul et al., 2024; Ucpinar Durmaz et al., 2024). For instance, blending PLA with PBAT or PBS improves its flexibility and reduces gas permeability, addressing inherent limitations such as brittleness and poor moisture resistance (Zhang et al., 2019). Simultaneously, the integration of natural active agents, including essential oils (e.g., thymol and cinnamaldehyde (CI)), phenolic compounds, and plant extracts, has resulted in active packaging solutions capable of antimicrobial activity, antioxidant protection, ethylene scavenging, and controlled release, all of which contribute to food quality preservation and shelf-life extension (Oladzadabbasabadi et al., 2022; Suwanamornlert et al., 2020; Yang et al., 2020).

The functional design of bioplastic-based packaging must also address food-specific preservation needs, consumer safety, and regulatory requirements. For example, bakery

Figure 9.1
Circular economy model illustrating sustainable material flows relevant to bioplastic packaging systems.

products demand high moisture barrier films to prevent staling. Similarly, fresh fruits and vegetables benefit from materials with selective gas permeability that can modulate in-package atmospheres (Gigante et al., 2023). In all cases, bioplastics must meet stringent food contact standards, such as those outlined in EU Regulation No. 10/2011 or FDA 21 CFR, and remain compatible with commercial-scale processing methods, including extrusion, thermoforming, and coating (Stoica et al., 2022, 2024). Despite their promise, bioplastics face several practical challenges. These include higher production costs compared to fossil-derived polymers, variable degradation behaviors depending on environmental conditions, and occasionally inferior mechanical or barrier properties. Nevertheless, the literature increasingly supports their successful application across diverse food sectors. Numerous studies have demonstrated that well-formulated bioplastic packaging can preserve sensory and nutritional quality, prevent microbial spoilage, and extend shelf life in categories such as fresh produce, dairy, bakery items, and meats (Bumbudsanpharoke et al., 2022; Yang et al., 2020; Signori et al., 2009). Innovations such as microperforated films, multilayer composites, and bioactive compound-loaded scaffolds are rapidly narrowing the performance gap between bioplastics and conventional plastics (Alshehri et al., 2024; Winotapun et al., 2023; Xiao et al., 2021). Importantly, the growing societal and policy momentum behind sustainable packaging is encouraging the adoption of bioplastics. National and regional regulations targeting single-use plastics, extended producer responsibility policies, and the rise of environmentally conscious consumer behavior are aligning in favor of bio-based alternatives (Xanthos and Walker, 2017). Consumers often perceive bioplastics as environmentally superior, safer, and more natural for food contact. This convergence of technical innovation, policy support, and public demand is positioning bioplastics as more than a material substitute; they are emerging as enablers of a more resilient and responsible food system.

This chapter presents a comprehensive and critical overview of the current landscape of bioplastics for food packaging. It begins by outlining the chemistry, classification, and performance characteristics of major bioplastics systems. Subsequent sections explore their sector-specific applications, with in-depth case studies on fruits and vegetables, bakery products, and high-protein foods, highlighting the influence of bioplastics on preservation outcomes under real-world conditions. This chapter draws on insights from materials science, microbiology, food engineering, and sustainability analysis. It aims to clarify the opportunities and challenges of bioplastics. The chapter also envisions their role in shaping the next generation of packaging technologies. These systems are informed by technological advances, designed to support food safety, and promote environmental stewardship in a rapidly evolving global food supply chain.

9.2 Types of bioplastics used in food packaging

As the food packaging industry seeks more sustainable practices, a diverse range of bioplastics has become essential. These materials are known for their environmental benefits and also for their practical applications. These materials include PLA, PHAs, PBS, PBAT, starch-based thermoplastic blends (TPS), and polycaprolactone (PCL). PLA has emerged as a fundamental material in the packaging of perishables such as fresh produce, salads, and takeaway containers. Appreciated for its clear appearance and effective barrier against oxygen, PLA plays a crucial role in preserving freshness and extending the shelf life of food products (Taechutrakul et al., 2024). Derived from renewable plant starch, typically corn or sugarcane, PLA exemplifies the shift toward sustainable materials in bioplastics and underscores the industry's efforts to reduce carbon footprints through innovative packaging solutions. Moving from PLA, starch-based blends are tailored for versatility and are used extensively across a diverse range of packaging applications. These blends are particularly favored in the production of items such as lightweight shopping bags. The adaptability of starch-based plastics stems from their formulation, which often incorporates other biodegradable polymers such as PLA or PCL (Hubackova et al., 2013; Martinez Villadiego et al., 2022). This combination enhances the blends' functional properties, notably improving water resistance and mechanical strength, which are critical for maintaining the integrity of packaging under various environmental conditions. Additionally, the use of starch, a natural and abundant material, in these blends highlights the industry's move toward utilizing more renewable resources, thereby reducing reliance on fossil fuels and minimizing the overall environmental footprint of packaging solutions (Surendren et al., 2022).

Another significant class of bioplastics, PHAs are uniquely produced through the microbial fermentation of sugars or lipids. This process leverages natural bacterial processes, which convert organic substrates into polymeric materials (Carvalho et al., 2014). Due to its biodegradation potential and origins from renewable resources, PHAs represent a compelling shift toward reducing dependency on petrochemical plastics. The production of PHAs is versatile; it can be tailored through the fermentation process to have specific properties by altering the types of feedstock and fermentation conditions (Colombo et al., 2016). This flexibility allows for the production of PHA variants that are suitable for a broad range of applications in the food packaging industry. PHAs' ability to create an effective moisture barrier makes it particularly suitable for packaging applications. This property is critical in maintaining the product's integrity and extending its shelf life by effectively blocking out humidity. Furthermore, PHAs' natural origins and eventual biodegradability under certain conditions make it a preferred choice in scenarios where environmental impact is a concern.

PBS is synthesized from renewable crops such as sugarcane or corn, making it an environmentally favorable option in the realm of bioplastics. PBS is particularly noted for its heat resistance and mechanical strength, which are characteristics that make it stand out as a robust alternative to traditional plastics (Rajgond et al., 2024). These properties allow PBS to perform exceptionally well in applications that demand durability under heat, such as microwaveable trays and disposable utensils. The material's capacity to withstand high temperatures not only enhances its usability but also extends its applicability across various food packaging formats that require sterilization or hot-fill processes. Moreover, the use of PBS contributes to a reduced carbon footprint, thanks to its basis in renewable resources. This aspect is increasingly important as industries push toward greater sustainability and lower environmental impact. PBS's compatibility with existing plastic manufacturing and recycling processes further bolsters its status as a practical, sustainable alternative to petroleum-based plastics. Its ability to integrate seamlessly into the current recycling infrastructure makes PBS an attractive choice for manufacturers looking to meet eco-friendly standards without compromising on quality or performance (Aliotta et al., 2022).

Moreover, PBAT is highly valued for its excellent biodegradability and is primarily utilized in the production of compostable bags and agricultural films. This bioplastic is particularly favored for applications that demand a delicate balance between environmental responsibility and performance. When blended with other bioplastics, PBAT significantly enhances the flexibility and durability of products. This improvement in mechanical properties allows it to withstand the physical demands of usage in agricultural settings, such as mulch films, which require durability against environmental stresses while ensuring they do not persist in the environment long after their useful life (Roy et al., 2024). Furthermore, PBAT's ability to degrade under commercial composting conditions aligns with global sustainability goals, helping to reduce waste accumulation and support a circular economy. Its versatility and eco-friendly profile make PBAT an essential material in efforts to replace conventional plastics with more sustainable alternatives in sensitive applications.

PCL, although synthetic and not derived from natural resources, is highly valued within the bioplastics industry for its unique properties. PCL is distinguished by its slow degradation rate, which makes it an ideal additive in bioplastic blends to enhance their performance, especially in products that require durability over an extended period (Dash and Konkimalla, 2012). Its inclusion in such blends allows for the creation of materials that maintain structural integrity under stress and usage yet still offer an environmentally friendly end-of-life option through biodegradability. PCL's ability to be processed at relatively low temperatures and its compatibility with other bioplastics make it a versatile choice for improving the usability and functional range of eco-conscious products. It is particularly effective in applications such as medical implants, biodegradable packaging, and even in three-dimensional printing filaments, where its precise melting point can be leveraged for detailed and custom manufacturing. Moreover, PCL's biodegradability is activated under composting conditions, aligning with sustainable disposal practices that prevent long-term environmental accumulation (Gürler et al., 2023). By enhancing the performance and lifespan of bioplastic blends, PCL plays a crucial role in bridging the gap between traditional durability in plastics and the ecological benefits of biodegradables. Its continued use and development in the bioplastics sector underscore its pivotal role in the advancement of sustainable material technologies. Together, these bioplastics are paving the way for a new era in food packaging. By balancing functional needs with ecological impacts, they offer viable solutions that cater to both consumer

Table 9.1 Comparative overview of common bioplastics utilized in food packaging.

Bioplastics	Source	Biodegradability	Key properties	Limitations	Typical applications	Reference
PLA	Derived from renewable resources like corn or sugarcane through fermentation	Industrial composting conditions	Transparent, rigid, and good printability	Brittleness and low thermal resistance	Rigid trays, bottles, and films	Han et al. (2018)
TPS	Sourced from starch-rich plants such as corn, potato, or cassava	Home compostable	Cost-effective, hydrophilic, easily blendable	Poor moisture barrier and weak mechanical properties	Films, loose-fill packaging, and coated papers	Rahardiyan et al. (2023)
PHA	Produced via bacterial fermentation of sugars or lipids	Home compostable, including marine environments	Biocompatible, good barrier properties, and biodegradable in various environments	High production cost and slow crystallization rate	Coatings and single-use packaging items	Adak et al. (2024)
PBS	Synthesized from bio-based succinic acid and 1,4-butanediol	Industrial composting conditions	Flexible, heat-resistant, and good elongation at break	Lower oxygen barrier properties	Food trays, bags, and cling films	Barletta et al. (2022)
PBAT	Fossil-based copolymer of adipic acid, terephthalic acid, and 1,4-butanediol	Industrial composting conditions	Flexible and good impact strength	Not fully bio-based	Compostable bags and flexible wraps	Roy et al. (2024)
PCL	Synthesized via ring-opening polymerization of ε-caprolactone	Slow in ambient conditions; faster in composting environments	Tough, flexible, and low melting point	Higher cost and slower degradation in natural environments	Controlled-release systems and blend component in packaging	Barletta et al. (2022)

PBAT, polybutylene adipate terephthalate; *PBS*, polybutylene succinate; *PCL*, polycaprolactone; *PHAs*, polyhydroxyalkanoates; *PLA*, polylactic acid; *TPS*, thermoplastic starch.

preferences and environmental imperatives. A comparative summary of the most widely studied bioplastics for food packaging applications is presented in Table 9.1, outlining their origin, environmental characteristics, and suitability for various packaging functions.

9.3 Properties and characteristics of key bioplastics

9.3.1 Polylactic acid

PLA is a leading bioplastic, recognized for its biodegradability and renewable origins, primarily from corn starch and sugarcane. Structurally, PLA is an aliphatic polyester composed of repeating lactic acid units, as shown in Fig. 9.2A, which allows it to be synthesized through ring-opening polymerization or direct polycondensation of lactic acid. This molecular architecture underpins its desirable thermal and mechanical properties and enables its decomposition under industrial composting conditions into carbon dioxide and water, positioning PLA as a key material in sustainable packaging. Thermally, PLA is adaptable; its processing parameters, such as the glass transition and melting temperatures, can be finely tuned by adjusting the molecular weight and the composition of the polymer. This flexibility is advantageous for manufacturing processes, allowing PLA to be easily molded into films or fibers and making it suitable for both high-temperature applications and products requiring barrier properties against moisture and gases. PLA's barrier effectiveness

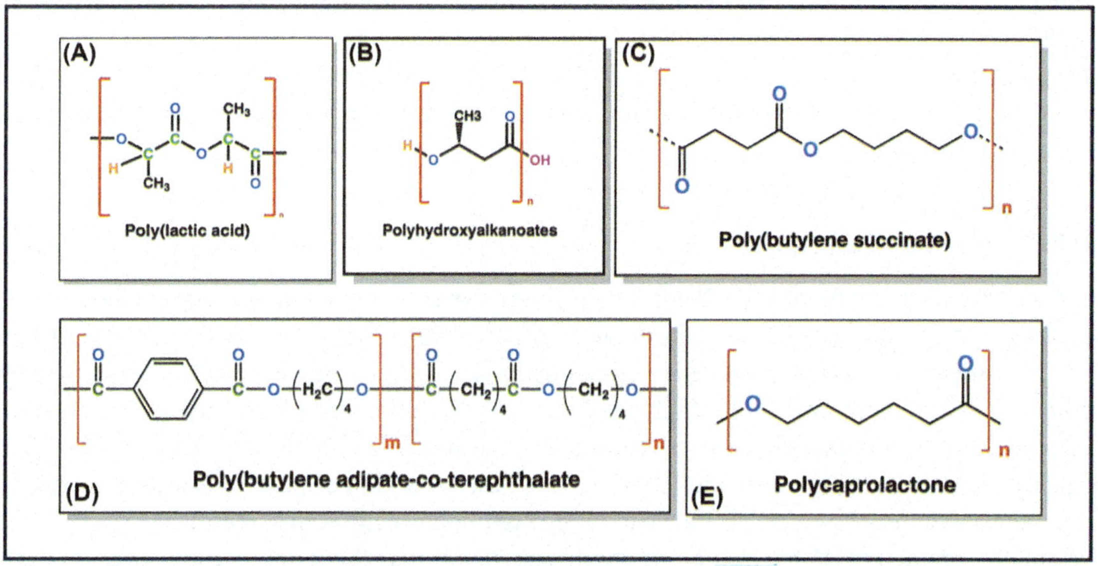

Figure 9.2
(A) Chemical structure of poly(lactic acid). (B) Chemical structure of polyhydroxyalkanoates. (C) Chemical structure of polyhydroxyalkanoates. (D) Chemical structure of poly(butylene adipate-co-terephthalate). (E) Chemical structure of polycaprolactone.

against oxygen, carbon dioxide, and water vapor is critical for packaging applications where product integrity and shelf life are paramount. Advances in polymer technology have enabled the development of PLA formulations that provide barrier properties that can meet or even exceed those of many traditional plastics (Huang et al., 2014). In terms of mechanical properties, PLA exhibits high rigidity and strength, which makes it comparable to conventional plastics like PET. This robustness enables its use across a variety of applications, including rigid packaging and disposable items. However, to overcome its inherent brittleness and improve impact resistance, PLA is often modified through copolymerization or blended with other biocompatible polymers, enhancing its versatility and applicability in more demanding scenarios (Alshehri et al., 2024). The processability of PLA is also a significant benefit, as it can be processed on standard plastic manufacturing equipment without the need for new technologies or significant modifications to existing facilities. This ease of use extends to various manufacturing techniques, including extrusion, thermoforming, and injection molding, making PLA highly valued for its ease of integration into current production processes (Chang et al., 2022; Gonon et al., 2023). Although PLA has many benefits, its hydrolytic stability poses challenges, particularly in humid environments where its structural integrity might be compromised over time. Ongoing research aims to enhance the hydrolytic stability of PLA without sacrificing its biodegradability, ensuring that it remains a practical option for long-term applications (Tripathi et al., 2021). As PLA continues to evolve, its applications have expanded beyond packaging to include biomedical devices, agricultural films, and even automotive interiors, showcasing its potential to replace traditional plastics in a wide array of industries. The approval of US Food and Drug Administration (FDA) of PLA for food contact further underscores its safety and versatility, reinforcing its status as a pivotal material in the shift toward more sustainable material solutions (Wang et al., 2021).

9.3.2 Thermoplastic starch

TPS stands out in the bioplastics landscape primarily due to its excellent biodegradability and its derivation from renewable resources such as corn and potatoes (Hubackova et al., 2013). As industries increasingly prioritize sustainability, TPS's ability to decompose under industrial composting conditions highlights its role in mitigating plastic waste issues. This biodegradability aligns TPS with environmental sustainability goals and even makes it an appealing choice for single-use applications where the reduction of long-term waste is crucial. The mechanical properties of TPS have evolved significantly through scientific advancements. Originally characterized by limited strength and flexibility, TPS now incorporates various plasticizers and biopolymers that enhance these properties (Mohd Nizam et al., 2021). Research indicates that additives like glycerol improve TPS's tensile strength and elasticity, making it suitable for a wider range of packaging applications that require durable materials. These enhancements have propelled TPS into competitive realms traditionally dominated by conventional plastics, offering a viable alternative without the environmental toll (Tarique et al., 2021). Thermal properties are equally critical for the application of TPS in food packaging. The modification of starch through the addition of plasticizers lowers its processing temperature, enabling it to be molded and extruded with less energy compared to synthetic polymers. This thermal adaptability makes TPS more energy-efficient in production and expands its application spectrum to include products that must endure varying temperature conditions (Abedi et al., 2025). Despite its many advantages, TPS's sensitivity to moisture remains a challenge; however, strides in technology have led to improved water resistance and barrier properties. By blending TPS with synthetic polymers or applying advanced chemical treatments, researchers have significantly enhanced its barrier against moisture and oxygen, essential for preserving the quality and extending the shelf life of food products (Dong et al., 2024). The processability of TPS has been substantially improved by integrating compatibilizers, which facilitate better blending with other polymers. This compatibility enhancement allows for more uniform material properties and eases manufacturing processes, enabling TPS to meet diverse commercial requirements more effectively. The expanding bioplastics market reflects a shift toward more sustainable packaging solutions, with TPS at the forefront due to its low environmental impact and versatility in applications ranging from agricultural films to protective food packaging. Its ability to adapt to various consumer needs while supporting sustainability initiatives positions TPS as a key player in the future development of bioplastics.

9.3.3 Polyhydroxyalkanoates

PHAs represent a promising class of biopolymers due to their biodegradability and versatile physical properties, which closely mirror those of traditional petrochemical plastics. Derived naturally through bacterial synthesis, PHAs accumulate in microbial cells under nutrient-limited conditions with excess carbon sources, serving as both energy reserves and carbon storage compounds (Zhou et al., 2023). This dual functionality highlights their ecological role and potential industrial applications, spanning from packaging to biomedical devices. The biocompatibility and biodegradability of PHAs make them especially appealing for medical applications, such as sutures and implants, and for sustainable packaging solutions. These polymers degrade into carbon dioxide and water in natural composting conditions, thereby contributing significantly to waste reduction without releasing toxic residues. This aspect is crucial for applications where environmental conservation and sustainable material cycles are prioritized (Zhou et al., 2023). PHAs are characterized by a variety of mechanical properties depending on their monomeric compositions. Short-chain length PHAs, such as poly(3-hydroxybutyrate) (PHB), are noted for their stiffness and brittleness due to their highly crystalline structure. The general chemical structure of PHAs, represented here by PHB, is shown in Fig. 9.2B. PHB consists of repeating hydroxybutyrate units linked via ester bonds, forming a linear aliphatic polyester backbone. This structural regularity contributes to its thermal processability and mechanical strength. To improve flexibility and impact resistance, PHB is often copolymerized with other monomers such as 3-hydroxyvalerate, resulting in materials like poly(3-hydroxybutyrate-co-3-hydroxyvalerate). These modifications lower crystallinity and enhance toughness, thereby broadening the applicability of PHAs. Medium-chain length PHAs, on the other hand, are more elastomeric and suitable for items requiring higher flexibility and elongation before breaking (Naser et al., 2021). Thermal properties of PHAs also vary; for instance, PHB has a melting point that makes it suitable for thermal processing akin to conventional plastics, which facilitates its use in common manufacturing processes such as injection molding and extrusion. The ability to process PHAs using conventional plastic manufacturing technologies without requiring new machinery or infrastructure represents a significant economic and logistical advantage (Mai et al., 2024). Despite their benefits, the widespread adoption of PHAs is hampered by their high production costs compared to that of conventional plastics. The cost barriers arise from the need for pure cultures, high-purity substrates, and controlled fermentation processes. However, advancements in using mixed microbial cultures and waste-based feedstocks are paving the way for more cost-effective production methods. These approaches not only reduce costs but also align with sustainable practices by valorizing waste streams, thereby contributing to a circular economy.

9.3.4 Polybutylene succinate

PBS is an important bioplastic due to its ability to break down naturally and its mechanical properties, which are comparable to traditional plastics. PBS can be made from both renewable resources, like corn and sugarcane, and petrochemical sources. The chemical structure of PBS is shown in Fig. 9.2C, illustrating its aliphatic polyester backbone, which provides flexibility and enables biodegradation under industrial composting conditions. This dual origin increases its appeal as a sustainable and flexible material. Its ability to degrade in various environments makes it a strong option for eco-friendly products such as agricultural films, disposable packaging, and other single-use items (Aliotta et al., 2022). PBS has a good combination of strength, flexibility, and toughness, making it suitable for a wide range of uses. However, it can sometimes be brittle. To overcome this, PBS is often mixed with other polymers or plasticizers to improve its resistance to impact and cracking. These modifications enhance its performance, enabling it to work well in both long-lasting products and items designed for short-term use (Gigli et al., 2016; Rajgond et al., 2024). In terms of heat resistance, PBS performs well in conventional plastic processing. Its melting point allows it to be molded and shaped through common techniques like injection molding and extrusion. This means PBS can easily fit into existing manufacturing systems, making it a practical replacement for traditional plastics without requiring major equipment changes (Chang et al., 2022; Gigli et al., 2016). The production of PBS is also becoming more sustainable. When PBS is made from bio-based sources such as corn or sugarcane, it further reduces its environmental impact. Recent advancements in production technology have improved efficiency and decreased the ecological footprint of PBS manufacturing. The ability to use renewable raw materials enhances PBS's reputation as an environmentally friendly option (Rafiqah et al., 2021; Rajgond et al., 2024). PBS is also finding new uses in areas like medicine. Its compatibility with the human body makes it suitable for temporary medical implants and drug delivery systems. Additionally, its ability to blend with other materials allows it to serve many industries, making it a highly adaptable bioplastic (Freire et al., 2024). With ongoing research aimed at improving its properties and reducing costs, PBS is set to grow in importance. Its potential to replace traditional plastics in various applications supports the global shift toward more sustainable materials, ensuring PBS will play a key role in the future of bioplastics.

9.3.5 Poly(butylene adipate-co-terephthalate)

PBAT has rapidly gained attention in the field of sustainable materials due to its unique combination of flexibility, biodegradability, and adaptability. This aliphatic–aromatic copolyester is synthesized by polycondensation of adipic acid, 1,4-butanediol, and terephthalic acid, giving it a dual nature that combines the strength of rigid polyesters with the degradation characteristics of flexible aliphatic components. The chemical structure of PBAT is shown in Fig. 9.2D, illustrating the repeating units of terephthalate and adipate segments within the polymer backbone, which are primarily responsible for its mechanical flexibility and biodegradability. PBAT stands out as one of the few commercially viable biodegradable polymers that can fully decompose under industrial composting conditions. Certified under standards such as ASTM D6400 and EN 13432, PBAT has demonstrated a degradation rate exceeding 90% in 3 months (Apicella et al., 2024). This rapid breakdown occurs through enzymatic hydrolysis, followed by microbial metabolism of its degradation by-products. This makes PBAT particularly suitable for applications like compostable packaging and agricultural mulch films, where postuse environmental integration is a priority (Roy et al., 2024). However, biodegradation is influenced by the polymer's chemical structure. Studies show that the proportion of terephthalate units impacts the hydrolysis rate; too many aromatic components can slow down the process. Nevertheless, even with a moderate level of aromatic content, PBAT maintains sufficient degradability to meet global composting standards (Hu et al., 2022). PBAT has similar properties to low-density polyethylene (LDPE), a common flexible plastic. With a tensile strength of approximately 21 MPa and an elongation at break exceeding 670%, PBAT demonstrates exceptional flexibility, which is critical for applications requiring stretchable films (Li et al., 2022). These properties make PBAT suitable for use in shopping bags, food wraps, and disposable packaging materials. However, PBAT's mechanical performance can be tailored through various techniques. Adjustments to the ratio of adipate and terephthalate units, as well as blending with other biodegradable polymers like PLA, can enhance its tensile strength and stiffness (Li et al., 2022; Roy et al., 2024).

Additionally, incorporating nanofillers such as clay and cellulose nanofibers has been shown to significantly boost the mechanical performance of PBAT-based films (Fourati et al., 2021). PBAT's thermal properties make it highly processable using conventional plastic manufacturing techniques, including extrusion, injection molding, and blow molding. The polymer has a melting temperature of around 125°C and maintains thermal stability up to 350°C. This stability enables manufacturers to integrate PBAT seamlessly into existing production lines without requiring significant adjustments. Additionally, PBAT's moderate crystallinity aids in balancing processing efficiency and material performance. The inclusion of long-chain branching agents can further enhance its melt strength, enabling the production of ultrathin, durable films suitable for packaging (Dong et al., 2024; Martinez Villadiego et al., 2022). Beyond conventional packaging, PBAT is making inroads into specialized applications. In active food packaging, PBAT films infused with antimicrobial agents have been developed to extend the shelf life of fresh produce. Meanwhile, its biocompatibility is opening doors to medical applications, including temporary implants and drug delivery systems. However, one challenge that persists is PBAT's cost, which remains higher than that of traditional plastics. To address this, researchers are exploring cost-reduction strategies such as using renewable feedstocks and optimizing polymer formulations (Signori et al., 2009). PBAT exemplifies how bioplastics can bridge the gap between functionality and environmental responsibility. With ongoing research aimed at improving its properties and reducing costs, PBAT is poised to play a pivotal role in replacing conventional plastics across industries. Its ability to adapt to both flexible and durable applications while maintaining biodegradability underscores its importance in the global push for more sustainable materials.

9.3.6 Polycaprolactone

PCL is a synthetic yet biodegradable polyester that plays a significant role in bioplastics due to its adaptable and complementary properties (Hubackova et al., 2013). Although it is not derived from renewable sources, its exceptional compatibility with other biodegradable polymers makes it essential for creating composites and blends. The chemical structure of PCL is shown in Fig. 9.2E, illustrating its linear aliphatic polyester backbone composed of repeating ε-caprolactone units. This structure contributes to PCL's flexibility, low melting point, and ability to undergo controlled hydrolytic degradation. By enhancing both mechanical performance and biodegradability, PCL serves as a key material in sustainable applications (Gürler et al., 2023). One of PCL's most valuable attributes is its semicrystalline structure, combined with a low melting point of around 60°C. This feature allows for efficient processing at reduced energy consumption, making PCL suitable for manufacturing processes such as extrusion, injection molding, and thermoforming (Aoyama et al., 2021). Its processability ensures that it can be easily integrated into standard plastic production without requiring special modifications. Thermally, PCL's low processing temperature offers energy-saving benefits, which further support its use in environmentally conscious manufacturing.

However, its moderate thermal stability may limit its application in high-temperature environments, prompting ongoing research into enhancing its thermal performance through polymer blending (Xiao et al., 2021). Additionally, PCL's mechanical profile includes excellent flexibility and high elongation at break, often exceeding 700%. This flexibility allows PCL to withstand significant mechanical stress, making it ideal for both stretchable and impact-resistant packaging (Bahramian et al., 2025). However, PCL's hydrophobic nature can limit its effectiveness in environments where rapid biodegradation is required. To overcome this limitation, it is often blended with more hydrophilic polymers or chemically modified to increase its water sensitivity and improve degradation rates. Despite its hydrophobicity, PCL has demonstrated effective biodegradability through hydrolytic degradation in moist or compostable environments. Its slower degradation rate compared to polymers like PLA is an advantage in applications that require controlled, gradual breakdown over time (Bahramian et al., 2025; Gürler et al., 2023). In food packaging, PCL is rarely used on its own due to its relatively slow breakdown. Instead, it is frequently blended with materials like PLA and starch to improve the flexibility, toughness, and water resistance of biodegradable films (Hubackova et al., 2013; Shahverdi et al., 2022). These enhancements make it particularly valuable in packaging applications that demand extended usability while maintaining eco-friendliness. In addition, PCL's compatibility with bioactive agents has paved the way for advanced packaging technologies, such as antimicrobial films, which aim to extend the shelf life of perishable goods (Bahramian et al., 2025). Ultimately, PCL contributes to sustainability by enhancing the performance of other biodegradable materials in packaging. Its versatility, ease of processing, and biodegradation under controlled conditions demonstrate its importance in the development of high-performance bioplastic solutions. Despite being synthetic, PCL remains a critical material for advancing environmentally conscious packaging technologies.

9.4 Applications in food packaging

The adoption of bioplastics in food packaging has attracted increasing scholarly and industrial interest due to their potential to reduce environmental impact and improve food preservation. This section examines the application of bioplastics across various food categories, with a focus on their functional performance and material suitability.

9.4.1 Meat and poultry products

Meat and poultry products are highly perishable, necessitating packaging solutions that offer excellent barrier properties against oxygen and moisture to prevent spoilage and extend shelf life. Bioplastics can be engineered into films and trays that provide the necessary protective barriers. Additionally, the incorporation of natural antioxidants and antimicrobial agents into bioplastic films has shown promise in further enhancing the shelf life and safety of meat

Table 9.2 Applications of bioplastics in meat and poultry packaging.

Bioplastic system	Functional additives	Processing method	Target product	Key functional benefits	References
PLA/PBAT/starch composite	Cinnamaldehyde (CI), tea polyphenols (TP)	Extrusion	Soy protein-based meat analogues	Reduced *Escherichia coli* and *Staphylococcus aureus* by 3.6 and 4.1 log CFU/g, respectively; moisture retention; texture stability	Wang et al. (2022)
PCL-based film	AgNPs (1%), hairy cellulose nanocrystals (2%)	Extrusion	Tilapia (fish)	Improved strength and elasticity; reduced surface roughness; lowered pH, total volatile base nitrogen, thiobarbituric acid reactive substances; delayed spoilage	Ahmadi et al. (2024)
PBAT/TPS blend (30/70 and 40/60)	Nitrite (1%–5%)	Blown-film extrusion	Vacuum-packaged pork	Lower O_2 and H_2O vapor permeability; improved color stability; reduced microbial load; safe residual nitrite levels	Katekhong et al. (2022)
PBS/PBSA multilayer film	Talc (10% in some formulations)	Vacuum packaging	Raw chicken, turkey, and smoked turkey	Adequate barrier properties; lower light transmission; maintained quality over 5–15 days	Vytejčková et al. (2017)
PHB/PCL blend	Organoclay (Cloisite 30B, 10A); activated with Nisin	Casting + activation treatment	Processed meats (sliced ham)	Improved tensile and thermal stability; antimicrobial activity after nisin activation; extended shelf life	Correa et al. (2017)

CFU, colony-forming unit; *PBAT*, polybutylene adipate terephthalate; *PBS*, polybutylene succinate; *PBSA*, poly(butylene-succinate-co-adipate); *PCL*, polycaprolactone; *PHB*, poly(3-hydroxybutyrate); *PLA*, polylactic acid; *TPS*, thermoplastic starch.

products. A summary of recent studies on bioplastic-based packaging solutions for meat and poultry products, highlighting their composition, functional additives, processing methods, and preservation benefits, is presented in Table 9.2.

Wang et al. (2022) investigated the preservation efficacy of antimicrobial packaging films on soy protein-based meat analogues stored at 4°C over a 10-day period. In their study, CI and tea polyphenols were incorporated into PLA/PBAT/starch composite films using an extrusion method to fabricate active packaging. Among the tested variants, the CI-loaded PLA/PBAT film exhibited the most significant antimicrobial performance, reducing *Escherichia coli* and *Staphylococcus aureus* populations by 3.6 and 4.1 log colony-forming unit (CFU)/g, respectively, by the end of the storage period. The film also effectively retained moisture and suppressed texture degradation, thereby mitigating common spoilage issues such as surface drying and hardness increase. These findings highlight the potential of CI-functionalized PLA/PBAT packaging as a promising approach for improving the microbial safety and quality retention of plant-based meat analogues during refrigerated storage. Furthermore, the incorporation of silver nanoparticles (AgNPs) and hairy cellulose nanocrystals (HCNs) into PCL-based films for active food packaging has been explored by Ahmadi et al. (2024) to address challenges associated with nanoparticle migration and fish preservation. Their study demonstrated that combining 1% AgNPs with 2% HCNs in the film matrix significantly improved mechanical strength and elasticity while reducing the surface roughness responsible for excessive Ag ion migration. When used for packaging tilapia, these composite films effectively lowered pH, total volatile base nitrogen, thiobarbituric acid reactive substances, and peroxide value levels, indicating delayed spoilage. In addition, microbiological assessments revealed reduced bacterial growth compared to controls, confirming the antimicrobial efficacy of the active packaging. This research highlights the potential of biopolymer-based nanocomposite films to provide both functional and safety advantages in preserving highly perishable foods such as fish. In another study, Katekhong et al. (2022) developed a functional bioplastic packaging film based on TPS and PBAT, incorporating nitrite (1%–5%) to enhance meat preservation during storage (Fig. 9.3). Using blown-film extrusion, they produced PBAT/TPS blend films at 30/70 and 40/60 ratios, which demonstrated modified microstructures and improved interfacial compatibility due to nitrite-induced disruption of starch granules and interactions with PBAT's carbonyl groups. The films effectively reduced oxygen and water vapor permeability while maintaining smooth, compact film matrices. When applied to vacuum-packaged pork, the nitrite-containing films significantly improved product color stability by promoting nitrosyl myoglobin formation and retained lipid integrity during 12 days of storage. Microbiological analysis revealed a marked reduction in total viable count, lactic acid bacteria, and yeast and mold levels in comparison to the control. Despite a reduction in mechanical strength of up to ~40%, the films delivered antimicrobial activity and color stabilization while maintaining residual nitrite levels below regulatory limits, offering a promising alternative to direct nitrite additives in meat processing.

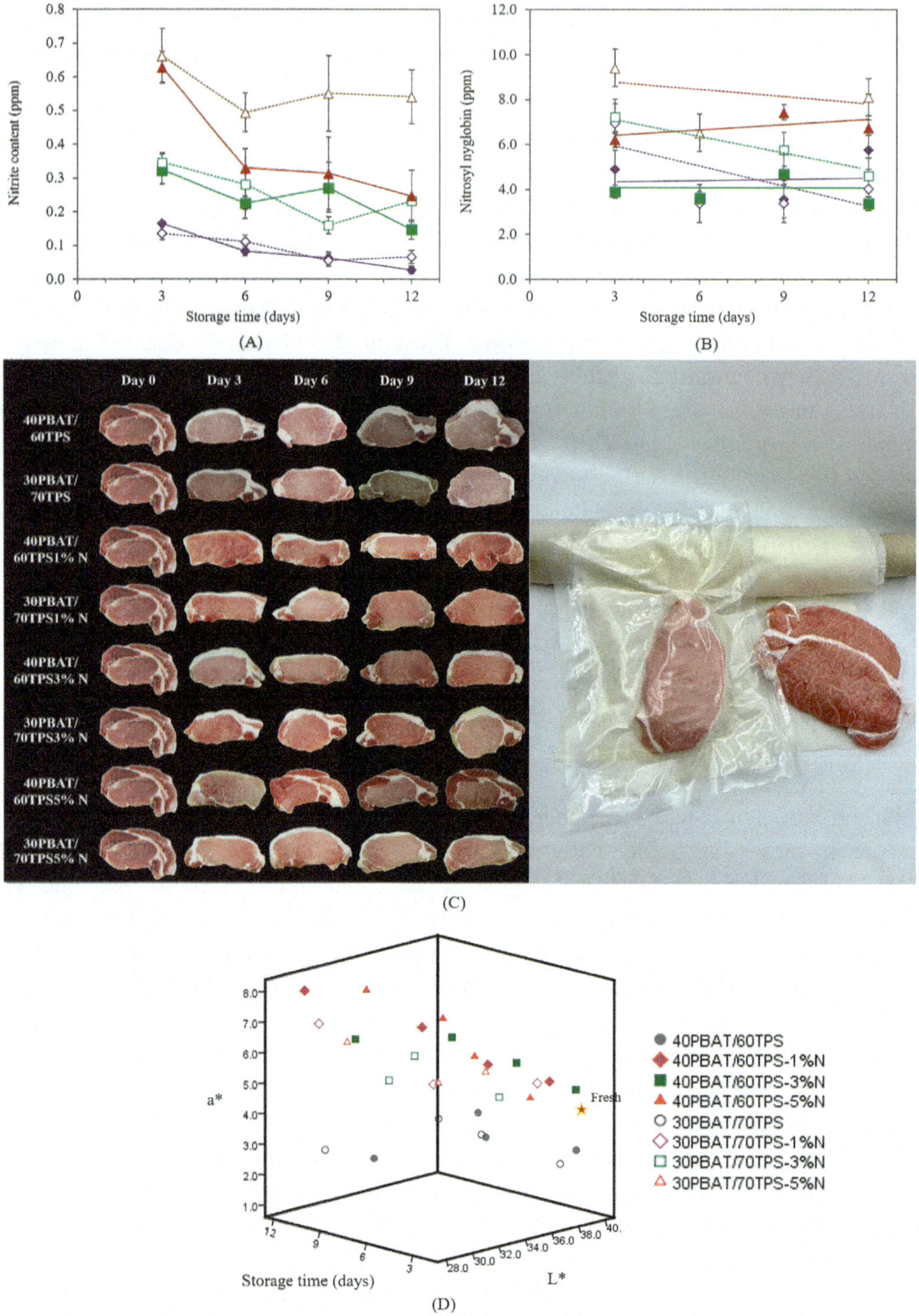

Figure 9.3
Appearance of packaged pork during 12 days of storage time.

Also, an evaluation of PBS-based and poly(butylene-succinate-co-adipate) (PBSA)-based multilayer films for poultry packaging has been conducted by Vytejčková et al. (2017), focusing on their functional performance compared to conventional polyamide/polyethylene (PA/PE) materials. Two formulations, PBSA/(90% PBS + 10% PBSA) and PBSA/(80% PBS + 10% PBSA + 10% talc), were applied in vacuum packaging of raw chicken, raw turkey, and smoked turkey meat. While these bioplastic films exhibited higher water vapor transmission rates and oxygen permeability (OP) than PA/PE, they still provided adequate protection. Transmittance values were significantly lower than those of PA/PE, indicating superior light barrier properties. Despite the differences in mechanical and barrier characteristics, no significant deterioration in meat quality, assessed through pH, water activity, microbial load, color, and volatile compound profiles, was observed throughout the respective shelf lives of 5 and 15 days. The study concluded that PBS-based films derived from renewable resources are viable alternatives to conventional multilayer plastics for short-to-medium shelf life poultry products, offering sustainable packaging solutions without compromising food safety or quality. Moreover, Correa et al. (2017) developed and characterized biodegradable films based on PHB and PCL, incorporating organoclay nanocomposites (Cloisite 30B and 10A) for application in active packaging of processed meats. Their study assessed mechanical, thermal, and water vapor barrier properties of the resulting films, alongside the films' capability to be activated with the antimicrobial peptide nisin. Although the addition of organo-clays improved the polymer blend's thermal stability and tensile strength, it did not impart direct antimicrobial activity when incorporated into the matrix. However, subsequent activation of the PHB/PCL films with nisin cm, under optimized conditions of 4000 IU/cm^3, 40°C, and 10 minutes, produced films effective in inhibiting *Lactobacillus* plantarum CRL691, a spoilage model organism, in vacuum-packed sliced ham. The findings demonstrate that nisin-loaded PHB/PCL films, with or without organoclay reinforcements, are viable for extending the shelf life of processed meat products and present a promising route toward fully biodegradable, antimicrobial food packaging materials.

Collectively, these studies underscore the growing potential of bioplastic-based films, particularly those enhanced with antimicrobial agents or nanocomposites, as viable alternatives to conventional meat packaging materials. By offering tailored barrier properties, microbial inhibition, and preservation of physicochemical attributes, these biodegradable films align with both sustainability goals and stringent food safety standards. While formulation optimization and regulatory validation remain important next steps, the demonstrated effectiveness of various bioplastic systems and nanocomposites reinforces their readiness for broader implementation in the meat and poultry industry.

9.4.2 Bakery products

The shelf life of bakery products is strongly influenced by microbial contamination, moisture dynamics, and starch retrogradation (Suwanamornlert et al., 2020). Bread, in particular, is

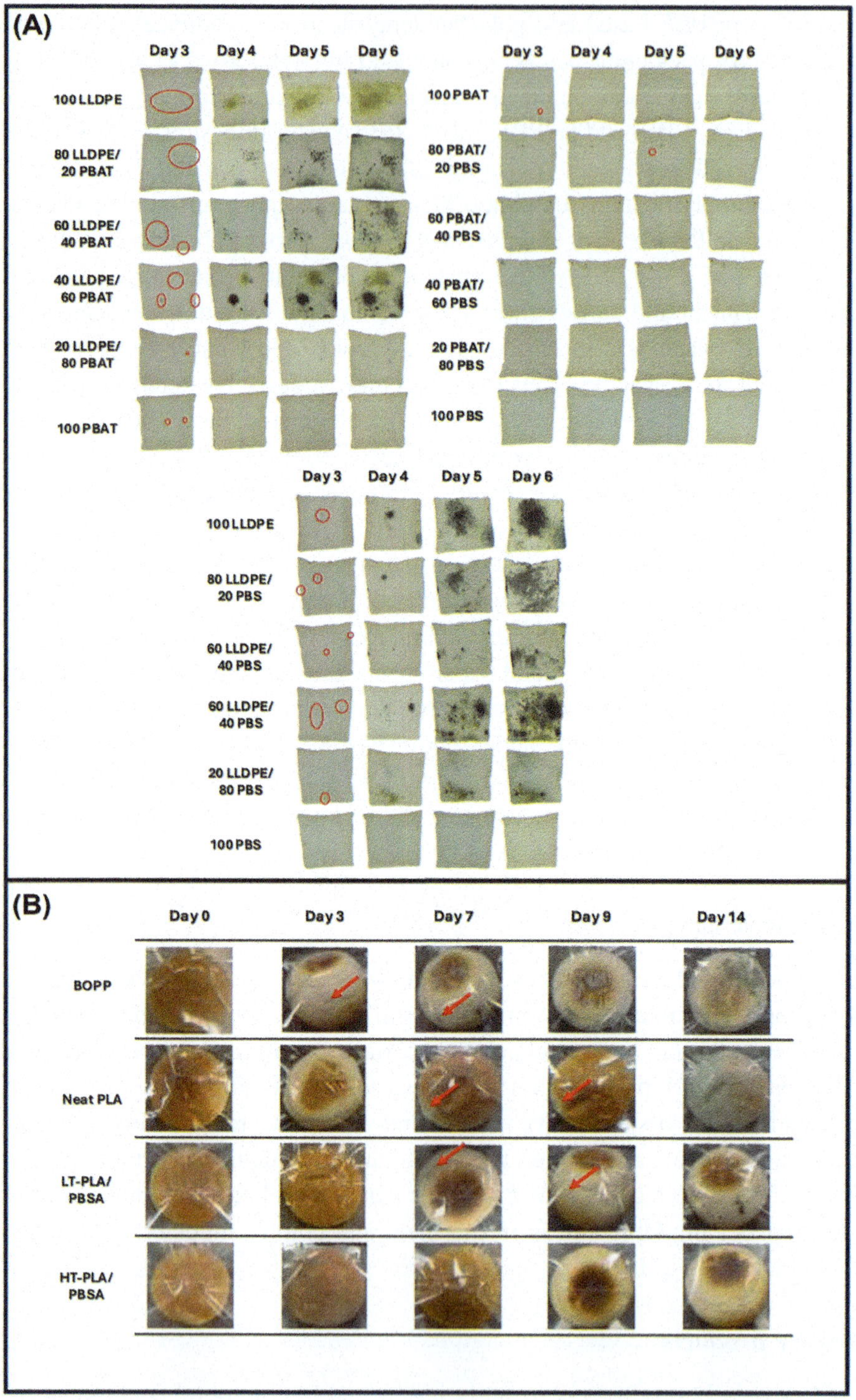
(A)
Day 3
Day 4
Day 5
Day 6
100 LLDPE
80 LLDPE/
20 PBAT
60 LLDPE/
40 PBAT
40 LLDPE/
60 PBAT
20 LLDPE/
80 PBAT
100 PBAT
Day 3
Day 4
Day 5
Day 6
100 PBAT
80 PBAT/
20 PBS
60 PBAT/
40 PBS
40 PBAT/
60 PBS
20 PBAT/
80 PBS
100 PBS
Day 3
Day 4
Day 5
Day 6
100 LLDPE
80 LLDPE/
20 PBS
60 LLDPE/
40 PBS
60 LLDPE/
40 PBS
20 LLDPE/
80 PBS
100 PBS
(B)
Day 0
Day 3
Day 7
Day 9
Day 14
BOPP
Neat PLA
LT-PLA/
PBSA
HT-PLA/
PBSA

susceptible to fungal growth and crumb staling during ambient storage. Bioplastic-based packaging has emerged as a functional and sustainable approach to mitigating these issues by offering tailored barrier properties and incorporating bioactive agents (Piroonpan et al., 2025). Recent studies have demonstrated the efficacy of various biopolymer blends and antimicrobial systems in extending the shelf life of bread while also addressing moisture regulation and packaging performance. Bumbudsanpharoke et al. (2022) evaluated bio-based films composed of PBAT, PBS, and linear low-density polyethylene (LLDPE) for their suitability in bread packaging (Fig. 9.4A). Their findings revealed that film composition significantly influenced morphological structure, permeability characteristics, and overall preservation performance. PBAT/PBS blends formed compact and smooth morphologies with effective interfacial adhesion, whereas blends containing LLDPE exhibited phase separation and fibrous structures, resulting in reduced mechanical strength. Water vapor permeability (WVP) and OP were found to vary notably with blend ratios. Specifically, PBAT/PBS films with >40% PBS exhibited elevated WVP and reduced OP, which contributed to mold growth suppression up to 6 days, in contrast to LLDPE/PBAT or LLDPE/PBS films that showed fungal growth within 3 days. Texture profile analysis further indicated that crumb hardness increased linearly with WVP during early storage (day 3), as demonstrated by high R^2 and F values. By day 6, the correlation shifted toward a second-order polynomial model, corresponding with the increased influence of starch retrogradation. Bread stored in PBAT/PBS films with optimized PBS content not only showed delayed microbial growth but also better maintained water activity levels, highlighting the role of bioplastics in controlling the internal package microenvironment. These findings underscore the effectiveness of well-designed biopolymer films in preserving the textural and microbiological quality of bread during storage. Suwanamornlert et al. (2020) developed antifungal active packaging films based on PLA/PBSA blends with 3 wt.% (LT-PLA/PBSA) and 6 wt.% (HT-PLA/PBSA)

Figure 9.4

(A) Visual comparison of bread stored in different biopolymer blend films, linear low-density polyethylene (LLDPE)/poly(butylene adipate-co-terephthalate) (PBAT), PBAT/poly(butylene succinate) (PBS), and LLDPE/PBS, highlighting changes in appearance over time under ambient storage at 25°C. (B) Progression of yeast and mold development in packaged bread during storage at 25°C ± 2°C and 75% relative humidity.

(A) Reprinted from Bumbudsanpharoke, N., Wongphan, P., Promhuad, K., Leelaphiwat, P., Harnkarnsujarit, N., 2022. Morphology and permeability of bio-based poly(butylene adipate-co-terephthalate) (PBAT), poly (butylene succinate) (PBS) and linear low-density polyethylene (LLDPE) blend films control shelf-life of packaged bread. Food Control 132, 108541. https://doi.org/10.1016/j.foodcont.2021.108541, with permission from Elsevier Science Ltd. (B) Reprinted from Suwanamornlert, P., Kerddonfag, N., Sane, A., Chinsirikul, W., Zhou, W., Chonhenchob, V., 2020. Poly(lactic acid)/poly(butylene-succinate-co-adipate) (PLA/PBSA) blend films containing thymol as alternative to synthetic preservatives for active packaging of bread. Food Packaging and Shelf Life 25, 100515. https://doi.org/10.1016/j.fpsl.2020.100515, with permission from Elsevier Science Ltd.

thymol content. Produced via blown-film extrusion, these films demonstrated potent antifungal efficacy against *Aspergillus* and *Penicillium* spp., both in vitro and in bread packaging trials (Fig. 9.4B). Bread packaged in commercial biaxially oriented polypropylene and neat PLA films exhibited mold growth by days 3 and 6, respectively, while LT-PLA/PBSA and HT-PLA/PBSA films delayed mold appearance until days 7 and 9, respectively. Yeast and mold counts after 9 days were markedly lower in HT-PLA/PBSA (<1.00 log CFU/g) compared to those in LT-PLA/PBSA (5.38 log CFU/g), neat PLA (5.83 log CFU/g), and BOPP (7.13 log CFU/g), confirming thymol's superior bioactivity, particularly at higher concentrations. In-package CO_2 levels further supported this microbial suppression. While BOPP packaging reached 12.30% CO_2 by day 9, indicating accelerated microbial respiration, HT-PLA/PBSA maintained CO_2 levels below 2.33%, with minimal fluctuations in O_2 levels. Bread moisture content (MC) declined from 43% to 35% over 9 days across all films, but higher WVP in PLA-based films facilitated moisture redistribution and reduced water activity (aw), which, in turn, inhibited mold proliferation. Although the crust MC was higher in BOPP due to its low WVP, this resulted in elevated aw and promoted microbial growth. Textural analysis confirmed that crumb hardness and chewiness increased during storage, largely driven by moisture migration and starch retrogradation. However, the packaging material had minimal effect on these textural trends, and the inclusion of thymol helped maintain microbiological safety without compromising mechanical performance. The HT-PLA/PBSA films extended bread shelf life up to 9 days, tripling that of commercial BOPP, while offering a biodegradable, bioactive alternative. Piroonpan et al. (2025) introduced a multifunctional PLA-based film enhanced with a cyclic amine-functionalized multibranched PLA (mPLA-PPR), providing dual antioxidant and antimicrobial activity for bakery applications. Incorporation of 2 wt.% mPLA-PPR into PLA significantly improved thermal stability, ductility, and resistance to high-dose electron beam sterilization (10 and 25 kGy), outperforming both neat PLA and conventional stabilizers such as Tinuvin 770. The film exhibited robust antibacterial performance, reducing *E. coli* growth by 78.24% within 24 hours, compared to just 8.51% inhibition by neat PLA. In bread storage trials, PLA/mPLA-PPR packaging effectively suppressed mold growth for 6–7 days, while visible fungal colonies developed in neat PLA within 3 days. This performance was attributed to the heterocyclic amine moieties in PPR, which disrupted microbial membranes. Compared to other antifungal systems, such as PLA–chiian (CS) nanoparticles or cellulose acetate with sodium propionate, the mPLA-PPR film delivered equal or superior preservation benefits. Gas exchange analysis indicated improved headspace stability in mPLA-PPR packages, indirectly curbing microbial metabolism. These findings position mPLA-PPR as a potent nucleating agent that enhances PLA's structural, oxidative, and microbial resistance characteristics, making it highly suitable for active bakery packaging under both ambient and sterilized storage conditions. Further advancing bioplastic applications, Srisa and Harnkarnsujarit (2020) developed cast-extruded PLA/PBAT films containing 2%–10% *trans*-cinnamaldehyde, evaluated for both textural and microbial performance in bread packaging. Two blend ratios,

PLA60/PBAT40 and PBAT60/PLA40, were tested. The PBAT-rich films exhibited greater release of CI, resulting in enhanced antifungal efficacy against *Penicillium* spp. and *Aspergillus niger*, extending bread shelf life up to 21 days, significantly longer than the 3–4 days observed with polypropylene films. CI acted both as a plasticizer and a microbial inhibitor. Water activity decreased from 0.874 in fresh bread, and films with higher CI content maintained lower crumb hardness over time (1.40–1.98 N per day in PBAT/PLA-10%) due to delayed starch retrogradation and lipid crystal dissolution. X-ray diffraction analysis confirmed an increase in crystallinity from 19.0% in fresh bread to 24.7% after storage, with PBAT/PLA films promoting higher crystallinity due to greater active compound migration. Films with ≥5% CI completely suppressed yeast and mold growth for 21 days. Slight yellowing was noted in 10% films due to compound migration, but visual and organoleptic quality remained acceptable at 2%–5% concentrations. These findings highlight the role of CI as a powerful antimicrobial and structural modulator, enabling PLA/PBAT films to serve as both protective and functional packaging for bread. Overall, these studies demonstrate that bioplastics, when blended and functionalized appropriately, can rival or surpass conventional plastics in bakery packaging. From passive moisture and gas control to active microbial inhibition, combinations of PLA, PBSA, PBAT, and tailored bioactive agents (e.g., thymol, CI, and mPLA-PPR) provide holistic preservation strategies. These biodegradable solutions not only align with sustainability imperatives but also offer shelf-life extension without compromising product quality or safety, positioning them as viable candidates for commercial bakery packaging systems.

9.4.3 Fruits and vegetables

Fresh fruits and vegetables require packaging that permits adequate respiration while protecting against microbial contamination and physical damage. To address these needs, various bioplastics have been investigated for their potential to create breathable, active, or modified atmosphere packaging (MAP) solutions that extend the shelf life of fresh produce. Zhang et al. (2019) examined the effectiveness of biodegradable PLA/PBAT composite films in preserving passion fruit during 21 days of ambient storage at 20°C. The bioplastic films were evaluated against PE films and unpackaged controls by tracking physiological loss, textural degradation, compositional changes, and sensory attributes. While both PLA/PBAT and PE delayed ripening and metabolic activity, PLA/PBAT more effectively preserved volatile aroma compounds and overall flavor, as confirmed by electronic nose, electronic tongue, and gas chromatography–mass spectrometry-MS analysis. Although PE outperformed PLA/PBAT in reducing shrinkage and water loss, the bioplastic films offered superior retention of ascorbic acid and acidity, suggesting promising potential for maintaining fruit quality in commercial applications. However, further optimization of film parameters is recommended. Building on active release strategies, Yang et al. (2020) developed CI-loaded porous microspheres using a PHA and CS matrix, intended for use in antimicrobial food

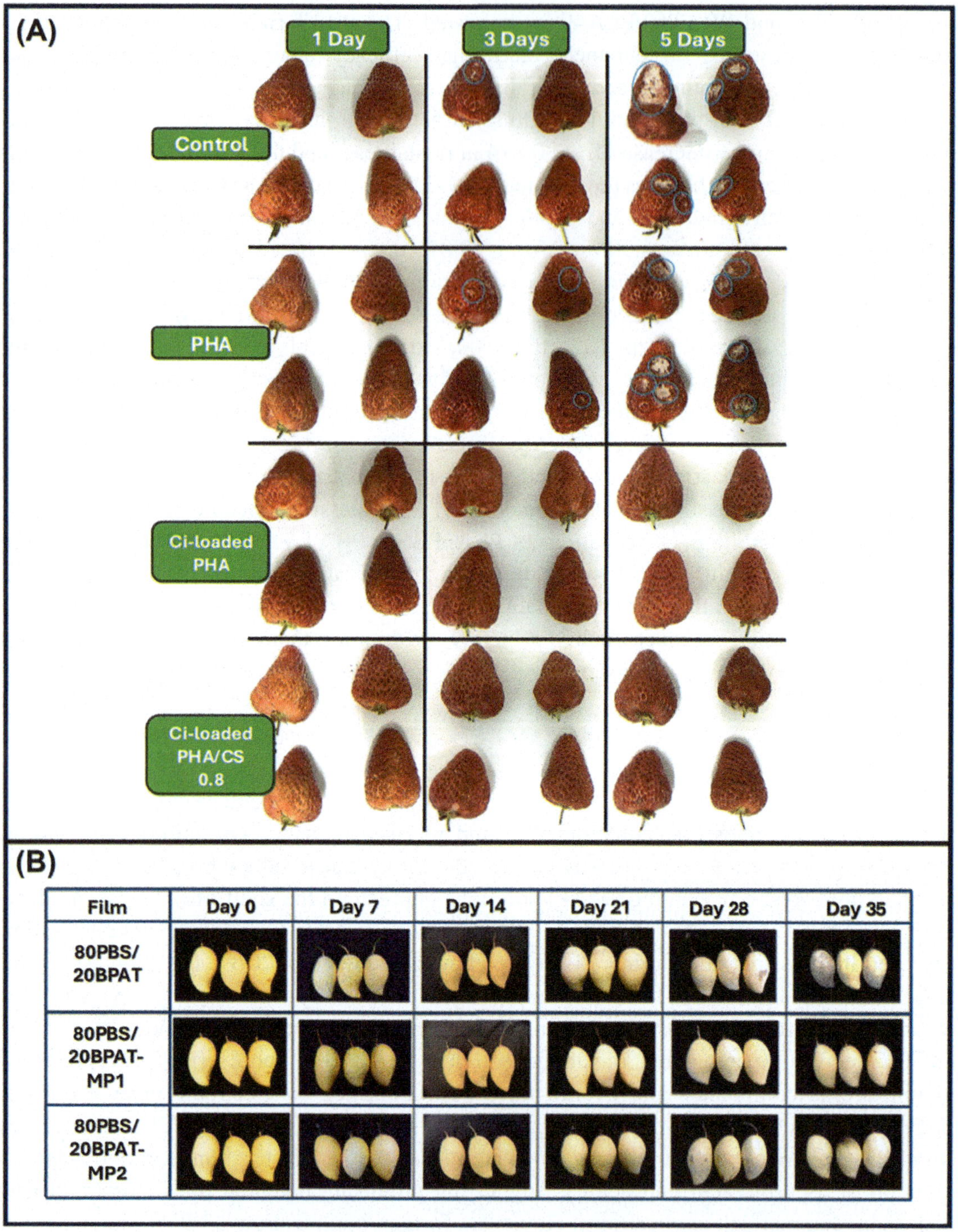
(A)
1 Day
3 Days
5 Days
Control
PHA
Ci-loaded PHA
Ci-loaded PHA/CS 0.8
(B)
Film
Day 0
Day 7
Day 14
Day 21
Day 28
Day 35
80PBS/ 20BPAT
80PBS/ 20BPAT-MP1
80PBS/ 20BPAT-MP2

packaging. Designed for controlled release, these microspheres exhibited tunable delivery profiles ranging from 7 to 60 days and demonstrated broad-spectrum antibacterial activity against both Gram-positive and Gram-negative bacteria. When applied in the preservation of strawberries, a highly perishable fruit prone to rapid microbial spoilage, the Ci-loaded PHAs and PHA/CS scaffolds significantly extended the freshness of strawberries compared to both unpackaged controls and neat PHA films. Visual observations (Fig. 9.5A) revealed noticeable decay in the control and PHA-only groups by day 3, while strawberries packaged with Ci-loaded scaffolds maintained acceptable color and aroma even after 5 days. Quantitative measurements confirmed that the Ci-loaded scaffolds slowed degradation in critical quality parameters, including firmness, titratable acidity, total soluble solids, and ascorbic acid content. Notably, the Ci-treated samples retained significantly more firmness and nutritional value, indicating reduced respiration and metabolic activity. This effect was attributed to the antimicrobial activity of CI, which suppressed microbial-induced spoilage pathways. These findings confirm that Ci-loaded PHA/CS microspheres not only offer structural and functional advantages but also effectively preserve the quality attributes of strawberries, highlighting their potential as active bioplastic packaging systems for high-value fresh produce. In terms of MAP, Zhang et al. (2025) developed PBAT/PCL blend films that naturally adjust the internal gas composition during cherry tomato storage. Incorporating 20% PCL into PBAT significantly improved film stiffness and generated a CO_2/O_2 permeability ratio exceeding 8.1 at 5°C, substantially higher than that of conventional PE films. The resulting films effectively slowed respiration, minimized nutritional loss, and preserved firmness and soluble solids under refrigerated conditions. Their ability to regulate gas without external flushing makes them highly suitable for passive MAP applications, while their biodegradability aligns with sustainability objectives. Further advancing MAP applications, Winotapun et al. (2023) introduced microperforated PBS/PBAT films enhanced with CO_2 laser treatment to improve gas transmission properties for mango preservation (Fig. 9.5B). Incorporating 20 wt.% PBAT

Figure 9.5

(A) Visual condition of strawberries in 1, 3, and 5 days of storage under different treatments: untreated control, with plain polyhydroxyalkanoate (PHA) scaffold, and with cinnamaldehyde-loaded PHA and PHA/chitosan (CS) scaffolds HT-polylactic acid (PLA)/poly(butylene-succinate-co-adipate) (PBSA). (B) Progressive visual changes in mangoes during storage under varying packaging conditions.

(A) Reprinted from Yang, S., Miao, Q., Huang, Y., Jian, P., Wang, X., Tu, M., 2020. Preparation of cinnamaldehyde-loaded polyhydroxyalkanoate/chitosan porous microspheres with adjustable controlled-release property and its application in fruit preservation. Food Packaging and Shelf Life 26, 100596. https://doi.org/10.1016/j.fpsl.2020.100596, with permission from Elsevier Science Ltd. (B) Reprinted from Winotapun, C., Tameesrisuk, M., Sirirutbunkajal, P., Sungdech, P., Leelaphiwat, P., 2023. Enhancing gas transmission rate of PBS/PBAT composite films: a study on microperforated film solutions for mango storage. ACS Omega. 9 (3), 3469–3479. https://doi.org/10.1021/acsomega.3c06999, with permission from Elsevier Science Ltd.

into PBS yielded a mechanically robust composite, and precision laser perforation significantly increased oxygen and carbon dioxide permeability. Two microperforated variants, PBS80/PBAT20-MP1 and MP2, demonstrated controlled internal atmosphere conditions, maintaining O_2 levels at 14.2% and 16.7% and CO_2 concentrations at 6% and 4%, respectively, throughout 33 days of storage. In contrast, mangoes packed in nonperforated films showed a rapid drop in O_2 to 2% and a sharp rise in CO_2 to 22% by day 2, leading to noticeable fermentative odor and freshness loss by day 5. By comparison, mangoes stored in the microperforated films retained stable levels of total soluble solids, color, and weight, with the PBS80/PBAT20-MP1 film extending shelf life by up to 35 days. These results highlight the role of microperforation in modulating gas exchange and preventing anaerobic conditions, validating it as a practical strategy to enhance the functionality of biodegradable films for high-respiration fruits like mango. Complementing these findings, Ucpinar Durmaz et al. (2024) investigated PBS-based films reinforced with alkaline-treated halloysite nanotubes (aHal) to enhance antimicrobial and ethylene-scavenging functionality. Films with 5 wt.% aHal demonstrated the highest ethylene removal, up to 80%, and displayed strong antibacterial activity against *E. coli*. Packaging tests using sliced apples and tomatoes under ambient conditions showed significantly improved preservation performance, particularly for extruded films, which exhibited superior mechanical flexibility and vapor barrier properties compared to solvent-cast variants. These results confirm the potential of PBS/aHal composite films for extending the shelf life of ethylene-sensitive produce in a sustainable and functional packaging format. Collectively, these studies illustrate the versatility of bioplastics in fresh produce packaging, from passive MAP systems to active antimicrobial and ethylene-scavenging platforms. Each material, whether PLA/PBAT blends, PHA-based microspheres, PBS/PBAT films, or nanotube-reinforced PBS, offers targeted functionality that can be tailored to specific postharvest challenges. Continued innovation in structure–property optimization and integration of natural additives will be essential for translating these materials into scalable, commercially viable packaging solutions that meet both shelf life and environmental performance criteria.

9.5 Advantages of bioplastics in food packaging

Bioplastics have emerged as a promising alternative to conventional fossil-derived plastics, particularly in the context of food packaging, where sustainability, functionality, and consumer safety converge. One of the most widely recognized advantages of bioplastics is their potential to reduce environmental impact. Life cycle assessment studies have consistently shown that bioplastics derived from renewable feedstocks, such as corn starch, sugarcane, or cellulose, exhibit a significantly lower carbon footprint and reduced nonrenewable energy usage compared to petroleum-based plastics. For instance, PLA and starch-based polymers have demonstrated greenhouse gas emission reductions of up to 70% during production when compared to PET, provided that agricultural practices and energy inputs are optimized (Sousa et al., 2021). This distinction is effectively illustrated in Fig. 9.6, which contrasts the environmental and material profiles of bioplastics and petroleum-based plastics. Bioplastics are made from renewable sources, are non-toxic, compostable, and capable of full biodegradation, making them suitable for circular economy models. In contrast, conventional plastics, derived from nonrenewable petroleum, are associated with high production demands, persistent environmental pollution, and contributions to global warming and aquatic ecosystem degradation. The figure visually reinforces the comparative sustainability advantages of bioplastics, particularly in their reduced carbon footprint and absence of long-term ecological harm. Furthermore, Table 9.3 summarizes the major advantages of commonly used bioplastics in food packaging.

Beyond their origin, many bioplastics offer the added benefit of end-of-life biodegradability. Materials such as PLA, PBS, PBAT, and PHAs can degrade under industrial composting conditions, and in some cases, under home composting or aquatic environments. PHAs, in particular, have demonstrated full biodegradation in marine and soil settings, a property that is particularly valuable given rising concern over microplastic pollution in natural ecosystems (Gigli et al., 2016; Hubackova et al., 2013; Rajgond et al., 2024). While biodegradability does not exempt these materials from the need for responsible waste management, it provides an alternative disposal route in regions with established composting infrastructure and supports

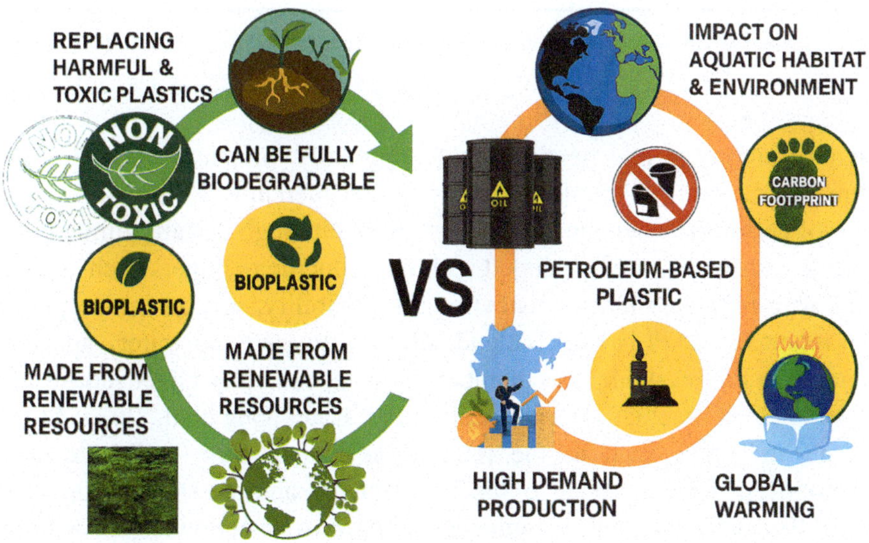

Figure 9.6
Comparison between bioplastics and petroleum-based plastics in terms of resource origin, environmental impact, and end-of-life behavior (Agarwal et al., 2023).
Reprinted from Agarwal, A., Shaida, B., Rastogi, M., Singh, N.B., 2023. Food packaging materials with special reference to biopolymers: properties and applications. Chemistry Africa 6(1), 117–144. https://doi.org/10.1007/s42250-022-00446-w.

Table 9.3 Key advantages of major bioplastics in food packaging.

Bioplastic	Key advantages	References
PLA	Derived from renewable feedstocks; good transparency; high stiffness; suitable for thermoforming; and compostable under industrial conditions	Chang et al. (2022) and Gonon et al. (2023)
TPS	Low cost; excellent biodegradability (including home composting); easy blending with other polymers; and derived from abundant starch sources	Fourati et al. (2021), Katekhong et al. (2022), Martinez Villadiego et al. (2022), and Rahardiyan et al. (2023)
PHAs	Fully biodegradable in various environments (including marine); good barrier properties; biocompatible; and suitable for active packaging	Carvalho et al. (2014), Colombo et al. (2016), and Mai et al. (2024)
PBAT	High flexibility; good impact strength; industrial compostability; excellent blending compatibility (e.g., with PLA); and suitable for films and bags	Aliotta et al. (2022); Chang et al. (2022), Freire et al. (2024), and Rajgond et al. (2024)
PCL	Excellent flexibility and elongation; low melting point (energy-efficient processing); slow, controlled biodegradation for longer shelf life; and compatible with nanocomposites and bioactive agents	Ahmadi et al. (2024), Dash and Konkimalla (2012), Gürler et al. (2023), and Hubackova et al. (2013)

PBAT, polybutylene adipate terephthalate; *PCL*, polycaprolactone; *PHAs*, polyhydroxyalkanoates; *PLA*, polylactic acid; *TPS*, thermoplastic starch.

efforts to reduce environmental persistence. Functionally, bioplastics can be engineered to meet the demands of food packaging through polymer blending, structural modification, and incorporation of performance-enhancing agents. PLA, although initially limited by brittleness and low thermal resistance, can be blended with more ductile polymers such as PBAT or PBS to yield films with improved mechanical strength, flexibility, and barrier properties. Several studies have reported that PLA/PBAT blends possess oxygen and water vapor transmission rates comparable to LDPE, while also maintaining compostability under industrial conditions (Bumbudsanpharoke et al., 2022; Correa et al., 2017; Katekhong et al., 2022; Li et al., 2022). Nanocomposite strategies, such as the integration of nanoclay, halloysite nanotubes, or cellulose nanofibers, further enhance tensile strength and barrier performance without compromising the environmental advantages of the base biopolymer. In addition to their structural versatility, bioplastics are particularly amenable to functionalization for active and intelligent packaging. The ability to incorporate antimicrobial agents, antioxidants, ethylene scavengers, or pH-sensitive indicators into biodegradable films has opened new avenues in food safety and shelf-life extension (Alshehri et al., 2024; Correa et al., 2017; Taechutrakul et al., 2024). Another important consideration is consumer perception and market alignment.

Growing awareness of plastic pollution and climate change has increased demand for packaging that aligns with environmentally responsible values. Consumer studies have shown that biodegradable and bio-based packaging materials are generally well-received, often perceived as safer and more natural. This positive perception can enhance brand image and offer competitive advantages in increasingly sustainability-driven markets. Moreover, bioplastics enable integration into circular economy models, particularly when produced from agricultural residues or industrial by-products, further decoupling material use from fossil carbon dependency. While some limitations remain, such as production cost, processability, and limited composting infrastructure, the collective advantages of bioplastics are clear. They present a viable route toward reducing dependency on fossil resources, curbing greenhouse gas emissions, and enabling packaging solutions that are both technically competent and environmentally restorative. With continued innovation in polymer science, scaling of bio-based feedstocks, and alignment of policy and infrastructure, bioplastics are well-positioned to play a central role in the future of sustainable food packaging.

9.6 Challenges and future prospects

Although bioplastics have emerged as a promising alternative to conventional petrochemical-based materials in food packaging, their broader adoption is hindered by a number of persistent challenges. One of the most pressing issues lies in their economic viability. The production of bioplastics, especially those derived through microbial fermentation or chemical synthesis from renewable feedstocks, remains more expensive than that of traditional plastics. This cost disparity is largely due to the high price of raw materials, limited economies of scale, and the technological complexity of production processes. Moreover, the reliance on agricultural inputs such as corn, sugarcane, or potatoes has sparked ongoing debate regarding competition with food supply and land use, raising ethical and environmental concerns about long-term scalability. Performance limitations also continue to constrain the applicability of many bioplastics in packaging formats that require prolonged shelf life, high mechanical durability, or resistance to moisture and temperature fluctuations. Materials such as thermoplastic starch are known for their biodegradability but are often too sensitive to humidity. PLA, while praised for its clarity and stiffness, tends to be brittle and susceptible to hydrolytic degradation in humid conditions. Others, like PCL, degrade too slowly under natural conditions, complicating their disposal. These material drawbacks restrict bioplastics from replacing conventional plastics in more demanding applications, especially in food systems that require strict barrier protection. Another major challenge is the disconnect between the designed biodegradability of bioplastics and the real-world infrastructure required to support their end-of-life management. Although several bioplastics are certified as compostable, this often refers to industrial composting conditions that are not widely available in many regions. The lack of sufficient facilities, along with inconsistent waste segregation practices, results in many compostable plastics ending up in landfills or contaminating recycling streams. Compounding this is the public confusion surrounding bioplastic labeling; terms like "biodegradable," "bio-based," and "compostable" are frequently misunderstood, leading to improper disposal and undermining the environmental benefits these materials are designed to deliver. On a regulatory level, fragmented standards and inconsistent certification frameworks across different countries present yet another barrier. Manufacturers are often required to navigate a complex landscape of local and international requirements, which can slow innovation and market entry. In parallel, environmental concerns have been raised about the potential for microplastic generation during incomplete degradation, particularly in marine ecosystems. These concerns have added another layer of scrutiny to the deployment of bioplastics, especially in regions with limited environmental regulation and enforcement. Nevertheless, the future of bioplastics in food packaging remains encouraging. Research continues to advance the development of novel polymer blends and nanocomposites with tailored performance characteristics. Scientists are exploring the incorporation of natural fillers, essential oils, and bioactive compounds to create multifunctional packaging with enhanced barrier, antimicrobial, and mechanical properties. Innovations in synthetic biology and green chemistry are also unlocking new pathways to produce biopolymers more efficiently and sustainably, using nonfood biomass or engineered microbial systems. Moreover, the global momentum toward a circular economy has placed bioplastics at the forefront of sustainable material strategies. As industries shift from linear to regenerative models, bioplastics offer the potential to close material loops through composting, enzymatic degradation, or even chemical recycling. These developments are being complemented by policy instruments such as bans on single-use plastics, incentives for sustainable materials, and producer responsibility frameworks that are beginning to redefine market dynamics. Ultimately, the success of bioplastics in transforming food packaging systems will depend on the convergence of scientific innovation, infrastructure development, regulatory coherence, and public engagement. While the path ahead is complex, it is also rich with opportunity. With coordinated action, bioplastics can evolve from niche alternatives into mainstream materials that redefine not only the packaging industry but also our broader relationship with plastics and sustainability.

9.7 Conclusion

The growing demand for sustainability in the food industry positions bioplastics as a promising alternative to conventional plastics. These materials, including PLA, TPS, PHAs, PBS, PBAT, and PCL, combine biodegradability, renewable origins, and compatibility with modern processing methods, making them viable solutions for reducing environmental impact without compromising functionality. Beyond passive packaging, recent advancements have enabled bioplastics to provide active roles in preservation, such as controlling gas exchange and inhibiting microbial growth, as demonstrated in applications for meat, bakery products, and fresh produce. Despite their potential, challenges remain. High production costs, limited processing adaptability, and insufficient composting infrastructure hinder large-scale adoption. Addressing these barriers through innovations in polymer science, bio-refining, and improved waste management systems will be critical. Regulatory frameworks and consumer demand for eco-friendly packaging further drive this transition. In essence, bioplastics represent a convergence of material innovation and environmental responsibility. Continued research and commercialization can transform food packaging into a sustainable, circular system. Bioplastics are not merely alternatives but critical enablers of a future food packaging system that integrates safety, functionality, and sustainability.

References

Abedi, E., Gahruie, H.H., Hosseini, S.M.H., Zhang, P., 2025. Impact of hydrodynamic cavitation-assisted microbubble with N2, CO2, and air on pasting, morphological, and thermal properties of starches from different botanical origins. International Journal of Biological Macromolecules 301, 140253. Available at https://doi.org/10.1016/j.ijbiomac.2025.140253.

Adak, S., Kayalvizhi, R., Bishai, M., Jacob, S., Kundu, D., 2024. Advancements in microbial production of polyhydroxyalkanoates (PHA) from wastes for sustainable active food packaging: an eclectic review. Biocatalysis and Agricultural Biotechnology 60, 103288. Available at https://doi.org/10.1016/j.bcab.2024.103288.

Agarwal, A., Shaida, B., Rastogi, M., Singh, N.B., 2023. Food packaging materials with special reference to biopolymers: properties and applications. Chemistry Africa 6 (1), 117–144. Available at https://doi.org/10.1007/s42250-022-00446-w.

Ahmadi, N., Ahari, H., Anvar, A., Khosravi-Darani, K., Gharachorloo, M., 2024. Polycaprolactone (PCL)-based films integrated with hairy cellulose nanocrystals and silver nanoparticles for active Tilapia packaging applications. Food Chemistry: X 22, 101490. Available at https://doi.org/10.1016/j.fochx.2024.101490.

Aliotta, L., Seggiani, M., Lazzeri, A., Gigante, V., Cinelli, P., 2022. A brief review of poly (butylene succinate) (PBS) and its main and applications. Polymer 14, 844.

Alshehri, A.A., Elsherief, M.F., Devecioglu, D., Salama, M.A., Sakr, H., Abdin, M., et al., 2024. Development and characterization of bioactive polyvinyl alcohol / chitosan multilayer-based films loaded with tea tree oil nanoemulsion to extend the shelf life of red grapes. Biocatalysis and Agricultural Biotechnology 58, 103206. Available at https://doi.org/10.1016/j.bcab.2024.103206.

Aoyama, T., Uto, K., Shimizu, H., Ebara, M., Kitagawa, T., Tachibana, H., et al., 2021. Development of a new poly-ε-caprolactone with low melting point for creating a thermoset mask used in radiation therapy. Scientific Reports 11 (1), 20409. Available at https://doi.org/10.1038/s41598-021-00005-2.

Apicella, A., Malafeev, K.V., Scarfato, P., Incarnato, L., 2024. Generation of microplastics from biodegradable packaging films based on PLA, PBS and their blend in freshwater and seawater. Polymers 16 (16), 2268. Available at https://doi.org/10.3390/polym16162268.

Bahramian, B., Abedi-Firoozjah, R., Kiani-Salmi, N., Ebrahimi, A., Oladzadabbasabadi, N., Moradiyan, M.H., et al., 2025. Development of polycaprolactone-based electrospun nanofiber incorporated lemon beebrush essential oil-loaded metal-organic frameworks as a novel active food packaging for meat preservation. Food Control 168, 110981. Available at https://doi.org/10.1016/j.foodcont.2024.110981.

Barletta, M., Aversa, C., Ayyoob, M., Gisario, A., Hamad, K., Mehrpouya, M., et al., 2022. Poly(butylene succinate) (PBS): materials, processing, and industrial applications. Progress in Polymer Science 132, 101579. Available at https://doi.org/10.1016/j.progpolymsci.2022.101579.

Bumbudsanpharoke, N., Wongphan, P., Promhuad, K., Leelaphiwat, P., Harnkarnsujarit, N., 2022. Morphology and permeability of bio-based poly(butylene adipate-co-terephthalate) (PBAT), poly(butylene succinate) (PBS) and linear low-density polyethylene (LLDPE) blend films control shelf-life of packaged bread. Food Control 132, 108541. Available at https://doi.org/10.1016/j.foodcont.2021.108541.

Carvalho, G., Oehmen, A., Albuquerque, M.G.E., Reis, M.A.M., 2014. The relationship between mixed microbial culture composition and PHA production performance from fermented molasses. New Biotechnology 31 (4), 257–263. Available at https://doi.org/10.1016/j.nbt.2013.08.010.

Chang, F.L., Hu, B., Huang, W.T., Chen, L., Yin, X.C., Cao, X.W., et al., 2022. Improvement of rheology and mechanical properties of PLA/PBS blends by in-situ UV-induced reactive extrusion. Polymer 259, 125336. Available at https://doi.org/10.1016/j.polymer.2022.125336.

Colombo, B., Sciarria, T.P., Reis, M., Scaglia, B., Adani, F., 2016. Polyhydroxyalkanoates (PHAs) production from fermented cheese whey by using a mixed microbial culture. Bioresource Technology 218, 692–699. Available at https://doi.org/10.1016/j.biortech.2016.07.024.

Correa, J.P., Molina, V., Sanchez, M., Kainz, C., Eisenberg, P., Massani, M.B., 2017. Improving ham shelf life with a polyhydroxybutyrate/polycaprolactone biodegradable film activated with nisin. Food Packaging and Shelf Life 11, 31–39. Available at https://doi.org/10.1016/j.fpsl.2016.11.004.

Dash, T.K., Konkimalla, V.B., 2012. Polymeric modification and its implication in drug delivery: poly-ε-caprolactone (PCL) as a model polymer. Molecular Pharmaceutics 9 (9), 2365–2379. Available at https://doi.org/10.1021/mp3001952.

Dong, M., Mastroianni, G., Bilotti, E., Zhang, H., Papageorgiou, D.G., 2024. Biodegradable starch-based nanocomposite films with exceptional water and oxygen barrier properties. ACS Sustainable Chemistry and Engineering 12 (29), 11056–11066. Available at https://doi.org/10.1021/acssuschemeng.4c04198.

Fourati, Y., Tarrés, Q., Delgado-Aguilar, M., Mutjé, P., Boufi, S., 2021. Cellulose nanofibrils reinforced PBAT/TPS blends: mechanical and rheological properties. International Journal of Biological Macromolecules 183, 267–275. Available at https://doi.org/10.1016/j.ijbiomac.2021.04.102.

Freire, N.F., Cordani, M., Aparicio-Blanco, J., Fraguas Sanchez, A.I., Dutra, L., Pinto, M.C.C., et al., 2024. Preparation and characterization of PBS (polybutylene Succinate) nanoparticles containing cannabidiol (CBD) for anticancer application. Journal of Drug Delivery Science and Technology 97, 105833. Available at https://doi.org/10.1016/j.jddst.2024.105833.

Gigante, V., Aliotta, L., Ascrizzi, R., Pistelli, L., Zinnai, A., Batoni, G., et al., 2023. Innovative biobased and sustainable polymer packaging solutions for extending bread shelf life: a review. Polymers 15 (24), 4700. Available at https://doi.org/10.3390/polym15244700.

Gigli, M., Fabbri, M., Lotti, N., Gamberini, R., Rimini, B., Munari, A., 2016. Poly(butylene succinate)-based polyesters for biomedical applications: a review in memory of our beloved colleague and friend Dr. Lara Finelli. European Polymer Journal 75, 431–460. Available at https://doi.org/10.1016/j.eurpolymj.2016.01.016.

Gonon, H., Srisa, A., Promhuad, K., Chonhenchob, V., Bumbudsanpharoke, N., Jarupan, L., et al., 2023. PLA thermoformed trays incorporated with cinnamaldehyde and carvacrol as active biodegradable bakery packaging. Food Packaging and Shelf Life 38, 101123. Available at https://doi.org/10.1016/j.fpsl.2023.101123.

Gurunathan, M.K., Navasingh, R.J.H., Selvam, J.D.R., Čep, R., 2025. Development and characterization of starch bioplastics as a sustainable alternative for packaging. Scientific Reports 15 (1), 15264. Available at https://doi.org/10.1038/s41598-025-00221-0.

Gürler, N., Pekdemir, M.E., Torğut, G., Kök, M., 2023. Binary PCL–waste photopolymer blends for biodegradable food packaging applications. Journal of Molecular Structure 1279, 134990. Available at https://doi.org/10.1016/j.molstruc.2023.134990.

Han, J.W., Ruiz-Garcia, L., Qian, J.P., Yang, X.T., 2018. Food packaging: a comprehensive review and future trends. Comprehensive Reviews in Food Science and Food Safety 17 (4), 860–877. Available at https://doi.org/10.1111/1541-4337.12343.

Hu, H., Tian, Y., Wang, J., Zhang, R., Zhu, J., 2022. Enhanced degradation and gas barrier of PBAT through composition design of aliphatic units. Polymer Degradation and Stability 195, 109795. Available at https://doi.org/10.1016/j.polymdegradstab.2021.109795.

Huang, H.D., Ren, P.G., Xu, J.Z., Xu, L., Zhong, G.J., Hsiao, B.S., et al., 2014. Improved barrier properties of poly (lactic acid) with randomly dispersed graphene oxide nanosheets. Journal of Membrane Science 464, 110–118. Available at https://doi.org/10.1016/j.memsci.2014.04.009.

Hubackova, J., Dvorackova, M., Svoboda, P., Mokrejs, P., Kupec, J., Pozarova, I., et al., 2013. Influence of various starch types on PCL/starch blends anaerobic biodegradation. Polymer Testing 32 (6), 1011–1019. Available at https://doi.org/10.1016/j.polymertesting.2013.05.008.

Katekhong, W., Wongphan, P., Klinmalai, P., Harnkarnsujarit, N., 2022. Thermoplastic starch blown films functionalized by plasticized nitrite blended with PBAT for superior oxygen barrier and active biodegradable meat packaging. Food Chemistry 374, 131709. Available at https://doi.org/10.1016/j.foodchem.2021.131709.

Li, C., Cui, Q., Li, Y., Zhang, K., Lu, X., Zhang, Y., 2022. Effect of LDPE and biodegradable PBAT primary microplastics on bacterial community after four months of soil incubation. Journal of Hazardous Materials 429, 128353. Available at https://doi.org/10.1016/j.jhazmat.2022.128353.

Mai, J., Kockler, K., Parisi, E., Chan, C.M., Pratt, S., Laycock, B., 2024. Synthesis and physical properties of polyhydroxyalkanoate (PHA)-based block copolymers: a review. International Journal of Biological Macromolecules 263 (Pt 1), 130204. Available at https://doi.org/10.1016/j.ijbiomac.2024.130204.

Martinez Villadiego, K., Arias Tapia, M.J., Useche, J., Escobar Macías, D., 2022. Thermoplastic starch (TPS)/polylactic acid (PLA) blending methodologies: a review. Journal of Polymers and the Environment 30 (1), 75–91. Available at https://doi.org/10.1007/s10924-021-02207-1.

Mohd Nizam, N.H., Mohammad Rawi, N.F., Mhd Ramle, S.F., Abd Aziz, A., Abdullah, C.K., Rashedi, A., et al., 2021. Physical, thermal, mechanical, antimicrobial and physicochemical properties of starch based film containing aloe vera: a review. Journal of Materials Research and Technology 15, 1572–1589. Available at https://doi.org/10.1016/j.jmrt.2021.08.138.

Naser, A.Z., Deiab, I., Defersha, F., Yang, S., 2021. Expanding poly(lactic acid) (pla) and polyhydroxyalkanoates (phas) applications: a review on modifications and effects. Polymers 13 (23), 4271. Available at https://doi.org/10.3390/polym13234271.

Oladzadabbasabadi, N., Mohammadi Nafchi, A., Ariffin, F., Karim, A.A., 2022. Chapter 11 - Plant extracts as packaging aids. In: Mir, S.A., Manickavasagan, A., Shah, M.A. (Eds.), Plant Extracts: Applications in the Food Industry. Academic Press, pp. 225–268. Available at https://doi.org/10.1016/B978-0-12-822475-5.00001-6.

Piroonpan, T., Haema, K., Chirachanchai, S., Pasanphan, W., 2025. A dual antioxidant and antimicrobial bioplastic film endowed with amine-branched poly(lactic acid): bioactive packaging for electron beam sterilization and food shelf-life extension. International Journal of Biological Macromolecules 291, 139093. Available at https://doi.org/10.1016/j.ijbiomac.2024.139093.

Rafiqah, S.A., Khalina, A., Harmaen, A.S., Tawakkal, I.A., Zaman, K., Asim, M., et al., 2021. A review on properties and application of bio-based poly(butylene succinate). Polymers 13 (9), 1436. Available at https://doi.org/10.3390/polym13091436.

Rahardiyan, D., Moko, E.M., Tan, J.S., Lee, C.K., 2023. Thermoplastic starch (TPS) bioplastic, the green solution for single-use petroleum plastic food packaging – a review. Enzyme and Microbial Technology 168, 110260. Available at https://doi.org/10.1016/j.enzmictec.2023.110260.

Rajgond, V., Mohite, A., More, N., More, A., 2024. Biodegradable polyester-polybutylene succinate (PBS): a review. Polymer Bulletin 81 (7), 5703–5752. Available at https://doi.org/10.1007/s00289-023-04998-w.

Roy, S., Ghosh, T., Zhang, W., Rhim, J.-W., 2024. Recent progress in PBAT-based films and food packaging applications: a mini-review. Food Chemistry 437 (Pt 1), 137822. Available at https://doi.org/10.1016/j.foodchem.2023.137822.

Shahverdi, M., Seifi, S., Akbari, A., Mohammadi, K., Shamloo, A., Movahhedy, M.R., 2022. Melt electrowriting of PLA, PCL, and composite PLA/PCL scaffolds for tissue engineering application. Scientific Reports 12 (1), 19935. Available at https://doi.org/10.1038/s41598-022-24275-6.

Shlush, E., Davidovich-Pinhas, M., 2022. Bioplastics for food packaging. Trends in Food Science & Technology 125, 66–80. Available at https://doi.org/10.1016/j.tifs.2022.04.026.

Signori, F., Coltelli, M.B., Bronco, S., 2009. Thermal degradation of poly(lactic acid) (PLA) and poly(butylene adipate-co-terephthalate) (PBAT) and their blends upon melt processing. Polymer Degradation and Stability 94 (1), 74–82. Available at https://doi.org/10.1016/j.polymdegradstab.2008.10.004.

Sousa, A.F., Patrício, R., Terzopoulou, Z., Bikiaris, D.N., Stern, T., Wenger, J., et al., 2021. Recommendations for replacing PET on packaging, fiber, and film materials with biobased counterparts. Green Chemistry 23 (22), 8795–8820. Available at https://doi.org/10.1039/d1gc02082j.

Srisa, A., Harnkarnsujarit, N., 2020. Antifungal films from trans-cinnamaldehyde incorporated poly(lactic acid) and poly(butylene adipate-co-terephthalate) for bread packaging. Food Chemistry 333, 127537. Available at https://doi.org/10.1016/j.foodchem.2020.127537.

Stoica, D., Alexe, P., Ivan, A.S., Moraru, D.I., Ungureanu, C.V., Stanciu, S., et al., 2022. Biopolymers: global carbon footprint and climate change. In: Nadda, A.K., Sharma, S., Bhat, R. (Eds.), Biopolymers. Springer Series on Polymer and Composite Materials. Springer, Cham, pp. 35–54. Available at https://doi.org/10.1007/978-3-030-98392-5_3.

Stoica, M., Bichescu, C.I., Crețu, C.M., Dragomir, M., Ivan, A.S., Podaru, G.M., et al., 2024. Review of bio-based biodegradable polymers: smart solutions for sustainable food packaging. Foods 13 (19), 3027. Available at https://doi.org/10.3390/foods13193027.

Surendren, A., Mohanty, A.K., Liu, Q., Misra, M., 2022. A review of biodegradable thermoplastic starches, their blends and composites: recent developments and opportunities for single-use plastic packaging alternatives. Green Chemistry 24 (22), 8606–8636. Available at https://doi.org/10.1039/d2gc02169b.

Suwanamornlert, P., Kerddonfag, N., Sane, A., Chinsirikul, W., Zhou, W., Chonhenchob, V., 2020. Poly(lactic acid)/poly(butylene-succinate-co-adipate) (PLA/PBSA) blend films containing thymol as alternative to synthetic preservatives for active packaging of bread. Food Packaging and Shelf Life 25, 100515. Available at https://doi.org/10.1016/j.fpsl.2020.100515.

Taechutrakul, S., Piroonpan, T., Pasanphan, W., 2024. Active film strips to extend the shelf life of fruits: multibranched PLA-gallic acid as an antioxidant/oxygen scavenger in a case study of bananas (Musa AAA group). Journal of Food Engineering 364, 111794. Available at https://doi.org/10.1016/j.jfoodeng.2023.111794.

Tarique, J., Sapuan, S.M., Khalina, A., 2021. Effect of glycerol plasticizer loading on the physical, mechanical, thermal, and barrier properties of arrowroot (Maranta arundinacea) starch biopolymers. Scientific Reports 11 (1), 13900. Available at https://doi.org/10.1038/s41598-021-93094-y.

Tripathi, N., Misra, M., Mohanty, A.K., 2021. Durable polylactic acid (PLA)-based sustainable engineered blends and biocomposites: recent developments, challenges, and opportunities. ACS Engineering Au 1 (1), 7–38. Available at https://doi.org/10.1021/acsengineeringau.1c00011.

Ucpinar Durmaz, B., Ugur Nigiz, F., Aytac, A., 2024. Active packaging films based on poly(butylene succinate) films reinforced with alkaline halloysite nanotubes: production, properties, and fruit packaging applications. Applied Clay Science 259, 107517. Available at https://doi.org/10.1016/j.clay.2024.107517.

Vytejčková, S., Vápenka, L., Hradecký, J., Dobiáš, J., Hajšlová, J., Loriot, C., et al., 2017. Testing of polybutylene succinate based films for poultry meat packaging. Polymer Testing 60, 357–364. Available at https://doi.org/10.1016/j.polymertesting.2017.04.018.

Wang, Y., Qin, B., Xia, G., Choi, S.H., 2021. FDA's poly (lactic-Co-glycolic acid) research program and regulatory outcomes. AAPS Journal 23 (4), 92. Available at https://doi.org/10.1208/s12248-021-00611-y.

Wang, L., Xu, J., Zhang, M., Zheng, H., Li, L., 2022. Preservation of soy protein-based meat analogues by using PLA/PBAT antimicrobial packaging film. Food Chemistry 380, 132022. Available at https://doi.org/10.1016/j.foodchem.2021.132022.

Winotapun, C., Tameesrisuk, M., Sirirutbunkajal, P., Sungdech, P., Leelaphiwat, P., 2023. Enhancing gas transmission rate of PBS/PBAT composite films: a study on microperforated film solutions for mango storage. ACS Omega 9 (3), 3469–3479. Available at https://doi.org/10.1021/acsomega.3c06999, pubs.acs.org/journal/acsodf.

Xanthos, D., Walker, T.R., 2017. International policies to reduce plastic marine pollution from single-use plastics (plastic bags and microbeads): a review. Marine Pollution Bulletin 118 (1-2), 17–26. Available at https://doi.org/10.1016/j.marpolbul.2017.02.048.

Xiao, X., Jiang, X., Yang, S., Lu, Z., Niu, C., Xu, Y., et al., 2021. Solvent evaporation induced fabrication of porous polycaprolactone scaffold via low-temperature 3D printing for regeneration medicine researches. Polymer 217, 123436. Available at https://doi.org/10.1016/j.polymer.2021.123436.

Yang, S., Miao, Q., Huang, Y., Jian, P., Wang, X., Tu, M., 2020. Preparation of cinnamaldehyde-loaded polyhydroxyalkanoate/chitosan porous microspheres with adjustable controlled-release property and its application in fruit preservation. Food Packaging and Shelf Life 26, 100596. Available at https://doi.org/10.1016/j.fpsl.2020.100596.

Zhang, R., Lan, W., Ding, J., Ahmed, S., Qin, W., He, L., et al., 2019. Effect of PLA/PBAT antibacterial film on storage quality of passion fruit during the shelf-life. Molecules (Basel, Switzerland) 24 (18), 3378. Available at https://doi.org/10.3390/molecules24183378.

Zhang, J., Wang, Y., Zhang, Y., Guo, J., Li, Y., Dong, T., et al., 2025. Preparation of PBAT/PCL films for passive modified atmosphere packaging of cherry tomatoes. Packaging Technology and Science 38 (4), 309–322. Available at https://doi.org/10.1002/pts.2874.

Zhou, W., Bergsma, S., Colpa, D.I., Euverink, G.J.W., Krooneman, J., 2023. Polyhydroxyalkanoates (PHAs) synthesis and degradation by microbes and applications towards a circular economy. Journal of Environmental Management 341, 118033. Available at https://doi.org/10.1016/j.jenvman.2023.118033.

CHAPTER 10

Main characteristics of bioplastics for food packaging applications

Maryam Azizi-lalabadi[1], Hananeh Yazdanbakhsh[1], and Shima Jafarzadeh[2,3]

[1]Research Center of Oils and Fats, Health Technology Institute, Kermanshah University of Medical Sciences, Kermanshah, Iran, [2]Center for Sustainable Bioproducts, Deakin University, Geelong Waurn Ponds Campus, Geelong, VIC, Australia, [3]School of Engineering, Deakin University, Geelong Waurn Ponds Campus, Geelong, VIC, Australia

10.1 Introduction

Food packaging, a crucial aspect of food technology, plays a significant role in protecting and preserving various food types (Jabeen et al., 2015). Food needs to be preserved from the time of production to consumption, and packaging is one of the things necessary for preserving food. The storage time of packaged food varies from a few days to more than a year based on the nature of the product. Therefore, the materials used in the packaging must have suitable stability so that the shelf life of the food does not suffer. In any case, food spoils over time, even if the protection is perfect. So, the main purpose of packaging is not to prevent the food from spoiling but to delay the spoilage of the food as much as possible. In other words, packaging does not improve product quality but only affects its shelf life (Jariyasakoolroj et al., 2020). The food industry is the largest end-user of the packaging sector, accounting for approximately 40% of produced plastic (Narancic et al., 2020). With economic abundance, petrochemical plastics (polyvinyl chloride, polypropylene, polyethylene terephthalate, polyethylene, polyamide, and polystyrene) (Flores-Morales et al., 2017; Ibrahim et al., 2021) have been predominantly utilized as packaging materials due to their favorable properties, including excellent barrier performance against O_2, carbon dioxide, oxygen, and aroma compounds, great tensile strength, and good tear resistance, high thermal sealing capability, easy access, and low cost (Asgher et al., 2020). However, they also present several drawbacks, such as a very low water vapor transmission rate, and their most significant disadvantage is that they are non-biodegradable, contributing to environmental pollution. The pollution caused by plastic is very high and has harmful effects on human health, so this problem is considered a big challenge. Recycling can be one of the approaches to reduce the impact of plastic pollution (Jafarzadeh et al., 2024; Karmakar, 2022; Siddiqui et al., 2024). Those recycled plastics can affect the quality. However, the recycling process requires

Bioplastics for Food Packaging. DOI: https://doi.org/10.1016/B978-0-443-41654-5.00011-1

thorough grading, washing, grinding, mixing, and extruding, which leads to degradation and limits polymer recycling. In addition, acceptance and demand due to chemical impurities are still lower than polymers in various applications. Therefore, one of the challenges of society is to reduce the amount of durable and non-biodegradable packaging materials such as glass, metal, and mostly plastic in order to find new solutions. The search for suitable alternatives with favorable packaging characteristics is continuously under study, and the reduction of these wastes can be achieved by developing new environmentally friendly packaging systems. One of the methods can be to replace synthetic plastic with biological plastic. Nowadays, consumers' awareness of preserving and protecting the environment has increased significantly and replaced synthetic plastics with bioplastics (Cruz et al., 2022; European Commission, 2011). Finally, considering the nonrenewable nature and waste disposal issues associated with petroleum, the concept of bioplastics has emerged.

According to the definition of the European Bioplastic Organization, the word bioplastic refers to plastics that originate from plants and are also renewable and recyclable. Those bioplastics are divided into three main types: natural polymers or polymers derived from biomass, synthetic polymers or polymers that are chemically synthesized from renewable sources, and microbial polymers or polymers derived from microorganisms (Rahman et al., 2019). In general, bioplastics are materials that are completely or partially obtained from biomass and substrates such as potatoes, barley, bananas, corn, sugarcane, among others. Bioplastics have two important features: (1) they can be extracted and (2) they can be decomposed, and these plastics can break down water, methane, and carbon dioxide through the enzymatic activity of microorganisms. Bioplastics can indeed be composed of materials such as starch and cellulose, which are derived from renewable biomass. In addition to these, bioplastics may also include other natural polymers like polylactic acid (PLA), polyhydroxyalkanoate (PHA), polyhydroxybutyrate, polynucleotide, polypeptide, and polysaccharide. It is important to note that the exact composition can vary based on the type of bioplastic and its intended application (Boccalon and Gorrasi, 2022; Chowdhury et al., 2023). Some plastics are biological but not biodegradable, for example, bio-derived polyethylene, which is physically and chemically similar to traditional polyethylene, but is known as a bioplastic because its monomers are produced from corn (Jariyasakoolroj et al., 2020). Biodegradable packaging materials are those that undergo the fermentation process by bacteria, fungi, and yeast, and they can also be used as fertilizer or humus during composting. Biodegradation depends on many factors, including environmental conditions such as temperature, humidity, the material and its intended use. Bio-made polymers cannot provide sufficient functional properties. Currently, bioplastics are 1% of the 320 million tons of plastics produced per year. Of course, the obtained statistics show that the use of bioplastics increases annually (Rahman et al., 2019). These plastics have disadvantages; for example, their mechanical inhibitory properties are poor and their production cost is high. A bioplastic must have good mechanical, thermal, and chemical resistance; this feature cannot always be available. To solve the problem of poor mechanical properties, reinforcing materials such as fillers, compatibilizers, plasticizers, and nanoparticles are used. The study of recyclable and biodegradable plastics is an interesting and emerging field in packaging science, but extensive research is needed to improve their performance, mechanical, thermal, and physical properties, and commercial use, which may be possible in the next few years (Rahman et al., 2019; Siddiqui et al., 2024).

10.2 Importance of material characteristics in food packaging

Food packaging is a crucial method for protecting food. It plays a significant role in ensuring the safety and preservation of food and beverages. In comparison, there are concerns regarding packaging, such as the potential release of harmful chemical gases from materials like PHA when heated. Food packaging must prioritize preserving the essential characteristics of food, including temperature, color, taste, and texture. Packaging offers numerous advantages; for instance, effective packaging can prevent breakage and vibration, as well as protect against the penetration of water, pollutants, and bacteria. All these features contribute to ensuring the safe shelf life of products (Abedi-Firoozjah et al., 2025; Elkhayat et al., 2024). The materials utilized in packaging significantly influence its quality. The primary materials employed in packaging are polymers, which can be categorized into two types: synthetic and natural polymers, also known as biopolymers. Synthetic polymers possess advantageous properties, such as transparency, flexibility, heat-sealing capability, and high strength. The extensive use of polymers is attributed not only to their favorable mechanical and thermal characteristics but also to their stability, which plays a crucial role in their application. However, the repeated use of synthetic polymers has led to environmental issues, contributing to approximately 20% of all municipal solid waste. Another concern with synthetic polymers was their detrimental impact on underground water, prompting the consideration of biopolymers (Gong et al., 2024). Biopolymers serve as an alternative to synthetic polymers. They are biodegradable and can be broken down into natural substances by microorganisms. Nevertheless, they come with drawbacks, such as low moisture barrier, poor thermal and mechanical resistance, and high cost. To enhance the properties of biopolymers certain materials must be added as improvers; these materials include nanomaterials, enzymes, antimicrobials, antioxidants, stabilizers, plasticizers, and colorants (Fig. 10.1) (Chen et al., 2024).

In recent years, nanotechnology has advanced methods to enhance the quality of food packaging, transforming the packaging industry. Nanomaterials are defined as materials with at least one dimension in the nanoscale range (1–100 nm) (Gupta et al., 2024). These materials are noted for their unique properties, the most significant of which is their ability to enhance food safety. They can serve as enhancers in the packaging sector. Among their beneficial features are resistance to gas and moisture, antimicrobial activity, and mechanical strength. Silver, titanium dioxide, and zinc oxide are examples of nanoparticles that possess antimicrobial properties. On the other hand, concerns exist regarding these materials,

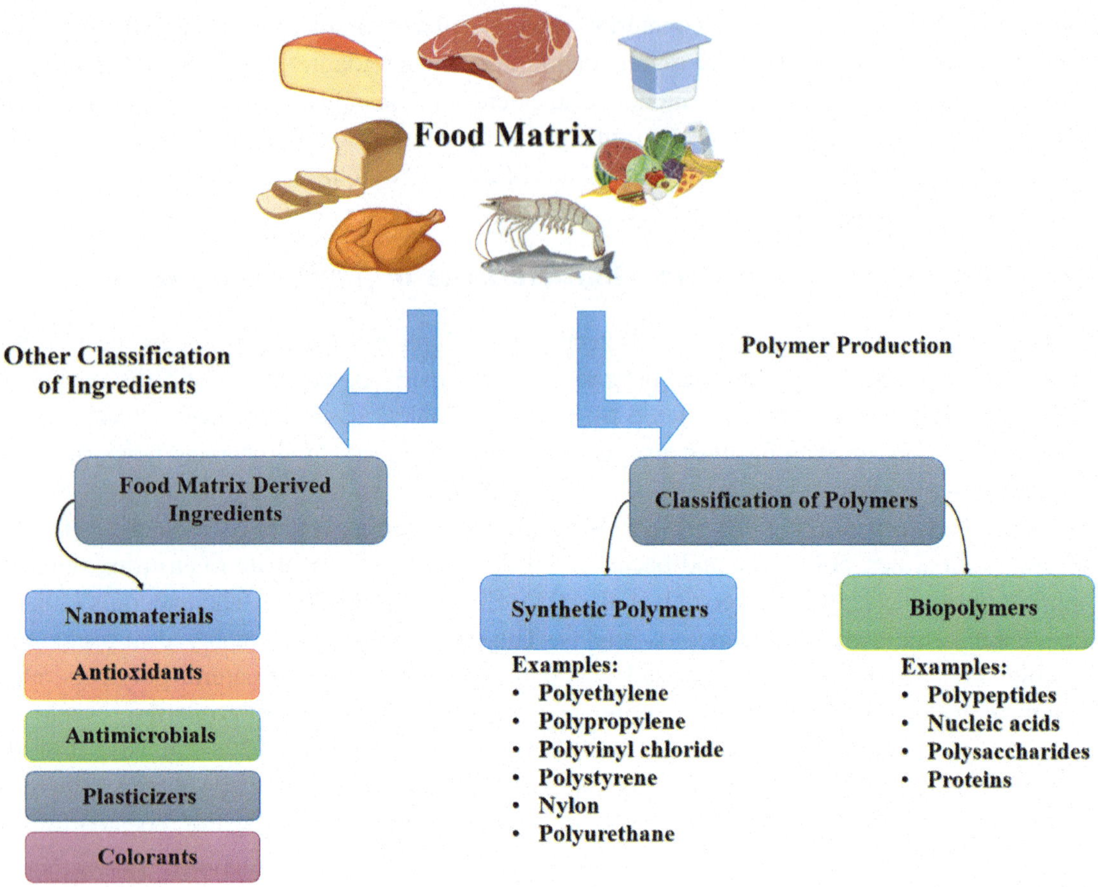

Figure 10.1

Overview of food matrix components and polymer classifications relevant to food packaging. The diagram shows food-derived ingredients alongside synthetic polymers and biopolymers used in packaging materials.

No permission required.

particularly their interaction with the environment. One of the most significant issues is the potential for migration from packaging materials to the food they contain (Gupta et al., 2024; Jafarzadeh et al., 2023). Other enhancing substances include enzymes that are crucial for biochemical processes, such as those utilized in the packaging industry, including laccase, which is present in bacteria, fungi, insects, and plants. Laccase serves as an effective option for creating transverse connections between the polymer matrix and fillers. Additionally, laccase can be incorporated into smart packaging sensors (Azizi-Lalabadi et al., 2023). However, enzymes have drawbacks; for instance, they may accelerate reactions, some of which could be detrimental (Abedi-Firoozjah et al., 2023; Alizadeh Sani et al., 2024).

Other materials utilized in the packaging industry include antimicrobial agents that enhance shelf life, ensure food safety, and inhibit the growth of microorganisms in food, fruits, and vegetables. Given that spoilage occurs rapidly, these substances can effectively prevent deterioration (Upadhyay et al., 2024). Antioxidants are also among the enhanced materials employed in packaging, particularly for meat. They can improve the stability of packaging and function as free radical scavengers; in the meat industry, they are directly incorporated into meat packaging to eliminate free radicals (Barzan et al., 2024). Stabilizers represent another additive used in the packaging sector. These materials stabilize intermolecular forces and prevent molecular dispersion. In fact, stabilizers are employed to enhance strength (Kane et al., 2024). Another additive found in food packaging is glycerol, which serves as a softener. Glycerol is a widely used plasticizer for polymers that enhances flexibility, diminishes intermolecular forces, boosts packing strength, and facilitates molding and shaping (Benitez et al., 2024). Colors are not the least important additives in food packaging, particularly in smart packaging. Colors are incorporated into the packaging to observe color changes. Various colors are utilized in packaging; for instance, anthocyanins are among the most significant color substances in the packaging industry. The addition of colorants to packaging materials, beyond facilitating color change, influences physical properties such as mechanical strength, solubility, permeability, and optical and thermal characteristics (Razmjoo et al., 2022; Xu et al., 2024). In the following sections, the key properties of food packaging are thoroughly evaluated, including mechanical properties, barrier properties, thermal properties, resistance and durability, biodegradability and compostability characteristics, and antimicrobial and antioxidant properties.

10.2.1 Mechanical properties

Mechanical properties are the characteristics that dictate how materials behave; they reflect the response of materials to applied forces, which varies for each material and is significantly influenced by factors such as temperature, applied force, deformation, and deformation rate. These properties are crucial for protecting food during storage and distribution. Numerous factors impact the mechanical properties of biopolymer layers, including the strength and type of interactions, as well as their quantity (Alizadeh Sani et al., 2022; Khezerlou et al., 2021). This was a general definition of mechanical properties; however, defining these properties is not an easy task and requires samples and experiments. Numerous mechanical properties assist in the design of biopolymer materials. Many beneficial changes in the packaging industry, particularly in active and intelligent packaging, stem from the enhancement of mechanical properties (Udayakumar et al., 2021). Evaluating the mechanical properties of food packaging is essential, as it enables the distribution of food products without potential losses during transportation. Packaging materials must possess mechanical strength and flexibility (Shah et al., 2023). The most critical mechanical properties include tensile strength and flexibility, which are explained respectively. There are various types, but the most significant ones that are assessed are tensile strength and flexibility. Tensile strength refers to the maximum load that a material can endure without breaking during stretching and is measured as a force against the surface. In essence, it represents the maximum stress that the material can tolerate before failure, indicating how far a polymer material can stretch. The chemical composition of packaging materials influences tensile strength, and an effective package is one whose materials possess strength against stretching and tearing while also being able to withstand stretching up to the standard level. The final tensile strength is typically determined by conducting tensile and stress tests in unit strain, which corresponds to the peak of the stress–strain curve, a compressive property that is affected by factors such as surface presence or absence, temperature, and more (Shah et al., 2023). Flexibility is the ability of a material to change its elastic shape and return to its original shape by removing the applied stress. It is an important feature because it allows materials to return to their original state when a force is applied to them. Being flexible in the food packaging industry has many advantages, such as formability, lightweight, heat-sealing ability, and so on; sealing is an important advantage of flexible packaging. The flexibility of packaging, owing to its lightweight nature, results in reduced energy consumption during transportation (Bamps et al., 2023). The following are some examples of mechanical features in different packagings.

The integration of certain types of compounds in the packaging may increase or decrease their mechanical properties (Table 10.1). For example, according to Khezerlou et al.'s (2021) report on the mechanical properties of whey protein layers, adding alpha-tocopherol nanoemulsion to whey reduces its flexibility because nanoparticles have emulsifying properties and soften the material, as well as adding alpha-tocopherol nanocapsules reduces tensile strength because alpha-tocopherol is a type of vitamin E and reduces the strength of polymer layers. Also, the addition of lecithin-based nanocapsules creates softer materials with lower tensile strength. Indeed, adding lecithin as an emulsion agent to nanoparticles and packaging materials softens the packaging material, and as a result, the tensile strength decreases (Khezerlou et al., 2021). The combination of essential oils weakens the network structure by breaking the intermolecular interactions. According to the report of Azizi and Almasi (2018), the combination of thyme extract nanoliposomes within whey protein layers alters the mechanical properties, resulting in reduced tensile strength due to the addition of essential oils and extracts, which significantly impacts the moisture content of the material. Another report by Fathi and Ranjbarian (2018) stated that the addition of 5% oil nanoemulsion increases the tensile strength (Ranjbaryan et al., 2019). Increasing tensile strength is due to the antioxidant properties. According to the studies of Jamróz et al. (2018), adding lavender to starch reduces its tensile strength, and finally, this reduction causes a decrease in mechanical resistance. Starch by itself leads to an increase in tensile strength, but since lavender is an essential oil, it increases the adhesion force, ultimately causing a decrease in tensile strength (Jamróz et al., 2018). As regards the studies of Oladzadabbasabadi et al. (2023), the composition of carbon dots (CDs) obtained from biomass can be a reinforcing

Table 10.1 Effect of different polymers on mechanical properties.

Polymer	Bioactive materials	Concentration	Production method	Tensile strength (TS)	Elongation at break	Effect on polymer matrix	References
Cassava starch	Lycopene	0%	Casting	3.09 ± 0.10	134.59 ± 2.69	Increasing TS due to better solubility and antioxidant properties	Assis et al. (2017)
Chitosan	Epigallocatechingallat	–	Casting	6.44 ± 0.28	22.5 ± 4.3	Increase due to strong mechanical properties	Chen et al. (2016)
Chitosan	Green tea extract	50%	Casting	5.03 ± 2.06	28	Increasing TS due to the increase in distance between molecular chains and decrease in hydrogen bonding	Liang et al. (2017)
HPMC	Nisin	5%	CCDS (constant cell disruptor system)	37.0 ± 2.5	2.6 ± 0.7	Reduction of TS due to emulsifying properties	Imran et al. (2012)
HPMC	Lecithin	30%	Solution	22.5 ± 2.8	75.7 ± 3.4	Reduction due to emulsifying properties	Imran et al. (2012)
Polylactic acid	Rosemary	1.5%	Cast-extruded	3880 ± 80	3700 ± 80	Increased TS due to antioxidant properties	Cerdá-Gandia et al. (2024)
Cassava starch	Bixin	5%	–	8.95 ± 1.32	15.55 ± 1.14	Bixin can create a three-dimensional structure and also has high strength so that it can increase the TS	Chen et al. (2016)
Chitosan	Cinnamaldehyde	0.6%	Casting	37.42 ± 4.02	14.5 ± 2.9	Due to the strong molecular structure, TS increases	Chen et al. (2016)
Quinoa protein/ chitosa	Thymol	10%	Chilean	2.9 ± 0.5	98 ± 1	Due to the possibility of chemical reactions, TS decreases	Robledo et al. (2018)

No permission required.

Table 10.2 Carbon nanotube sources with different properties.

Polymer	Carbon sources	Concentration	Methods of obtaining	Main properties	Effect on polymer matrix	References
Gelatin	Potato peel	74%	Carbonization	Polymer hardening	Gelatin/carbon dots have anti-UV, antioxidant, and antimicrobial properties	Oladzadabbasabadi et al. (2023)
Cellulose	Glucose	4.1%	Hydrothermal	UV-blocking and antimicrobial	Reduction of tensile strength due to lack of stability	Oladzadabbasabadi et al. (2023)
Silk sericin protein	Silkworm cocoon	54.7%	Hydrothermal	Due to their elasticity and tensile strength, they improve mechanical properties.	Anticounterfeiting, antioxidant, and antimicrobial	Oladzadabbasabadi et al. (2023)
Gelatin	Enoki mushroom	6.4%	Hydrothermal	Antimicrobial	Polymer hardening	
Polyaniline	Citric acid	1%	Hydrothermal	Strong fluorescence	TS increases due to hydrogen bonding	Robledo et al. (2018)
Polyvinyl alcohol	Citric acid	1 –0.5%	Hydrothermal	Thermal stability	TS increases due to the creation of hydrogen bonds and a decrease in humidity	Robledo et al. (2018)
Polyethylene glycol	Glucose	1%	Carbonization or hydrothermal	Strong fluorescence	Increasing TS and flexibility due to the creation of coherent molecular networks	Robledo et al. (2018)
Polycarbonate	Citric acid	0.1%–0.5%	Hydrothermal	Thermal stability and Strong fluorescence	Increase of TS and flexibility due to the formation of hydrogenous bonds and maintenance of flexibility	Robledo et al. (2018)
Polycarbonate	Glucose	1%	Hydrothermal	Thermal stability and Strong fluorescence	Increased flexibility and TS	Robledo et al. (2018)

TS, Tensile strength; *UV*, ultraviolet.

No permission required.

agent that increases tensile strength and flexibility because it can cause the integration of mechanical properties. CDs play one of the most vital roles and has wide applications in the packaging industry (Table 10.2). One of its applications is to increase mechanical properties.

10.2.2 Barrier properties

The barrier properties of packaging materials are essential for protecting certain products, such as moisture-sensitive foods and electronic devices. This feature refers to the packaging's ability to resist the absorption of light, moisture, and oxygen. The barrier properties are necessary to maintain the quality and safety of food. Packaging systems with optimal barrier properties protect their contents against gases, water vapor, light, and aromatic compounds (Azimi-Salim et al., 2022; Marano et al., 2022). Packaging materials should exhibit low permeability to oxygen and moisture vapor to increase shelf life. The inhibitory property is very important for packaging materials because they are hydrophilic and not effective at inhibiting water vapor. Consequently, exposure to water vapor reduces the quality of food and its shelf life. One of the notable bioplastics that are highly resistant to water vapor and gas is polyglycolic acid. The inhibitory properties of polyglycolic acid could be enhanced by incorporating additives such as nanoclays, pectin, and others. Indeed, these materials positively influence the enhancement of inhibitory characteristics (Kumari et al., 2024). The permeability of food packaging can change the color, taste, and texture of food due to the occurrence of oxidation. Therefore, improving the resistance of food packaging can extend the shelf life of food and preserve its properties (Azizi-Lalabadi et al., 2021; Tyagi et al., 2021). Weak barrier properties in biopolymers confine their usability in the packaging industry. In addition to polyglycolic acid, which has been mentioned, polyethylene terephthalate is also an effective inhibitor against oxygen. Polyhydroxybutyrate is another biopolymer that has strong barrier properties. However, due to its weak mechanical strength, it is not widely applied in the food packaging industry. In the following, the inhibition characteristics of packaging materials against various factors are examined.

The oxygen barrier is determined by oxygen permeability, which measures the amount of oxygen that penetrates a unit area over time through a packaging material at a specific temperature and relative humidity. Among synthetic polymers, polyvinyl alcohol (PVA) serves as an excellent inhibitor against oxygen. In contrast, starch, a biopolymer, exhibits the lowest permeability to oxygen. However, compared to PVA, its permeability remains significantly higher. So, it can be stated that the permeability of synthetic polymers is lower than that of biopolymers. To address the issue of starch's oxygen permeability, Patil et al. (2021) concluded that mixing starch with PVA enhances the inhibitory properties of starch, which is due to the creation of strong hydrogen bonds that restrict the penetration of oxygen. Calculation of water vapor permeability in the packaging system is crucial. Numerous standards exist for determining water vapor permeability. Many polymers, especially polar ones, exhibit nonideal behavior for water vapor penetration. Due to the frequent exposure of products to moisture, humidity is a critical factor that should be considered. Native starches are naturally very hydrophilic, rendering their films highly sensitive to moisture. According to the studies of Trinh et al. (2021), the addition of nanoclays as nanofillers improves permeability to moisture because nanofillers alter the path of water diffusion. Gas barrier food packaging is designed to limit the transfer of gases into or out of the food packaging. This technology relies on chain motion to prevent oxidation-induced spoilage loss of gas diffusion within the polymer. Therefore, it can only occur in the amorphous region, making the crystallinity of polymers significantly beneficial for gas impermeability (Qin et al., 2021). Additionally, moisture control or microbial growth plays an important role in maintaining the quality, freshness and shelf life of food. The diffusion of gas inside the polymer depends on the movement of the chain (Liu et al., 2024). According to the studies of Tuyet Thury et al., in 2021, they were able to reduce gas transmission by adding cellulose nanofibers, chitosan nanowires, and sunflower oil to polyethylene terephthalate due to the interactions of these crystalline nanomaterials. Chitosan, a natural polymer derived from chitin, possesses antibacterial and antioxidant properties. When in nanoparticle form, its increased surface area enhances its ability to prevent the penetration of gases such as oxygen and carbon dioxide into the packaging. These features preserve the freshness and quality of food, while sunflower oil acts as a moisture barrier and aids in creating a natural shield against gas penetration. Furthermore, the fats in this oil can inhibit gas exchange by forming a layer on the package's surface. Therefore, the combination of these two substances in a nano and natural form creates an effective barrier against gases that protect against oxidation and reduce the quality of products (Tyagi et al., 2021).

Aromatic compounds are composed of organic compounds such as carbon, hydrogen, and oxygen and, in some cases, nitrogen and sulfur derivatives. These are volatile compounds that are extracted from different parts of plants, including bark, seeds, flowers, bark, and fruit. Aromatic compounds, or arenes, typically refer to organic compounds characterized by benzene and cyclically conjugated structures. According to research by Moeini et al. (2022), chitosan, a biopolymer widely utilized across various fields, exhibits weak inhibition against aromatic compounds, which can also harm the food matrix. Consequently, incorporating nanoparticles into aromatic compounds and integrating them with chitosan enhanced its inhibitory effects because of the antioxidant properties of nanoparticles. Light barrier food packaging plays an important role in maintaining the quality and safety of food products. This type of packaging utilizes materials or coatings specifically designed to block light, especially harmful ultraviolet (UV) rays that can reduce the quality of food. These rays can negatively impact the color, taste, aroma, and nutritional value of the food. Common materials used in light barrier packaging include aluminum foil, opaque films, and specialized polymers, all aimed at shielding products from light exposure. The advantages of light-blocking polymers and biopolymers include increasing shelf life, maintaining freshness, and protecting sensitive nutrients such as vitamins. It is especially important for products such as dairy products, oils, and beverages that are highly susceptible to light degradation (Marangoni Júnior et al., 2020). Based on Khalid and Arif's (2022) research, different methods have been proposed to improve the light-blocking properties of biopolymers used in food packaging. The addition of nanoparticles, such as zinc oxide or titanium dioxide, can improve the blocking properties against light and UV. Multilayered films that combine proteins, lipids, and polysaccharides can further increase light inhibition (Khalid and Arif, 2022). According to Bhaskar et al. (2023), incorporating active substances such as antioxidants or photosensitive pigments can reduce the harmful effects of light. These materials can function as natural filters or can improve light and moisture resistance through surface modification of biopolymer layers via cross-linking or coating with light-resistant materials such as polyethylene glycol (Bhaskar et al., 2023). In Fig. 10.2, a general schematic of the barrier properties and durability characteristics of polymers is exhibited.

10.2.3 Impact resistance and durability

Impact resistance and durability are two important features in the biopolymer-based packaging industry. Packaging should be able to protect food from physical damage and environmental conditions while also being sustainable and environmentally friendly

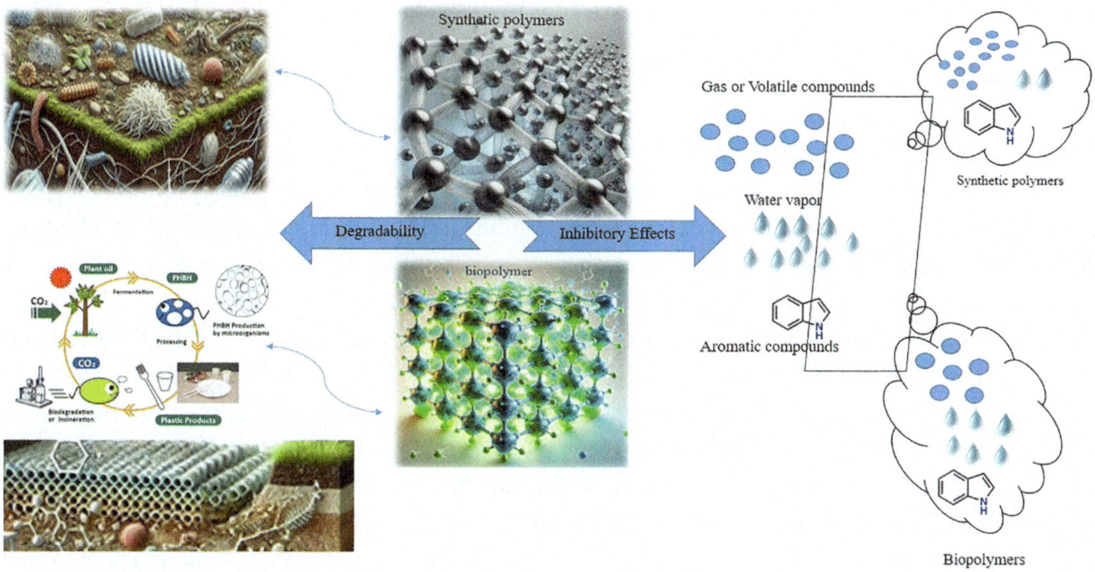

Figure 10.2

Schematic presentation of barrier properties and degradation characteristics in biopolymers and synthetic polymers.

No permission required.

(Azizi-Lalabadi et al., 2020; Blanco et al., 2020). Impact resistance refers to the packaging's ability to absorb energy and prevent biopolymer chain failure from pressure and impact (Fig. 10.2). A number of factors influence this characteristic, including hardness, which indicates a material's resistance to deformation; toughness, which reflects a material's capacity to absorb energy before failure; ductility, which denotes the ability to deform without breaking; and the microscopic structure that reveals the distribution of particles. Charpy and Izod tests are also conducted to measure resistance. Durability refers to the ability of packaging to withstand environmental conditions such as humidity, temperature, oxygen, and other environmental factors. It also depends on various aspects, including corrosion resistance, wear resistance, thermal stability, and quality of construction and raw materials. Tensile, compression, and fatigue tests are employed to assess durability. Biopolymers tend to be more sensitive to environmental factors than synthetic polymers (Blanco et al., 2020). Enhanced impact resistance and durability offer advantages, such as maintaining quality by preventing moisture or gas ingress, thereby reducing waste to food damage during transport. This also increases durability in harsh conditions. Since biopolymers used in the packaging industry, like PLA and PHA, are generally more brittle than synthetic polymers, it is essential to find ways to improve this property (Basavegowda and Baek, 2021; Blanco et al., 2020). A series of solutions were considered to improve corrosion resistance and durability properties (Basavegowda and Baek, 2021; Wu et al., 2021). For instance, Basavegowda and Baek (2021) found that adding flexible polymers, such as PLA with polybutylene succinate, can increase impact resistance. Another method involved combining polycaprolactone (PCL) to improve impact resistance. Since the microscopic structure affects the impact resistance and durability, nanotechnology plays an important role in enhancing the properties of biopolymers by altering their microscopic structure. In the study, incorporating cellulose nanofibers into biopolymers strengthens the matrix and better distributes stresses, while adding nanoclay and silicate nanoparticles prevents gas and moisture penetration, ultimately increasing the durability of biopolymers (Basavegowda and Baek, 2021). Softeners are additive chemical agents that enhance impact resistance and prevent moisture penetration through the interaction of softener and biopolymer molecules. According to research by Muobom et al. (2020), softener citrates exhibit a favorable combination that renders the biopolymer resistant to water and moisture, thereby enhancing its durability. Studies by Li et al. (2022) indicate that after conducting a tensile test, the applications of nanocellulose (including nanofibers or nanocellulose crystals) in these composites is a primary method for strengthening bio-packaging due to its high mechanical strength, biodegradability, and suitable thermal properties. Additionally, research by Basavegowda and Baek (2021) et al. shows that through impact and tension tests, the combination of biopolymers with inorganic nanoparticles (such as metal oxides) can significantly improve packaging durability. Biswas et al. (2022) conducted similar research, indicating that techniques such as reactive extrusion and surface modification can enhance the internal structure of biopolymers and raise impact resistance by creating a unified structure, ensuring uniform material distribution, enabling the formation of multiple layers, and improving heat resistance. Utilizing advanced methods can also enhance impact resistance and durability; according to Jospeh et al., chemical modifications like cross-linking or the addition of specific functional groups can significantly improve the impact resistance and durability of biopolymers by reducing permeability and enhancing thermal stability (Joseph and George, 2023). Additionally, refining the production process is another approach to enhance impact resistance and durability. Basavegowda and his colleagues found that the stretching or orientation process, which adjusts the direction of molecules during the production of biopolymer films, can increase strength. For instance, controlled polymerization yields polymers with suitable molecular weights that exhibit better impact resistance and durability (Basavegowda and Baek, 2021). In conclusion, biopolymers typically demonstrate good impact resistance and durability, which can be attributed to their properties, such as flexibility and resilience. However, their performance can vary significantly depending on the specific type of biopolymer and its formulation. Enhancements can also be realized through blending with other materials or through chemical modifications to optimize their mechanical properties. It is essential to conduct specific tests to assess the impact resistance and durability according to the intended application, as environmental factors can also play a vital role.

10.2.3.1 Biodegradability and compostability

In recent years, there has been a growing focus on biodegradability in food packaging. This characteristic allows materials to decompose naturally after use without causing harm to the environment. Biodegradability refers to the ability of a substance to be broken down by microorganisms such as bacteria and fungi. This process may occur in different environments, such as soil, water, or landfills, and the time required for decomposition can differ. Utilizing biodegradable materials in food packaging can aid in reducing plastic pollution and conserving natural resources. These materials are usually derived from renewable sources such as plants and naturally decompose after use. Although biodegradability offers advantages, it also presents challenges. Some biodegradable materials may only decompose under certain conditions, and without those conditions, they may need more time to decompose. Additionally, the production costs of these materials may exceed those of traditional plastics (Chen et al., 2024). Compostability is a subset of biodegradability, referring to a material's ability to decompose under specific industrial or home composting conditions. Compostable materials must break down within a certain time and under specific conditions, such as high temperature, humidity, and microorganisms. These materials usually decompose more quickly than biodegradable ones, transforming into compost that enriches soil (Nazareth et al., 2022). The following will examine more extensive examples of biodegradable and compostable packaging in the food industry. Fig. 10.3 illustrates an overview of biodegradable and compostable food packaging materials and their degradation pathways.

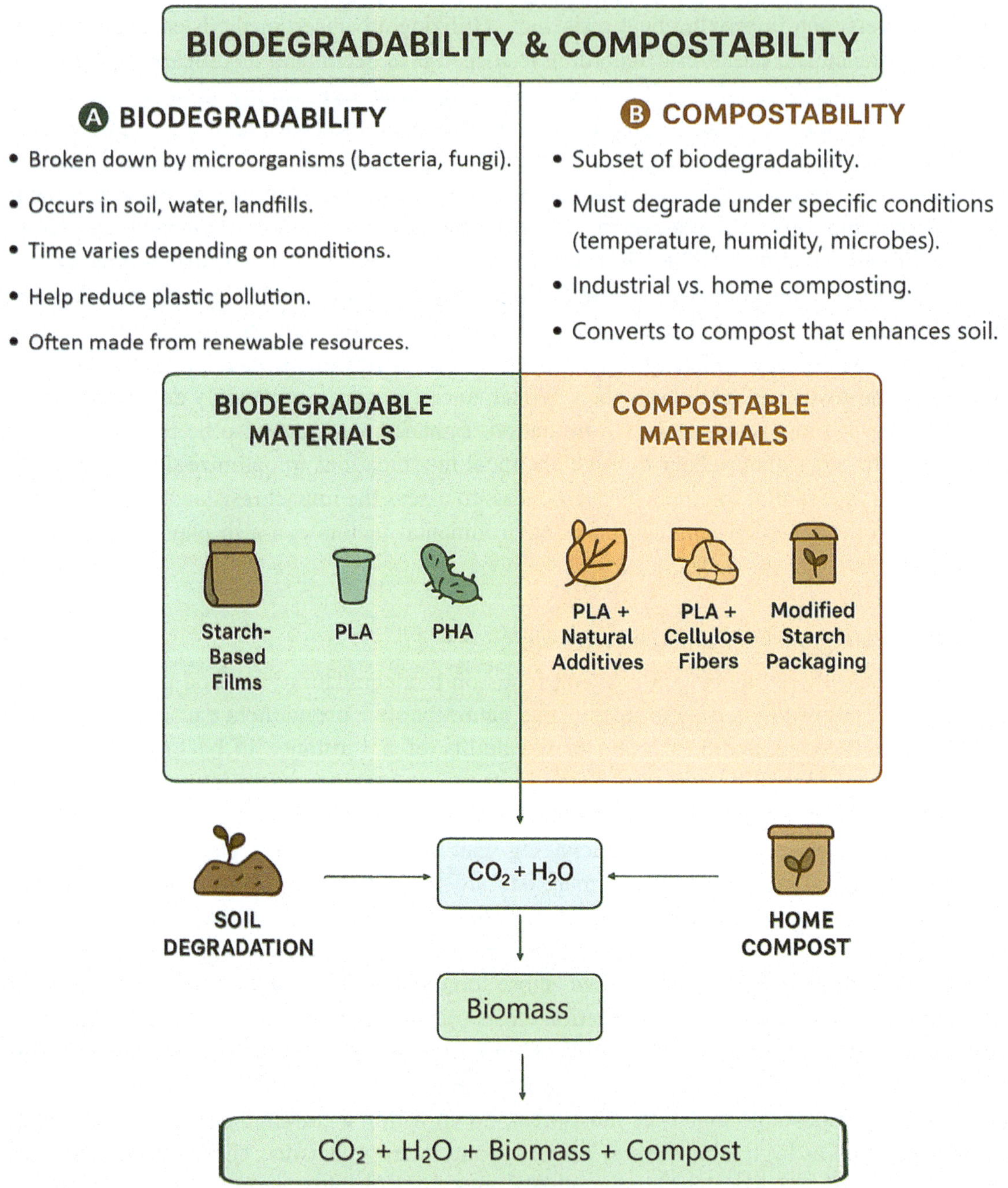

Figure 10.3

Overview of biodegradable and compostable food packaging materials and their degradation pathways. The figure compares biodegradable and compostable polymers, highlights key examples such as starch films, polylactic acid (PLA), polyhydroxyalkanoate (PHA), and PLA–cellulose blends, and illustrates their breakdown into CO_2, water, and biomass under soil or composting conditions.

No permission required.

Many brands use ingredients such as starch and vegetable proteins. This type of packaging is especially common for low-fat and natural products. For example, research by Surendren et al. (2022) indicates that packaging made from corn starch has gained popularity in the food industry due to its natural ability to degrade and minimize environmental harm. PLA is one of the widely used biodegradable biopolymers in food packaging. A recent innovation involves improving PLA properties by incorporating biological additives that accelerate the decomposition rate and ameliorate resistance to various environmental conditions. These improvements will also contribute to reducing energy consumption during production. For example, according to the studies of Kalita et al. (2021) show that PLA packaging containing natural materials such as corn flour or vegetable fibers can decompose in home composts, and the decomposition rate increases in air environments. PHA is one of the biopolymers produced by microorganisms in certain environments. Recent innovations in this field have enabled PHA to decompose in different environments such as water, soil, and sea. For example, research by Gegia et al. (2021) exhibits that food packaging made from PHA has improved chemical compositions. It is now possible for these substances to decompose even in industrial environments or lakes, which were previously unfeasible (Gioia et al., 2021). One of the innovative ways to improve compostability is to combine biodegradable biopolymers with natural materials such as starch or plant proteins. These compounds can increase compostability and biodegradability properties while simultaneously improving the efficiency of food packaging. For example, according to Momeni et al. (2023), the use of cellulose fibers in combination with PLA can create packages that decompose under home composting conditions while maintaining adequate strength. A notable advancement in biodegradable packaging involves modified starch, which can completely decompose in natural environments. These materials, combined with natural additives or biopolymers, can produce biodegradable and compostable packaging, according to research by Gutiérrez and Valencia (2021), requiring less energy during the production process.

10.2.4 Thermal properties

The thermal properties of biopolymers are important in the application of food packaging and other industries. Good thermal properties prevent the decomposition of biopolymers; those with good thermal characteristics, such as PLA, can endure high temperatures without breaking down. This capability enables them to tolerate processes, that is, extrusion and molding. Thermal properties play a vital role in determining the performance and application of packaging materials. The presence of this feature in biopolymers is necessary for maintaining structural integrity and preventing their degradation at high temperatures, thereby preserving food quality. It can also be formed to create stable edible films (Shah et al., 2024). Biopolymers show different thermal properties that are influenced by their structure and chemical composition. The thermal properties of biopolymers include the glass transition temperature (T_g), which is the temperature at which the biopolymer changes from a hard, glassy state to a soft, elastic state. This temperature is different for each biopolymer. This temperature indicates the change in the mobility of the molecular chains of the polymer. In biopolymers, T_g is influenced by the structure of polymers, molecular weight and the presence of side groups or additives. This features are crucial in the design and application of polymers, especially for food packaging. Additionally, T_g plays an important role in determining the performance, durability, and function of biopolymers across various industries. T_g varies among different biopolymers, that is, PLA has a T_g of approximately 65°C–55°C, while PHA ranges from −35°C to 10°C (Shah et al., 2024). Moreover, the melting temperature (T_m) is the temperature at which a biopolymer melts, indicating its thermal stability at elevated temperatures. Specifically, it is the temperature at which the biopolymers transition from a crystalline or semicrystalline state to a liquid or amorphous state. Several processes, such as extrusion, injection, and three-dimensional printing, depend on T_m, which can also influence the mechanical and biological properties of the polymers. Each biopolymer has a specific melting temperature. For example, PLA melts at around 150°C–180°C, and PCL melts at approximately 60°C–65°C (Fambri and Migliaresi, 2022).

Additionally, PLA has a thermal degradation temperature of approximately 200°C–250°C, while that of cellulose exceeds 250°C. Specific heat capacity refers to the amount of heat needed to raise the temperature of 1 g of a substance by 1°C, playing a crucial role in heat storage or conduction. Thermal conductivity measures a material's ability to conduct heat, and biopolymers usually show low thermal conductivity, making them suitable for insulation applications. Thermal stability refers to a material's ability to maintain its mechanical and chemical properties at high temperatures. For instance, biopolymers such as chitosan and polycarbonate exhibit greater resistance to thermal degradation. These characteristics enable biopolymers to be used in a wide range of applications, including biodegradable packaging, medicine, and insulation (Perera et al., 2023). The thermal properties of biopolymers play an important role in determining their functionality. Alongside biodegradability, these properties positively influence packaging, necessitating solutions to enhance them. A series of methods exist to improve thermal properties, facilitating the use of biopolymers in broader applications and at higher temperatures, as detailed below. It is possible to improve the thermal properties of biopolymers through changes in chemical structure, production processes, and the incorporation of reinforcing materials. Several methods have been proposed to improve the thermal properties, and chemical structure modification is one of these methods. For example, the copolymerization process is an important method for altering the thermal properties of biopolymers, as it can significantly affect their thermal properties. In this process, two or more monomers with different properties are polymerized together to obtain polymers with combined attributes that help improve T_g and thermal stability. According to the research by Taherimehr et al. (2021), the copolymerization of PLA with PCL enhances T_g, ultimately reducing brittleness and increasing flexibility.

One of the ways to increase the crystallinity is to improve the thermal properties. Crystallization is a process in which the molecular chains of the biopolymer transition from an irregular or amorphous state to a regular and crystalline structure under the appropriate conditions, which are beyond T_m and above T_g. Factors such as molecular weight and additives like softeners affect crystallization. The process of crystallization encompasses the steps of nucleation, crystal growth, and the cessation of growth. Crystallization significantly affects the thermal properties of biopolymers. Biopolymers' thermal stability refers to the ability of these materials to maintain their structure and physicochemical properties at high temperatures. These characteristics are linked to their chemical composition. Indeed, the stronger the covalent bonds, the more stable the biopolymers tend to be, particularly those with a higher crystalline structure. These biopolymers exhibit greater heat stability, and certain additives can also enhance their heat resistance. In general, it improves their thermal properties. For example, it increases T_g and T_m, increases thermal stability, reduces thermal expansion, and increases the resistance of biopolymers to thermal degradation. According to the study of Dong et al., the incorporation of nanoparticles such as silica in chitosan increases crystallinity and ultimately improves thermal properties (Dong et al., 2022). Another approach to improve thermal properties is to combine biopolymers. For example, research by Shah et al. (2024) indicates that the combination of PLA with polybutylene adipate terephthalate increases the thermal properties of biopolymers. Creating composite structures is another method to improve thermal properties. According to Mary et al. (2022), incorporating carbon fibers into PLA improves the thermal characteristics of biopolymers. Adding plasticizers to biopolymers can also improve their thermal properties. Plasticizers can reduce the T_g and increase the flexibility and softness of the biopolymer. This reduction in T_g can enhance the resistance of the biopolymer to high temperatures. Zhang et al. (2023) investigates the addition of a natural plasticizer, such as citric acid, to improve the thermal properties of chitosan. Incorporating thermal stabilizers like antioxidants into biopolymers, especially at high temperatures, can positively affect their thermal properties, preventing oxidation and thermal degradation of biopolymers. As regards the research of de Lima Silva et al. (2021), it was found that adding total antioxidant capacity to PVA increases its thermal stability. Adding nanoparticles can ameliorate the thermal properties of the biopolymer. They can increase the thermal resistance of the biopolymer while also leading to a decrease in T_g. Additionally, they can act as a protective layer, preventing thermal degradation of the biopolymer at elevated temperatures. Hakeem et al. (2020) evaluated the addition of ZnO into cellulose, which improves thermal properties due to its antioxidant characteristics and the formation of a protective layer on the antioxidant surface (Hakeem et al., 2020). T_g and T_m of different biopolymers have been evaluated in Table 10.3.

10.2.5 Antimicrobial and antioxidant properties

Biopolymers, as a sustainable alternative to traditional plastics in food packaging, have gained significant attention due to their biodegradability, food safety, and ability to enhance functional properties. One of the most important advantages of these materials is their

Table 10.3 Glass transition temperature (T_g) and melting temperature (T_m) of various biopolymers.

Biopolymer	Properties	Formula	T_g	T_m	References
Polylactic acid (PLA)	Biodegradable, strong mechanical properties, good thermal performance, dyeability, and suitable for food and medicine packaging	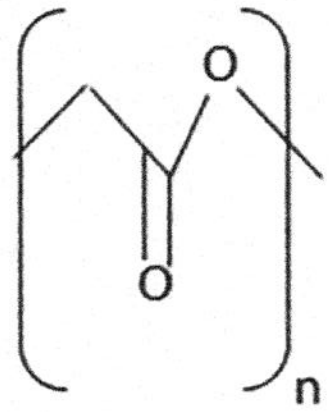	60	160–180	Patel et al. (2022)
Polyhydroxyl butyrate (PHB)	Biodegradable, high resistance to moisture, and strong mechanical properties, thermal properties, and chemical resistance	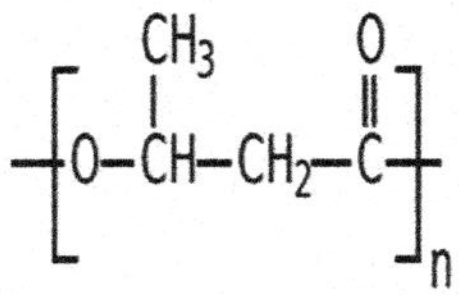	0	170–180	Patel et al. (2022)
Polyhydroxyalkanoate (PHA)	Biodegradable, good thermal properties, used in food packaging and medical items, printability, and biological diversity	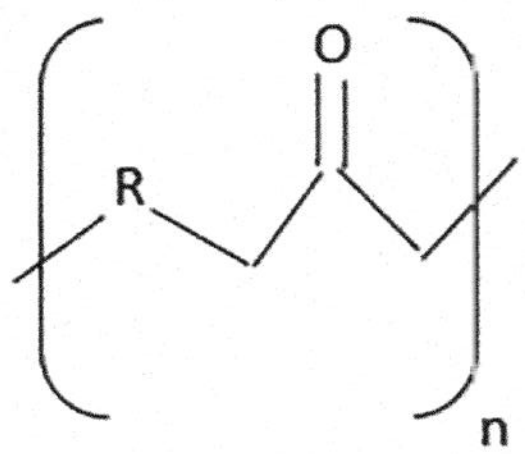	40–60	170–180	Ishak et al. (2024)
Starch	Biodegradable and a source of carbohydrates	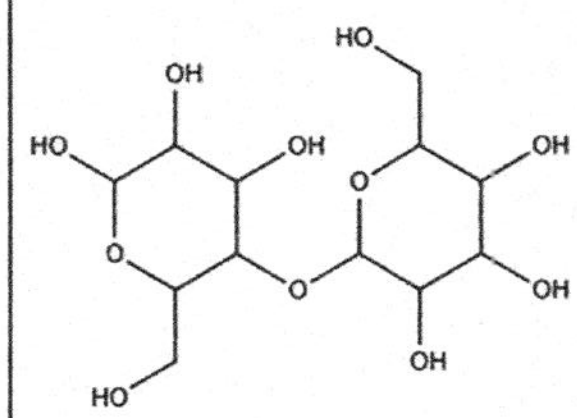	50	70–150	Othman et al. (2024)
Natural rubber (NR)	Elastic resistant to water and humidity	$(C_5H_8)_n$	0	200	Chueangchayaphan et al. (2024)

Cellulose	Biodegradable and resistant to chemicals	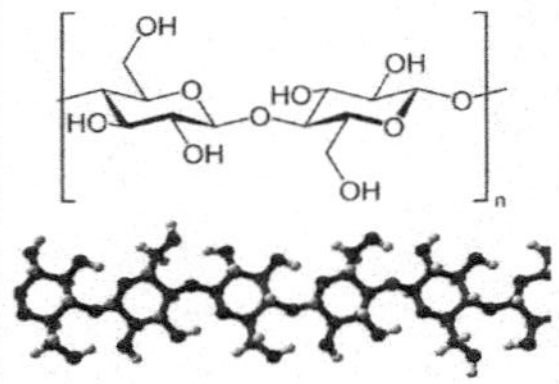	160	250–350	Liu et al. (2021)
Casein	Protein (milk index protein)	$C_{81}H_{125}N_{23}O_{39}P$	70–90	220–230	Sahin et al. (2023)
Chitosan	Biodegradable and antibacterial	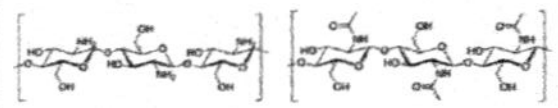	150	215	Karydis-Messinis et al. (2024)
Polyglycolic acid (PGA)	Suitable for use in medicine	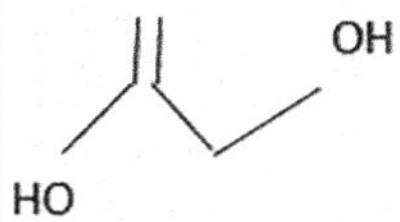	35–40	220	Wang et al. (2021)
Bio-polyurethanes	Elastic and high wear resistance	Variable	50–100	150–200	Chen et al. (2024)

No permission required.

antioxidant and antimicrobial properties, which can play a crucial role in extending the shelf life of food, preventing spoilage, and preserving its color, flavor, and nutritional value (Azizi-Lalabadi et al., 2020; Moeini et al., 2021). Biopolymers in food packaging can play a significant role in preserving quality and extending the shelf life of products due to their antioxidant properties (Azizi-Lalabadi et al., 2020; Moeini et al., 2021).

Oxidation is one of the main factors that degrade food quality, leading to color changes, loss of nutritional value, unpleasant flavors, and the formation of harmful compounds. Biopolymer-based packaging, by incorporating natural antioxidant materials, can prevent this process. These compounds work by neutralizing free radicals, preventing lipid peroxidation, and reducing the effect of metal ions that act as catalysts for oxidative reactions, thereby enhancing the stability of food products. Additionally, the use of such packaging can reduce the need for chemical additives and contribute to consumer health and environmental sustainability. On the other side, the antimicrobial properties of biopolymers in food packaging play a crucial role in inhibiting the growth of harmful microorganisms, including bacteria, fungi, and yeasts, which ultimately enhances the shelf life and safety of food products. Biopolymers naturally possess antimicrobial agents that either block microbial metabolic functions or damage their cell structures. This effect can be further improved through the incorporation of natural compounds, such as essential oils from plants (like cinnamon, rosemary, and garlic), metal nanoparticles (including silver and zinc), or plant extracts (Fig. 10.4). These substances work by disrupting essential processes in microorganisms like, interference with membrane permeability and cell wall, thereby slowing their growth and preventing food contamination. Utilizing antimicrobial biopolymers in food packaging minimizes the reliance on chemical preservatives, thus promoting consumer health and mitigating negative effects on the environment (Deshmukh and Gaikwad, 2024). There are a number of solutions that can improve the antimicrobial and antioxidant properties, as explained below.

Essential oils are low-molecular-weight plant metabolites that possess various properties, including antimicrobial and antioxidant effects. Cinnamon essential oil is recognized as one of the widely used essential oils in the packaging industry, imparting significant antimicrobial features to biopolymers. For instance, according to research conducted by Mohammadi et al. (2020), an active packaging system was designed using chitosan nanofibers and cinnamon essential oil in the production of whey protein films. The results of this study demonstrated the inhibitory effect of cinnamon essential oil on harmful bacteria such as *Staphylococcus aureus* and *Escherichia coli* (Mohammadi et al., 2020). Phenolic compounds present in rosemary oil have the ability to inhibit bacteria, fungi, and molds. This property leads to a reduction in the growth of microorganisms in food and increases its safety. According to research done by Moieni et al. (2021), the addition of rosemary essential oil to cellulose significantly reduces the number of *Pseudomonas* bacteria. This protective effect is attributable to the strong antimicrobial activity of phenolic compounds, which can affect

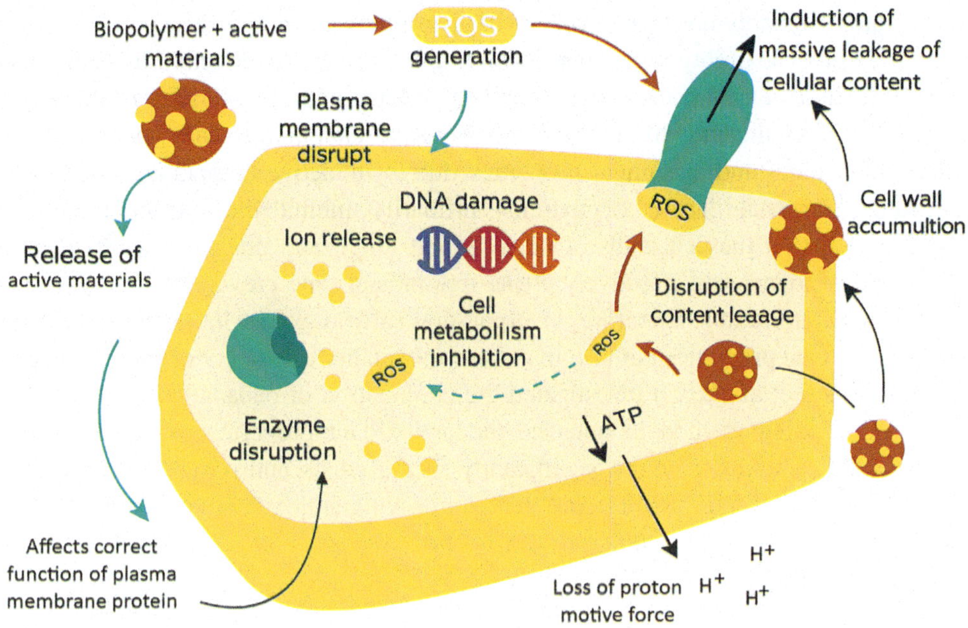

Figure 10.4

Antimicrobial and antioxidant characteristic schematics of active compound (essential oils, plant extracts, and nanoparticles).
No permission required.

the cell membranes of microorganisms and disrupt their metabolic processes. As a result, this action leads to a decrease in microbial growth in food, thereby preserving its quality and safety (Moeini et al., 2021). Sodium alginate is used as an effective material in cherry packaging. According to research conducted by Sun et al. (2021), adding quercetin peel extract to sodium alginate prevents oxidative degradation and can inhibit the growth of harmful bacteria such as *Salmonella* and *E. coli*. This combination not only enhances antioxidant properties but also strengthens the antimicrobial properties of the packaging. These positive effects are due to the ability of the phenolic compounds in quercetin extract to inhibit the metabolic activities of microorganisms and prevent oxidative damage to the food. As a result, the use of sodium alginate along with quercetin extract can significantly improve the quality and safety of food products (Sun et al., 2021). Polyphenols in green tea extract prevent the oxidation of fats and proteins and increase the shelf life of food products. Additionally, due to the presence of polyphenols in green tea, its antioxidant properties are enhanced. Moreover, these polyphenols also exhibit antimicrobial properties, inhibiting the growth of bacteria such as *E. coli* and *Listeria*. According to research by Sabaghi et al. (2021), the antioxidant properties of green tea are enhanced when added to chitosan due to the polyphenol compounds. Furthermore, the inhibition of *E. coli* and *Listeria* also increases when green tea extract is added to chitosan due to its antimicrobial properties (Sabaghi et al., 2021).

Gelatin is a natural biopolymer that is widely used in the packaging industry to produce edible films. Due to its high flexibility and water absorption capacity, it is effective in food packaging. According to research carried out by Ezati et al. (2023b), the addition of saffron extract to gelatin can effectively inhibit the growth of bacteria such as *Pseudomonas* and *Penicillium.* Saffron extract has antibacterial and antifungal properties due to its active compounds such as Crocin, Picroxine, and Zeroranal. These compounds primarily inhibit the metabolic activities of microorganisms, disrupt the cell walls of bacteria and fungi, and prevent microbial growth. In addition, these compounds can effectively create free radicals and prevent oxidative damage that can lead to food deterioration. Therefore, incorporating saffron extract into gelatin films not only increases antimicrobial properties but also maintains the quality and safety of food products (Ezati et al., 2023b). Zhang et al. (2020) investigated the conjugation of banana tannins to inulin using the free radical grafting method to enhance the antioxidant properties of inulin. The results showed that adding tannins to inulin significantly increased its antioxidant activity. They also found that the optimal ratio of tannin to inulin for improving antioxidant properties was 1:1. This study demonstrates that adding polyphenols to biopolymers can improve their antioxidant properties (Zeng et al., 2020).

10.2.6 Safety concerns of bioplastics

10.2.6.1 Sustainability aspects

Biopolymer-based food packaging is suggested as a sustainable alternative to petroleum-derived plastics, emphasizing biodegradability, renewable resources, and minimizing environmental impacts. Biopolymers, derived from renewable plant materials like starch, cellulose, and PLA, naturally decompose in the environment, reducing plastic waste in landfills and oceans. The production of these materials also has a lower carbon footprint, as they emit fewer greenhouse gases than traditional plastics. Additionally, biopolymers generally possess antimicrobial properties, enhance the shelf life of food products, and decrease food waste. Given the increasing consumer demand for environmentally friendly products, biopolymer packaging supports sustainability goals by minimizing plastic pollution, increasing recycling opportunities, and promoting a circular economy. However, to achieve widespread adoption of these materials, challenges such as performance, cost, and waste management must be addressed (Perera et al., 2023).

10.2.6.2 Degradation and migration impacts

Food packaging plays a crucial role in maintaining product quality safety and extending shelf life; however, environmental challenges, migration of chemical compounds, and safety concerns are some of the fundamental issues in this field. Bioplastics, including PLA and PHA, are attracting interest as sustainable alternatives to conventional plastics because of their biodegradability and production from renewable resources. These materials are engineered to degrade under specific conditions, such as elevated temperature and humidity. However, due to environmental constraints, this process can significantly slow down or even stop in natural environments like soil and open waters. Starch-based bioplastics have higher degradation rates in moist environments, but this process is reduced dramatically in dry or cold conditions (Perera et al., 2023). Therefore, bioplastics can persist for extended periods in certain environmental conditions and lead to pollution, similar to synthetic plastics. During the degradation process of these materials, volatile organic compounds and organic acids may be released, negatively impacting soil and water quality. Furthermore, bioplastics designed for industrial conditions may not fully degrade in natural environments or landfills, resulting in microplastics that can enter the food chain and cause environmental and toxic effects on living organisms (Fig. 10.5). The stability of bioplastics in aquatic ecosystems is also challenging. Even if these materials undergo physical degradation, residual chemical compounds may remain in the water, contaminating aquatic habitats. Marine organisms may ingest these particles, consequently suffering from physiological damage that can lead to bioaccumulation and the transfer of these compounds throughout the food chain. Certain bioplastics exhibit properties like moisture absorption and structural alterations under specific conditions, influencing their stability and performance in natural environments (Perera et al., 2023). Additionally, these materials are often combined with various additives to enhance mechanical strength, thermal stability, and antimicrobial characteristics. Some of these additives, like metal nanoparticles and antimicrobial compounds, may be released during the degradation of plastics, negatively impacting natural ecosystems and human health. The entry of nanoparticles into the environment can result in biological changes in living organisms and create long-term effects in ecosystems (Thakur and Satapathy, 2024).

Despite the environmental challenges, bioplastics have notable advantages over petroleum-based plastics. These materials are primarily produced from renewable resources such as corn starch, potatoes, and other polysaccharides, reducing dependence on fossil fuels and limiting the carbon footprint associated with plastic production. Furthermore, the production process of bioplastics typically generates fewer greenhouse gas emissions, which can play an essential role in mitigating climate change. Innovative biodegradability methods, designing biopolymers capable of degradation under various environmental conditions, and employing effective recycling technologies are critical solutions to enhance bioplastic sustainability. Strengthening recycling processes, developing innovative bioplastic packaging, and optimizing the conditions for the biodegradation of these materials reduce their environmental impact and improve their efficiency and safety in food packaging (Perera et al., 2023).

It should be noted that the migration of chemical compounds from biopolymer packaging to food is a critical issue in food safety and product quality, influenced by various factors. This process, particularly in biological packaging, occurs when multiple compounds within the polymer migrate into the food, potentially affecting its sensory, nutritional, and safety characteristics. The most significant compounds that may migrate from biopolymer packaging

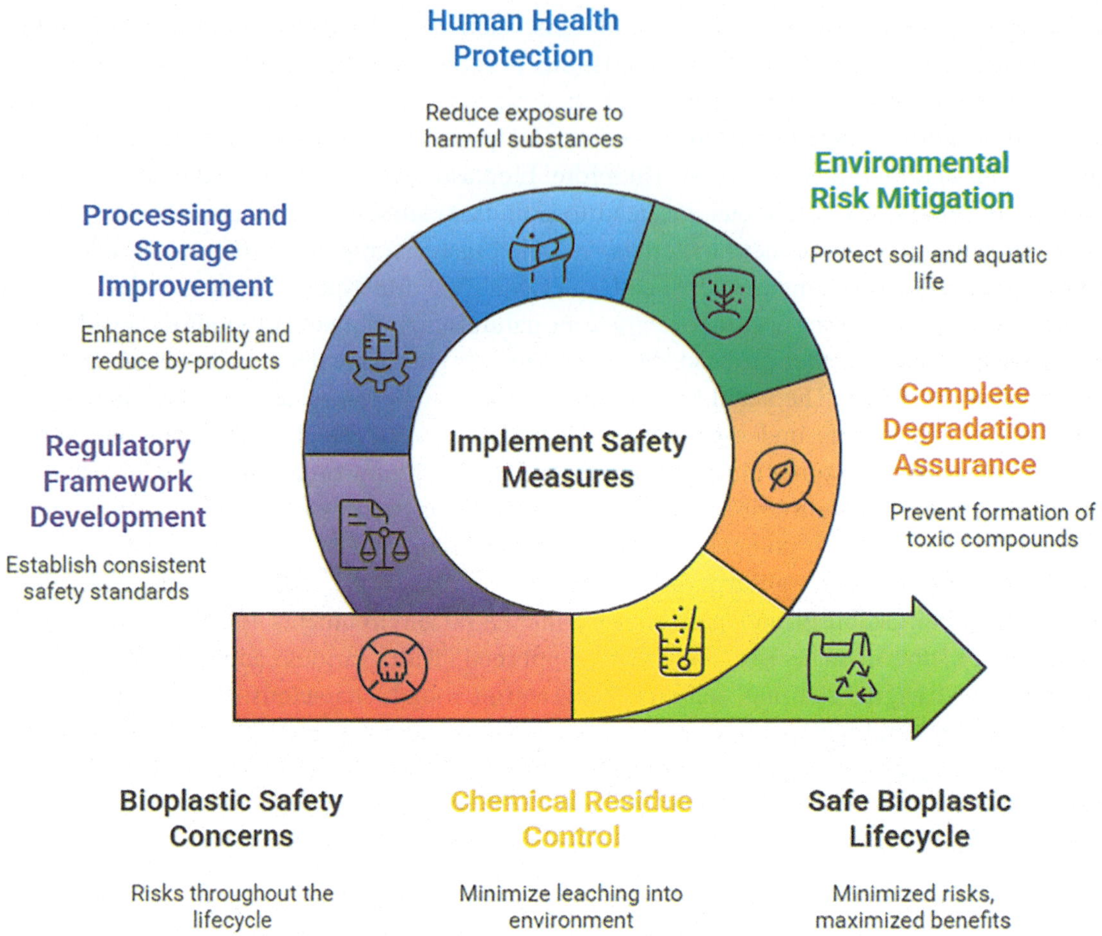

Figure 10.5

Key safety concerns of biodegradable plastics. The diagram highlights potential risks including chemical leaching, formation of micro/nanoplastics, environmental toxicity, and human health impacts arising from degradation products and additives.
No permission required.

to food include residual monomers from the production process, chemical additives such as plasticizers, antimicrobial agents, nanoparticles, and degradation products from the chemical breakdown of the polymers. Specifically, in polymers like PLA, which have lower thermal stability, high temperatures can lead to chemical degradation or hydrolysis, releasing migratory substances into the food. Other compounds that may migrate from biological packaging to food include bioactive substances such as plant extracts, plant essences, or polyphenols utilized in active packaging to improve antioxidant and antimicrobial properties. If transferred uncontrollably, these compounds can change sensory properties such as taste, aroma, and color of the food and may also interfere with the metabolic processes of microorganisms present in the food, consequently reducing quality and increasing microbial contamination (Thakur and Satapathy, 2024; Perera et al., 2023). Various factors influence the extent of migration of chemical compounds from packaging, with critical ones being the type of food, storage temperature, contact time of the packaging with food, and environmental conditions. Specifically, increased temperature (such as using packaging in the microwave) can accelerate migration and chemical changes. Additionally, specific polymers like PLA can become unstable at high temperatures and undergo thermal and chemical degradation, leading to the release of undesirable compounds and their migration into food. Furthermore, biopolymer packaging may degrade under specific conditions, such as humid environments or exposure to sunlight, releasing new chemical compounds that can be transferred to food (Thakur and Satapathy, 2024). Clay has a layered structure that can also slow the migration rate of monomers, additives, and nanoparticles. According to the research of Thakur and Satapathy (2024), clay was added to a biopolymer containing nanoparticles, preventing nanoparticle migration. Chitosan, a natural biopolymer, is extracted from the fats of insects and the exoskeletons of crabs and shrimp. This material is used in food packaging as a protective layer to reduce chemical compounds' migration into food. Chitosan is well-known for its antibacterial and antifungal properties, which can play a significant role in maintaining the quality and safety of food products. According to research conducted by Aresta et al. (2024), adding chitosan to packaging materials or using chitosan bags for packaging meat can significantly prevent the migration of chemical substances into food. Multilayer packaging can effectively avoid the migration of chemical compounds into food. For instance, in another research by Zhao et al. (2024), the use of a PLA layer as the primary layer, in combination with a polyethylene terephthalate layer, in the packaging of meat or dairy products, can significantly reduce the migration of chemical substances from plastic layers into food. This approach is particularly beneficial for high-fat or acidic food products, which are more prone to absorbing migratory substances due to their enhanced affinity for such compounds. The synergistic effect of using multiple layers with distinct barrier properties can provide a more effective solution for preserving the safety and quality of food, thereby extending its shelf life (Zhao et al., 2024). Silver nanoparticles (AgNPs) can prevent the migration of chemical compounds from packaging into food. Thakur and Satapathy (2024), assessed the addition of silver nanoparticles to PVA and can prevent the migration of chemical substances. This characteristic of AgNPs is attributed to their notable antibacterial and antifungal properties, which can enhance the safety and quality of food products (Thakur and Satapathy, 2024). Polyphenols play a role in preventing the migration of chemical compounds by creating protective layers. For example, according to research conducted by Niu et al. (2024), adding green tea extract containing polyphenols to PLA can hinder the migration of free radicals. This ability of polyphenols is attributed to their antioxidant properties, which can help preserve the quality and safety of food products (Niu et al., 2024). In short, using certain additives in the polymer matrix can address the migration concerns in biopolymer packaging.

10.2.7 Optical properties

The quality of food products is significantly influenced by the extent of their protection against UV radiation and visible light, as these radiations can induce undesirable photochemical reactions in food components. The optical properties of packaging materials, including color, transparency, and the degree of UV transmission and absorption, play a crucial role in determining the level of protection for food products. Exposure of food items to UV radiation resulted in lipid oxidation, degradation of vitamins, loss of antioxidant properties, and undesirable alterations in color and flavor. To mitigate these harmful effects, food packaging should be designed with engineered properties that minimize the transmission of harmful rays. Incorporating polymers with UV absorbers, mineral nanoparticles like titanium dioxide (TiO_2), zinc oxide (ZnO), and multilayer structures engineered to reflect and absorb light represent effective strategies for developing light and UV-resistant packaging. These approaches not only protect the product's nutritional value but also enhance its shelf life and prevent undesirable physicochemical changes (Jafarzadeh and Jafari, 2021). The parameter of optical properties commonly assessed is transparency, which significantly influences the visual appearance of packaging (Abdullah et al., 2022). Exposure of food products to light, remarkably UV light, can damage their structure and texture. This degradation occurs via photolysis and photooxidation reactions, where molecules in the food are influenced by light, producing reactive oxygen species and free radicals. These unstable molecules interact with other compounds in the food, leading to chemical changes that ultimately develop unpleasant flavors and odors (Ezati et al., 2023a). Certain additives that can improve the optical properties are discussed in Table 10.4.

Sodium polyacrylate is an acrylic polymer known for its moisture-absorbing properties and ability to form a gel in water. This material is primarily used in various industries, especially coatings and packaging. When combined with other polymers, such as pullulan, it can enhance clarity and antifogging properties. According to research conducted by Ghaani et al. (2024), combining pullulan with sodium polyacrylate results in a clear and effective coating under humid conditions. This coating significantly improves the visual display of food in polypropylene packaging and enhances its moisturizing properties. Furthermore, this combination effectively increases sealing properties and improves the sealing strength of packaging without compromising quality at high temperatures, which is particularly important for maintaining the quality and safety of food products (Ghaani et al., 2024). Various nanocomposites significantly affect the optical properties of food packaging materials. For instance, according to reports by Rezghi Rami et al. (2024), adding silver nanoparticles or TiO_2 to chitosan improved the surface transparency of packaging films. These combinations enhanced optical properties, increasing the clarity and brilliance of the packaging while simultaneously improving antifogging features and protection against UV radiation. Such modifications in chitosan-based packaging are particularly beneficial for food preservation applications, leading to improved appearance and extended shelf life of food items under

Table 10.4 The influence of various compounds in biopolymer matrices on optical properties.

Biopolymer	Optical properties	Added component	Effect on optical properties	References
PVA	PVA has high transparency and light transmittance but weak UV barrier properties.	Clove essential oil	Reduces transparency and decreases UV light transmission	Zhao et al. (2023)
Chitosan	Moderate transparency with good UV barrier properties due to its polymeric structure and interaction with light	TiO_2	Increases opacity and improves UV shielding	
Starch	High transparency but poor UV barrier properties	ZnO	Enhances UV blocking and slightly reduces transparency	Mallakpour et al. (2021)
Cellulose	Good transparency and optical clarity but exhibit limited UV barrier properties	AgNPs	Increases UV blocking and slightly increase optically	M'sakni and Alsufyani (2023)
Sodium alginate	Good transparency but limited UV-blocking ability	Essential oil (lemon)	Enhances color and reduces UV light penetration	Zhang et al. (2025)
Agar	High transparency and good light transmission but exhibit limited UV-blocking properties	Polyphenols (ellagic acid)	Improves UV resistance and reduces light transmission	Farasati Far et al. (2024)
Gelatin	Good transparency but poor UV-blocking properties	Ginkgo bilob extract	Increases UV barrier properties	Kim et al. (2021); Rangaraj et al. (2021)
PLA	High transparency but limited UV protection	Carbon quantum dots	Reduces UV transmission	Singh et al. (2024)
PBS	Good transparency but limited UV-blocking properties	TiO_2	Enhances UV blocking and slightly reduces transparency	Varghese et al. (2023)
Xanthan gum	Moderate transparency and high UV transmission	Ocimum tenufloriam	Reduces UV transmission and enhances color stability	Annu et al. (2021)

PLA, polylactic acid; *UV*, ultaviolet.

various storage conditions (Rezghi Rami et al., 2024). According to research conducted by Saeed et al. (2024), adding ZnO nanoparticles to chitosan and PVA nanocomposite films significantly improved the optical properties of these films. This combination increased surface transparency and enhanced the optical features of the films, ultimately contributing to improved visual quality of the packaging and expanding its applications in food packaging (Saeed et al., 2024). Essential oils incorporated into biopolymers enhance packaging films' optical and color properties, boosting their visual appeal and making them more vibrant. Additionally, the inclusion of TiO_2 in biopolymer films leads to a decrease in brightness index (L^*) and an increase in redness (a^*) and yellowness (b^*) indexes, which ultimately causes the films to become darker, yellow, and redder in color. These changes indicate that TiO_2 contributes to the enhancement of the optical and color characteristics of the films. For instance, according to reports by Noorian et al. (2022), adding TiO_2 and essential oils to cassava starch-based biopolymers results in darkening, yellow, and reddening packaging colors while simultaneously increasing its transparency. The extract of *Ocimum tenuiflorum*, which contains phytochemical compounds such as benzene rings, carbonyl compounds, and polyphenols, can reduce light transmission in the UV range. These compounds assist in creating a protective barrier against UV radiation by enhancing the optical absorption process in the UV region. According to reports by Annu et al. (2021), adding *O. tenuiflorum* extract to PVA significantly improves the protective properties against UV radiation, which contributes to reducing fat oxidation in food products (Annu et al., 2021). Antioxidant compounds added to polymer films can enhance their optical properties. These compounds can prevent UV radiation transmission and strengthen light absorption properties. Additionally, they may reduce the films' transparency, increase their opacity, and avoid color change in the films due to light and environmental factors. For example, according to reports by Annu et al. (2021), adding clove extract to PVA, which acts as an antioxidant compound, results in reduced transparency and decreased light transmission in the UV range.

10.2.8 Energy efficiency in biopolymer production

Energy efficiency in biopolymer packaging production involves strategies designed to minimize energy consumption and enhance production processes, facilitating the replacement of synthetic polymers with bio-based materials (Fig. 10.6). The biopolymer production process, encompassing various stages like extraction, purification, extrusion, and molding, requires a considerable amount of energy. To reduce energy consumption in these processes, the use of methods such as two-stage extrusion and employing high-energy-efficient equipment can significantly contribute to energy reduction. These strategies not only optimize energy consumption but can also offer substantial economic benefits. Other important methods for improving energy efficiency include the use of renewable resources, such as solar, wind, or biomass energy, to supply electricity and heat in factories. These resources can significantly reduce the consumption of fossil fuels and decrease the negative environmental

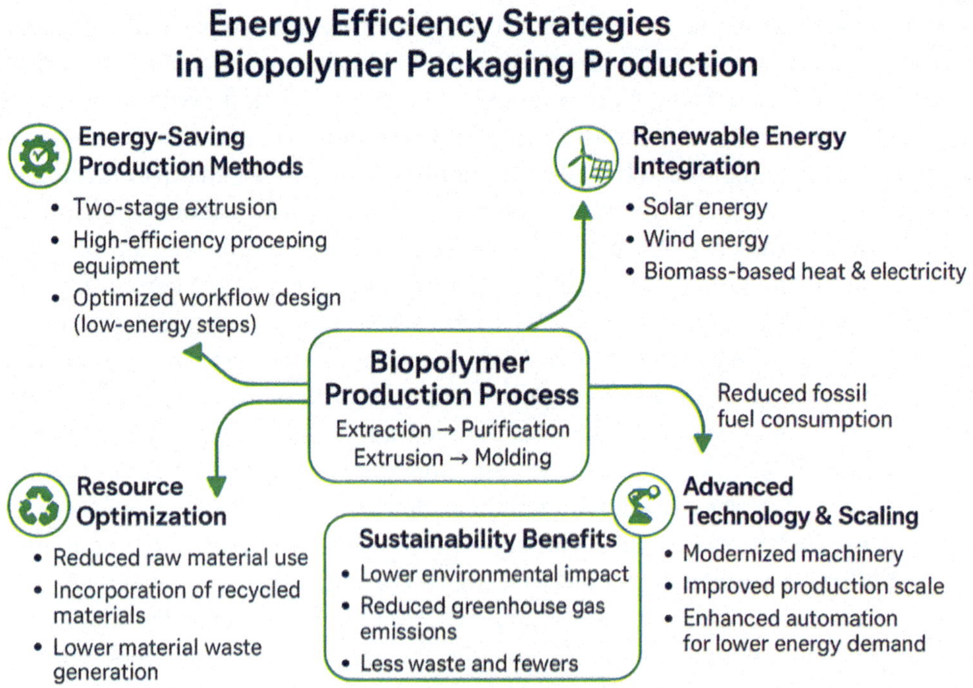

Figure 10.6

Key strategies for improving energy efficiency in biopolymer packaging production, including optimized processing, renewable energy use, resource optimization, and advanced technologies. *No permission required.*

impacts of production. Additionally, reducing the use of raw materials, utilizing recycled materials, and designing production processes that consume the least energy are other methods for improving energy efficiency. Furthermore, improving production scale and using more advanced equipment can also reduce energy consumption in biopolymer packaging production. Finally, focusing on design and recycling processes in biopolymer packaging production reduces environmental impacts and enhances their sustainability, as these processes utilize renewable raw materials and diminish waste and by-products (D'Almeida and de Albuquerque, 2024).

One of the biopolymers that offer many environmental benefits but consumes a significant amount of energy is PHA, with an energy consumption of 79 MJ/kg, which needs to be optimized. Based on the reports presented by Stublić et al. (2024), the use of sustainable methods in the extraction and purification of PHA plays a significant role in optimizing energy consumption in the production process of this biopolymer. One of these methods is the use of bio-based solvents instead of traditional chemical solvents, which leads to reducing environmental impacts and increasing the efficiency of the extraction process. Furthermore, the use of novel and more efficient technologies in the separation and purification stages, such as microfiltration processes and optimized chromatography, has resulted in a 30% reduction in energy consumption compared to conventional methods (Stublić et al., 2024). The twin-screw extrusion method is recognized as the most common type of extrusion for starch-based polymers. According to the research by do Val Siqueira et al. (2021), the twin-screw extrusion method enables precise control of temperature and pressure and reduces processing time, which results in lower energy consumption. The microwave radiation method is a very low-energy physical method that can alter the performance of native starch; for example, according to reports by Zhang et al. (2024), microwave radiation applied to native starch enhances the gelatinization process and leads to reduced energy consumption in starch processing. Optimization of the chitosan purification and structural modification process can reduce energy consumption and make production more sustainable. Another example by Iñiguez-Moreno et al. (2024) demonstrated that employing enzyme extraction methods with proteolytic enzymes for protein removal and chitosan extraction led to a reduction in energy requirements. UV radiation can have various effects on the properties of biopolymers. For example, according to research by Said et al. (2024), UV radiation can influence PLA, since it can accelerate chemical processes and polymerization, thereby reducing the time and energy required for PLA production. In conclusion, the energy efficiency of biopolymer-derived energy storage is closely linked to the stability of the materials and their capacity to maintain performance under diverse environmental conditions.

10.3 Conclusion and future trends

Food packaging is one of the key strategies for preserving the quality, safety, and shelf life of food products. In this field, polymers are used as the primary raw materials in packaging production, which can be divided into two categories: synthetic polymers and biopolymers. Synthetic polymers are primarily produced from petroleum derivatives and have raised significant concerns due to their high stability, non-biodegradability, and negative environmental impacts. In contrast, biopolymers are introduced as sustainable alternatives to plastic packaging due to their biodegradability and ecological compatibility. These polymers are derived from renewable resources such as starch, proteins, and microorganisms, and they are compostable. Among the most commonly used biopolymers in food packaging are PHA, polyhydroxybutyrate, polypeptides, PLA, and starch-based biopolymers. Biopolymers have garnered significant attention across various industries because of their biodegradability, utilization of renewable resources, reduction of greenhouse gas emissions, environmental compatibility, and compostability. These materials are recognized as sustainable alternatives to synthetic plastics by reducing dependence on petroleum resources and minimizing environmental pollution. Despite the numerous advantages of biopolymers, some disadvantages must be considered. One of the main issues is the high energy consumption during the production process, which may reduce environmental efficiency. Additionally, some of these polymers require specific conditions for biodegradability and may not fully decompose in natural environments. The production costs of these materials are still higher than those of synthetic plastics, which can create limitations for their widespread use. Furthermore, biopolymers' mechanical and barrier properties may impose restrictions, especially in industrial applications. Due to their natural structures, many biopolymers exhibit weaker mechanical properties compared to synthetic plastics, which can reduce the strength and durability of products such as packaging and industrial components. Moreover, certain biopolymers, like PHA, may have lower barrier properties against gases, moisture, or other substances, negatively impacting their performance in food packaging or other sensitive applications. Assuredly, the properties of biopolymers can be improved through various methods, such as adding antioxidants and antimicrobial components, polymer blending, incorporating essential oils, adding plasticizers, and using other additives. These strategies can enhance the performance of biopolymers in applications like food packaging.

The future of bioplastics in food packaging is poised for significant growth due to the global shift toward sustainable practices and consumer demand for eco-friendly solutions. Bioplastics, especially those derived from renewable resources such as plant starches, sugars, and agricultural by-products, will be crucial in addressing the plastic waste crisis as technological advancements continue. Innovations in material science will enable the development of bioplastics with improved mechanical properties, making them suitable for a broader range of packaging applications, including food products with varying shelf life. Packaging materials will also be designed more effectively, capable of providing sustainability, preserving food freshness, and extending the shelf life of products through active and smart packaging technologies such as moisture control and oxygen barriers. Additionally, they are expected to possess more substantial mechanical and barrier properties. Moreover, as bioplastic production costs decrease due to improvements in manufacturing processes such as fermentation and enzyme treatment, they are anticipated to become cost-competitive alternatives to synthetic plastics. Overall, the future of bioplastics in food packaging is bright, with a continued focus on innovation, scalability, and environmental responsibility.

References

Abdullah, J., Cai, M.A., Hafeez, Q., Wang, S., Farooq, Q., Huang, W., et al., 2022. Biopolymer-based functional films for packaging applications: a review. Frontiers in Nutrition 9, 1000116. Available at https://doi.org/10.3389/fnut.2022.1000116.

Abedi-Firoozjah, R., Salim, S.A., Hasanvand, S., Assadpour, E., Azizi-Lalabadi, M., Prieto, M.A., et al., 2023. Application of smart packaging for seafood: a comprehensive review. Comprehensive Reviews in Food Science and Food Safety 22 (2), 1438–1461. Available at https://doi.org/10.1111/1541-4337.13117.

Abedi-Firoozjah, R., Azimi-Salim, S., Afrah, A., Moradiyan, M.H., Assadpour, E., Sarlak, Z., et al., 2025. Comparison of the pH-and NH3-sensitivity of chitosan/polyvinyl alcohol smart films containing anthocyanins or betacyanins for monitoring fish freshness. Carbohydrate Polymer Technologies and Applications 9, 100639.

Alizadeh Sani, M., Tavassoli, M., Salim, S.A., Azizi-lalabadi, M., McClements, D.J., 2022. Development of green halochromic smart and active packaging materials: TiO2 nanoparticle- and anthocyanin-loaded gelatin/κ-carrageenan films. Food Hydrocolloids 124, 107324. Available at https://doi.org/10.1016/j.foodhyd.2021.107324.

Alizadeh Sani, M., Priyadarshi, R., Zhang, W., Khezerlou, A., Rhim, J.-W., 2024. Innovative application of laccase enzyme in food packaging. Trends in Food Science & Technology 151, 104623. Available at https://doi.org/10.1016/j.tifs.2024.104623.

Annu, A., Ali, S., Ahmed, 2021. Eco-friendly natural extract loaded antioxidative chitosan/polyvinyl alcohol based active films for food packaging. Heliyon 7 (3), e06550. Available at https://doi.org/10.1016/j.heliyon.2021.e06550.

Aresta, A.M., De Vietro, N., Mancini, G., Zambonin, C., 2024. Effect of a chitosan-based packaging material on the domestic storage of "ready-to-cook" meat products: evaluation of biogenic amines production, phthalates migration, and in vitro antimicrobic activity's impact on Aspergillus niger. Separations 11 (6), 159. Available at https://doi.org/10.3390/separations11060159.

Asgher, M., Qamar, S.A., Bilal, M., Iqbal, H.M.N., 2020. Bio-based active food packaging materials: sustainable alternative to conventional petrochemical-based packaging materials. Food Research International 137, 109625. Available at https://doi.org/10.1016/j.foodres.2020.109625.

Assis, R.Q., Lopes, S.M., Costa, T.M.H., Flôres, S.H., Rios, Ad.O., 2017. Active biodegradable cassava starch films incorporated lycopene nanocapsules. Industrial Crops and Products 109, 818–827. Available at https://doi.org/10.1016/j.indcrop.2017.09.043.

Azimi-Salim, S., Azizi-Lalabadi, M., Tavassoli, M., Alizadeh Sani, M., 2022. Design of nanocomposite packaging based on gelatin biopolymer containing titanium dioxide nanoparticles and saffron extract for use in food packaging. Journal of Food Science and Technology (Iran) 18 (121), 25–37. Available at https://doi.org/10.52547/fsct.18.121.3.

Aziz, S.G.G., Almasi, H., 2018. Physical characteristics, release properties, and antioxidant and antimicrobial activities of whey protein isolate films incorporated with thyme (Thymus vulgaris L.) extract-loaded Nanoliposomes. Food and Bioprocess Technology 11 (8), 1552–1565. Available at https://doi.org/10.1007/s11947-018-2121-6.

Azizi-Lalabadi, M., Garavand, F., Jafari, S.M., 2021. Incorporation of silver nanoparticles into active antimicrobial nanocomposites: release behavior, analyzing techniques, applications and safety issues. Advances in Colloid and Interface Science 293, 102440. Available at https://doi.org/10.1016/j.cis.2021.102440.

Azizi-Lalabadi, M., Ehsani, A., Ghanbarzadeh, B., Divband, B., 2020. Polyvinyl alcohol/gelatin nanocomposite containing ZnO, TiO_2 or ZnO/TiO_2 nanoparticles doped on 4A zeolite: microbial and sensory qualities of packaged white shrimp during refrigeration. International Journal of Food Microbiology 312, 108375. Available at https://doi.org/10.1016/j.ijfoodmicro.2019.108375.

Azizi-Lalabadi, M., Alizadeh-Sani, M., Divband, B., Ehsani, A., McClements, D.J., 2020. Nanocomposite films consisting of functional nanoparticles (TiO2 and ZnO) embedded in 4A-Zeolite and mixed polymer matrices (gelatin and polyvinyl alcohol). Food Research International 137, 109716. Available at https://doi.org/10.1016/j.foodres.2020.109716.

Azizi-Lalabadi, M., Rahimzadeh-Sani, Z., Feng, J., Hosseini, H., Jafari, S.M., 2023. The impact of essential oils on the qualitative properties, release profile, and stimuli-responsiveness of active food packaging nanocomposites. Critical Reviews in Food Science and Nutrition 63 (13), 1822–1845. Available at https://doi.org/10.1080/10408398.2021.1971154.

Azizi-Lalabadi, M., Rafiei, L., Divband, B., Ehsani, A., 2020. Active packaging for Salmon stored at refrigerator with polypropylene nanocomposites containing 4A zeolite, ZnO nanoparticles, and green tea extract. Food Science & Nutrition 8 (12), 6445–6456. Available at https://doi.org/10.1002/fsn3.1934.

Bamps, B., Buntinx, M., Peeters, R., 2023. Seal materials in flexible plastic food packaging: a review. Packaging Technology and Science 36 (7), 507–532. Available at https://doi.org/10.1002/pts.2732.

Barzan, G., Sacco, A., Giovannozzi, A.M., Portesi, C., Schiavone, C., Salafranca, J., et al., 2024. Development of innovative antioxidant food packaging systems based on natural extracts from food industry waste and

Moringa oleifera leaves. Food chemistry 432, 137088. Available at https://doi.org/10.1016/j.foodchem.2023.137088.

Basavegowda, N., Baek, K.-H., 2021. Advances in functional biopolymer-based nanocomposites for active food packaging applications. Polymers 13 (23), 4198. Available at https://doi.org/10.3390/polym13234198.

Benitez, J.J., Florido-Moreno, P., Porras-Vázquez, J.M., Tedeschi, G., Athanassiou, A., Heredia-Guerrero, J.A., et al., 2024. Transparent, plasticized cellulose-glycerol bioplastics for food packaging applications. International Journal of Biological Macromolecules 273, 132956. Available at https://doi.org/10.1016/j.ijbiomac.2024.132956.

Bhaskar, R., Zo, S.M., Narayanan, K.B., Purohit, S.D., Gupta, M.K., Han, S.S., 2023. Recent development of protein-based biopolymers in food packaging applications: a review. Polymer Testing 124, 108097. Available at https://doi.org/10.1016/j.polymertesting.2023.108097.

Biswas, M.C., Jony, B., Nandy, P.K., Chowdhury, R.A., Halder, S., Kumar, D., et al., 2022. Recent advancement of biopolymers and their potential biomedical applications. Journal of Polymers and the Environment 30 (1), 51–74. Available at https://doi.org/10.1007/s10924-021-02199-y.

Blanco, I., Ingrao, C., Siracusa, V., 2020. Life-cycle assessment in the polymeric sector: a comprehensive review of application experiences on the Italian scale. Polymers 12 (6), 1212. Available at https://doi.org/10.3390/polym12061212.

Boccalon, E., Gorrasi, G., 2022. Functional bioplastics from food residual: potentiality and safety issues. Comprehensive Reviews in Food Science and Food Safety 21 (4), 3177–3204. Available at https://doi.org/10.1111/1541-4337.12986.

Cerdá-Gandia, R., Agüero, Á., Arrieta, M.P., Fenollar, O., 2024. Effect of different porous size of porous inorganic fillers on the encapsulation of rosemary essential oil for PLA-based active packaging. Polymers 16 (18), 2632. Available at https://doi.org/10.3390/polym16182632.

Chen, H., Hu, X., Chen, E., Wu, S., McClements, D.J., Liu, S., et al., 2016. Preparation, characterization, and properties of chitosan films with cinnamaldehyde nanoemulsions. Food Hydrocolloids 61, 662–671. Available at https://doi.org/10.1016/j.foodhyd.2016.06.034.

Chen, X., Feng, J., Han, S., Bai, Y., Wang, P., Liu, H., et al., 2024. Cellulose nanocrystal-modified bio-based aqueous polyurethane coating agent for kraft paper packaging. Journal of Applied Polymer Science 141 (32), e55745. Available at https://doi.org/10.1002/app.55745.

Chowdhury, M.A., Hossain, N., Badrudduza, M.D., Rana, M.M., 2023. Development and characterization of natural sourced bioplastic for food packaging applications. Heliyon 9 (2), e13538. Available at https://doi.org/10.1016/j.heliyon.2023.e13538.

Chueangchayaphan, W., Nooun, P., Ummarat, N., Chueangchayaphan, N., 2024. Eco-friendly biocomposite foam from natural rubber latex and rice starch for sustainable packaging applications. Express Polymer Letters 18 (1), 27–40. Available at https://doi.org/10.3144/expresspolymlett.2024.3.

Cruz, R.M.S., Krauter, V., Krauter, S., Agriopoulou, S., Weinrich, R., Herbes, C., et al., 2022. Bioplastics for food packaging: environmental impact, trends and regulatory aspects. Foods 11 (19), 3087. Available at https://doi.org/10.3390/foods11193087.

D'Almeida, A.P., de Albuquerque, T.L., 2024. Innovations in food packaging: from bio-based materials to smart packaging systems. Processes 12 (10), 2085. Available at https://doi.org/10.3390/pr12102085.

de Lima Silva, I.D., de Moraes, L.E.P.T., Caetano, V.F., De Andrade, M.F., Hallwass, F., Brito, A.M.S.S., et al., 2021. Development of antioxidant active PVA films with plant extract of Caesalpinia ferrea Martius. Lwt 144, 111215.

Deshmukh, R.K., Gaikwad, K.K., 2024. Natural antimicrobial and antioxidant compounds for active food packaging applications. Biomass Conversion and Biorefinery 14 (4), 4419–4440. Available at https://doi.org/10.1007/s13399-022-02623-w.

Dong, W., Su, J., Chen, Y., Xu, D., Cheng, L., Mao, L., et al., 2022. Characterization and antioxidant properties of chitosan film incorporated with modified silica nanoparticles as an active food packaging. Food chemistry 373 (Pt A), 131414. Available at https://doi.org/10.1016/j.foodchem.2021.131414.

do Val Siqueira, L., Arias, C.I.L.F., Maniglia, B.C., Tadini, C.C., 2021. Starch-based biodegradable plastics: methods of production, challenges and future perspectives. Current Opinion in Food Science 38, 122–130. Available at https://doi.org/10.1016/j.cofs.2020.10.020.

Elkhayat, A., El-Shafey, F., Mohamed, E., Fouad, A., Shamiea, M., Hassabo, A., 2024. Eco-friendly biological materials for healthy food packaging. Journal of Textiles, Coloration and Polymer Science 0 (0). Available at https://doi.org/10.21608/jtcps.2024.259041.1275.

European Commission, 2011. Communication from the Commission to the European Parliament, the Council, the Economic and Social Committee and the Committee of the Regions.

Ezati, P., Khan, A., Rhim, J.W., Roy, S., Hassan, Z.U., 2023. Saffron: perspectives and sustainability for active and intelligent food packaging applications. Food and Bioprocess Technology 16 (6), 1177–1196. Available at https://doi.org/10.1007/s11947-022-02949-3.

Ezati, P., Khan, A., Priyadarshi, R., Bhattacharya, T., Tammina, S.K., Rhim, J.W., 2023. Biopolymer-based UV protection functional films for food packaging. Food Hydrocolloids 142, 108771. Available at https://doi.org/10.1016/j.foodhyd.2023.108771.

Fambri, L., Migliaresi, C., 2022. Crystallization and thermal properties. In: Auras, R.A., Lim, L.-T., Selke, S.E.M., Tsuji, H. (Eds.), Poly(Lactic Acid): Synthesis, Structures, Properties, Processing, Applications, and End of Life. Wiley, pp. 135–151. Available at https://doi.org/10.1002/9781119767480.ch8.

Farasati Far, B., Jahanbakhshi, M., Jameie, L., Zolfigol, F., Taromi, P., Ertas, Y.N., 2024. Chitosan-graft-pomegranate extract hydrogel: a dual-functional pad for antibacterial and antioxidant enhancement for shelf life extension in food packaging. ACS Applied Polymer Materials 6 (16), 9545–9558. Available at https://doi.org/10.1021/acsapm.4c01240.

Flores-Morales, P., Campos-Requena, V.H., Gatica, N., Munoz, C., Pérez, M.A., Rivas, B.L., et al., 2017. Teaching polymer science in the department of polymers at the University of Concepción, Chile: a brief history. Journal of Chemical Education 94 (11), 1702–1713. Available at https://doi.org/10.1021/acs.jchemed.7b00212.

Ghaani, M., Soltanzadeh, M., Carullo, D., Farris, S., 2024. Development of a biopolymer-based anti-fog coating with sealing properties for applications in the food packaging sector. Polymers 16 (12), 1745. Available at https://doi.org/10.3390/polym16121745.

Gioia, C., Giacobazzi, G., Vannini, M., Totaro, G., Sisti, L., Colonna, M., et al., 2021. End of life of biodegradable plastics: composting versus re/upcycling. ChemSusChem 14 (19), 4167–4175. Available at https://doi.org/10.1002/cssc.202101226.

Gong, S., Li, B., Liu, H., Xu, C., Wang, J., Fan, Y., et al., 2024. Facile preparation of eco-friendly plastic from rosin modified microcrystalline cellulose for food packaging. Industrial Crops and Products 221, 119370. Available at https://doi.org/10.1016/j.indcrop.2024.119370.

Gupta, R.K., Guha, P., Srivastav, P.P., 2024. Investigating the toxicological effects of nanomaterials in food packaging associated with human health and the environment. Journal of Hazardous Materials Letters 5, 100125. Available at https://doi.org/10.1016/j.hazl.2024.100125.

Gutiérrez, T.J., Valencia, G.A., 2021. Reactive extrusion-processed native and phosphated starch-based food packaging films governed by the hierarchical structure. International Journal of Biological Macromolecules 172, 439–451. Available at https://doi.org/10.1016/j.ijbiomac.2021.01.048.

Hakeem, M.J., Feng, J., Nilghaz, A., Ma, L., Seah, H.C., Konkel, M.E., et al., 2020. Active packaging of immobilized zinc oxide nanoparticles controls Campylobacter jejuni in raw chicken meat. Applied and Environmental Microbiology 86 (22), e01195-20. Available at https://doi.org/10.1128/aem.01195-20.

Ibrahim, N.I., Shahar, F.S., Sultan, M.T.H., Shah, A.U.M., Safri, S.N.A., Mat Yazik, M.H., 2021. Overview of bioplastic introduction and its applications in product packaging. Coatings 11 (11), 1423. Available at https://doi.org/10.3390/coatings11111423.

Imran, M., Revol-Junelles, A.M., René, N., Jamshidian, M., Akhtar, M.J., Arab-Tehrany, E., et al., 2012. Microstructure and physico-chemical evaluation of nano-emulsion-based antimicrobial peptides embedded in bioactive packaging films. Food Hydrocolloids 29 (2), 407–419. Available at https://doi.org/10.1016/j.foodhyd.2012.04.010.

Iñiguez-Moreno, M., Santiesteban-Romero, B., Melchor-Martínez, E.M., Parra-Saldívar, R., González-González, R.B., 2024. Valorization of fishery industry waste: chitosan extraction and its application in the industry. MethodsX 13, 102892. Available at https://doi.org/10.1016/j.mex.2024.102892.

Ishak, K.A., Zahid, N.I., Velayutham, T.S., Khyasudeen, M.F., Annuar, M.S.M., 2024. Corroborative studies on chain packing characteristics of biological medium-chain-length poly-3-hydroxyalkanoates with different monomeric composition. International Journal of Biological Macromolecules 269, 131973. Available at https://doi.org/10.1016/j.ijbiomac.2024.131973.

Jabeen, N., Majid, I., Ahmad Nayik, G., Yildiz, F., 2015. Bioplastics and food packaging: a review. Cogent Food & Agriculture 1 (1), 1117749. Available at https://doi.org/10.1080/23311932.2015.1117749.

Jafarzadeh, S., Jafari, S.M., 2021. Impact of metal nanoparticles on the mechanical, barrier, optical and thermal properties of biodegradable food packaging materials. Critical Reviews in Food Science and Nutrition 61 (16), 2640–2658. Available at https://doi.org/10.1080/10408398.2020.1783200.

Jafarzadeh, S., Forough, M., Kouzegaran, V.J., Zargar, M., Garavand, F., Azizi-Lalabadi, M., et al., 2023. Improving the functionality of biodegradable food packaging materials via porous nanomaterials. Comprehensive Reviews in Food Science and Food Safety 22 (4), 2850–2886. Available at https://doi.org/10.1111/1541-4337.13164.

Jafarzadeh, S., Golgoli, M., Azizi-Lalabadi, M., Farahbakhsh, J., Forough, M., Rabiee, N., et al., 2024. Enhanced carbohydrate-based plastic performance by incorporating cerium-based metal-organic framework for food packaging application. International Journal of Biological Macromolecules 265 (Pt 1), 130899. Available at https://doi.org/10.1016/j.ijbiomac.2024.130899.

Jamróz, E., Juszczak, L., Kucharek, M., 2018. Investigation of the physical properties, antioxidant and antimicrobial activity of ternary potato starch-furcellaran-gelatin films incorporated with lavender essential oil. International Journal of Biological Macromolecules 114, 1094–1101. Available at https://doi.org/10.1016/j.ijbiomac.2018.04.014.

Jariyasakoolroj, P., Leelaphiwat, P., Harnkarnsujarit, N., 2020. Advances in research and development of bioplastic for food packaging. Journal of the Science of Food and Agriculture 100 (14), 5032–5045. Available at https://doi.org/10.1002/jsfa.9497.

Joseph, A.M., George, B., 2023. Cross-linking biopolymers for biomedical applications. In: Thomas, S., Ajitha, A.R., Jose Chirayil, C., Thomas, B. (Eds.), Handbook of Biopolymers. Springer Nature, Singapore, pp. 1135–1172. Available at https://doi.org/10.1007/978-981-19-0710-4_42.

Kalita, N.K., Sarmah, A., Bhasney, S.M., Kalamdhad, A., Katiyar, V., 2021. Demonstrating an ideal compostable plastic using biodegradability kinetics of poly(lactic acid) (PLA) based green biocomposite films under aerobic composting conditions. Environmental Challenges 3, 100030. Available at https://doi.org/10.1016/j.envc.2021.100030.

Kane, A., Diallo, Y., Tamba, A., Diakhaté, P.A., Sagna, C.G., Mbodji, H., et al., 2024. Frequency and diversity of stabilizers, thickeners and gelling agents used as food additives in food products sold on Dakar markets. OALib 11 (01), 1–13. Available at https://doi.org/10.4236/oalib.1111147.

Karmakar, G.P., 2022. Regeneration and recovery of plastics. Encyclopedia of Materials: Plastics and Polymers 1–4, 634–651. Available at https://doi.org/10.1016/B978-0-12-820352-1.00045-6.

Karydis-Messinis, A., Kyriakaki, C., Triantafyllou, E., Tsirka, K., Gioti, C., Gkikas, D., et al., 2024. Development and physicochemical characterization of edible chitosan–casein hydrogel membranes for potential use in food packaging. Gels 10 (4), 254. Available at https://doi.org/10.3390/gels10040254.

Khalid, M.Y., Arif, Z.U., 2022. Novel biopolymer-based sustainable composites for food packaging applications: a narrative review. Food Packaging and Shelf Life 33, 100892. Available at https://doi.org/10.1016/j.fpsl.2022.100892.

Khezerlou, A., Tavassoli, M., Alizadeh Sani, M., Mohammadi, K., Ehsani, A., McClements, D.J., 2021. Application of nanotechnology to improve the performance of biodegradable biopolymer-based packaging materials. Polymers 13 (24), 4399. Available at https://doi.org/10.3390/polym13244399.

Kim, B.K., Lee, H.S., Yang, H.S., Song, K.B., 2021. Development of ginkgo (Ginkgo biloba) nut starch films containing cinnamon (cinnamomum zeylanicum) leaf essential oil. Molecules 26 (20), 6114. Available at https://doi.org/10.3390/molecules26206114.

Kumari, S., Debbarma, R., Nasrin, N., Khan, T., Taj, S., Bhuyan, T., 2024. Recent advances in packaging materials for food products. Food Bioengineering 3 (2), 236–249. Available at https://doi.org/10.1002/fbe2.12096.

Li, J., Zhang, F., Zhong, Y., Zhao, Y., Gao, P., Tian, F., et al., 2022. Emerging food packaging applications of cellulose nanocomposites: a review. Polymers 14 (19), 4025. Available at https://doi.org/10.3390/polym14194025.

Liang, J., Yan, H., Zhang, J., Dai, W., Gao, X., Zhou, Y., et al., 2017. Preparation and characterization of antioxidant edible chitosan films incorporated with epigallocatechin gallate nanocapsules. Carbohydrate Polymers 171, 300–306. Available at https://doi.org/10.1016/j.carbpol.2017.04.081.

Liu, B., Sun, F., Zhu, P., Wang, K., Peng, L., Zhuang, Y., et al., 2024. Preparation of multi-barrier and multi-functional paper-based materials by chitosan, ethyl cellulose and green walnut husk biorefinery products for sustainable food packaging. International Journal of Biological Macromolecules 278 (Pt 2), 134557. Available at https://doi.org/10.1016/j.ijbiomac.2024.134557.

Liu, Y., Ahmed, S., Sameen, D.E., Wang, Y., Lu, R., Dai, J., et al., 2021. A review of cellulose and its derivatives in biopolymer-based for food packaging application. Trends in Food Science and Technology 112, 532–546. Available at https://doi.org/10.1016/j.tifs.2021.04.016.

M'sakni, N.H., Alsufyani, T., 2023. Part B: improvement of the optical properties of cellulose nanocrystals reinforced thermoplastic starch bio-composite films by ex situ incorporation of green silver nanoparticles from Chaetomorpha linum. Polymers 15 (9), 2148. Available at https://doi.org/10.3390/polym15092148.

Mallakpour, S., Sirous, F., Hussain, C.M., 2021. A journey to the world of fascinating ZnO nanocomposites made of chitosan, starch, cellulose, and other biopolymers: progress in recent achievements in eco-friendly food packaging, biomedical, and water remediation technologies. International Journal of Biological Macromolecules 170, 701–716. Available at https://doi.org/10.1016/j.ijbiomac.2020.12.163.

Marangoni Júnior, L., Alves, R.M.V., Moreira, C.Q., Cristianini, M., Padula, M., Anjos, C.A.R., 2020. High-pressure processing effects on the barrier properties of flexible packaging materials. Journal of Food Processing and Preservation 44 (11), e14865. Available at https://doi.org/10.1111/jfpp.14865.

Marano, S., Laudadio, E., Minnelli, C., Stipa, P., 2022. Tailoring the barrier properties of PLA: a state-of-the-art review for food packaging applications. Polymers 14 (8), 1626. Available at https://doi.org/10.3390/polym14081626.

Mary, S.K., Koshy, R.R., Arunima, R., Thomas, S., Pothen, L.A., 2022. A review of recent advances in starch-based materials: bionanocomposites, pH sensitive films, aerogels and carbon dots. Carbohydrate Polymer Technologies and Applications 3, 100190. Available at https://doi.org/10.1016/j.carpta.2022.100190.

Moeini, A., Germann, N., Malinconico, M., Santagata, G., 2021. Formulation of secondary compounds as additives of biopolymer-based food packaging: a review. Trends in Food Science & Technology 114, 342–354. Available at https://doi.org/10.1016/j.tifs.2021.05.040.

Moeini, A., Pedram, P., Fattahi, E., Cerruti, P., Santagata, G., 2022. Edible polymers and secondary bioactive compounds for food packaging applications: antimicrobial, mechanical, and gas barrier properties. Polymers 14 (12), 2395. Available at https://doi.org/10.3390/polym14122395.

Mohammadi, M., Mirabzadeh, S., Shahvalizadeh, R., Hamishehkar, H., 2020. Development of novel active packaging films based on whey protein isolate incorporated with chitosan nanofiber and nano-formulated cinnamon oil. International Journal of Biological Macromolecules 149, 11–20. Available at https://doi.org/10.1016/j.ijbiomac.2020.01.083.

Momeni, S., Craplewe, K., Safder, M., Luz, S., Sauvageau, D., Elias, A., 2023. Accelerating the biodegradation of poly(lactic acid) through the inclusion of plant fibers: a review of recent advances. ACS Sustainable Chemistry & Engineering 11 (42), 15146–15170. Available at https://doi.org/10.1021/acssuschemeng.3c04240.

Muobom, S.S., Umar, A.-M.S., Soongseok, Y., Brolin, A.-P., 2020. A review on plasticizers and eco-friendly bioplasticizers: biomass sources and market. International Journal of Engineering Research 9 (5), 1138–1144.

Narancic, T., Cerrone, F., Beagan, N., O'Connor, K.E., 2020. Recent advances in bioplastics: application and biodegradation. Polymers 12 (4), 920. Available at https://doi.org/10.3390/POLYM12040920.

Nazareth, M.C., Marques, M.R.C., Pinheiro, L.M., Castro, Í.B., 2022. Key issues for bio-based, biodegradable and compostable plastics governance. Journal of Environmental Management 322, 116074. Available at https://doi.org/10.1016/j.jenvman.2022.116074.

Niu, J., Shang, M., Li, X., Sang, S., Chen, L., Long, J., et al., 2024. Health benefits, mechanisms of interaction with food components, and delivery of tea polyphenols: a review. Critical Reviews in Food Science and Nutrition 64 (33), 12487–12499. Available at https://doi.org/10.1080/10408398.2023.2253542.

Noorian, S., Mohammadi Nafchi, A., Bolandi, M., Jokar, M., 2022. Effects of nano-titanium dioxide and mentha piperita essential oil on physicochemical, mechanical, and optical properties of cassava starch film. Starch - Stärke 74 (9–10), 2200090. Available at https://doi.org/10.1002/star.202200090.

Oladzadabbasabadi, N., Dheyab, M.A., Nafchi, A.M., Ghasemlou, M., Ivanova, E.P., Adhikari, B., 2023. Turning food waste into value-added carbon dots for sustainable food packaging application: a review. Advances in Colloid and Interface Science 321, 103020. Available at https://doi.org/10.1016/j.cis.2023.103020.

Othman, Siti Hajar, Shapi'i, Ruzanna Ahmad, Ronzi, Nur Diana Arisya, 2024. Starch biopolymer films containing chitosan nanoparticles: A review. Carbohydrate Polymers 329, 121735. Available at https://doi.org/10.1016/j.carbpol.2023.121735.

Patel, M.K., Hansson, F., Pitkänen, O., Geng, S., Oksman, K., 2022. Biopolymer blends of poly(lactic acid) and poly(hydroxybutyrate) and their functionalization with glycerol triacetate and chitin nanocrystals for food packaging applications. ACS Applied Polymer Materials 4 (9), 6592–6601. Available at https://doi.org/10.1021/acsapm.2c00967.

Patil, S., Kumar Bharimalla, A., Mahapatra, A., Dhakane-Lad, J., Arputharaj, A., Kumar, M., et al., 2021. Effect of polymer blending on mechanical and barrier properties of starch-polyvinyl alcohol based biodegradable composite films. Food Bioscience 44, 101352. Available at https://doi.org/10.1016/j.fbio.2021.101352.

Perera, K.Y., Jaiswal, A.K., Jaiswal, S., 2023. Biopolymer-based sustainable food packaging materials: challenges, solutions, and applications. Foods 12 (12), 2422. Available at https://doi.org/10.3390/foods12122422.

Qin, P., Wu, L., Li, B., Li, N., Pan, X., Dai, J., 2021. Superior gas barrier properties of biodegradable PBST vs. PBAT copolyesters: a comparative study. Polymers 13 (19), 3449. Available at https://doi.org/10.3390/polym13193449.

Rahman, R., Sood, M., Gupta, N., Bandral, J.D., Hameed, F., Ashraf, S., 2019. Bioplastics for food packaging: a review. Int. J. Curr. Microbiol. Appl. Sci 8 (3), 2311–2321.

Rangaraj, V.M., Rambabu, K., Banat, F., Mittal, V., 2021. Natural antioxidants-based edible active food packaging: an overview of current advancements. Food Bioscience 43, 101251. Available at https://doi.org/10.1016/j.fbio.2021.101251.

Ranjbaryan, S., Pourfathi, B., Almasi, H., 2019. Reinforcing and release controlling effect of cellulose nanofiber in sodium caseinate films activated by nanoemulsified cinnamon essential oil. Food Packaging and Shelf Life 21, 100341. Available at https://doi.org/10.1016/j.fpsl.2019.100341.

Razmjoo, F., Sadeghi, E., Alizadeh-Sani, M., Noroozi, R., Azizi-Lalabadi, M., 2022. Fabrication and application of functional active packaging material based on carbohydrate biopolymers formulated with Lemon verbena/Ferulago angulata extracts for the preservation of raw chicken meat. Journal of Food Processing and Preservation 46 (10), e16830. Available at https://doi.org/10.1111/jfpp.16830.

Rezghi Rami, M., Forouzandehdel, S., Aalizadeh, F., 2024. Enhancing biodegradable smart food packaging: fungal-synthesized nanoparticles for stabilizing biopolymers. Heliyon 10 (18), e37692. Available at https://doi.org/10.1016/j.heliyon.2024.e37692.

Robledo, N., Vera, P., López, L., Yazdani-Pedram, M., Tapia, C., Abugoch, L., 2018. Thymol nanoemulsions incorporated in quinoa protein/chitosan edible films: antifungal effect in cherry tomatoes. Food Chemistry 246, 211–219. Available at https://doi.org/10.1016/j.foodchem.2017.11.032.

Sabaghi, M., Hoseyni, S.Z., Tavasoli, S., Mozafari, M.R., Katouzian, I., 2021. Strategies of confining green tea catechin compounds in nano-biopolymeric matrices: a review. Colloids and Surfaces B: Biointerfaces 204, 111781. Available at https://doi.org/10.1016/j.colsurfb.2021.111781.

Saeed, A., Guizani, I., Hanash, F.E., Asnag, G.M., Al-Harthi, A.M., Alwafi, R., et al., 2024. Enhancing optical, structural, thermal, electrical properties, and antibacterial activity in chitosan/polyvinyl alcohol blend with ZnO nanorods: polymer nanocomposites for optoelectronics and food/medical packaging applications. Polymer Bulletin 81 (13), 11645–11670. Available at https://doi.org/10.1007/s00289-024-05270-5.

Sahin, D., Aksoy, P., Ucpinar Durmaz, B., Aytac, A., 2023. Improvement of polyvinyl alcohol/casein blend film properties by adding cellulose nanocrystals. Journal of Vinyl and Additive Technology 29 (6), 1002–1014. Available at https://doi.org/10.1002/vnl.21979.

Said, N.S., Olawuyi, I.F., Lee, W.Y., 2024. Tailoring pectin-PLA bilayer film for optimal properties as a food pouch material. Polymers 16 (5), 712. Available at https://doi.org/10.3390/polym16050712.

Shah, Y.A., Bhatia, S., Al-Harrasi, A., Afzaal, M., Saeed, F., Anwer, M.K., et al., 2023. Mechanical properties of protein-based food packaging materials. Polymers 15 (7), 1724. Available at https://doi.org/10.3390/polym15071724.

Shah, Y.A., Bhatia, S., Al-Harrasi, A., Oz, F., Khan, M.H., Roy, S., et al., 2024. Thermal properties of biopolymer films: insights for sustainable food packaging applications. Food Engineering Reviews 16 (4), 497–512. Available at https://doi.org/10.1007/s12393-024-09380-8.

Siddiqui, S.A., Yang, X., Deshmukh, R.K., Gaikwad, K.K., Bahmid, N.A., Castro-Muñoz, R., 2024. Recent advances in reinforced bioplastics for food packaging – a critical review. International Journal of Biological Macromolecules 263, 130399. Available at https://doi.org/10.1016/j.ijbiomac.2024.130399.

Singh, A.K., Itkor, P., Lee, M., Saenjaiban, A., Lee, Y.S., 2024. Synergistic integration of carbon quantum dots in biopolymer matrices: an overview of current advancements in antioxidant and antimicrobial active packaging. Molecules 29 (21), 5138. Available at https://doi.org/10.3390/molecules29215138.

Stublić, K., Ranilović, J., Ocelić Bulatović, V., Kučić Grgić, D., 2024. Advancing sustainability: utilizing bacterial polyhydroxyalkanoate for food packaging. Processes 12 (9), 1886. Available at https://doi.org/10.3390/pr12091886.

Sun, X., Zhang, H., Wang, J., Dong, M., Jia, P., Bu, T., et al., 2021. Sodium alginate-based nanocomposite films with strong antioxidant and antibacterial properties enhanced by polyphenol-rich kiwi peel extracts bio-reduced silver nanoparticles. Food Packaging and Shelf Life 29, 100741. Available at https://doi.org/10.1016/j.fpsl.2021.100741.

Surendren, A., Mohanty, A.K., Liu, Q., Misra, M., 2022. A review of biodegradable thermoplastic starches, their blends and composites: recent developments and opportunities for single-use plastic packaging alternatives. Green Chemistry 24 (22), 8606–8636. Available at https://doi.org/10.1039/d2gc02169b.

Taherimehr, M., YousefniaPasha, H., Tabatabaeekoloor, R., Pesaranhajiabbas, E., 2021. Trends and challenges of biopolymer-based nanocomposites in food packaging. Comprehensive Reviews in Food Science and Food Safety 20 (6), 5321–5344. Available at https://doi.org/10.1111/1541-4337.12832.

Thakur, V., Satapathy, B.K., 2024. Migration Concerns of Biopolymer-Based Food Packaging. In: Kumar, S., Mukherjee, A., Katiyar, V. (Eds.), Agro-Waste Derived Biopolymers and Biocomposites: Innovations and Sustainability in Food Packaging. Wiley, pp. 421–450. Available at https://doi.org/10.1002/9781394175161.ch15.

Trinh, B.M., Chang, C.C., Mekonnen, T.H., 2021. Facile fabrication of thermoplastic starch/poly (lactic acid) multilayer films with superior gas and moisture barrier properties. Polymer 223. Available at https://doi.org/10.1016/j.polymer.2021.123679.

Tyagi, P., Salem, K.S., Hubbe, M.A., Pal, L., 2021. Advances in barrier coatings and film technologies for achieving sustainable packaging of food products – a review. Trends in Food Science and Technology 115, 461–485. Available at https://doi.org/10.1016/j.tifs.2021.06.036.

Udayakumar, G.P., Muthusamy, S., Selvaganesh, B., Sivarajasekar, N., Rambabu, K., Banat, F., et al., 2021. Biopolymers and composites: properties, characterization and their applications in food, medical and pharmaceutical industries. Journal of Environmental Chemical Engineering 9 (4), 105322. Available at https://doi.org/10.1016/j.jece.2021.105322.

Upadhyay, P., Zubair, M., Roopesh, M.S., Ullah, A., 2024. An overview of advanced antimicrobial food packaging: emphasizing antimicrobial agents and polymer-based films. Polymers 16 (14), 2007. Available at https://doi.org/10.3390/polym16142007.

Varghese, S.A., Phothisarattana, D., Srisa, A., Laorenza, Y., Jarupan, L., Bumbudsanpharoke, N., et al., 2023. Novel eco-friendly antimicrobial UV-blocking PBAT/PBS/TiO2 nanocomposite films for improved shelf-life of bananas. Food Bioscience 55, 102993. Available at https://doi.org/10.1016/j.fbio.2023.102993.

Wang, X., Gou, C., Gao, C., Song, Y., Zhang, J., Zhai, D., et al., 2021. Production of a novel non-toxic γ-PGA/casein composite hydrogel using MTG and optimization by response surface methodology. Grain & Oil Science and Technology 4 (2), 71–80. Available at https://doi.org/10.1016/j.gaost.2021.04.001.

Wu, F., Misra, M., Mohanty, A.K., 2021. Challenges and new opportunities on barrier performance of biodegradable polymers for sustainable packaging. Progress in Polymer Science 117, 101395. Available at https://doi.org/10.1016/j.progpolymsci.2021.101395.

Xu, M., Fang, D., Kimatu, B.M., Lyu, L., Wu, W., Cao, F., et al., 2024. Recent advances in anthocyanin-based films and its application in sustainable intelligent food packaging: a review. Food Control 162, 110431. Available at https://doi.org/10.1016/j.foodcont.2024.110431.

Zeng, X., Du, Z., Ding, X., Zhao, Y., Jiang, W., 2020. Preparation, characterization and in vitro hypoglycemic activity of banana condensed tannin–inulin conjugate. Food & Function 11 (9), 7973–7986. Available at https://doi.org/10.1039/d0fo01652g.

Zhang, J., Zhang, J., Huang, X., Zhai, X., Li, Z., Shi, J., et al., 2025. Lemon-derived carbon quantum dots incorporated guar gum/sodium alginate films with enhanced the preservability for blanched asparagus active packaging. Food Research International 202, 115736. Available at https://doi.org/10.1016/j.foodres.2025.115736.

Zhang, L., Zhao, J., Li, F., Jiao, X., Yang, B., Li, Q., 2024. Effects of amylose and amylopectin fine structure on the thermal, mechanical and hydrophobic properties of starch films. International Journal of Biological Macromolecules 282 (Pt 5), 137018. Available at https://doi.org/10.1016/j.ijbiomac.2024.137018.

Zhang, W., Roy, S., Assadpour, E., Cong, X., Jafari, S.M., 2023. Cross-linked biopolymeric films by citric acid for food packaging and preservation. Advances in Colloid and Interface Science 314, 102886. Available at https://doi.org/10.1016/j.cis.2023.102886.

Zhao, C.C., Lin, Q.B., Xie, C.H., Liu, Y.Q., Zhong, H.N., Gu, W.Y., et al., 2024. Screening and safety assessment of migrating substances released from biodegradable packaging materials into milk. Food Control 166, 110755. Available at https://doi.org/10.1016/j.foodcont.2024.110755.

Zhao, Z., Liu, H., Tang, J., He, B., Yu, H., Xu, X., et al., 2023. Pork preservation by antimicrobial films based on potato starch (PS) and polyvinyl alcohol (PVA) and incorporated with clove essential oil (CLO) Pickering emulsion. Food Control 154, 109988. Available at https://doi.org/10.1016/j.foodcont.2023.109988.

CHAPTER 11

Biodegradation and circular economy practices for certain bioplastics

Mehran Ghasemlou[1,2,3], Benu Adhikari[2,4], and Fugen Daver[5]

[1]Center for Sustainable Bioproducts, Deakin University, Geelong Waurn Ponds Campus, Geelong, VIC, Australia, [2]School of Science, STEM College, RMIT University, Melbourne, VIC, Australia, [3]Department of Chemical Engineering, The University of Melbourne, Parkville, VIC, Australia, [4]Centre for Advanced Materials and Industrial Chemistry (CAMIC), RMIT University, Melbourne, VIC, Australia, [5]School of Engineering, RMIT University, Bundoora, VIC, Australia

11.1 Introduction

Plastics are deeply integrated into the fabric of modern society. Their lightweight, mechanical strength, and versatility have made them indispensable across virtually every sector, from extending food shelf life and enabling safe transport to reducing fuel consumption through lightweight vehicle components and improving energy efficiency as thermal insulators (Ghasemlou et al., 2025). This linear (take–make–waste) economic model currently produces around 380 million tons (Mt) annually, resulting in huge amounts of plastic waste (Fig. 11.1).

Despite these undeniable benefits, their ubiquity and resistance to degradation have turned plastics into a significant threat to ecosystems and human well-being. The accumulation of plastic waste in landfills, oceans, and terrestrial ecosystems, combined with the inefficiencies of current recycling systems, has created an urgent demand for sustainable alternatives. Strategies to mitigate the impact of plastic waste include reducing plastic use through eco-design, reconfiguring supply chains to promote circularity, and developing alternatives with lower life-cycle footprints. A promising solution is to use bioplastics, which are derived from renewable and/or biodegradable sources. Ideally, bioplastics offer similar functionality to fossil-based plastics but with a lower carbon footprint (Rosenboom et al., 2022). Estimates suggest that replacing two-thirds of conventional plastics with bioplastics could reduce greenhouse gas emissions by 241–316 Mt CO_2-equivalent annually (Hao et al., 2024). The field of bioplastic research is filled with conceptual and practical ambiguities. The prefix "bio" is commonly used to imply sustainability and environmental protection, but such marketing has proven counterproductive. Consumers frequently assume that all bioplastics are biomass-derived, rapidly degradable and environmentally benign. The terminology surrounding "bioplastic," "biodegradable," and "compostable" is neither harmonized nor consistently

Bioplastics for Food Packaging. DOI: https://doi.org/10.1016/B978-0-443-41654-5.00004-4

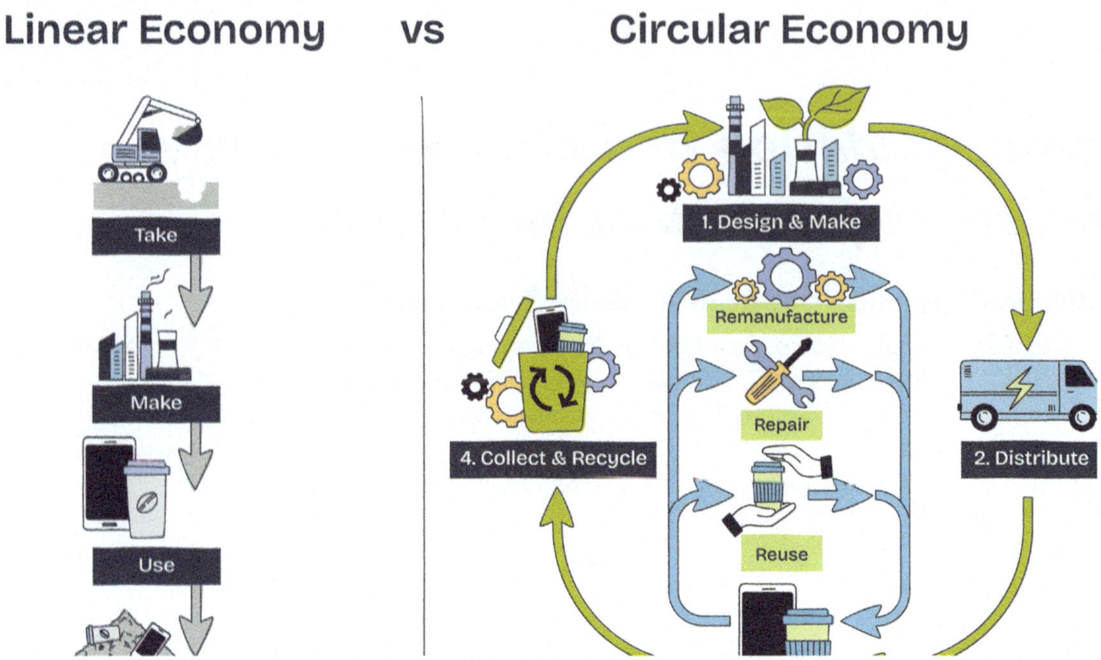

Figure 11.1
Schematic comparison of linear and circular economic models. The linear economy has a one-way pathway of *take–make–use–dispose*, resulting in significant waste generation. In contrast, the circular economy is designed as a closed-loop system that integrates reuse, repair, remanufacturing, and recycling across the product life cycle, thereby extending material value, reducing resource extraction, and minimizing waste.
The image is obtained under a license from Shutterstock.com.

applied across scientific, industrial, and policy contexts (Goel et al., 2021). This lack of clarity has fostered widespread misconceptions, most notably the assumption that all bioplastics are derived from biomass, degrade rapidly in nature, or are inherently safer for the environment. This chapter reviews the terminologies of bioplastics, identifies the most prevalent misconceptions, and examines how these shape consumer behavior, industrial practices, and policy design. Finally, it highlights the role of certification and labeling schemes in validating environmental claims and guiding responsible disposal. Understanding this information is critical to ensure that bioplastics contribute meaningfully to a sustainable circular economy.

11.2 Terminologies and misconceptions about bioplastics

The term bioplastic refers to a broad class of polymers that are either bio-based, biodegradable, or exhibit both attributes (Yuan et al., 2021). The terminology surrounding bioplastics does not have clear boundaries, which causes ambiguity, even within scientific and industrial communities. The absence of clear and harmonized definitions not only impedes informed policy and research discussions but also contributes to widespread misconceptions among consumers. In many cases, this ambiguity is strategically exploited to support greenwashing practices, where environmental claims are overstated or unfounded (Nazareth et al., 2019). One of the most common strategies is the unqualified use of the prefix "bio" to suggest ecological superiority and to appeal to environmentally conscious buyers. However, this label has become increasingly counterproductive (Ghasemlou et al., 2024). The public now widely associates terms such as "bioplastic" or "biodegradable plastic" with environmental safety, recyclability, or rapid degradation, regardless of the material's actual chemical properties or degradation pathways. For example, a pervasive misconception is that all bioplastics are derived from biomass and naturally degrade within a short period, irrespective of environmental conditions (Wang et al., 2021). Such misunderstanding frequently originates from deceptive advertisements that pairs visuals of agricultural feedstocks (e.g., corn, sugarcane, or leaves) with fragmented bottles or cups to promote the idea that the sustainability of a product is governed by its feedstock origin or disposal route, rather than by its full life-cycle impacts or postconsumer management (Dilkes-Hoffman et al., 2019). In some cases, this perception has driven consumer choices toward bio-based but non-biodegradable polymers (such as bio-polyethylene terephthalate (PET) or bio-polyethylene (bio-PE)) over fossil-based biodegradable alternatives. Other studies, however, have shown that consumers care less about whether a product is made from biomass and/or has a lower production carbon footprint if it is labeled "degradable" (which could indeed be oxo-degradable). Dispelling the false equivalence between bioplastics and biodegradability remains a pressing challenge in both science communication and policy development (Hao et al., 2024).

The term "biodegradable" itself is full of inconsistencies. Although several definitions exist, none universally establishes quantitative criteria for biodegradability labeling. A commonly used definition states that a material is biodegradable if it can be broken down by naturally occurring microorganisms into CO_2 under aerobic conditions, or into CO_2 and methane under anaerobic conditions. However, this definition omits critical parameters such as temperature, microbial diversity, moisture availability, degradation timescale, and extent of mineralization. Without any specification for time frame, virtually any material could qualify as biodegradable, whether it takes hundreds or even millions of years to fully degrade (Zhu and Wang, 2020). In the absence of environmental context and standardized test conditions, the term becomes largely scientifically meaningless and highly susceptible to misuse.

Another common misconception is the belief that "biodegradable" and "compostable" are interchangeable. In fact, compostable plastics are a specific subset of biodegradable materials that must be decomposed within a defined period under industrial composting conditions (typically 58°C and high humidity). For instance, polylactic acid (PLA) is often marketed as biodegradable, yet in common environmental settings such as soil or seawater, where temperatures remain far lower than 58°C, it shows little to no degradation. Only under tightly controlled thermophilic conditions, PLA can likely meet compostability standards (Taufik et al., 2020). Similarly, polycaprolactone, poly (butylene adipate-co-terephthalate) (PBAT), polyhydroxyalkanoates (PHAs), and poly (butylene succinate) (PBS), despite being labeled biodegradable, often require long periods of more than 6 months to degrade significantly in natural environments, and may accumulate in landfills or oceans if mismanaged. Paper-based products (such as paper cups or cardboard) are often viewed as more eco-friendly because they are plant-based materials. However, these products do not necessarily decompose faster than bioplastics because they are often coated and/or laminated with other synthetic compounds and do not always meet biodegradation requirements.

Another frequently cited concern is that using food crops for bioplastic feedstocks could threaten food security by diverting land and resources away from agriculture. While this issue merits attention in the broader context of bioeconomy ethics, current data suggest the actual land-use impact of bioplastics is minimal (Carus and Dammer, 2013). Farmland dedicated to bioplastics production remains below 0.06% of total agricultural land worldwide. Even in a hypothetical scenario where all 380 million tons of global fossil-based plastics are replaced with bioplastics, land use would rise only to ~1.5% (Blesin et al., 2017). Moreover, emerging innovations in second-generation feedstocks such as agricultural waste, algae, and CO_2-derived polymers offer scalable alternatives that do not compete with food production (Herrmann et al., 2022).

11.3 Biodegradation of plastics and/or bioplastics

The biodegradation of plastics and bioplastics is a multistep, biologically mediated process governed by a combination of polymer properties, environmental conditions, and microbial activity (Ghasemlou et al., 2022). As illustrated in Fig. 11.2, this process can be broadly divided into three sequential phases: biofilm formation and enzymatic depolymerization, biofragmentation, and bio-assimilation.

In this first phase, free-living bacteria and fungi from soil, aquatic systems, or the atmosphere colonize the polymer surface. These microorganisms secrete extracellular polymeric substances, which are sticky biopolymers that facilitate adhesion and form a biofilm on the plastic surface. Within this microenvironment, cells release a suite of hydrolytic enzymes (such as lipases, proteases, and hydrogenases), which preferentially cleave the amorphous regions of the polymer matrix (Chen and Patel, 2012). These enzymatic reactions break down long-chain macromolecules into shorter oligomers and monomers, making them more accessible for microbial uptake. The second phase, which is termed as biofragmentation,

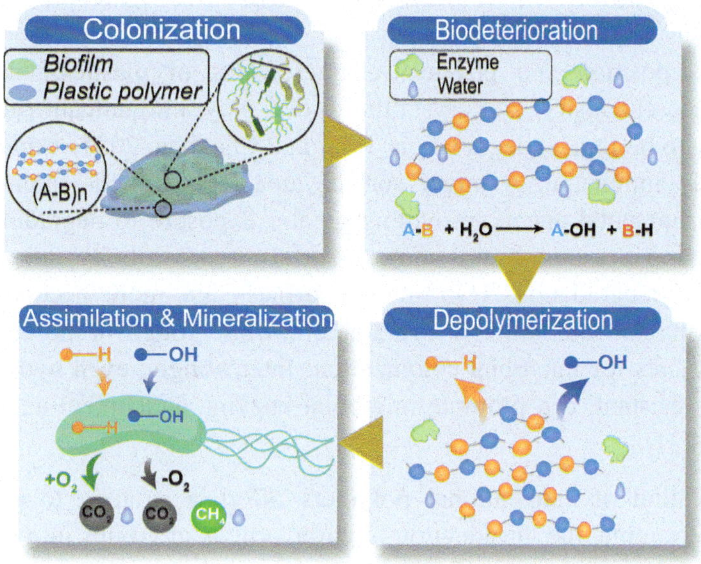

Figure 11.2

Schematic illustration of a typical biodegradation process of polymeric materials. The process begins with surface colonization and biofilm formation on the polymer, followed by enzymatic biodeterioration in the presence of water, leading to bond cleavage. Subsequent depolymerization generates lower-molecular-weight fragments, which are assimilated by microorganisms and ultimately mineralized into end products such as CO_2, CH_4, and water under aerobic or anaerobic conditions. *No permission required.*

involves further enzymatic degradation by a broader consortium of microorganisms, leading to a gradual reduction in molecular weight and the generation of low-molecular-weight intermediates. As fragmentation proceeds, the polymer structure becomes increasingly porous and chemically accessible, enabling more enzymatic penetration. The final phase, which is termed as bio-assimilation, begins when degradation products are sufficiently small to be transported across microbial cell membranes. Once internalized, these fragments enter central metabolic pathways, such as glycolysis and the citric acid cycle. Under aerobic conditions, the products are ultimately oxidized to CO_2 and water, while anaerobic conditions yield a mixture of CO_2 and methane. When the polymer's carbon content is completely converted into these simple metabolites, the biodegradation process is complete. Despite its apparent simplicity, plastic biodegradation is affected by a complex interplay of physical, chemical, and biological factors, which can either accelerate or hinder the rate of degradation. These factors are generally grouped into three domains: polymer characteristics, environmental conditions, and microbial function. Understanding these factors is essential not only for evaluating the environmental fate of bioplastics but also for guiding the rational design of polymers with tailored degradation profiles (Kim et al., 2023a).

11.3.1 Polymer characteristics

The chemical composition and physical characteristics of plastics, such as molecular architecture, surface chemistry, and crystalline order, play a significant role in determining their susceptibility to biological degradation (Kaur and Chauhan, 2024). Among these, surface area is particularly influential. A larger surface area increases the interface between the polymer and environmental agents, providing greater exposure to degrading agents such as enzymes, oxygen, and moisture. In this context, surface hydrophilicity and roughness are also key variables in biodegradation. Hydrophilic surfaces are more accessible to microbial enzymes, facilitating the initiation of degradation, while rough or porous surfaces create additional contact sites for microbial colonization. Interestingly, even hydrophobic surfaces, although initially resistant, can promote microbial enzyme attachment and persistence over time (Ali et al., 2021).

Chemical modification of conventional polymers offers a strategy to enhance microbial accessibility. For example, the introduction of carboxylic acid groups into traditional plastics such as PE has been shown to promote microbial assimilation (Filiciotto and Rothenberg, 2021). More generally, bio-based plastics that contain polar functional groups (e.g., such as esters, acids, or alcohols) are more susceptible to hydrolytic attack because these groups promote hydrophilic interactions and facilitate water diffusion into the polymer matrix. Similarly, incorporating amine or amide groups supports biodegradability by providing enzymatically cleavable sites, while the presence of halogens, due to their inherent toxicity and chemical stability, tends to hinder microbial colonization and enzymatic activity. In contrast, aliphatic groups, which are less reactive and more chemically inert, offer fewer sites for enzymatic cleavage and thus reduce the overall rate of biodegradation.

Another critical factor influencing biodegradation is the degree of crystallinity. Highly crystalline polymers have a tightly packed molecular arrangement that creates a physical barrier to the penetration of water, oxygen, and enzymes. This structural rigidity impedes microbial access and slows down degradation rates. Conversely, amorphous regions, with their looser molecular packing, are more vulnerable to enzymatic attacks and often degrade more rapidly. Polymers with low glass transition and melting temperatures are also typically more prone to biodegradation, as they tend to soften under ambient environmental conditions, allowing microbial and enzymatic agents to access and act on the polymer matrix more readily.

The chemical complexity of a polymer backbone further shapes its interaction with microbial communities. Plastics with simpler, linear chemical structures are generally easier to degrade because their bonds can be cleaved more efficiently by microbial enzymes. In contrast, complex molecular architectures with strong covalent bonds and aromatic rings resist degradation, requiring additional abiotic processes such as hydrolysis, oxidation, or ultraviolet (UV)-assisted breakdown to reduce their molecular size before microbial uptake can occur.

Molecular weight is another important determinant. High-molecular-weight polymers are generally more resistant to microbial degradation because large polymer chains cannot be directly internalized by microbial cells. Instead, microbes depend on extracellular enzymes to fragment the polymer into low-molecular-weight oligomers and monomers, which can then be metabolized. Natural polymers such as starch and cellulose are highly biodegradable due to their hydrophilicity, relatively simple structure, and the widespread presence of specific depolymerizing enzymes. In contrast, many synthetic biodegradable polymers must first undergo abiotic depolymerization via hydrolysis, oxidation, or photodegradation before becoming bioavailable. Based on these mechanistic insights, it has been hypothesized that even some fossil-based polymers, such as low-density polyethylene (LDPE), may exhibit limited biodegradation under specific conditions owing to their amorphous, branched structure. However, branching alone is insufficient to ensure biodegradability, particularly in the absence of cleavable functional groups. The case of PLA illustrates the interplay between crystallinity and degradation: PLA in its amorphous state degrades more readily than its semicrystalline counterpart, which resists enzymatic attack due to tighter molecular packing. Annealing PLA postinjection molding increases its crystallinity, thereby reducing its biodegradability. A study by Itävaara et al. (2002) demonstrated increased biodegradation of amorphous PLA, while Kolstad et al. (2012) reported minimal degradation of semicrystalline PLA under landfill conditions, but up to 37% biodegradation in amorphous PLA over the same period. This difference is attributable to the fact that semicrystalline PLA, with its longer chains and higher molecular weight, requires initial hydrolytic depolymerization at high temperatures before microbial metabolism can proceed (Bátori et al., 2018).

11.3.2 Environmental conditions

Environmental conditions, including geographical location, climatic factors, and air pollution, exert a profound effect on the mechanisms and rates of plastic degradation (Andrady et al., 2022). Multiple studies have demonstrated that plastic fragmentation on beaches can occur within months to a few years, primarily due to intense solar radiation and extensive environmental exposure. Key environmental drivers such as temperature, humidity, oxygen availability, and sunlight intensity interact in complex ways to accelerate or decelerate the breakdown of plastic materials. High temperatures generally increase microbial metabolic activity and accelerate hydrolytic cleavage of polymeric chains. This is achieved through increased diffusion of oxygen and the promotion of free radical formation. In contrast, lower temperatures slow microbial respiration and enzymatic degradation processes, thereby extending the environmental persistence of plastics. The role of oxygen is also critical: in aerobic environments, oxidative reactions between oxygen- and carbon-centered radicals (during the initial steps of polymer degradation) are essential to advance the breakdown of plastics (Queste et al., 2013). Microorganisms, particularly aerobic bacteria and fungi, use oxygen-dependent enzymatic pathways to efficiently break polymer chains. However, under anaerobic conditions, both microbial diversity and metabolic rates are reduced, which substantially limit the overall degradation rate.

Moisture content is another critical element for biodegradation because it facilitates the transport of enzymes and other degrading agents to the plastic surface. Water also aids in dissolving plastic additives and enables enzymatic access to polymer backbones. As a result, plastics degrade more rapidly in humid or water-submerged environments, where water acts as both a transport medium and a reaction facilitator. UV radiation from sunlight plays a dual role in the biodegradation of plastics. It can initiate photochemical chain scission via photooxidation, thereby enhancing biodegradability by creating surface cracks and reducing molecular weight. In this context, Kitamoto et al. (2011) reported increased plastic degradation with greater sunlight exposure, attributing the phenomenon to enhanced photooxidative cleavage of molecular bonds. Likewise, Pischedda et al. (2019) observed that higher temperatures intensified polymer chain mobility, further facilitating enzymatic degradation. UV-induced photodegradation typically targets C–H bonds, particularly in systems with conjugated bonds (e.g., aromatic rings). Although aromatic groups contribute to polymer rigidity and exhibit radical-scavenging properties that suppress biodegradation, their structures are also vulnerable to UV-induced chain breakage (Gewert et al., 2015). The integration of aliphatic ester groups into such polymers has been shown to improve their biodegradability (e.g., in PBAT, a semiaromatic biodegradable polyester) (Shah et al., 2014). However, UV radiation may also initiate photo-cross-linking reactions, especially in conjugated systems, which can hinder degradation by increasing polymer rigidity. From a broader perspective, UV-mediated photodegradation not only prepares plastics for microbial attack by disrupting structural integrity but may also generate toxic by-products that inhibit microbial colonization. Despite these trade-offs, plastics exposed directly to sunlight tend to exhibit faster degradation than those shielded from it. Therefore, a synergistic combination of high temperature, abundant oxygen, and high humidity levels is generally required to promote efficient plastic degradation. For example, PLA, despite being a biodegradable polymer, requires high temperatures to initiate the hydrolysis of its molecular structure and the subsequent reduction in molecular weight necessary for microbial uptake (Lyu et al., 2007). Under low-temperature aerobic conditions (e.g., those found in home composting systems), PLA degrades very poorly. Indeed, Rudnik and Briassoulis (2011) observed no significant degradation of a 30 μm thick PLA film over an 11-month period under home composting conditions. In contrast, Itävaara et al. (2002) reported a 400% increase in biodegradation rate and a near-elimination of the lag phase when PLA was subjected to industrial composting, which maintains high temperatures and microbial activity levels. The different outcomes observed in these two studies show the critical role of environmental design in shaping plastic degradation pathways and highlight the importance of aligning material properties with realistic end-of-life conditions.

11.3.3 Role of microorganisms

Microorganisms such as bacteria, fungi, and algae play a central role in the biodegradation of plastics. Their ability to colonize polymer surfaces, depolymerize long-chain molecules, and assimilate the resulting monomers is essential for the mineralization of synthetic and bio-based plastics. To date, more than 90 microbial genera capable of degrading plastics have been isolated from soils, suggesting that terrestrial environments offer significant potential for plastic breakdown when conditions are favorable. In contrast, although marine ecosystems are rich in microbial diversity, less than 0.1% of marine microbes are thought to possess plastic-degrading capabilities. This limited functional diversity may partly explain the slower degradation rates observed in aquatic environments, where low temperatures, oxygen limitation, and nutrient scarcity further prevent microbial activity. Microbial specificity towards certain plastic types has also been observed, particularly for polyesters such as PLA and PET. These microorganisms secrete hydrolytic enzymes that target ester bonds, enabling more effective breakdown under suitable environmental conditions. Importantly, abiotic stressors (e.g., UV radiation, mechanical abrasion, thermal aging, and chemical exposure) can serve as priming steps for microbial attack by introducing oxidation and chain scission, thereby reducing molecular weight and increasing surface roughness. Recent studies have highlighted the role of rhizospheric and plant growth–promoting microorganisms in accelerating plastic degradation. For example, the inoculation of PLA with rhizosphere microbes or burial in dairy wastewater sludge rich in nitrogen has led to substantial degradation, with one report showing complete disintegration of PLA sheets within 15 days. These findings suggest that the surrounding biotic environment can be tuned or engineered to enhance biodegradation rates (Pattanasuttichonlakul et al., 2018). However, the expanding diversity of polymeric materials used in consumer products poses an evolutionary challenge for the global microbial community. Unlocking the potential of natural and engineered microbial consortia, in concert with the rational design of more degradable polymers, is a promising frontier for mitigating plastic pollution. This synergy between microbiology and materials science is likely to play a pivotal role in closing the loop on plastic waste and advancing the goals of a circular bioeconomy.

11.4 Compostable bioplastics

Composting is a biological process that typically operates under aerobic conditions and converts organic biomass into nutrient-rich soil enrichment. In industrial composting facilities, bacteria and fungi decompose organic feedstock at high temperatures (50°C–60°C) (Nazareth et al., 2022). This thermophilic window is critical for removing heat-sensitive pathogens, fly larvae, and weed seeds. To ensure effective pathogens reduction, the US Environmental Protection Agency recommends maintaining a minimum temperature of 40°C for 5 days, with temperatures >55°C for at least 4 hours (Fig. 11.3).

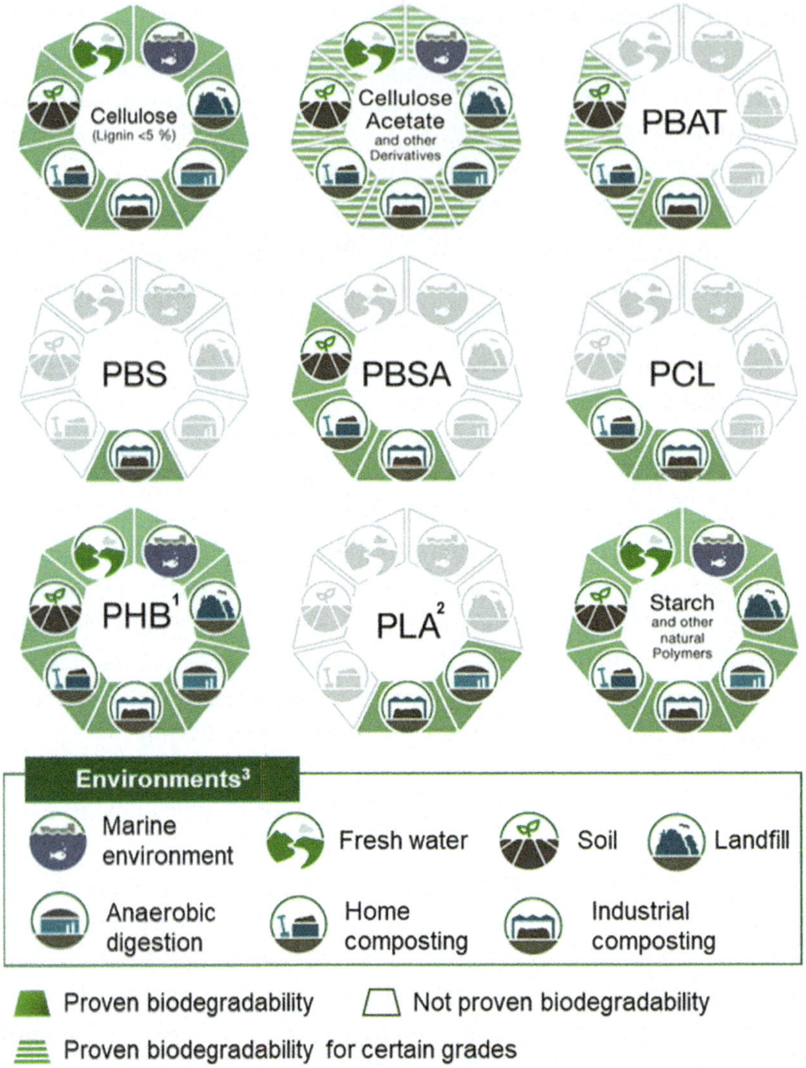

Figure 11.3

The biodegradability of major bioplastic materials in marine, freshwater, soil, landfill, anaerobic digestion, home composting, and industrial composting environments based on the information from market. Colored segments indicate environments in which the material demonstrates biodegradation, whereas grayed segments show conditions where biodegradation has not been demonstrated. The biodegradation conditions for each environment are as follow: Marine environment: temperature of 30°C, 90% biodegradation within a maximum of 6 months; Freshwater: temperature of 21°C, 90% biodegradation within a maximum of 56 days; Soil: temperature of 25°C, 90% biodegradation within a maximum of 2 years; Landfill: no European standard specifications are available; anaerobic digestion: two thermophilic (52°C) and mesophilic (37°C) conditions, 50% biodegradation within 2 months; home composting: temperate of 28°C,

(Continued)

Biodegradation of plastic substrates under these conditions converts organic carbon to CO_2, which (despite the benefits of composting) may contribute to global warming (Song et al., 2009). As such, industrial composting has also been identified as a main contributor to greenhouse gas emissions. That said, methane emissions are typically low, and nitrous oxide emissions are even lower, minimizing the overall climate impact under proper aerobic conditions. Despite these concerns, composting offers three main environmental benefits: (1) it helps sequester carbon in soil; (2) nitrogen content in compost serves as an organic fertilizer, enriching soil nutrient levels; and (3) it improves soil structure and acts as a natural conditioner. Plastic is deemed compostable when three criteria are met: (1) it disintegrates under microbial action; (2) it undergoes complete mineralization (i.e., conversion into carbon dioxide, methane, water, inorganic compounds, or biomass under aerobic conditions); and (3) it achieves a mineralization rate compatible with the composting process. Therefore, while some plastics may be biodegradable, they are not necessarily compostable. For instance, if a plastic leaves toxic residues after biodegradation, it cannot be classified as compostable. Although the terms *biodegradable* and *compostable* are often used interchangeably, a clear distinction between these two terminologies is important to understand their end-of-life fate. The fundamental difference lies in that composting is a human-driven, managed, and accelerated form of biodegradation and is conducted under defined environmental conditions. In contrast, biodegradation occurs naturally without external intervention. Under appropriate control, certain biodegradable plastics may also fulfill compostability requirements (Kim et al., 2023b). Hence, for the sake of brevity, the two labels may be used in parallel, but they are not synonymous. The bioplastics industry broadly categorizes compostable products into two groups: (1) industrial-composting products, which decompose at ~58°C, and (2) home-composting products, which decompose at lower temperatures (< 35°C) (Ghasemlou et al., 2024). Home composting processes are generally slower due to lower thermal input and less frequent aeration/mixing. Moreover, poorly managed home composting systems may result in methane production due to localized anaerobic conditions (Kale et al., 2007). However, in general, all home-compostable bioplastics are expected to degrade under industrial composting conditions. TÜV Austria offers two distinct certifications for compostable plastics. Each certification addresses two different end-of-life environments. The *OK Compost Home* certification indicates that a bioplastic is suitable for decomposition in home composting conditions, which typically involve lower and less consistent temperatures compared to industrial settings. In contrast, the *OK Compost* certification is limited to plastics/bioplastics

90% biodegradation within a maximum of 12 months and industrial composting: temperature of 58°C, 90% biodegradation within a maximum of 6 months.

Reproduced with permission from Ghasemlou, M., Barrow, C.J., Adhikari, B., 2024. The future of bioplastics in food packaging: an industrial perspective. Food Packaging and Shelf Life 43 101279. https://doi.org/10.1016/j.fpsl.2024.101279.

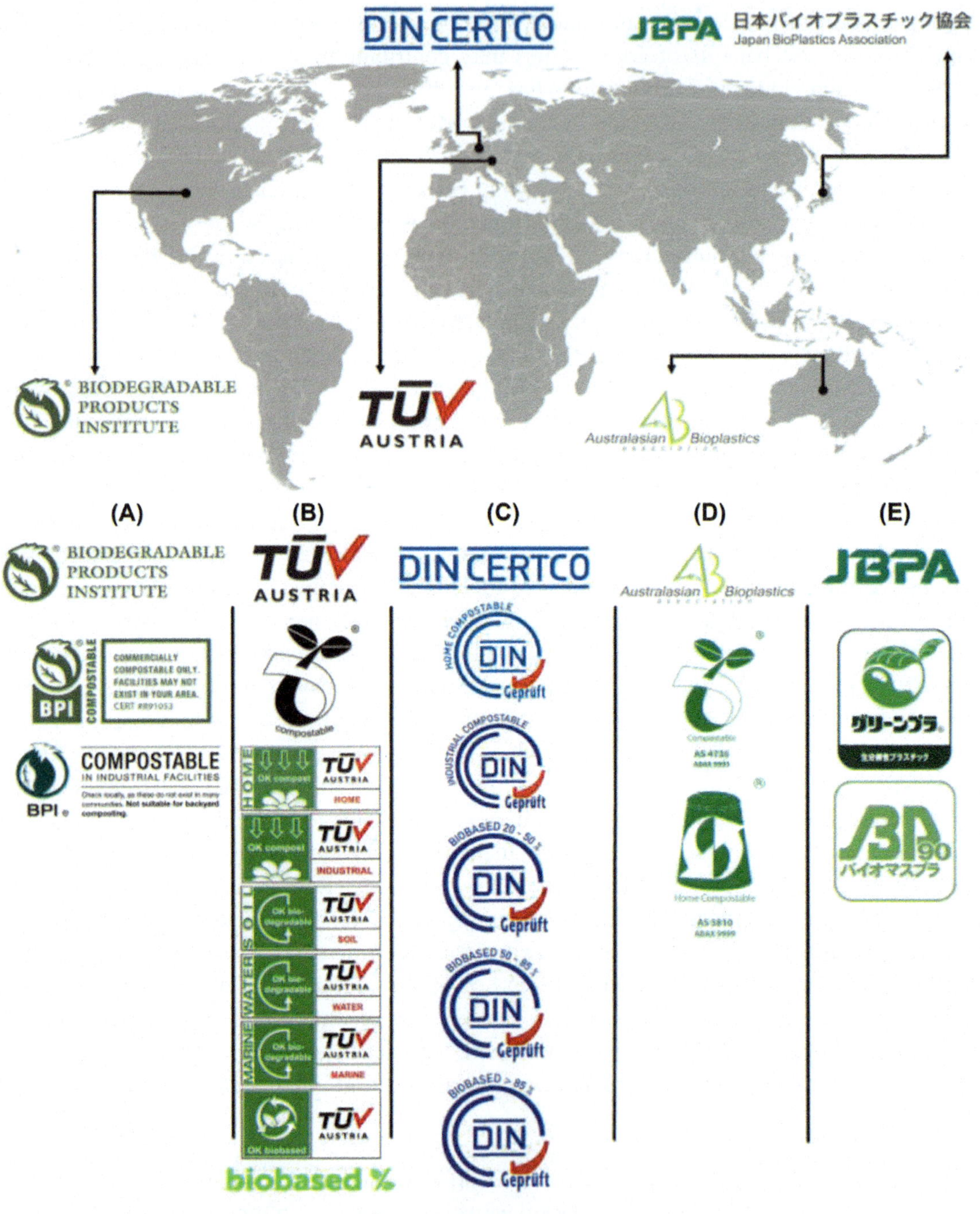

Figure 11.4

Global overview of certification and labeling schemes for bioplastics. (A) In North America, the Biodegradable Products Institute (BPI) certifies compostable products for industrial composting

(*Continued*)

that can fully or partially biodegrade only under the controlled, high-temperature conditions of industrial composting facilities. The distinction between these two certifications is critical for ensuring proper disposal pathways and for achieving the environmental benefits often attributed to compostable bioplastics. Misinterpretation of these two labels may result in improper disposal, potentially undermining the intended ecological advantages (Melchor-Martínez et al., 2022) (Fig. 11.4).

11.5 Testing methodologies for biodegradable and compostable bioplastics

International standard organizations, such as ASTM (American Society for Testing and Materials) and ISO (International Organization for Standardization), European Committee for Standardization (CEN) in collaboration with various national agencies, have developed detailed guidelines, specifications, and performance criteria to ensure that a material, product, process, or service is suitable for its intended use. These standards are essential for facilitating international trade by establishing a shared technical language among buyers, sellers, and regulatory bodies. Conformity assessment serves as the mechanism through which manufacturers and other stakeholders demonstrate that a product, process, system, or individual complies with a defined standard. This process underpins consumer confidence by verifying product reliability, safety, and quality. Specification standards for bioplastics are typically provided by ISO, CEN, and ASTM. These standards are primarily designed to assess the degradability of plastics through the action of naturally occurring microorganisms (e.g., bacteria, fungi, and algae) in environmental conditions that simulate those found in real-world conditions. However, current methodologies focus largely on abiotic factors (e.g., pH, water content, and carbon-to-nitrogen ratio), while often neglecting biotic variables such as microbial community profiles. Moreover, significant variability in abiotic conditions, test

facilities. (B) In Europe, TÜV Austria provides a suite of "OK" certifications, including *OK compost HOME*, *OK compost INDUSTRIAL*, *OK biodegradable SOIL*, *OK biodegradable WATER*, *OK biodegradable MARINE*, and *OK biobased*. (C) DIN CERTCO (Germany) certifies products under EN 13432 and International Organization for Standardization (ISO) standards, with labels for home compostability, industrial compostability, and bio-based carbon content. (D) In Australia and New Zealand, the Australasian Bioplastics Association (ABA) certifies products to AS 4736 (industrial compostability) and AS 5810 (home compostability). (E) In Japan, the Japan BioPlastics Association (JBPA) oversees the *GreenPla* and *BiomassPla* logos, the latter requiring >90% bio-based carbon content. Map lines delineate study areas and do not necessarily depict accepted national boundaries. *Reprinted with permission from Melchor-Martínez, E.M., Macías-Garbett, R., Alvarado-Ramírez, L., Araújo, R.G., Sosa-Hernández, J.E., Ramírez-Gamboa, D. et al., 2022. Towards a circular economy of plastics: an evaluation of the systematic transition to a new generation of bioplastics. Polymers 14 (6), 1203. https://doi.org/10.3390/polym14061203.*

durations (ranging from 30 to 180 days), and test scales (spanning from 100- mL to 180-L vessels) contribute to low reproducibility and limited predictive power of these assays. This inconsistency reflects the inherent complexity of biodegradation processes, which are governed by intricate interactions between the polymer's molecular structure and both abiotic and biotic environmental factors (Briassoulis and Dejean, 2010). This section aims to briefly summarize the current international standards and evaluate their relevance in the context of bioplastics testing. One common method for assessing aerobic biodegradation involves quantifying the cumulative CO_2 released from microbial oxidation of the test material. According to biodegradability criteria, the amount of CO_2 generated from the plastic must exceed 90% of that released from a reference substrate (typically cellulose) within a maximum period of 6 months (Kijchavengkul and Auras, 2008). To ensure accurate comparison between materials, CO_2 emissions must be measured daily over this period. CO_2 quantification can be achieved via gas trapping in an alkaline solution or an absorption column. Alternatively, more advanced techniques such as gas chromatography or infrared (IR) gas analysis may be used, although these methods require rigorous calibration (Hao et al., 2024). Despite their widespread adoption, these methods have inherent limitations. The percentage of biodegradability determined through CO_2 evolution does not account for the fraction of organic carbon assimilated into microbial biomass, which may not yet have been mineralized to CO_2. Another complicating factor is the *priming effect*—a phenomenon whereby the addition of external organic substances alters microbial decomposition rates of native organic matter in soil or compost (Bingeman et al., 1953). This effect can confound results because the measured CO_2 may not originate solely from plastic degradation. In response to this challenge, the revised ISO 14855-2 standard recommends using vermiculite (i.e., an inert clay mineral less susceptible to the priming effect) combined with compost extract as the composting matrix. A range of international test standards now exist for assessing aerobic biodegradation of plastics under different environmental conditions. These include ISO 14855 and ASTM D5338 for controlled composting, ISO 14851 and ISO 14852 for aqueous media, and ISO 17556 for soil environments. This section briefly reviews several of these standards used for assessing biodegradability and compostability of bioplastics and evaluates their relevance and limitations in the context of environmental performance.

11.5.1 American Society for Testing and Materials standards for assessing biodegradable and compostable plastics

To provide reliable assessments for biodegradable materials, the ASTM Committee on Plastics has developed a suite of test standards tailored to simulate various environmental disposal scenarios. One such standard is ASTM D5988-2003, titled *Standard Test Method for Determining Aerobic Biodegradation in Soil of Plastic Materials or Residuals After Composting*. This method specifically targets plastic materials that do not inhibit microbial activity in soil and compost environments. Its primary goal is to assess the extent and rate of aerobic biodegradation of plastic materials, particularly those containing biodegradable additives, under controlled laboratory conditions that simulate contact with soil or soil–compost mixtures. The biodegradation rate in this method is quantified by measuring cumulative CO_2 release over time. A plastic or bioplastic is typically considered biodegradable under this protocol if at least 70% of the material degrades within 6 months. The testing is conducted under tightly controlled conditions, including a constant room temperature (25°C–27°C), soil pH of 6–8, water content of 50%–55%, and a carbon-to-nitrogen (C/N) ratio of 15:1, which is adjusted using an ammonium phosphate solution. These parameters are designed to simulate environmental conditions conducive to biodegradation. However, this protocol does not fully capture the complexity of real soil environments, and its relevance to natural field conditions is limited. Furthermore, it does not address all potential safety concerns associated with the use of the tested materials.

ASTM D6400 provides a broader definition of biodegradability, defining a biodegradable plastic as one whose breakdown is driven by the metabolic activity of microorganisms such as bacteria, fungi, and algae. Under this definition, certain biopolymers such as PLA do not qualify as biodegradable because their principal decomposition pathway is hydrolysis rather than microbial oxidation. Although ASTM D6400 includes guidance on compostable plastics, it does not specify the type of composting environment, or the time frame required for complete degradation. This lack of specificity enables some materials to carry a "compostable" label even if they decompose more slowly than naturally compostable substances such as food waste or kraft paper. To claim compostability under ASTM D6400, a material must meet three distinct criteria: (1) Physical disintegration—visible breakdown of the plastic during composting; (2) Biodegradation—conversion of organic carbon into CO_2; and (3) Absence of toxicity—the final compost must not adversely affect soil or aquatic life (Hao et al., 2024). If these three benchmarks (i.e., physical, biochemical, and toxicological) are met, the material may be marketed as compostable. These requirements are verified in controlled laboratory-scale composting tests conducted in sealed vessels under thermophilic conditions. The protocol involves at least 12 vessels: three blanks, three negative controls (e.g., LDPE), three positive controls (e.g., cellulose), and three vessels containing the test material. Each vessel is filled with a composting mixture and receives a constant flow of humidified, CO_2-free air for 45–180 days at a constant temperature of 58°C. CO_2 evolution is measured either through titration or direct analysis using IR sensors or gas chromatography. For criterion I, the material must disintegrate such that less than 10% of its original dry weight remains after sieving through a 2-mm mesh. For criterion II, the cumulative organic carbon converted into CO_2 (also known as mineralization) must reach at least 60% of the theoretical value for homopolymers or random copolymers, and 90% for block copolymers, segmented copolymers, blends, or formulations containing low-molecular-weight additives. To meet criterion III, the concentration of heavy metals in the plastic must be less than 50% of the thresholds listed in 40CFR§503.13 and the final compost must pass a series of toxicity assessments.

These include aquatic toxicity tests with the rotifer *Brachionus* (Guide E1440), plant germination tests using cress seeds, plant growth tests following OECD Guideline 208, and earthworm toxicity tests in accordance with OECD Guideline 207 (Kale et al., 2007). In order to pass the toxicity test, the results from compost containing the test material must not differ significantly from those of blank compost. Prior to these evaluations, the final compost must be prepared according to test method D5951. Test method ASTM D5988, also referred to as the "soil contact test" can be also used as a rapid screening tool before conducting the more rigorous ASTM D5338 composting test. In this method, samples and reference materials are mixed into a soil matrix and placed in air-sealed vessels (e.g., desiccators) with each vessel containing only one type of material. These vessels are stored in the dark at 21°C (mesophilic conditions) for 30–60 days and CO_2 evolution is measured by titration. A positive outcome, defined as >60% or >90% conversion of total organic carbon into CO_2, suggests that the material is likely to biodegrade under higher-temperature composting conditions. However, a negative result should be followed up with the more definitive ASTM D5338 test.

Several other ASTM standards involve exposing materials to composting environments but are not specifically included in specification ASTM D6400 for identifying compostable materials. For example, test method ASTM D5929 evaluates biodegradability in municipal solid waste composting conditions at 40°C by measuring oxygen uptake and CO_2 generation over 45 days. The total biodegradability is calculated by comparing cumulative gas evolution to the theoretical maximum based on the test material's organic carbon content. ASTM D6954 provides a tier-based framework to compare and rank the degradation performance of polymers exposed to thermal and photo-oxidative aging, followed by biodegradation under different environmental conditions. In Tier 1, materials are exposed to thermal or photooxidation based on their intended use. In Tier 2, the aged fragments from Tier 1 are subjected to biometer tests such as D5338 or D5988 to quantify CO_2 evolution in simulated environments: soil (20°C–30°C), landfill (20°C–35°C), or compost (30°C–65°C). The 60% and 90% thresholds for organic carbon conversion, as defined in D6400, also apply here. In Tier 3, the residual materials are subjected to the same toxicity tests used in D6002 Tier 2.

11.5.2 European and International Organization for Standardization standards

The ISO has established rigorous methods for evaluating the aerobic biodegradability of plastics in soil environments. ISO 17556 provides a standardized procedure that quantifies either the oxygen demand or the CO_2 evolution as a function of microbial activity. To ensure accuracy and reproducibility, this test operates under optimized conditions: soil temperatures between 20°C and 25°C, a carbon-to-nitrogen (C/N) ratio of 40:1, water-holding capacity between 40% and 60%, and a pH range of 6–8. These parameters simulate typical soil conditions and enable reliable assessments of the material's ultimate biodegradation potential under aerobic settings. While European and ISO standards for composting environments remain relatively limited, EN 13432 stands out as a comprehensive specification. This standard applies to all forms of packaging and packaging components, setting it apart from ASTM standards in scope. EN 13432 establishes a four-pillar framework for evaluating compostability: (1) characterization, (2) biodegradability, (3) disintegration, and (4) compost quality or ecotoxicity. Together, these criteria reflect a similar scope to the three criteria defined by ASTM D6400, albeit with greater specificity and clarity.

The *characterization* phase involves a detailed analysis of the material's composition, including its organic carbon content, presence of hazardous substances (e.g., heavy metals), and levels of total dry and volatile solids. The content of volatile solids (defined as components that volatilize upon heating) must reach at least 50% of the material's dry weight. This is typically measured via ignition loss at 550°C, as detailed in ASTM D5338. In addition, heavy metal concentrations must remain below the thresholds specified in EN 13432. For *biodegradability*, ISO 14855 offers a closely aligned test method to ASTM D5338 yet introduces key differences. Unlike ASTM D5338, ISO 14855 does not require negative control vessels, which reduces the total number of test vessels from 12 to 9. It also permits optional calculation of biodegradability based on weight loss, providing a complementary metric to the standard carbon dioxide evolution measurement. Importantly, ISO 14855 adopts a more stringent threshold: biodegradation is considered successful when at least 90% of the theoretical or maximum CO_2 release from the reference material is achieved, and both the test and reference samples have reached a biodegradation plateau. The *disintegration* criterion emphasizes testing in pilot- or full-scale composting systems, as outlined in ISO 16929. This contrasts with ASTM D6400, which focuses solely on controlled laboratory-scale testing. Despite these methodological differences, the acceptance criterion is consistent: following 12 weeks of composting, <10% of the material's original dry mass should remain, based on screening through a 2 mm sieve. Lastly, the *compost quality and ecotoxicity* criterion aims to confirm that the degraded material does not impair the functional properties or safety of the compost. Parameters such as pH, density, dry and volatile solids, and salt content are measured to assess physical and chemical integrity. To address ecological impacts, EN 13432 incorporates the plant growth assay based on OECD Guideline 208. In this test, compost samples containing the test material are compared against control compost for both germination rates and biomass accumulation. Any statistically significant inhibition indicates potential ecotoxicity and precludes certification.

11.5.3 Testing methodologies for marine biodegradability

Understanding how plastics and bioplastics behave in marine environments is a critical component of environmental sustainability assessment. While many bioplastics marketed as biodegradable or compostable perform well under industrial composting conditions, their fate in marine ecosystems is different. Marine biodegradability testing is therefore essential to determine whether these materials pose a long-term pollution risk if released into aquatic environments. However, the current landscape of standardized methods for evaluating marine biodegradation remains fragmented and underdeveloped. At present, only a limited number of international standards (primarily from ASTM and ISO) exist for this purpose. Critically, no CEN method currently addresses marine biodegradability, creating a regulatory gap for European stakeholders. The existing methods predominantly assess aerobic degradation, even though many marine environments are suboxic, hypoxic, or anoxic, raising questions about the ecological relevance of such tests. Among ASTM standards, ASTM D6691 is the most widely used. It evaluates the aerobic biodegradation of plastics in the marine environment, either with a defined microbial inoculum or using natural seawater columns. The method involves monitoring CO_2 evolution as the primary degradation endpoint, providing a quantitative measure of microbial activity. Another relevant standard, ASTM D7473-12, focuses on the physical disintegration of materials in simulated marine environments, complementing biodegradability testing by assessing how quickly plastics break down into smaller pieces. Historically, ASTM D7081provided one of the first guidelines for assessing the biodegradation of nonfloating plastics under marine conditions. However, it was officially withdrawn in 2014 due to scientific and methodological limitations and was not replaced. To fill this void, TÜV Austria (formerly Vinçotte) introduced the OK biodegradable MARINE certification scheme. Although based on the withdrawn ASTM D7081 protocol, the certification retains many of its core principles and continues to be used by manufacturers and researchers as a benchmark for assessing marine biodegradability. Developing reliable and reproducible methods for marine biodegradation remains a substantial scientific challenge. Marine systems are highly dynamic, characterized by fluctuating salinity, temperature, nutrient availability, microbial diversity, light exposure, and oxygen concentrations. These variations create heterogeneous microenvironments that affect both the rate and mechanism of biodegradation. For instance, different marine environments such as the open ocean, coastal sediments, and intertidal zones differ vastly in microbial community composition and physical conditions. Consequently, results from one marine zone may not translate directly to another, complicating standardization efforts. In addition, the complex ecological phenomena unique to marine systems (e.g., biofouling, where plastic debris becomes colonized by marine microorganisms, algae, and invertebrates) introduce further variables. Biofouling can alter a plastic's density, surface chemistry, and degradation pathway. While it may increase microbial colonization, it can also form protective layers that inhibit degradation, depending on the species involved and the structure of the biofilm. Despite its relevance, the influence of biofouling on plastic degradation remains under-explored in most existing test protocols.

Another complicating factor is the comparatively low biodegradation potential of seawater. Unlike compost or soil, seawater has both lower enzymatic activity and microbial biomass, resulting in slower degradation rates. However, plastics in marine environments can still release dissolved organic carbon, which fuels microbial metabolism and may initiate subtle but meaningful degradation processes over long time frames. Capturing these nuanced dynamics requires sensitive analytical techniques and longer test durations than typically used in laboratory standards.

Recognizing these complexities, several international initiatives have been launched to improve and harmonize testing protocols. One of the most comprehensive efforts is the Open-Bio project, funded by the European Commission under the FP7 program. This project focuses on the environmental performance of bio-based materials and includes a strong emphasis on marine biodegradability. It evaluates material behavior across three ecologically distinct marine zones: (1) the sandy eulittoral zone, or intertidal beach, where materials are exposed to fluctuating wet–dry cycles and variable microbial communities; (2) the sublittoral zone, representing the benthic interface between water and seafloor, where low oxygen conditions and fine sediments predominate; and (3) the pelagic zone, comprising the water column, where sunlight and microbial degradation interact with photochemical and physical processes.

11.6 Labeling and certification schemes for biodegradability

As global concern over plastic pollution intensifies, standardized certification and labeling schemes have become essential tools for validating the environmental claims of biodegradable and compostable bioplastics. These schemes serve both as technical quality control mechanisms and as instruments for guiding consumer behavior and regulatory enforcement. Across different regions, certification systems are anchored in internationally recognized standards, such as ASTM D6400 in North America and EN 13432 in Europe. As mentioned earlier, EN 13432 defines criteria for packaging products that require compliance with specific thresholds for heavy metals, volatile solids, biodegradation rate, and plant ecotoxicity. The certification process under EN 13432 is carried out by authorized bodies such as TÜV Austria in Belgium and DIN CERTCO in Germany. One of the most recognizable indicators of EN 13432 compliance is the "Seedling" logo, a consumer-facing label that has become widely adopted in countries including Belgium, Switzerland, Poland, and the United Kingdom (Fig. 11.4). When printed on packaging, this logo signals that a product has been independently verified as compostable under industrial conditions, supporting consumer confidence and enabling responsible disposal. Beyond Europe and North America, certification programs have emerged globally to align with these international benchmarks. In Malaysia, the SIRIM ECO 001 standard administered by Standards and Industrial Research Institute of Malaysia (SIRIM), Quality Assurance Services certifies biodegradable and compostable plastics and bioplastics. This certification integrates criteria from EN 13432, ASTM, and ISO standards, and forms part of a broader eco-labeling strategy aimed at promoting sustainable development and green consumerism. Products certified under this scheme display the SIRIM Eco-Label, signifying compliance with internationally harmonized biodegradability metrics. For packaging manufacturers, obtaining a "compostable" certification is not only a mark of environmental responsibility but also a critical step for acceptance into industrial composting facilities. Certification assures facility operators that the product will degrade effectively under composting conditions without leaving harmful residues. Two of the most influential certifiers in this space are the Biodegradable Products Institute (BPI) and DIN CERTCO. BPI, in collaboration with the US Composting Council (USCC), certifies products based on ASTM D6400 and ASTM D6868 through a network of accredited laboratories. Certified products are authorized to display the BPI Compostable logo, while a list of approved laboratories is available through the BPI website. Similarly, DIN CERTCO uses ASTM D6400 or the EN 13432 standard to evaluate and certify compostable materials, awarding the DIN CERTCO Compostable logo to qualifying products. Additional certifications are provided by organizations like TÜV Austria, which offers the "OK compost Industry" label for products proven to meet EN 13432 criteria under industrial composting conditions. The OK Compost Industry program is a cornerstone for certifying industrial compostability. To qualify for this scheme, products must meet EN 13432 benchmarks, including ≥90% degradation within 12 weeks under thermophilic conditions (55°C–60°C), with adequate moisture and aeration. Importantly, the material must leave no harmful residues and must pass ecotoxicity assessments to confirm safety for soil and microbial life. However, this certification only applies to industrial composting settings. Home composting, which typically occurs at lower temperatures and with less active aeration, requires separate certification such as "OK compost Home" to reflect the slower degradation dynamics in domestic systems.

Certification bodies such as SGS also provide biodegradability and ecotoxicity testing aligned with ISO standards. Although not always coupled with compostable logos, these assessments function similarly to ISO 9000 series certifications, validating conformity with global biodegradability protocols. Currently, certified compostable products by BPI and DIN CERTCO include a wide range of consumer and commercial goods: compostable bags and films, water bottles, trays, dishes, disposable utensils, and bio-based resins. These certifications play a key role in ensuring that such materials are not only functionally compostable but also widely accepted by composting infrastructure, enabling a truly circular material flow.

Regulatory efforts have also begun to reinforce these standards. In California, Assembly Bill No. 2147 (2006) titled "Solid Waste: Compostable Plastic Food and Beverage Containers" prohibits labeling any plastic item as "degradable," "biodegradable," or "compostable" unless it meets ASTM-defined specifications. Unlike earlier laws targeting plastic bags, this bill extends to all food and beverage containers and aims to curb misleading environmental claims. However, the law does not explicitly mandate which ASTM method (e.g., D5338 or D6400) must be used, which could be a persistent gap between legislative intent and technical enforcement. In specialized applications, such as agriculture, targeted certifications have emerged. In France, biodegradable mulching films can be certified under the NFU52-001 standard, which requires successful biodegradation in at least two environmental matrices, one of which must be soil. Likewise, TÜV Austria has developed the "OK Biodegradable Soil" certification for products used in horticulture and agriculture. This certification applies to both finished products and upstream components, such as intermediate materials and polymer blends. The criteria for OK biodegradable soil are stringent: each organic constituent present at over 1% must achieve at least 90% biodegradation (absolute or relative to a reference material) within 2 years, using test methods such as ISO 17556, ISO 11266, or ASTM D5988. Constituents below 1% are included unless their total contribution exceeds 5%, and ecotoxicity testing is waived only for constituents under 0.1%, provided their cumulative amount does not exceed 0.5% of total dry weight. This comprehensive protocol ensures environmental safety in soil ecosystems.

11.7 Recyclability of bioplastics

The exponential growth of plastic production over the past century has generated an unparalleled material revolution. However, this success story has come at a considerable environmental cost. With over 390 million tons of plastic produced annually, only a fraction is effectively recycled, while the rest accumulates in landfills, incinerators, or the natural environment (Soroudi and Jakubowicz, 2013). Bioplastics have emerged as a hopeful alternative to conventional petroleum-based plastics, offering the promise of reduced carbon footprints, renewable feedstocks, and enhanced environmental performance. These materials ranging from bio-based, non-biodegradable plastics such as bio-PE and bio-PET to biodegradable polymers such as PLA and PHAs are increasingly used across packaging, agriculture, textiles, and biomedical sectors. However, despite their proliferation and green branding, a critical challenge remains largely unresolved, which is the absence of harmonized, science-based end-of-life guidelines for bioplastics. Without clear regulatory frameworks and practical disposal routes, bioplastics risk undermining their own sustainability narrative and can become not a solution to plastic pollution, but a parallel contributor to it. Chemical recycling is increasingly recognized as a critical enabler of circularity in the plastics economy, particularly for bioplastics, which often face barriers to mechanical recycling due to their distinct chemistries and limited integration into existing infrastructure. Unlike mechanical recycling, which reprocesses plastics or bioplastics through physical means (e.g., melting and extrusion), chemical recycling aims to break down polymers into their original monomers or other low-molecular-weight compounds (Choi et al., 2023) (Fig. 11.5).

These products may then be purified and repurposed as virgin-like feedstock, intermediates for industrial use (e.g., solvents or plasticizers), or substrates for further chemical transformation (Soroudi and Jakubowicz, 2013). For bioplastics, whose molecular structures can differ substantially from conventional fossil-based plastics, this route offers a much-

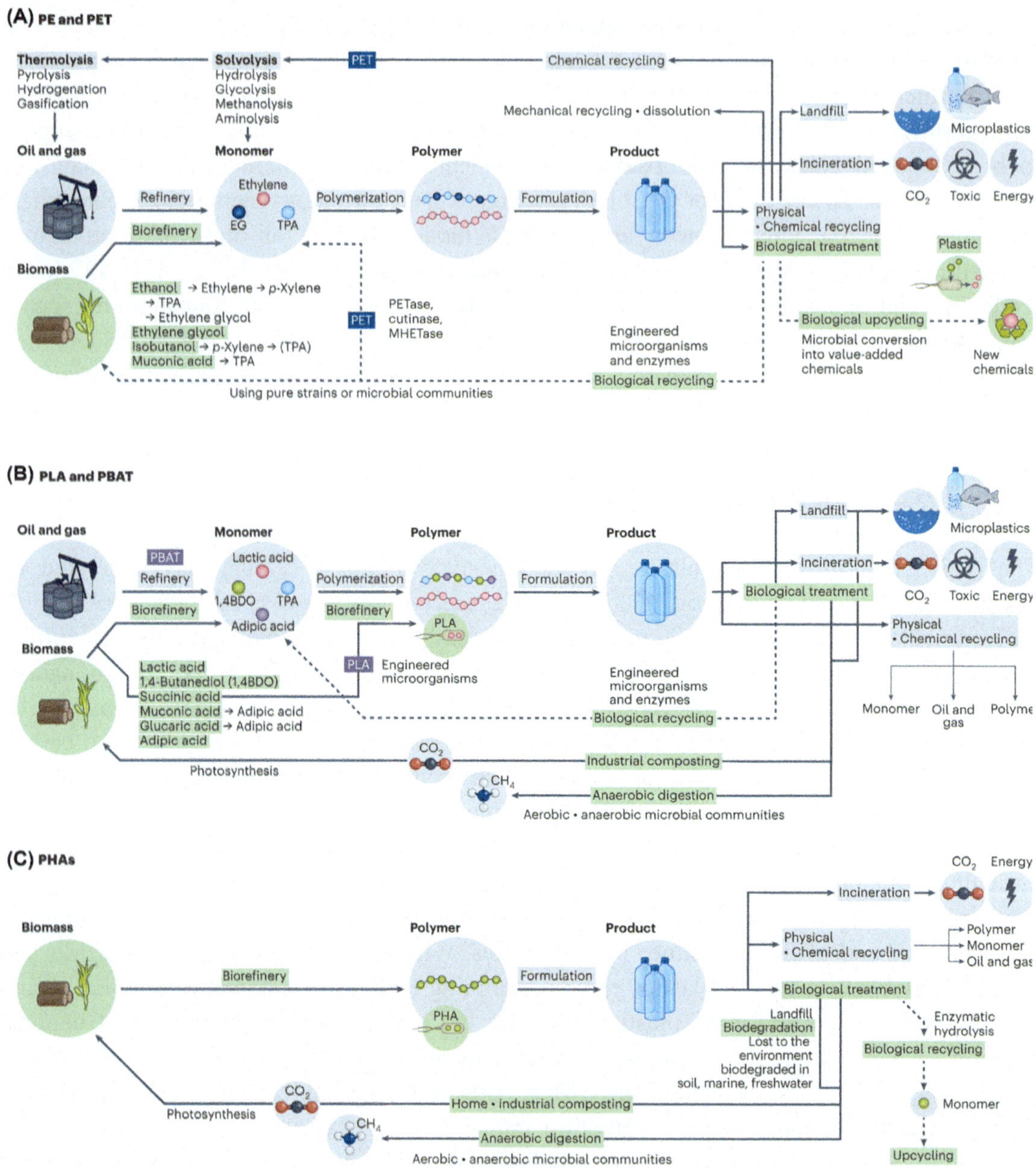

Figure 11.5

Life cycle pathways for fossil-based and biomass-derived plastics. Fossil-based routes rely on oil and gas refining to generate petrochemical monomers, whereas bio-based pathways utilize biomass in biorefineries, with carbon sourced from atmospheric CO_2 via photosynthesis. End-of-life options include mechanical and chemical recycling, enzymatic or microbially mediated biological recycling, industrial and home composting, anaerobic digestion under aerobic or anaerobic microbial

(*Continued*)

needed pathway for end-of-life valorization. Several bioplastics are well-suited for chemical depolymerization. For instance, PLA can be chemically hydrolyzed to recover lactic acid or methyl lactate, both of which are valuable building blocks for repolymerization or other applications. Similarly, PHAs, a family of microbial polyesters, can undergo methanolysis or pyrolysis to produce hydroxyalkanoic acid derivatives or short-chain esters. Some bio-based polyesters, such as PBS are also amenable to glycolysis, solvolysis, or enzymatic depolymerization under suitable conditions. Despite these advantages, the implementation of chemical recycling for bioplastics remains in its infancy, and several technological, economic, and environmental challenges must be addressed to unlock its full potential. A major technical challenge is to reduce energy efficiency. Depolymerization reactions often require high temperatures, catalysts, or prolonged reaction times, which make them energy-intensive and potentially carbon-positive unless powered by renewable energy. Achieving selective depolymerization without overdegradation of the target molecules is also difficult, particularly in the presence of additives, dyes, or multilayer film structures. Purification of the degradation products (particularly when starting from post-consumer waste) is another hurdle because residual contaminants may interfere with downstream applications or require costly separation steps. Another complication is the low availability of clean bioplastic waste for chemical recycling. Unlike PET or HDPE, which are widely used and collected at scale, bioplastics still occupy a small scale. The inconsistent labeling and insufficient consumer sorting results in bioplastic waste streams are mixed or misdirected into landfills and incinerators. Furthermore, life cycle assessments of chemical recycling for bioplastics are still sparse, and more data are needed to validate its environmental benefits over traditional end-of-life scenarios. Nevertheless, promising advances are beginning to emerge. Research in catalyst development is improving the selectivity and efficiency of depolymerization reactions at lower temperatures. Green solvents, such as ionic liquids and deep eutectic solvents, are also being explored to reduce process emissions and enhance product recovery. Looking ahead, the integration of hybrid recycling where mechanical, chemical, and organic routes coexist and complement one another represents a promising strategy. In such a system, bioplastics unsuitable for mechanical recycling could be chemically depolymerized, while compostable materials might be directed to industrial composters, and durable bio-based plastics returned back into mechanical cycles. Digital traceability tools, such as blockchain or QR-code labeling, could support this triage by identifying polymer types and directing waste

conditions, incineration with energy recovery, and landfill disposal. The figure shows that conventional plastics are predominantly managed through physical or chemical recycling and thermal treatments, while bio-based and biodegradable polymers enable biological treatment, depolymerization, and mineralization into CO_2 and CH_4, supporting more circular material flows and reduced microplastic persistence.

Reproduced with permission from Choi, S.Y., Lee, Y., Yu, H.E., Cho, I.J., Kang, M., Lee, S.Y., 2023. Sustainable production and degradation of plastics using microbes. Nature microbiology 8 (12), 2253–2276.

accordingly. Ultimately, achieving a circular economy for bioplastics will require innovation not only in materials and processes but also in infrastructure, regulation, and consumer engagement. Chemical recycling, despite its current limitations, holds the potential to become a cornerstone of this vision, and transform bioplastics from single-use commodities into regenerative materials within a truly sustainable cycle.

11.7.1 Chemical recyclability of bio-based and biodegradable bioplastics

PLA is currently the most widely used bio-based and biodegradable plastic. However, the biodegradation of PLA in environmental conditions is notoriously slow, which can lead to prolonged environmental persistence and pollution. As a result, PLA may pose waste management challenges similar to those of non-biodegradable plastics. In addition, the production of PLA precursors requires large amounts of agricultural land, raising concerns about competition with food crops, particularly in regions with limited resources. This aspect of PLA production could undermine the goals of a sustainable bioeconomy, especially in economically disadvantaged areas. Given these concerns, the widespread use of PLA must be accompanied by a thorough evaluation of its end-of-life options, particularly its recyclability. Identifying the most suitable recycling strategies and determining whether PLA can be effectively separated from other plastics for efficient recycling is crucial for minimizing its carbon footprint throughout its life cycle. One significant challenge lies in establishing an effective PLA waste collection system. Attempts to mix PLA into the well-established PET recycling stream have encountered resistance, primarily due to the difficulty in sorting PLA from PET. Despite their similar appearance, PLA and PET have distinct chemical properties, and even a small amount of PLA (as little as 1 wt.%) can severely impact the thermo-mechanical properties of recycled PET, forcing the resulting material to be downgraded for less demanding applications. As a result, there is a strong push from groups such as Plastics Recyclers Europe to develop separate recycling streams for biodegradable plastics to improve waste management efficiency. Chemical recycling (hydrolysis and alcoholysis) is an effective strategy for reprocessing PLA waste. Hydrolysis is the best option that allows PLA to be broken down into lactic acid with up to 95% conversion within 2 hours at temperatures between 160°C and 180°C. The efficiency of the hydrolysis process can be affected by factors such as polymer bond concentration, water content, and the presence of acidic hydrolysis products. Indeed, the hydrolysis of PLA is an autocatalytic process, meaning the carboxyl groups generated during hydrolysis further catalyze the reaction. However, the rate of hydrolysis is also impacted by polymer crystallinity, pH, temperature, absorbed water content, and the diffusion coefficient of degradation products. Although PLA can be hydrolyzed into lactic acid, further processing is required to convert it back into lactide, the key building block for PLA polymerization. This involves dehydrating the lactic acid, followed by polycondensation into oligomers and depolymerization to yield lactide. These additional steps complicate the process and make it less economically viable for closed-loop recycling, especially considering that dehydration and polycondensation can lead to an increased ratio of D-lactic acid over L-lactic acid due to partial racemization. A potential solution could involve partial hydrolysis, where PLA is broken down into oligomers that can be directly depolymerized into lactide, simplifying the recycling process. One of the significant advantages of chemical recycling over conventional biomass fermentation is the lower energy consumption. Producing 1 kg of lactic acid via hydrolysis requires only 14 MJ of energy, compared to the 55 MJ needed when obtaining it from corn biomass fermentation. This reduction in energy use makes hydrolysis a more sustainable option for PLA recycling. Alcoholysis is another viable chemical recycling method, where PLA reacts with alcohols such as methanol, ethanol, or propanol in the presence of catalysts (e.g., $Zn(Et)_2$ or $Zn(Pr)_2$) to produce lactate esters, such as methyl lactate, ethyl lactate, or propyl lactate. These lactate esters are valuable by-products with high market potential and can be further converted into lactide, closing a circular economy loop. Alcoholysis typically yields high-purity and high-yield lactate esters, offering significant advantages over other chemical recycling methods. However, the use of solvents in solvolysis processes also raises concerns regarding cost and environmental impact. Ideally, the solvents chosen for alcoholysis should be nontoxic, environmentally friendly, and mild enough to avoid the production of unwanted side products.

Mechanical recycling of PLA is also the most promising waste valorization strategy. However, to fully realize its potential, several challenges must be addressed. First, recycling plants need to accommodate a dedicated PLA stream, separate from other plastics, to prevent contamination and ensure the quality of the recycled material. Another key hurdle is the degradation of PLA's mechanical properties during mechanical recycling. Thermomechanical degradation is a common issue during the mechanical reprocessing of PLA, which leads to a reduction in its mechanical strength, thermal stability, and other performance characteristics. This can make recycled PLA less suitable for high-performance applications. However, this issue can be mitigated through the use of various additives. Functionalized clays, chain extenders, cross-linking agents, and organic fillers have all shown promise in reducing the degradation effects. Importantly, these additives are carefully selected to ensure they do not compromise the biodegradability of the recycled PLA. Among the most effective organic fillers are silk fibroin nanoparticles and functionalized chitosan, which not only act as nucleating agents to enhance PLA's properties but also help maintain the recyclate's thermal stability, tensile strength, toughness, and gas barrier performance. By incorporating these additives, it is possible to improve the quality of recycled PLA, making it more marketable and promoting its further integration into circular plastic systems (Piemonte et al., 2013). With PLA production reaching ~200,000 tons globally in 2018, establishing dedicated recycling streams and plants for PLA could soon become a viable and profitable solution. Recycling plants are typically considered economically feasible only when the global production of a polymer exceeds 200,000 tons, with annual processing capacities of 5000 to 18,000 tons per plant. Given the ongoing growth in PLA production, such infrastructure could be a practical step forward, facilitating the more efficient and sustainable recycling of PLA and supporting the broader goals of reducing plastic waste and promoting circularity in the plastic economy.

Short-chain PHAs such as polyhydroxybutyrate (PHB) thermally decompose into a spectrum of volatile products, including CO_2, H_2O, propene, ketene, acetaldehyde, β-butyrolactone, crotonic acid isomers, 3-butenoic acid, oligomers, and cyclic trimers. Under noncatalytic conditions, crotonic and 3-butenoic acid yields increase from ~40% to ~87%; with $Mg(OH)_2$, crotonic acid yields approach 100%. At higher temperatures, CO_2 and propene dominate, reflecting decarboxylation of crotonic acid. Degradation proceeds mainly through random cis-elimination chain scission, generating crotonic acid and its oligomers (Merchan et al., 2022). Hydrolysis is accelerated under acidic or basic conditions, particularly at amorphous chain-fold regions adjacent to lamellae. For PHB–Poly(hydroxyl valerate) copolymers in drug delivery, incorporation of basic groups has been used to tune hydrolysis and thus drug release rates. At high hydrolysis temperatures (175°C–220°C), PHB first depolymerizes to 3-hydroxybutyric acid and crotonic acid, which subsequently undergo dehydration and decarboxylation to propylene and CO_2.

11.8 Future developments and innovations

Biological recycling (the use of enzymes and microbes to depolymerize plastics/bioplastics into reusable intermediates) is likely the most compelling pathway for addressing the end-of-life challenges of bioplastics. The future of bioplastics is likely to be shaped by the convergence of synthetic biology and artificial intelligence (AI). Indeed, the integration of AI with bioreactor systems can deliver autonomous and self-optimizing recycling infrastructures (Ghasemlou et al., 2025). The first priority is to develop robust depolymerizing enzymes. Future generations of enzymes will need to combine high catalytic turnover with stability across a broad range of temperatures, pH, and contaminant loads. Progress in protein design and structural biology is beginning to reveal the motifs and domains that govern substrate recognition and bond cleavage. Using engineered microbial consortia can expand the scope of biological recycling. Instead of having one organism to perform all metabolic activities such as oxidizing polymers at the surface, depolymerizing them into oligomers, and mineralizing intermediates into high-value chemicals, communities of microbes can be programmed to do this all together. Looking further ahead, there is also scope for integrating biodegradation with other recycling technologies. In the longer term, bioplastics might even be designed at the molecular level with "traceable recycling codes" that make them more easily identifiable at end-of-life. These codes are not literal barcodes but rather chemical, structural, or digital signatures that link a product to a recycling pathway. As scientific knowledge and industrial practice converge, the end-of-life of bioplastics may come to be seen not as their greatest limitation but as their defining strength.

11.9 Conclusions

Bioplastics will not solve plastic pollution on their own, but they can play an important role when used in the right systems. Their environmental impact depends less on whether they are "biodegradable" in name, and more on whether they are designed and managed for recovery, recycling, or safe degradation. Current standards and certification schemes provide a starting point, but gaps remain between controlled laboratory tests and what happens in real environments, especially in the ocean. At the same time, recycling strategies are also emerging to close the loop, but challenges of scale, cost, and contamination still limit progress. Real progress will require a combination of material innovation, harmonized international standards, stronger policy frameworks, and clear communication with consumers. If these elements come together, bioplastics could become a meaningful part of a circular and more sustainable plastics economy.

References

Ali, S.S., Elsamahy, T., Al-Tohamy, R., Zhu, D., Mahmoud, Y.A.G., Koutra, E., et al., 2021. Plastic wastes biodegradation: mechanisms, challenges and future prospects. Science of the Total Environment 780, 146590. Available at https://doi.org/10.1016/j.scitotenv.2021.146590.

Andrady, A.L., Lavender Law, K., Donohue, J., Koongolla, B., 2022. Accelerated degradation of low-density polyethylene in air and in sea water. Science of the Total Environment 811, 151368. Available at https://doi.org/10.1016/j.scitotenv.2021.151368.

Bátori, V., Åkesson, D., Zamani, A., Taherzadeh, M.J., Sárvári Horváth, I., 2018. Anaerobic degradation of bioplastics: a review. Waste Management 80, 406–413. Available at https://doi.org/10.1016/j.wasman.2018.09.040.

Bingeman, C.W., Varner, J.E., Martin, W.P., 1953. The effect of the addition of organic materials on the decomposition of an organic soil. Soil Science Society of America Journal 17 (1), 34–38. Available at https://doi.org/10.2136/sssaj1953.03615995001700010008x.

Blesin, J.-M., Jaspersen, M., Wiebke, M., 2017. Boosting Plastics' image? Communicative challenges of innovative bioplastics. e-plastory 3.

Briassoulis, D., Dejean, C., 2010. Critical review of norms and standards for biodegradable agricultural plastics. Part I. Biodegradation in soil. Journal of Polymers and the Environment 18 (3), 384–400. Available at https://doi.org/10.1007/s10924-010-0168-1.

Carus, M., Dammer, L., 2013. Food or non-food: which agricultural feedstocks are best for industrial uses? Industrial Biotechnology 9 (4), 171–176. Available at https://doi.org/10.1089/ind.2013.1580.

Chen, G.Q., Patel, M.K., 2012. Plastics derived from biological sources: present and future: a technical and environmental review. Chemical Reviews 112 (4), 2082–2099. Available at https://doi.org/10.1021/cr200162d.

Choi, S.Y., Lee, Y., Yu, H.E., Cho, I.J., Kang, M., Lee, S.Y., 2023. Sustainable production and degradation of plastics using microbes. Nature microbiology 8 (12), 2253–2276.

Dilkes-Hoffman, L., Ashworth, P., Laycock, B., Pratt, S., Lant, P., 2019. Public attitudes towards bioplastics – knowledge, perception and end-of-life management. Resources, Conservation and Recycling 151, 104479. Available at https://doi.org/10.1016/j.resconrec.2019.104479.

Filiciotto, L., Rothenberg, G., 2021. Biodegradable plastics: standards, policies, and impacts. ChemSusChem 14 (1), 56–72. Available at https://doi.org/10.1002/cssc.202002044.

Gewert, B., Plassmann, M.M., Macleod, M., 2015. Pathways for degradation of plastic polymers floating in the marine environment. Environmental Sciences: Processes and Impacts 17 (9), 1513–1521. Available at https://doi.org/10.1039/c5em00207a.

Ghasemlou, M., Adhikari, B., Naebe, M., Barrow, C.J., 2025. Bioengineering strategies to promote plastic circularity. Nature Reviews Bioengineering 3 (5), 353–354. Available at https://doi.org/10.1038/s44222-025-00287-5.

Ghasemlou, M., Barrow, C.J., Adhikari, B., 2024. The future of bioplastics in food packaging: an industrial perspective. Food Packaging and Shelf Life 43, 101279. Available at https://doi.org/10.1016/j.fpsl.2024.101279.

Ghasemlou, M., Daver, F., Murdoch, B.J., Ball, A.S., Ivanova, E.P., Adhikari, B., 2022. Biodegradation of novel bioplastics made of starch, polyhydroxyurethanes and cellulose nanocrystals in soil environment. Science of the Total Environment 815, 152684. Available at https://doi.org/10.1016/j.scitotenv.2021.152684.

Goel, V., Luthra, P., Kapur, G.S., Ramakumar, S.S.V., 2021. Biodegradable/bio-plastics: myths and realities. Journal of Polymers and the Environment 29 (10), 3079–3104. Available at https://doi.org/10.1007/s10924-021-02099-1.

Hao, L.T., Ju, S., Hwang, D.K., Hwang, D.S., Ok, Y.S., Hwang, S.Y., et al., 2024. Optimizing bioplastics translation. Nature Reviews Bioengineering 2 (4), 289–304. Available at https://doi.org/10.1038/s44222-023-00142-5.

Herrmann, C., Rhein, S., Sträter, K.F., 2022. Consumers' sustainability-related perception of and willingness-to-pay for food packaging alternatives. Resources, Conservation and Recycling 181, 106219. Available at https://doi.org/10.1016/j.resconrec.2022.106219.

Itävaara, M., Karjomaa, S., Selin, J.F., 2002. Biodegradation of polylactide in aerobic and anaerobic thermophilic conditions. Chemosphere 46 (6), 879–885. Available at https://doi.org/10.1016/S0045-6535(01)00163-1.

Kale, G., Kijchavengkul, T., Auras, R., Rubino, M., Selke, S.E., Singh, S.P., 2007. Compostability of bioplastic packaging materials: an overview. Macromolecular bioscience 7 (3), 255–277. Available at https://doi.org/10.1002/mabi.200600168.

Kaur, R., Chauhan, I., 2024. Biodegradable plastics: mechanisms of degradation and generated bio microplastic impact on soil health. Biodegradation 35 (6), 863–892. Available at https://doi.org/10.1007/s10532-024-10092-3.

Kijchavengkul, T., Auras, R., 2008. Compostability of polymers. Polymer International 57 (6), 793–804. Available at https://doi.org/10.1002/pi.2420.

Kim, M.S., Chang, H., Zheng, L., Yan, Q., Pfleger, B.F., Klier, J., et al., 2023a. A review of biodegradable plastics: chemistry, applications, properties, and future research needs. Chemical Reviews 123 (16), 9915–9939.

Kim, H., Shin, G., Jang, M., Nilsson, F., Hakkarainen, M., Kim, H.J., et al., 2023. Toward sustaining bioplastics: add a pinch of seasoning. ACS Sustainable Chemistry and Engineering 11 (5), 1846–1856. Available at https://doi.org/10.1021/acssuschemeng.2c06247.

Kitamoto, H.K., Shinozaki, Y., X.h. Cao, T., Morita, M., Konishi, K., Tago, H., et al., 2011. Phyllosphere yeasts rapidly break down biodegradable plastics. AMB Express 1 (1), 1–11. Available at https://doi.org/10.1186/2191-0855-1-44.

Kolstad, J.J., Vink, E.T.H., De Wilde, B., Debeer, L., 2012. Assessment of anaerobic degradation of Ingeo™ polylactides under accelerated landfill conditions. Polymer Degradation and Stability 97 (7), 1131–1141. Available at https://doi.org/10.1016/j.polymdegradstab.2012.04.003.

Lyu, Schley, J., Loy, B., Lind, D., Hobot, C., Sparer, R., et al., 2007. Kinetics and time–temperature equivalence of polymer degradation. Biomacromolecules 8 (7), 2301–2310. Available at https://doi.org/10.1021/bm070313n.

Melchor-Martínez, E.M., Macías-Garbett, R., Alvarado-Ramírez, L., Araújo, R.G., Sosa-Hernández, J.E., Ramírez-Gamboa, D., et al., 2022. Towards a circular economy of plastics: an evaluation of the systematic transition to a new generation of bioplastics. Polymers 14 (6), 1203. Available at https://doi.org/10.3390/polym14061203.

Merchan, A.L., Fischöder, T., Hee, J., Lehnertz, M.S., Osterthun, O., Pielsticker, S., et al., 2022. Chemical recycling of bioplastics: technical opportunities to preserve chemical functionality as path towards a circular economy. Green Chemistry 24 (24), 9428–9449. Available at https://doi.org/10.1039/d2gc02244c.

Nazareth, M., Marques, M.R.C., Leite, M.C.A., Castro, Í.B., 2019. Commercial plastics claiming biodegradable status: is this also accurate for marine environments? Journal of Hazardous Materials 366, 714–722. Available at https://doi.org/10.1016/j.jhazmat.2018.12.05.

Nazareth, M.C., Marques, M.R.C., Pinheiro, L.M., Castro, Í.B., 2022. Key issues for bio-based, biodegradable and compostable plastics governance. Journal of Environmental Management 322, 116074. Available at https://doi.org/10.1016/j.jenvman.2022.116074.

Pattanasuttichonlakul, W., Sombatsompop, N., Prapagdee, B., 2018. Accelerating biodegradation of PLA using microbial consortium from dairy wastewater sludge combined with PLA-degrading bacterium. International biodeterioration & biodegradation 132, 74–83. Available at https://doi.org/10.1016/j.ibiod.2018.05.014.

Piemonte, V., Sabatini, S., Gironi, F., 2013. Chemical recycling of PLA: a great opportunity towards the sustainable development? Journal of Polymers and the Environment 21 (3), 640–647. Available at https://doi.org/10.1007/s10924-013-0608-9.

Pischedda, A., Tosin, M., Degli-Innocenti, F., 2019. Biodegradation of plastics in soil: the effect of temperature. Polymer Degradation and Stability 170, 109017. Available at https://doi.org/10.1016/j.polymdegradstab.2019.109017.

Queste, B.Y., Fernand, L., Jickells, T.D., Heywood, K.J., 2013. Spatial extent and historical context of North Sea oxygen depletion in August 2010. Biogeochemistry 113 (1-3), 53–68. Available at https://doi.org/10.1007/s10533-012-9729-9.

Rosenboom, J.G., Langer, R., Traverso, G., 2022. Bioplastics for a circular economy. Nature Reviews Materials 7 (2), 117–137. Available at https://doi.org/10.1038/s41578-021-00407-8.

Rudnik, E., Briassoulis, D., 2011. Degradation behaviour of poly(lactic acid) films and fibres in soil under Mediterranean field conditions and laboratory simulations testing. Industrial Crops and Products 33 (3), 648–658. Available at https://doi.org/10.1016/j.indcrop.2010.12.031.

Shah, A.A., Kato, S., Shintani, N., Kamini, N.R., Nakajima-Kambe, T., 2014. Microbial degradation of aliphatic and aliphatic-aromatic co-polyesters. Applied Microbiology and Biotechnology 98 (8), 3437–3447. Available at https://doi.org/10.1007/s00253-014-5558-1.

Song, J.H., Murphy, R.J., Narayan, R., Davies, G.B.H., 2009. Biodegradable and compostable alternatives to conventional plastics. Philosophical Transactions of the Royal Society B: Biological Sciences 364 (1526), 2127–2139. Available at https://doi.org/10.1098/rstb.2008.0289.

Soroudi, A., Jakubowicz, I., 2013. Recycling of bioplastics, their blends and biocomposites: a review. European Polymer Journal 49 (10), 2839–2858. Available at https://doi.org/10.1016/j.eurpolymj.2013.07.025.

Taufik, D., Reinders, M.J., Molenveld, K., Onwezen, M.C., 2020. The paradox between the environmental appeal of bio-based plastic packaging for consumers and their disposal behaviour. Science of the Total Environment 705, 135820. Available at https://doi.org/10.1016/j.scitotenv.2019.135820.

Wang, C., Yu, J., Lu, Y., Hua, D., Wang, X., Zou, X., 2021. Biodegradable microplastics (BMPs): a new cause for concern? Environmental Science and Pollution Research 28 (47), 66511–66518. Available at https://doi.org/10.1007/s11356-021-16435-4.

Yuan, X., Bank, M.S., Sonne, C., Ok, Y.S., 2021. Dual closed-loop chemical recycling support sustainable mitigation of plastic pollution. Matter 4 (4), 1095–1097. Available at https://doi.org/10.1016/j.matt.2021.03.014.

Zhu, J., Wang, C., 2020. Biodegradable plastics: green hope or greenwashing? Marine Pollution Bulletin 161 (Pt B), 111774. Available at https://doi.org/10.1016/j.marpolbul.2020.111774.

CHAPTER 12

Bioplastics for active and intelligent food packaging

Mehdi Jahanbakhshi[1], Bahareh Farasati Far[2], Caixia Wan[3], and Shima Jafarzadeh[4,5]

[1]School of Chemical Engineering, University of Tehran, Tehran, Iran, [2]Department of Chemistry, Iran University of Science and Technology, Tehran, Iran, [3]Department of Chemical and Biomedical Engineering, University of Missouri, Columbia, MO, United States, [4]Center for Sustainable Bioproducts, Deakin University, Geelong Waurn Ponds Campus, Geelong, VIC, Australia, [5]School of Engineering, Deakin University, Geelong Waurn Ponds Campus, Geelong, VIC, Australia

12.1 Introduction

Since the 1940s, traditional plastics have continued to be extensively utilized to replace items created using paper, glass, and metal owing to their versatility, affordability, and desirable characteristics (Thakur et al., 2018). In modern society, our existence is intricately linked to plastics within many different industries, including textiles, toy items, and automotive components, as well as electronic devices, agricultural applications, medical equipment, and packaging technology. The food industry represents the most significant users of the packaging field, representing around 40% of the total plastic manufactured (Narancic et al., 2020). Traditional plastics are synthetic materials made from petroleum-based polymers, characterized by their robustness and extremely slow degradation rates, leading to resource shortages and significant waste buildup. Furthermore, the burning of plastic waste results in significant emissions of carbon dioxide and various toxic compounds into the atmosphere, which leads to the phenomenon of global warming (Zheng et al., 2005). To address such environmental challenges, efforts have been directed over several decades on two main areas: recycling processes of current plastic items and the innovation of sustainable alternatives to substitute traditional plastics (Kale et al., 2007). Recycling serves as a careful approach to reducing the volume of plastic waste, aiming to substitute the traditional landfilling technique currently employed for plastic waste removal. Nevertheless, the process of recycling presents certain limitations. First, biological compounds frequently contaminate plastic materials, making their recycling unpracticable. Second, the process of recycling involves significant energy and time investment, and this can potentially elevate the carbon footprint of the item, thereby rendering it less advantageous. Nonetheless, given that the advancements in this area are not yielding optimal results, it is crucial to identify biodegradable alternatives that possess

Bioplastics for Food Packaging. DOI: https://doi.org/10.1016/B978-0-443-41654-5.00009-3

comparable properties to traditional plastics, along with developing greener methodologies, to significantly reduce environmental hazards (Peelman et al., 2013). In that context, the goal of developing sustainable materials to replace petroleum-derived plastics has resulted in the emergence of a novel category of materials known as bioplastics. The advancement of food packaging extends beyond simply improving production volume. A notable trend is emerging in the innovation of materials, characterized by the increasing importance of composting, biodegradable, and reusable packaging in large-scale environmental actions (Almeida et al., 2023).

Packaging serves as a critical mechanism to maintain the integrity of food products during transportation, storage, and final consumption. It reduces the rate of quality deterioration and enhances the efficiency of distribution and marketing processes. Packaging plays four key roles: it preserves the product, provides essential information, makes handling and use easier, and protects the contents (Han, 2005). Packages serve several key purposes: they protect products from external factors, communicate information through text, branding, and visual elements, and support modern consumer lifestyles by offering convenience, such as ready-made or heat-and-eat options, easy-open features, resealable designs, and improved usability. They also provide versatile containment for products of different shapes and sizes, ensuring practical efficiency throughout handling, storage, and use (Yam and Lee, 2012). The primary objectives within the food supply system are to ensure safe transportation and maintain the integrity of packed foods prior to consuming. Nevertheless, the degradation of characteristics associated with food quality arises over distribution and storage because of biological, chemical, and physical processes. The maintaining of food quality represents a major field of research, as it is closely associated with the broader objective of improving the quality of human life (Farasati Far et al., 2024). Ensuring the safety of food is essential, as it guarantees consumers from foodborne illnesses and enhances the operational efficacy of food industries, such as minimizing losses caused by microbial contamination of perishable items. At the organizational level, the evaluation of food quality typically involves systematic microbiological and chemical assessments that form the basis of standard testing protocols (Cruz-Romero et al., 2013).

Recently, two innovative concepts have significantly enhanced the development of advanced packaging aimed at ensuring better health and safety for food: the terms of active packaging and intelligent packaging. Active additives are components added to packaging to extend shelf life and maintain or improve food quality. These systems are purposely designed to include elements that can release beneficial compounds into the food environment or absorb unwanted substances from the product or its surroundings (European Commission, 2004). To enhance the efficacy of food packaging and provide extra functions, various active compounds may be integrated into the packaging system (Siddiqui et al., 2023). A variety of active packaging techniques have been extensively documented, including oxygen and ethylene scavengers, moisture regulators, carbon dioxide scavengers and emitters, antioxidant and antimicrobial controlled-release packages, and mechanisms designed to regulate the release or adsorption of flavors and odors (European Commission, 2004).

Intelligent packaging products refer to materials and items that determine the state of the food being packaged or the environment in which the food is placed (European Commission, 2004). In a more general way, intelligent packaging is characterized as the application of scientific principles and technological advancements that leverage the communication capabilities of packaging systems. This facilitates informed decision-making by observing variations in environments inside as well as outside and transferring the status of the packaged food product (Yam, 2012). In contrast to active packaging technology, smart packaging is not involved in direct actions for extending the freshness of food products. Intelligent packaging is designed to communicate essential information to the various stakeholders involved in the food supply process, including manufacturers, retailers, and buyers, regarding the quality of the food product (Restuccia et al., 2010). For instance, a smart packaging system can indicate the freshness of a food product or signal whenever its shelf life has passed; it can reveal the food's temperature through the use of thermochromic inks or microwave doneness indicators; and it can present the temperature history of the food using time–temperature indicators (TTIs) (Robertson, 2005). Furthermore, intelligent packaging serves as a method to evaluate the efficacy of active packaging systems (Kerry et al., 2006). In fact, active packaging serves as the agent that performs specific functions, whereas intelligent packaging acts as the sensor that detects and communicates information (Yam, 2012). Intelligent and active packaging may interact in a beneficial way to produce what is referred to as "smart" packaging, which represents a comprehensive packaging approach that integrates the advantages derived from both active and intelligent technologies (Fig. 12.1) (Vanderroost et al., 2014).

12.2 Active packaging technology

12.2.1 Mechanisms of active packaging (antimicrobial, oxygen scavenging, and moisture control)

Active packaging is the type of packaging that contains one or more active components that control the production of one or more gases, moisture, microorganisms, and so on. Active packaging therefore ensures the presence of antimicrobial agents, oxygen scavengers, and moisture regulators (Fig. 12.2). The outstanding advantages of the developed active packaging technologies include reducing to a minimum the factors that lead to food spoilage and deterioration (Alves et al., 2023).

12.2.1.1 Antimicrobial mechanisms

Antimicrobial mechanisms therefore find their application in the microbial context. In most cases, this involves the production of packaging, in which doses of substances that prevent or

Figure 12.1
Schematic diagram of active and intelligent packaging concepts.

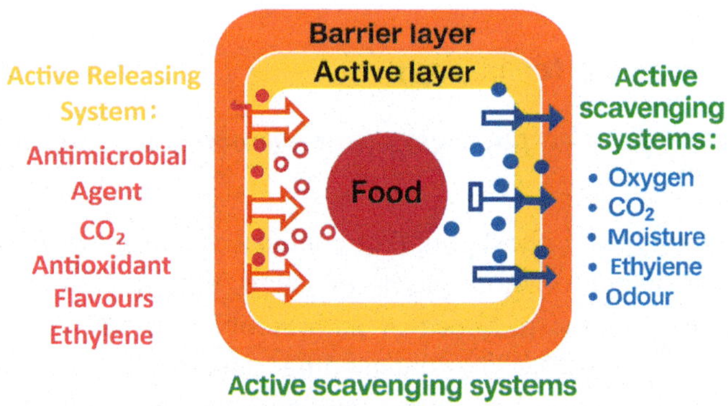

Figure 12.2
Mechanisms of active packaging systems.

kill microbes are introduced into the packaging structure. It is of significant importance for food safety and preservation (Upadhyay et al., 2024). This technology is more appropriate for products that are expected to be contaminated with microbes, such as meat, dairy, and ready-to-eat meals. One of the most popular groups of antimicrobial agents is silver nanoparticles. This was shown to have a broad-spectrum activity against *Escherichia coli* and *Staphylococcus aureus*. Silver nanoparticle-doped polyethylene as active antimicrobial packaging polymer materials reduced bacterial growth by 99% as a result of oxygen-releasing ions which alter microbial cell membrane activity, thereby prolonging the shelf life of some foods (Jafarzadeh et al., 2019).

Composite resins incorporating silver/zinc oxide nanoparticles have garnered recent attention, as they have shown to possess antimicrobial activity against important pathogens such as *Streptococcus mutans*, *Lactobacillus acidophilus*, and *Candida albicans*. The bond strength and biocompatibility of these materials were proven effective for antimicrobial packaging applications (Kachoei et al., 2021). ZnO nanoparticles have undergone extensive developmental research as antimicrobial materials. In broad-spectrum and biopolymer-coated hybrid packaging application materials, ZnO demonstrates high activity against various microorganisms, which are inhibitory against the growth of pathogens and improved mechanical strength. Great reductions in pathogens on food surfaces can reduce the public health concerns associated with foodborne illnesses (Jieying et al., 2024). In some instances, an oxygen scavenger and antimicrobial agent are combined, and this synergism aids in microbial control. Furthermore, its application can reduce microbial counts by 2 log colony-forming unit (CFU)/g, thus increasing control over spoilage and enhancing food quality (Han, 2005). In another study, when the ZnO nanoparticles were introduced into chitosan-based films, the films showed high antimicrobial activity against *E. coli* and *Brevibacterium lactofermentum*. Nevertheless, a film matrix dependence on performance and the significance of proper blending is also observed because of their interaction with different film matrices (Boura-Theodoridou et al., 2020). The results show that silver-doped ZnO nanoparticles exhibit excellent antimicrobial effects, outperforming both stand-alone ZnO and silver alone, and providing strong bactericidal activity and improved food safety for antimicrobial packaging applications (Mortezagholi et al., 2022).

12.2.1.2 Oxygen scavenging mechanisms

Oxygen scavenging is an important part of active packaging, which is especially meant for products susceptible to oxidative spoilage, including meats, oils, and dairy products. In addition, oxygen stimulates spoilage by promoting aerobic bacterial growth and the start of lipid oxidation, which contributes to off-flavors and rancidity. Oxygen scavengers normally achieve their intrinsic function by bringing down the oxygen via a chemical reaction to very low levels inside the package environment, typically from the ambient 21% to below 0.1%, hence slowing oxidizing processes (Dey and Neogi, 2019). Excellent iron-based oxygen scavengers, especially in active packaging. For example, these scavengers can be lowered to nearly zero oxygen in the package and extend the shelf life of oxygen sensitive food from

20% to 50% (Kumar et al., 2024b). For applications in dairy products, which are sensitive to oxidation, ascorbic acid has been incorporated into polylactic acid (PLA) films to create an oxygen scavenging effect. The oxygen content can be reduced to less than 0.5% within 48 hours with these films, which efficiently maintains product quality over extended periods of time (Dey and Neogi, 2019; Reshmy et al., 2021). Along with iron-based oxygen scavengers, innovative methods were developed, including using oxygen-absorbing nanoparticles doped in PLA sheets. These films not only extend the shelf life of oxidation sensitive items but also bring oxygen levels down to less than 0.5% (Kim et al., 2022).

12.2.1.3 Moisture control mechanisms

Because products like cereals, dry foods, and baked goods are more likely to experience textural deterioration and become more vulnerable to microbial growth, maintaining moisture control is a basic use for active packaging. However, moisture-absorbing films maintain humidity in the package inside which the product is kept, retaining the texture and quality of the product. By increasing water activity and extending the shelf life of sensitive products, starch-based films plasticized with glycerol bioplastics can absorb moisture up to twice as well as traditional packaging (Benitez et al., 2024; Kumar et al., 2024b). The films of bioplastic use materials and formulations that attract or repel water to effectively manage moisture in food packaging. Starch-based films are used extensively because of their capacity to regulate water activity. For instance, glycerol is frequently used to plasticize them in product formulations where water activity needs to be regulated, particularly in moisture-sensitive products like cereals and baked goods. For instance, cassava starch-based bioplastics have good moisture controlling properties under certain temperature and humidity conditions (Supardi and Karmini, 2023). The use of bioplastic films is on the rise as an eco-friendly alternative to petroleum-based films, which do not biodegrade. Because these films are compostable and made from nonplastic materials like sugarcane bagasse and cassava starch, they can help reduce the carbon footprint of plastic packaging, increase plastic recycling, and decrease landfill waste (Azmin et al., 2020). Moreover, bioplastics also help in minimizing waste food by improving their product quality through increased control of moisture, leading to reduced microbial contamination and textural degradation (Gaikwad et al., 2019). In contrast to traditional plastics, bioplastics degrade in response to changes in environmental conditions such as humidity and temperature, rendering them less consistent. In addition, from an economic perspective, scaling up could make it more problematic to increase the manufacturing cost of biofilms by incorporating two or more nanoparticles (Asgher et al., 2020).

To improve the water-resistant properties of biopolymer films produced from semolina, nano-kaolin has been incorporated, along with films manufactured from starch. These nanoparticle-reinforced films can be used to package products that need to be stored in a controlled environment to maintain quality and avoid microbiological contamination. One example is cheese. Researchers have found that by incorporating nano-kaolin into films, they can significantly decrease moisture transmission, which in turn improves the product's freshness and stability (Jafarzadeh et al., 2019). Furthermore, the gas barrier characteristics and water vapor transmission rates are enhanced in nanocomposites that combine silicon dioxide and zinc oxide ($ZnO–SiO_2$) with polymer films. The results showed that packed bread had a longer shelf life and fewer microbiological dangers (Al-Tayyar et al., 2020).

12.3 Role of bioplastics in active packaging solutions

Bioplastics offers a more environmentally friendly option than conventional plastics made from petroleum, particularly for use in active packaging. Bioplastics, which are made from sustainable materials, break down faster in nature. For example, bioplastics made from bacterial cellulose can break down by more than 90% within just 2 weeks in composting facilities‹ much faster than many synthetic polymers, which often take decades to decompose (Sulistyo et al., 2023).

They also work well with active ingredients that help keep food fresh. Bioplastics are especially useful in active packaging because they can easily incorporate antimicrobial agents, oxygen absorbers, and moisture controllers without losing their ability to biodegrade. Combining these active agents with biodegradable plastics offers a strong advantage: it supports sustainability while improving food safety, meeting both industry needs and consumer expectations (Masmoudi et al., 2016; Siddiqui et al., 2023; Swetha et al., 2023; Reshmy et al., 2021). The main ecological benefit of bioplastics is their availability, mainly in the active packaging field. Control of spoilage and growth of fungi and bacteria in foods is of vital importance to meet both consumer preference and industry requirements. Active agents with agents embedded in bioplastics may make bioplastics the item of choice in active packaging, impart sustainability, and provide food safety. New biodegradable polymer technologies based on PLA and cellulosic bioplastics have been efficiently degraded, and their use in short-lived products such as food packaging is being promoted (de Abreu et al., 2012) (Fig. 12.3).

One possible application of bioplastics to be highlighted is their presence as active agent carriers for the packaging of foods. PLA films were found to be 85% more effective against *Listeria monocytogenes* bacteria when essential oils like oregano and thyme were added. By inhibiting the growth of spoilage microbes, these films also reduced the storage time of perishable items by 60% (Varghese et al., 2020). Previous studies have shown that adding essential oils to PLA films can reduce *L. monocytogenes* by 2 logs within 2 weeks under refrigeration. This significant reduction helps maintain the quality of perishable foods and demonstrates how bioplastics can extend shelf life while reducing the need for synthetic preservatives (Hossain et al., 2019).

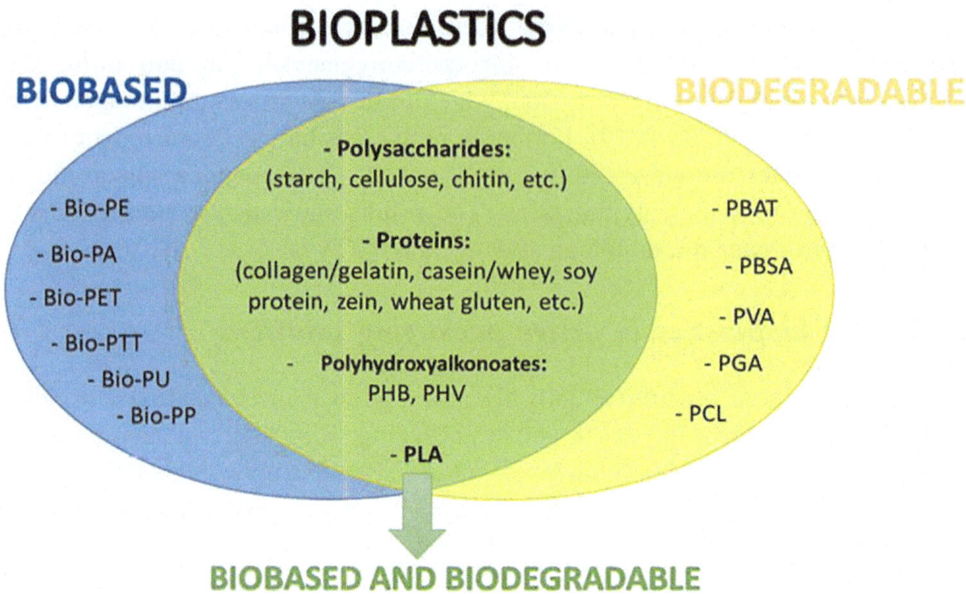

Figure 12.3

Classification of Bioplastics. This figure illustrates the categorization of bioplastics into three distinct groups: (1) bio-based but non-biodegradable polymers, (2) biodegradable polymers derived from fossil resources, and (3) polymers that are both bio-based and biodegradable.
Reshmy, R., Madhavan, A., Philip, E., Paul, S.A., Sindhu, R., Binod, P. et al., 2021. Sugarcane bagasse derived nanocellulose reinforced with frankincense (Boswellia serrata): physicochemical properties, biodegradability and antimicrobial effect for controlling microbial growth for food packaging application. Environmental Technology & Innovation 21, 101335. https://doi.org/10.1016/j.eti.2020.101335.

Alternatively, chitosan-based bioplastics incorporate ZnO nanoparticles as an active packaging feature. New biodegradable polymers made from PLA and cellulose-based bioplastics break down efficiently and are increasingly recommended for short-lived products like food packaging. Studies have also shown that films made from chitosan and zinc oxide can actively fight against harmful pathogens, offering added protection for packaged foods. Therefore, contamination with *S. aureus* and *E. coli* was reduced while recycling contaminated materials, thereby increasing the safety of a particular food product. In addition, these films also exhibited better oxygen and moisture barrier properties, which helped against oxidative and moisture-related spoilage, thus making the chitosan–ZnO bioplastics suitable for packaging products susceptible to bacterial contamination, such as poultry and fish (Ahmadi et al., 2021).

Regarding the response to the use of bioplastics in active packaging applications, it has been reported that there is a comparative study on the performance of bioplastics showing that phytochemical components in bioplastics have promising potential to impart and reduce preservative properties in food environmental impact and pollution than traditional materials

(Oluwasina and Awonyemi, 2021). Barriers such as zinc oxide and cerium oxide (CeO_2) film components result in reduced growth of microbes and lowering oxidation rates, ideal for extending the shelf life of perishable products. For products that need a long shelf life and sterile conditions, these barrier materials are very effective. They help keep the product safe while also innovatively supporting environmental sustainability (Figueroa-Lopez et al., 2024). Also, by adding cerium oxide nanoparticles to polyhydroxybutyrate films, oxygen levels can be lowered to less than 0.2% in just 48 hours, which means packaged foods can stay fresh for almost 30% longer. When tested against microorganisms that cause spoilage, these films also showed improved antibacterial activity (Figueroa-Lopez et al., 2023).

Bioplastics also support the shift toward a circular economy in the packaging industry. Because they are made from plant-based materials that can biodegrade or be composted, they help improve waste management and reduce reliance on fossil fuels. Studies have shown that nanocellulose films made from sugarcane bagasse and infused with frankincense also exhibit strong antibacterial effects. These films reduced microbial contamination by more than 90% and remained biodegradable under soil conditions for 45 days. In addition to improving food safety, these advances are better for the environment (Reshmy et al., 2021). The trend toward using bioplastics also responds to the global emphasis on eco-friendliness, as they can biodegrade easily. Such attributes are useful in the food industry where single-use packaging is common. With studies now proving that bioplastics can be used in food packaging biotechnologically, this idea fits well within the framework of green packaging ideas and innovation (Asgher et al., 2020).

Therefore, the use of active packaging methods in bioplastics offers a very efficient and environmentally friendly way to keep food fresh. By using antimicrobials, oxygen scavengers and moisture regulators, the packaging industry would be able to provide longer shelf life and cleaner labels, which consumers demand, in an environmentally friendly way through bioplastics. The use of bioplastics in active packaging is truly groundbreaking in food technology, for this is the means to safer, fresher, more sustainable food systems that will meet both the goals of sustainability and consumer health. Bioplastics has many uses in intelligent and active food packaging, as shown in Table 12.1. These uses include increasing food safety, prolonging the shelf life of the contents, and improving sustainability.

12.4 Intelligent packaging technology

12.4.1 Sensors, indicators, and smart labels in food packaging

Intelligent packaging technologies include the use of sensors, indicators, and smart labels to track and report the status of packaged foods, providing important information about freshness, safety, and quality. These allow for the immediate monitoring of any changes in the environment or any signs of spoilage, which is particularly important with perishable

Table 12.1 Applications of bioplastics in active and intelligent food packaging

No.	Active/ intelligent mechanism	Bioplastic material	Active agent incorporated	Target food product	Observed effect	Numerical data	Method of incorporation	Mechanical properties	Barrier properties	References
1	Antimicrobial packaging	Chitosan film	Oregano essential oil	Mozzarella cheese	Extended shelf life	Shelf life increased by 7 days at 4°C	Blending during film formation	Tensile strength decreased by 15%	Water vapor permeability increased	Bonilla et al. (2013)
2	Oxygen scavenging	PLA film	Ascorbic acid	Sliced apples	Reduced oxidation	Oxygen levels reduced to <0.5% within 48 h	Casting with active agent	Mechanical properties unchanged	Oxygen transmission rate decreased	Yildirim et al. (2017)
3	Moisture control	Starch-based film	Glycerol	Baked goods	Maintained texture	Water activity reduced by 20%	Glycerol as plasticizer	Increased flexibility by 25%	Moisture absorption increased about 15%	Jouki et al. (2013)
4	pH indicator	Gelatin film	Red cabbage extract	Fish fillets	Spoilage detection	Color change from purple to green between pH 6–8	Dye incorporated into film	Mechanical properties acceptable	Transparency maintained	Taha (2023)
5	Antimicrobial packaging	PLA film	Silver nanoparticles	Bread	Mold inhibition	Shelf life extended by 5 days at room temperature	Nanoparticles dispersed in matrix	Tensile strength increased by 10%	No significant change	Gan and Chow (2018)
6	Intelligent packaging (TTI)	PLA film	Thermochromic pigments	Perishable foods	Temperature monitoring	Visible color change above 5°C	Pigments mixed into PLA	Mechanical properties maintained	No effect on barrier properties	Park et al. (1994)
7	Antimicrobial packaging	Chitosan film	Nisin	Beef patties	Inhibited bacterial growth	Reduction of Listeria *monocytogenes* by 2 log CFU/g	Coating onto film surface	Tensile strength decreased by 5%	Slight increase in water vapor permeability	Zimet et al. (2019)

8	Oxygen scavenging	Starch-based film	Iron powder	Nuts	Prevented oxidation	Oxygen levels reduced to <0.1%	Iron embedded in film	Mechanical properties acceptable	Improved oxygen barrier	Miltz and Perry (2005)
9	Antimicrobial packaging	PLA/Cs blend	Cinnamon essential oil	Fresh produce	Reduced microbial load	Bacterial reduction by 3 log CFU/g over 10 days	Emulsified oil in polymer blend	Decreased tensile strength by 12%	Increased flexibility	Suryani et al. (2020)
10	Moisture control	Cellulose film	Silica gel particles	Dried foods	Reduced moisture content	Reduced water vapor transmission rates (WVTR) by nearly 4-fold (from 350 $g/m^2/day$ to 90 $g/m^2/day$) while maintaining mechanical stability	Particles embedded in film	Mechanical strength maintained	Enhanced moisture absorption	Rovera et al. (2020)
11	Intelligent packaging (gas sensor)	Chitosan film	Carbon dioxide indicator dye	Meat products	Spoilage detection	Color change observed with increased CO_2 levels	Dye incorporated during film formation	Mechanical properties acceptable	No impact on barrier properties	Mills (2005)
12	UV protection	PLA film	Titanium dioxide nanoparticles	Dairy products	Protected from UV light	UV transmission reduced by 95% at 300 nm wavelength	Nanoparticles dispersed in PLA	Tensile strength slightly decreased (5%)	Improved UV barrier properties	Cao et al. (2020)
13	Antimicrobial packaging	PHA film	Zinc oxide nanoparticles	Ready-to-eat meat	Antimicrobial effect	4-log reduction in bacteria over 12 days at 4°C	Nanoparticles incorporated in film	Increased tensile strength	Barrier properties enhanced (oxygen and moisture)	Hakeem et al. (2020)
14	Oxygen scavenging	Pectin-based film	Glucose oxidase enzyme	Cheese	Reduced oxygen levels	Oxygen reduced to <0.2% within 24 hours	Enzyme immobilized in film	Mechanical properties acceptable	Oxygen barrier improved	Cruz-Romero et al. (2013)

(*Continued*)

Table 12.1 (Continued)

No.	Active/ intelligent mechanism	Bioplastic material	Active agent incorporated	Target food product	Observed effect	Numerical data	Method of incorporation	Mechanical properties	Barrier properties	References
15	Intelligent packaging (pH indicator)	PLA film	Anthocyanins	Poultry products	Freshness indicator	Color change from red to yellow as pH increased from 5 to 8	Dye blended into film	Mechanical properties maintained	Transparency affected	Zhang et al. (2019)
16	Antimicrobial packaging	Starch-based film	Chitosan nanoparticles	Fresh fruits	Delayed ripening	Shelf life extended by 4 days at room temperature	Nanoparticles embedded in film	Mechanical strength increased by 20%	Improved barrier properties (gas and moisture)	Elsabee and Abdou (2013)
17	Moisture control	PHA Fflm	Calcium oxide	Snack foods	Moisture absorption	Moisture content reduced by 10%	Additive mixed into polymer	Mechanical properties unchanged	Moisture barrier enhanced by 15%	Dilkes-Hoffman et al. (2018)
18	Antimicrobial packaging	PLA film	Green tea extract	Oily fish	Inhibited Lipid Oxidation	Shelf life extended at 4°C	Extract incorporated into film	Tensile strength decreased	No significant change in barrier properties	Martins et al. (2018)
19	Intelligent packaging (RFID)	PLA substrate	RFID antenna	Various foods	Enhanced traceability	Data storage	Antenna printed on PLA	Mechanical properties acceptable	No impact on barrier properties	Machiels et al. (2021)
20	Antimicrobial packaging	Chitosan/ PVA blend	Grapefruit seed extract	Cooked ham	Reduced microbial growth	Reduction in *L. monocytogenes* over 15 days	Extract blended into film	Increased flexibility	Water vapor permeability increased	Wang et al. (2019)

PLA, polylactic acid; *TTIs*, time–temperature indicators; *PHA*, polyhydroxyalkanoate; *RFID*, radio frequency identification; *UV*, ultraviolet.

foods like meat, dairy, and produce. Normal packaging is just a barrier, but intelligent packaging interacts with the environment or the food itself and reports back to the consumer or retailer important information that can help reduce food waste and increase safety (Yousefi et al., 2019).

12.4.1.1 Sensors

Sensors play a key role in monitoring food by detecting changes in temperature shifts, gas release, or microbial activity. For example, TTIs record the temperature history of a product and show whether it has been exposed to unsafe conditions. This is especially important for cold-chain items, where even a small temperature change can spoil the product. Modern TTIs are highly accurate, with sensitivity as fine as ±0.2°C, greatly improving the reliability of cold-chain monitoring (Reshmy et al., 2021).

TTIs are these things, they are small tabs that change color over time and temperature. So, if the product is in the "danger zone," both the consumer and the retailer can tell if it is The other type of sensors are gas sensors, which detect changes in the concentration of oxygen, carbon dioxide, or ethylene, all of which may indicate that microbial activity is occurring, or that the fruit is ripening. Recent innovations in oxygen and gas sensors include flexible organic polymer-based gas sensors integrated into packaging. These sensors can detect trace amounts of gases like carbon dioxide with high sensitivity, with a detection limit as low as 10 ppm, ensuring early detection of spoilage (Li et al., 2023). The first are the oxygen sensors that they use in modified atmosphere packaging (MAP) packaging when the food is sensitive to oxygen because then oxidative spoilage is prevented (Chayavanich et al., 2020).

12.4.1.2 Indicators

Indicators provide an easy, visual way to check the quality or freshness of food. Freshness indicators, for example, change color in response to biochemical changes that occur as food spoils. Natural dyes like anthocyanins are often used to create these color-based indicators, allowing real-time detection of spoilage. In meat and fish packaging, these indicators respond to pH changes caused by spoilage gases, producing a measurable color shift. Their sensitivity can be as precise as 0.1 pH units, enabling accurate monitoring of food freshness (Balbinot-Alfaro et al., 2019).

These are especially good in fish and meat products, which will give off volatile amines when they start to spoil. These amines, when they reach a certain level of concentration, cause the freshness indicator to change color, informing the consumer of possible spoilage. Dual-channel spoilage indicators developed for fish packaging can visually detect spoilage within 24 hours, with clear color transitions from red to yellow as spoilage gases increase (Reshmy et al., 2021). In the same way, they use those ethylene indicators in the fresh produce packages. Ethylene is a natural plant hormone that speeds up ripening and since many fruits and veggies are ethylene sensitive, they can be used as an indicator of when they are approaching over-ripeness. pH indicators also are used in packaging materials for highly acidic foods because they can detect changes in acidity caused by microbial activity, which can lead to food spoilage and possible health hazards (Realini and Marcos, 2014). One of the main problems is being able to find and single out unique biochemical markers that will consistently and accurately indicate spoilage in all sorts of foods. Like how volatile amines are good indicators of spoilage in seafood, but not necessarily in fruits or dairy products. It takes much fine chemical engineering and testing against many food matrices to make sure that the indicators react only to these markers. The other problem is that these markers must be stable throughout the shelf life of the product and still be sensitive enough to the spoilage to not give false positives. However, according to some newer research, the use of pH-sensitive dyes with nanocomposite materials could improve the specificity and sensitivity of these indicators and provide a more accurate answer (Soltani Firouz et al., 2021; Reshmy et al., 2021).

12.4.1.3 Smart labels

Smart labels, such as QR codes and radio frequency identification (RFID) tags provide a digital interactive element for traceability and transparency in food packaging. QR codes, when scanned by a smartphone, provide consumers with detailed information on product origin, storage, handling practices, and even certifications or freshness data (Rahman et al., 2021). That way consumers know exactly what they are getting and can make a choice based on the facts and it makes the product more reliable. RFID tags are more common in logistics where they enable real-time tracking and monitoring of products throughout the supply chain. Interactive QR code systems, when paired with pH-sensitive freshness sensors, have been shown to improve spoilage detection accuracy by 70%, making them more effective for consumer use (Ibrahim et al., 2021; Reshmy et al., 2021). RFID tags can store large amounts of information, and they can be read automatically at checkpoints, eliminating human error, and allowing for much more accurate data on the condition of products. RFID can also talk to sensors and signals inside the package itself, thus RFID ensures a comprehensive method of tracking food safety (Chayavanich et al., 2020).

Smart labels improve food safety and quality through the ability to collect and share data continuously along the supply chain. Especially RFID tags, which can track a product's location and temperature at any given moment and send out an alert if the product is not being stored in its most ideal conditions (Pârvulescu et al., 2023). When RFID tags are combined with temperature and humidity sensors, they create "smart labels" that can detect and report potential quality risks, such as temperature abuse, throughout the supply chain. This level of monitoring allows companies to act early, by redirecting or rescuing products before they reach store shelves, helping to reduce waste and improve food safety (Corradini, 2018). As smart label technology continues to advance, wireless communication will become more sophisticated, making it easier and more accurate to track food quality in real time.

12.5 Role of bioplastics in developing intelligent packaging systems

Bioplastics will be a vital part in the creation of intelligent packaging systems, not only is it a green alternative to regular plastic, but it also enhances the performance of intelligent packaging. With consumer and industry focus shifting to sustainability, biodegradable plastics (PLA, polyhydroxyalkanoate (PHA), starch-based polymers) are preferred due to their ability to breakdown and because they are compatible with sensors and indicators (Swetha et al., 2023). With the combination of bioplastics and intelligent packaging components such as freshness indicators and time–temperature sensors, it is possible to have environmentally friendly packaging and also enhance food safety and quality monitoring (Jafarzadeh et al., 2024).

PLA is one of the most widely used bioplastics in intelligent packaging because it is compatible with many sensor technologies. PLA-based films can be designed with pH-sensitive dyes that act as freshness indicators, changing color with the acidity level changes, which are usually a sign of microbial activity. PLA also blends nicely with QR codes and RFID tags, offering a green solution to the traceability and transparency issues of food supply chains. Another advantage of bioplastics such as PLA for smart packaging is that they are compatible with natural sensors and indicators, making the packaging both smart and compostable at the end of its life (Jariyasakoolroj et al., 2020). In smart packaging made from bioplastics, the main challenge is balancing good functionality with biodegradability. Researchers are working to improve the strength and barrier properties of bioplastics by adding nanomaterials like cellulose nanofibers or clay nanoparticles. These additives can increase tensile strength, reduce moisture transfer, and make the materials more durable, important qualities for "intelligent" packaging. Another active area of research is finding ways to make bioplastics compatible with electronic components, such as RFID tags and chemical indicators, while still keeping them fully biodegradable. Fig. 12.4 shows schematic illustration of various smart packaging labels/systems. Several innovations including the use of completely biodegradable markers and sensors that break down with the bioplastic are being explored to fulfill the ecological needs of a cradle-to-cradle economy (Das et al., 2024).

12.6 Examples of intelligent packaging in bioplastics

Many examples show that bioplastics prove to be highly effective when used in intelligent packaging, as well as being environmentally friendly. One of the best examples is the PLA films into which natural and pH-sensitive dyes are incorporated to indicate freshness. With perishable products such as meat and seafood, any change of color on the PLA film can be an indication of

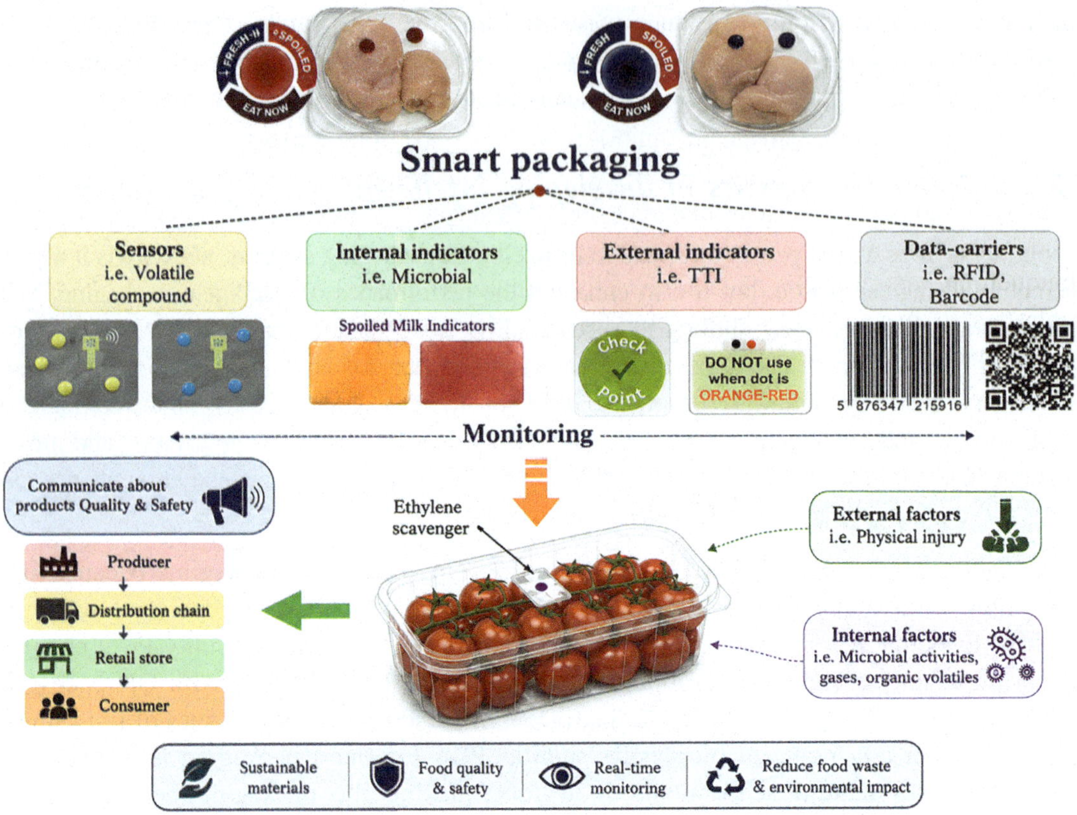

Figure 12.4
Schematic illustration of various smart packaging labels/systems.

spoilage by microbes that are known to change the pH of the food product. Naturally occurring dyes like anthocyanins from red cabbage or blueberries are commonly used for these purposes because they are compostable and nonleaching and their intensity varies based on the conditions inside of the package. This feature helps the consumers in determining the freshness of food, and also the company's products do not require preservatives because the freshness is always indicated passively and continuously (Păuşescu et al., 2022).

One other example is bioplastic packaging that allows air to circulate. The permeable nature of bioplastics like Bio-Flex keeps perishables fresh for longer by letting moisture escape. This is particularly beneficial for perishable foods such as fruits and vegetables, as it keeps the humidity in the packaging at an ideal level, reducing the likelihood of mold formation and extending the shelf life of the products.

Another example is drop-in bioplastics. Drop-in bioplastics like bio-polyethylene and bio-polypropylene are structurally identical to their fossil-derived counterparts but made from renewable resources. These materials can be recycled using existing infrastructure, promoting a circular economy while providing the same performance characteristics as traditional plastics (Haylock and Rosentrater, 2018). Multilayered films with nanocomposites is another example. Films that incorporate nanomaterials, such as nanoclays or silver nanoparticles, exhibit enhanced mechanical properties and barrier functions. These multilayered films not only provide better protection against oxygen and moisture but also possess antimicrobial properties that help in preserving food quality over extended periods (D'Almeida and de Albuquerque, 2024). These intelligent packaging solutions illustrate the potential of bioplastics to meet modern demands for sustainability while improving food safety and reducing waste in the food industry.

However, the generally lower barrier and mechanical properties of bioplastics compared with those of conventional plastics represent one limitation in the use of bioplastics in intelligent packaging (Fig. 12.5). This can be challenging for applications that require high moisture or oxygen barriers. In response, researchers are trying to combine nanocomposites such as clay or graphene oxide to increase barrier properties of bioplastics, while maintaining biodegradability (Rossa et al., 2022). The cost and scalability of bioplastics against conventional materials are other limitations to adoption in large-scale applications. Challenges of bioplastic packaging are being overcome by advances in bioplastic production, such as microbial fermentation to produce low-cost PHA, making intelligent bioplastic packaging a more viable and sustainable option for the food industry (Vicente et al., 2023).

Intelligent packaging through ethylene-sensitive indicators for fresh produce is also shown to hold great promise for starch-based bioplastics. Ethylene is a natural plant hormone that speeds up ripening of fruit and vegetables, but, at high levels, it encourages spoilage. The packaging can warn consumers and retailers that produce is becoming over ripe with ethylene indicators embedded in starch-based films. In the fresh produce supply chain, where timing is critical to reduce waste and maintain product quality, timing is this feature's greatest value. An added benefit is that starch-based films are also biodegradable, making them a good fit with environmentally friendly goals and not contributing to permanent plastic pollution. The use of this type of intelligent packaging is becoming popular for use in distribution centers and retail environment to optimize inventory by identification of produce batches that require immediate attention (Jafarzadeh et al., 2019).

Other innovative applications of bioplastic films in intelligent packaging are PHA films that are suitable for packaging temperature-sensitive products like dairy. Microbial fermentation yields PHA, a bioplastic that biodegrades naturally and is compatible with TTIs, making it suitable for use in products vulnerable to temperature fluctuations like milk and yogurt. TTIs embedded in PHA films change color in the presence of unsafe temperature conditions, allowing consumers and retailers to quickly ascertain whether a product has been stored properly. Cold-chain logistics especially can benefit from this app, where a small deviation in

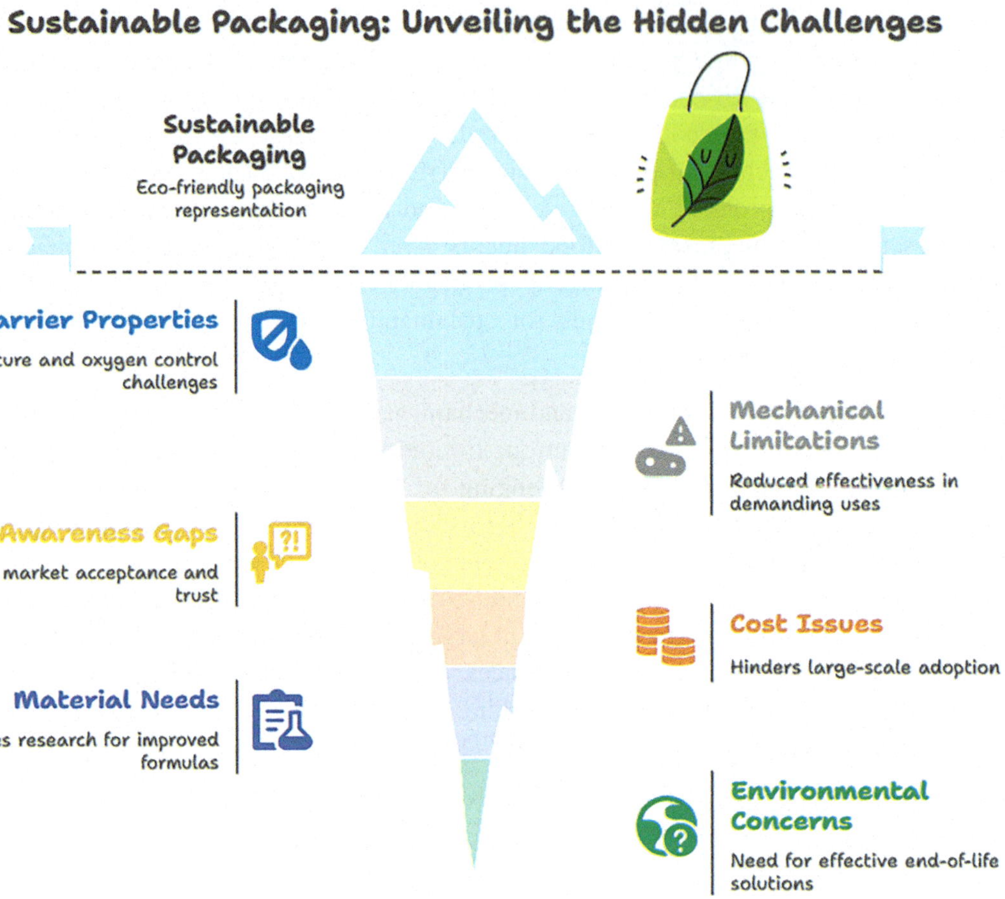

Figure 12.5
Limitation of bioplastics in intelligent packaging.

temperature can affect product quality and safety. PHA in this way shows how bioplastics combine environmental benefits with a high level of monitoring capacity for food safety and waste reduction (Paidari et al., 2024). The use of bioplastics in intelligent packaging is also consistent with the broader industry trend toward sustainability and resource efficiency in other specific applications. For instance, combining bioplastic packaging with smart labels—like QR codes or RFID tags—makes it possible for consumers to see detailed information about a product's origins, handling, and freshness. Most times, consumers' email is shared with consumers so they can scan the QR code on their smartphone and see data related to how the product was stored or even predicted shelf life. Consumer trust comes from this level of transparency, and requiring food traceability is one of its strengths. Using bioplastics with smart labeling reduces the industry's reliance on single-use plastics. The opportunity is offered to be interactive with consumers beyond the functionality of traditional packaging (Pereira et al., 2020).

Bioplastics such as PHA and PLA, which are being developed to biodegrade with their embedded sensors and indicators, are one last step toward reducing environmental impact even further. Bioplastic degradation has been focused on making sure the active or intelligent particulates in the packaging degrade at the end of the life cycle of the product to minimize any residual waste. This feature is critical for scaling intelligent bioplastic packaging in a circular economy by enabling waste reduction while retaining the functionality necessary for monitoring food quality and safety. While new formulations and manufacturing techniques are making bioplastics more durable, flexible, and degradable, these materials are poised to help create intelligent packaging systems that are both effective and environmentally responsible (de Abreu et al., 2012). Through these applications, intelligent packaging using bioplastics addresses multiple goals: keeping food safe, minimizing food waste, and increasing sustainability. Highlighting each how bioplastics improve the functionality of intelligent packaging by providing a responsible alternative to conventional materials. This type of approach facilitates participation by companies, retailers, and consumers in a system that gives equal priority to food safety and environmental stewardship, orienting a more sustainable future for food packaging (Almeida et al., 2023).

12.7 Performance and functional properties of active and intelligent packaging

12.7.1 Barrier properties (oxygen, moisture, and ultraviolet light)

For this reason, the performance of active and intelligent packaging largely depends on the barrier properties, as the barrier properties are crucial in maintaining the quality of the food as they prevent external factors from damaging the contents (Fig. 12.6). Effective packaging can extend the longevity and safety of packaged food products that are limited in their exposure to oxygen, moisture, and ultraviolet (UV) light (Reshmy et al., 2021). Robust barrier properties in the form of active and intelligent packaging applications of bioplastics have also been engineered based on combined use of advanced materials and nanotechnology. These barriers are critical for extending shelf life and ensuring food quality because each type of barrier targets a specific degradation pathway: this may be providing oxygen for fuel oxidation, providing moisture for nascent microbial growth, or exposing your food and ingredients to UV radiation in an effort to damage nutrients or add a new bit of appeal to your food or ingredients (Chandrababu et al., 2024).

12.7.2 Oxygen barrier

Oxidation and aerobic bacterial growth are two of the main reasons for food spoilage, and oxygen is one of them. Fats and oils of foods are especially prone to oxidative rancidity,

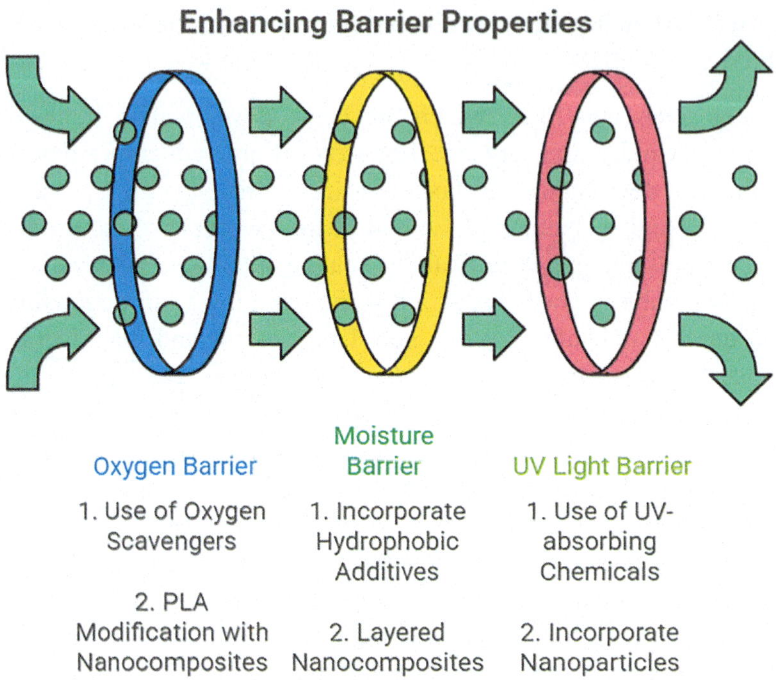

Figure 12.6
Barrier properties help maintain food quality by preventing external factors from damaging the contents.

negatively affecting flavor, aroma, and nutritional quality. Different oxygen scavenging agents or barrier-enhancing additives have been used to tailor these bioplastics to provide barrier properties to combat these effects. The moderate oxygen barrier properties of the bioplastic PLA have been demonstrated, and the addition of nanocomposites can greatly improve oxygen barrier properties. Adding clay nanoparticles or graphene oxide to PLA films has been shown to lower oxygen permeability by creating an additional tortuous path for oxygen molecules to take to avoid being able to pass through the package and get to the food (Ozdemir and Floros, 2004). The addition of nanocellulose and nanochitin to PLA films significantly decreased oxygen transmission rates, from 1118 to 13.10 cc/m^2 per day, according to recent studies. This resulted in an excellent oxygen barrier performance and the films maintained to be biodegradable (Nguyen et al., 2021).

Oxygen barriers are even more critical in MAP. By displacing oxygen with gases like nitrogen or carbon dioxide, MAP extends the shelf life of products like meats and cheeses. PLA or PHA films containing oxygen scavengers will reduce the oxygen levels in the packaging environment from 21% to nearly 0%. For example, Yildirim et al. have demonstrated that ascorbic acid or iron-based scavengers embedded in PLA can reach an oxygen concentration of less than 0.5% at 48 hours, effectively delaying the quality loss of a product (Yildirim et al., 2017). The use of bioplastic as the oxygen barrier shows its efficacy in maintaining the food quality, as bioplastic-based oxygen barrier will actively reduce oxidative stress to the food, but without adding preservatives to the food itself (Vilela et al., 2018). An innovative film consisting of polyisoprene and low-density polyethylene (LDPE) that is activated by UV light successfully reduced oxygen levels in packaged beef jerky by 16.72 mL O_2/g. This film successfully maintained the oxidative stability of the beef jerky for a period of 90 days (Gaikwad et al., 2020).

12.7.3 Moisture barrier

Another important property of active packaging is moisture control, especially for hydrated foods and products with high-fat content. Package moisture levels above 65% can encourage microbial growth and deteriorate product texture, especially in items that must remain crisp or dry. The incorporation of hydrophobic additives, or layered nanocomposites, often improves moisture barrier properties in bioplastics. For example, starch-based bioplastics are biodegradable and low-cost, but they naturally absorb water due to their hydrophilic nature. To reduce this moisture absorption, researchers have developed starch films coated or blended with hydrophobic compounds such as waxes or natural oils. These additions help make the material more moisture-resistant (Mane, 2016). Incorporating polyvinyl alcohol and vermiculite nanocomposites into starch-based films resulted in a 40% reduction in water vapor permeability under high-humidity conditions, highlighting their efficacy in packaging foods with high moisture content (Kim et al., 2018). Moisture control has also been demonstrated for enhanced bioplastics. Both montmorillonite clay and silica nanoparticle integration into bioplastic matrices like PLA or PHA have been shown to decrease water vapor transmission rates. The nanoparticles generate a labyrinthine effect within the film structure, preventing moisture molecules from permeating. A study showed that the incorporation of nano-kaolin into PLA films reduced moisture permeability by nearly 40%, improving shelf life of moisture-sensitive products such as cookies and crackers (Jafarzadeh et al., 2017). Carboxymethyl cellulose films that were inoculated with citric acid and aloe vera showed a 35% reduction in water vapor permeability, which helped to keep high-fat foods like ground beef intact and postponed their spoilage (Kanatt and Makwana, 2020). By controlling moisture transfer, these advanced bioplastic materials offer an effective solution to maintain product quality and prevent spoilage associated with high humidity.

12.7.4 Ultraviolet light barrier

Some foods, especially those rich in fats, vitamins, and natural pigments, can spoil quickly when exposed to UV light. UV rays can break down these sensitive components, causing color changes, off-flavors, and nutrient loss through photochemical reactions. This can seriously affect the quality and safety of the food (Ghandehari-Alavijeh et al., 2024). UV rays cause nutritional losses of certain vitamins, making products such as milk—which are sensitive to UV-induced nutrient degradation and quickly lose their vitamin concentration when exposed to UV treatment—taste different. In response, they make sure that active and intelligent packaging contains UV-blocking constituents to protect food from harmful light. Bioplastics formulated with UV-absorbing chemicals, such as titanium dioxide (TiO_2) or zinc oxide (ZnO), can help reduce the effects of UV exposure (Chaudhary et al., 2020). Bixin is a natural carotenoid that has recently been shown to be able to block up to 95% of UVA and 90% of UVB light. This means that food items like oils and dairy can be protected from photodegradation in a transparent and effective way (Stoll et al., 2021).

Addtionally, many investigations have also reported considerable enhancements in UV-blocking capability for PLA-based films containing TiO_2 nanoparticles, as an example (Zhang et al., 2024). The films that absorb UV rays can still be transparent to visible light for consumer viewing of the product while not compromising quality of sensitive foods. Another suitable additive is zinc oxide (ZnO), which has dual functionality in respect to blocking UV light and provide antimicrobial activity, which makes it particularly valuable for application of active packaging for meat and dairy. Research studies have shown that up to 99% of UV light can be blocked by most bioplastics with incorporated TiO_2 or ZnO nanoparticles, considerably improving the storage life of food products sensitive to UV while maintaining their quality over an extended period (Nikolic et al., 2021). Furthermore, chitosan/polyvinyl alcohol films functionalized with catechol demonstrated a 67.6% reduction in UV transmittance at 280 nm, ensuring prolonged shelf life for UV-sensitive products (Lei et al., 2021). This UV barrier functionality suits well with sustainability goals, as it minimizes food loss by retaining product quality with no extra chemical stabilizer or preservative and profiling them for commercial end-use.

12.7.5 Mechanical properties and shelf-life extension

Mechanical properties are among the most important features of packaging materials because they affect how well the packaging remains stable, flexible, and strong while protecting food. Good packaging must maintain its strength and durability throughout storage and distribution, without harming the environment. For bioplastics used in active and intelligent packaging, chemical or physical modifications are often made to improve their strength and durability so they can perform reliably during the entire food distribution process. Stronger bioplastics translate to better defense against physical damage and moisture loss, reducing product waste by increasing its storage life further along the supply chain (Reshmy et al., 2021).

12.7.6 Strength and durability

Mechanical strength may play a key factor for packaging in which barrier functions must be maintained through food distribution and storage phase. As an illustration, PLA and PHAs are two of the most important bioplastics in active and intelligent packaging owing to their biodegradability and renewability (Zheng et al., 2022). However, PLA and PHA films are typically more brittle than common plastics such as polyethylene terephthalate (PET) or LDPE/linear low-density polyethylene, preventing their use in certain types of packaging. To address this, researchers added nanomaterials like cellulose nanofibers, clay nanoparticles, or graphene oxide inside these bioplastics, which formed nanocomposites with the biopolymer and improved their strength and flexibility by a large margin. Research has reported the enhancement of tensile strength and elasticity characteristics when clay nanoparticles are included in PLA films, which makes it an appropriate candidate for food packaging applications in terms of required durability and impact resistance (Ozdemir and Floros, 2004).

In one example, scientists showed that PLA films with graphene oxide nanofillers have a 60% greater tensile strength than without these nanofillers, which gives the polymers much greater resistance to tearing and puncturing. It makes them suitable for the intelligent and active packaging, where the package integrity must be maintained against different environmental conditions preventing oxygen, moisture and/or contaminants penetration (Liesenfeld et al., 2024). For instance, a study demonstrated that adding graphene oxide to PLA increased the tensile strength from 57.98 to 62.18 MPa, an improvement of approximately 7.2%. This enhancement improves the material's resistance to tearing and puncturing, making it suitable for active and intelligent packaging applications where maintaining package integrity against environmental factors is crucial (Chieng et al., 2014). Mechanical strength development in bioplastics is crucial to preserving packaged food as the barrier properties turn ineffective if mechanical damage occurs like during transportation or handling (Jariyasakoolroj et al., 2020)

12.7.7 *Flexibility and adaptability*

Once more, flexibility is the other supplementary mechanical property, together with strength, that gives an active and intelligent packaging its broad applicability. It also allows the packaging to bend and adapt to different shapes and sizes of food products, making it suitable for a wide range of uses, from solid items like fruits and vegetables to liquids and powders. Starch-based bioplastics are flexible and cost-effective to manufacture when utilized independently; at this point, they tend to be brittle. Plasticizers such as glycerol are added to enhance the flexibility, or bioplastics can be blended with more flexible polymers like polycaprolactone. By making these changes, the starch-based films can stretch instead of breaking apart so that they are more useful for wrapping items like baked goods or fresh fruit and vegetables (Mane, 2016). This improvement also allows intelligent components, such as TTIs and freshness indicators, to be added to the packaging without weakening its structure. For example, PHA films that contain TTIs can be bent or handled without affecting how the indicators work. This makes it much easier to include intelligent features in a wide range of products, from dairy items to frozen foods. Bioplastics provide great flexibility for use in active and intelligent packaging; the packaging having kept itself intact throughout the packaging life cycle of the product, thereby extending shelf life by reducing its exposure to environmental hazards as well as mechanical stress (Yin and Woo, 2024).

12.7.8 Role in shelf-life extension

The enhancement of mechanical attributes in bioplastic wrap leads to longer periods of food storage because the tough protective barrier protects foods from physical harm and microbiological damage and spoiling. Tough bioplastics can retain their oxygen, moisture, and light barrier properties for a longer time after packing by preventing their packaging from ripping, puncturing, or deformation. It is crucial to this protection and in high-value or sensitive products such as meats, cheeses, and fresh produce that can spoil quickly if exposed to air or contaminants. When this happens, the authors stated in their study improvement of the mechanical properties of active bioplastics not only minimizes breakage but also stabilizes the atmospheric conditions inside the package, thereby preventing/delaying microbial growth and oxidative changes (Siciliano et al., 2024).

Shelf life can be further improved by adding nanoparticles to bioplastic materials, as these particles can provide antimicrobial effects. For instance, PHA films containing silver or zinc oxide not only maintain their mechanical strength but also prevent the growth of bacteria on the surface of food, helping reduce spoilage. These functional films are important in active packaging, where the main goal is to keep food fresh for longer.

Sustainable food packaging can be achieved by using biodegradable materials that are strong, flexible, and able to protect food from physical damage and microbial contamination. By combining bioplastics with active components, it becomes possible to reduce food waste, improve food safety, and move toward a more sustainable food system (Waldo-Mendoza et al., 2024). Improving the mechanical properties of bioplastics is important for both active and intelligent packaging. Stronger, more durable bioplastics ensure that the packaging can protect food effectively while remaining environmentally friendly. With advances in nanotechnology, polymer blending, and plasticizers, bioplastics are now being engineered to meet the demanding standards of the food packaging industry. These improvements make bioplastics a promising and sustainable alternative to traditional plastics (Fontecha-Umaña et al., 2020).

12.8 Interaction of bioplastics with active and intelligent packaging components

The interactions of bioplastics together with active and intelligent packaging parts help to develop dual aspiration: increase lifetime span and food quality preservation as well preserve the ecological balance. Active packaging is based on the introduction of agents that are able to change the composition of the environment existing in internal part of package (oxygen scavengers, antimicrobial agent, and moisture regulators), while intelligent packaging systems are those that contain sensor, indicator, or smart label to provide information about product status (Wagh et al., 2023). However, for these functions to be fulfilled by bioplastics, a very compatible bio-based material is expected to possess not only the compatibility with active and intelligent functionalities but also biodegradability and renewability (Gamage et al., 2024).

12.8.1 Compatibility with active agents

The incorporation of active agents in bioplastic matrices like PLA and PHA has shown high effectiveness generating packaging that actively participates in food preservation. Oxygen stands out among the major spoilage agents, driving both microbial activity and oxidation-related deterioration of flavor, aroma, and nutritional quality of food (Cruz et al., 2022). Oxygen scavenging agents are incorporated in PLA films for reducing the amount of oxygen within the package and decreased spoilage when stored under very low-oxygen environments.

This approach has been especially successful in developing bioplastics that extend the shelf life of oxygen-sensitive products, for example, cured meats and cheese, by increasing oxygen barrier properties without affecting biodegradability. Studies show these oxygen scavenging agents are still functional inside the PLA, as they effectively lower the amount of oxygen without compromising biodegradability (Siddiqui and Chand, 2022).

Another important active packaging factor is the antimicrobials, which reduce or suppress microbe growth on food surface essential for protection from foodborne diseases and spoilage. Chitosan is an antimicrobial derived from natural sources that could be mixed with PLA or PHA to form bioplastic possessing antimicrobial activity; this may be particularly advantageous for high-risk foods like fresh produce and meats (Jiang et al., 2023). Finally, metal oxide nanoparticles (nano-ZnO and nano-Ag) have been successfully incorporated into bioplastics with extended antimicrobial function. These stable nanoparticles are also good compatibility with the bioplastic matrix and, consequently, produce a sustained antimicrobial activity. Preservation of the mechanical performance and environmental advantages of bioplastics while not compromising these performances as active packaging systems is essential (Kumar et al., 2024a).

12.8.2 Integration with intelligent packaging components

The ability to work with intelligent packaging components, such as sensors and indicators, is essential for providing reliable information about the freshness and quality of food. With tools like TTIs, freshness indicators, and pH sensors built into the packaging, both consumers and supply-chain managers can monitor product conditions in real time. Bioplastics such as PLA and PHA work well with pH-sensitive dyes that change color when spoilage occurs. Microbial activity can alter the pH of acidic foods, and the color shift gives a clear visual warning. This is especially useful for seafood, where freshness is critical. The dyes are blended into the bioplastic matrix in a way that keeps them sensitive to pH changes without reducing the material's compostability, supporting both food safety and sustainability goals (Realini and Marcos, 2014).

Also, smart labels, RFID, tags and QR codes provide increased traceability and transparency of products on the supply chain. These digital components can be accommodated with PLA-based packaging, for example, to improve monitoring and tracking of product quality from production to retail. Specifically, RFID tags enable on the fly monitoring storage conditions, such as temperature, humidity, and so on, all of which can be very harmful to product quality. The tags, when integrated with sensors inside the bioplastic packaging, provide for an integrated approach to food quality management. However, adding traditional electronic components to biodegradable materials is challenging because most electronics do not break down along with the bioplastic. Current research is focused on creating electronic components that can work inside bioplastic materials and are also biodegradable or compostable. This would support the development of fully sustainable, traceable packaging systems (Pereira et al., 2020).

Although progress has been made in integrating active and intelligent components into bioplastics, they still must be uniformly distributed and environmentally compatible. To maintain both barrier and toughness, as well as antimicrobial activity, achieving an even distribution of active agents within the bioplastic matrix, such as antimicrobial nanoparticles or oxygen scavengers, is important. To address this, films, and therefore their functionality and durability, have been improved through exploration of techniques like electrospinning and solvent casting to achieve more uniform active agent incorporation (Jariyasakoolroj et al., 2020).

Another focus is in developing biodegradable sensors and indicators that match actual decomposition rates of bioplastics. Bioplastics offer environmental benefits, but traditional sensors, most electrically, RFID tags, contain significant amounts of traditional (not compostable) electronic components, limiting their advantages. As such, researchers are looking at biodegradable conductive polymers and organic electronic materials that break down into bioplastics to make the first zero-waste packaging that works in a circular economy. These innovations not only promote environmental sustainability, but intelligent packaging can also be composted without leaving traces of nondegradable residues, by which bioplastics could be maximally utilized as eco-friendly material for active and intelligent packaging systems (Zhai et al., 2022).

Essentially, bioplastics represent the core of the development of sustainable packaging solutions on two levels—active or intelligent components and interaction between bioplastics and active or intelligent components. While research continues to improve the compatibility and functionality of bioplastic-based packaging with food and the environment, these materials are set to increasingly come to the forefront in food preservation, safety, and environmental responsibility. In view of the increasing needs for both performance and sustainability in the modern food packaging industry, bioplastics' evolution to integrate seamlessly with active and intelligent components offers to advance (Tamjid et al., 2024).

12.9 Applications active and intelligent packaging in the food industry

Active food packaging indicates an advancement in the traditional packaging system, designed to enhance the sustainability of products through interactions with the food itself or with its immediate surroundings. European Regulation 450/2009/EC provides a comprehensive description for "active materials and articles." They are defined as "components that are intentionally incorporated to release or absorb compounds into or from the food that is packaged or into the environment related to the food" (European Commission, 2009). Traditional food packaging typically serves as passive barriers that preserve food from environmental contaminants; however, active packaging facilitates proactive measures through the packing material, hence enhancing food preservation. This active process may result from specific characteristics of the packing material or from active agents incorporated into the passive barrier to function as absorbers, emission sources, or releasers of substances essential for food preservation. The primary objective of active packaging is to inhibit microbiological and chemical contamination while preserving the physical and chemical composition attributes of food. There is a variety of commercial active packaging in the market, as given in Table 12.2. Activ-Film from CSP Technologies functions as a moisture absorber, particularly for fruits and vegetables, utilizing LDPE film. Tenderpac, supplied by Sealpac International, also serves as a moisture absorber but is designed for meat products, using PET trays. PEAKfresh and BIOPAC both act as ethylene scavengers. PEAKfresh targets fruits and vegetables and employs a film impregnated with a natural mineral, whereas BIOPAC is suited for fresh products contained in sachets made of porous material mixed with potassium permanganate. ATCO and SANDRY are both carbon dioxide absorbers, with ATCO providing fresh products packaging in film bags and SANDRY® designed for fruits, coffee, and fermented foods in sachet form. McAirlaid's CO_2Pad and FreshPax are carbon dioxide emitters; McAirlaid's CO_2Pad for fish, meat, and fruit products uses cellulose-based pads, while FreshPax targets processed and precooked foods with packets and films made from food-grade materials. Celox from W. R. Grace & Co. is an oxygen scavenger useful in beverages, providing can sealants and closure coatings. $ZERO_2$ from IPL Plastics is another oxygen scavenger, targeting fresh products with a multilayer film fused to injection-molded containers. Biomaster and Food-touch offer antimicrobial properties; Biomaster is used for chilled and frozen products in cool bags, while Food-touch suits all food products with various paper product forms. ATOX from Artibal serves as an antioxidant for cereal products,

Table 12.2 Key characteristics of active packaging systems on the market.

Industrial name	Concept	Applications	Materials and types
Activ-Film (www.csptechnologies.com)	Absorption of moisture	Fruits and vegetables	Low-density polyethylene (LDPE) film
Tenderpac (www.sealpacinternational.com)	Absorption of moisture	Meat products	Polyethylene terephthalate (PET) tray
PEAKfresh (www.peakfresh.com)	Ethylene scavenging	Fruits and vegetables	Film impregnated with a natural mineral
BIOPAC (www.biopac.com.au)	Ethylene scavenging	Fresh products	Sachet in porous material mixed with potassium permanganate
ATCO (www.emcotechnologies.co.uk)	CO_2 absorber	Fresh products	Film bags
SANDRY (www.hengsan.com)	CO_2 absorber	Fruits, coffee, and fermented food	Sachets
McAirlaid's CO_2Pad (www.mcairlaids.net)	CO_2 emitter	Fishes, meats, and fruit products	Cellulose-based pads
FreshPax (www.multisorb.com)	CO_2 emitter	Processed and precooked food	Packets and films realize with food-grade materials
Celox (www.grace.com)	Oxygen scavenging	Beverages	Cans sealants and closure coatings
ZERO$_2$ (www.ipl-plastics.com)	Oxygen scavenging	Fresh products	Multilayer film fused to injection-molded containers
Biomaster (www.biomasterprotected.com)	Antimicrobial activity	Chilled and frozen products	Cool bags
Food-touch (www.microbeguard.com)	Antimicrobial activity	All food products	Various forms of paper products
ATOX (www.artibal.com)	Antioxidant activity	Cereal products	Film coating containing oregano essential oils
Pure Temp (www.puretemp.com)	Phase-change Materials	Frozen food, cold storage	Palm oil–, coconut oil–, and soybean oil–based
Green Box (www.greenbox.it)	Phase-change Materials	Perishable products	Vegetable oil–based

employing a film coating containing oregano essential oils. Pure Temp and Green Box both provide phase change materials; Pure Temp for frozen food and cold storage using palm oil, coconut oil, and soybean oil-based materials, and Green Box for perishable products using vegetable oil-based materials.

The applications and principles of various intelligent packaging systems present on the market are discussed in Table 12.3. They include TTIs such as 3 M MonitorMark®, which uses a self-adhesive pad with a dye that changes color through a wick as a temperature threshold is exceeded. Fresh-Check and Insignia Deli Intelligent Labels also indicate temperature changes

Table 12.3 Key characteristics of intelligent packaging systems on the market.

Technology	Commercial Name	Application	Principle
Time-temperature indicators (TTIs)	3M MonitorMark (www.3m.com)	Backery, beverage, and meat	Self-adhesive pad for secondary packaging. Monitors temperature contact, not product quality. Contains blue fatty acid ester dye in a matrix. Dye is released upon phase transition of carrier. Triggered by temperature exceeding a threshold. Response is measured by dye migration distance.
	Fresh-Check®Temperature Intelligence (www.fresh-check.com)	All fresh products	Self-adhesive device employing a solid-state polymerization process that produces a polymer exhibiting vibrant coloration. As the active centre experiences fluctuations in temperature over time, it exhibits a progressive alteration in color, signaling the food's freshness. The active centre circle experiences a nonreversible darkening reaction.
	Insignia Deli Intelligent Labels (www.insigniatechnologies.com)	Chilled products	The alteration in color intensifies with variations or shifts within the established temperature parameters.
	OnVu (www.packworld.com)	Meats, fishes, and dairy products	The composition features a photochromic ink sourced from benzylpyridines, which is activated by ultraviolet (UV) light, leading to a conversion into a dark blue shade. Consequently, benzylpyridines demonstrate a progressive reduction in color intensity as time elapses, even under conditions of increased ambient temperatures.
	CoolVu Food (www.product.statnano.com)	Dairy products and beverages	Over a time period, an active region undergoes a transformation from a metallic coloration to a more subdued tone. Elevated storage temperatures enhance the rate of degradation.
	Smart dot (www.evigence.com)	Bakery and frozen products	The indicator demonstrates a color change from green to red when subjected to fluctuations in temperature.

(*Continued*)

Table 12.3 (Continued)

Technology	Commercial Name	Application	Principle
	WarmMark (www.deltatrak.com)	Shipping, storage, processing	Observation of adherence to temperature fluctuations. This pad contains blotter paper saturated with a reagent exhibiting a red hue.
	Cold Chain iToken (www.deltatrak.com)	Supply chain	The direct pull-tab mechanism guarantees a clear "ON" signal; it is suitable for use via barcode scanning gadgets.
	TempDot (www.deltatrak.com)	Seafood and meat	The indicator window confirms activating; the labels are appropriate for transporting and storing across a range of temperatures.
	Freshtag/Checkpoint (www.vitsab.com)	Meat, fish, and dairy products	The setup comprises two separate compartments: one filled with an enzyme solution as well as another containing the substrate solution, accompanied by a pH indicator. The enzymes break down the substrate in a controlled way, which lowers the pH and changes the color from green to red.
	OliTec (www.oli-tec.com)	Fresh products	This is a multilayer label engineered to evaluate the degrading profiles of food products subjected to specific storage conditions.
	TOPCRYO (www.cryolog.com)	Cold chain	The system demonstrates the ability to track cold chain regulation; the microbial label experiences a colorimetric transition from green to red.
	Traceo (www.cryolog.com)	Chilled food products	The basis is established inside a framework of microbiology. This label serves as a transparent adhesive medium that encapsulates distinct strains of lactic acid bacteria. The procedure results in a clear dual effect: an irreversible change from colorless to pink, alongside a simultaneous opacification reaction initiated by significant temperature fluctuations or upon getting the specified expiration date.
	eO (www.cryolog.com)	Cold chain	The basis is established within a microbiological framework. This is a bonding label formulated as a compact gel pad, mimicking the structure of flower petals, which experiences a color change from green

			to red. The noted change in color indicates a variation in pH due to the increase of lactic acid bacteria.
	Keep-it (www.keep-it.com)	Fresh products, especially fish	The composition of the indicator is carefully crafted for different products, illustrating the gradual reduction in the product's residual shelf life as time progresses. When the compound is subjected to elevated temperatures, the indicator will demonstrate rapid activity. When the substance is kept at reduced temperatures, the indicator will demonstrate a gradual shift. Upon the indicator attaining a value of zero, the product has exceeded its threshold for edibility.
	FreshCode (www.freshcodelabel.com)	Poultry products	The indicator's core white section is impregnated with a smart ink that detects the discharge of volatile gases linked with chicken rotting in modified environment packaging. The material is considered unsuitable for ingestion when the indicator displays a fully black hue.
	Tempix (www.tempix.com)	Cold chain	During the cold chain procedure, the black section of the indicator provides evidence that the product has been able to retain the required temperature continuously. If the product has been handled at any point in time when temperatures were higher than what is suggested, the black bar will be removed from the monitor.
Freshness indicators	Raflatac (www.upmraflatac.com)	Poultry	The base consists of a nanolayer of silver that interacts with hydrogen sulfide, a by-product of cysteine breakdown.
	RipeSense (www.ripesense.co.nz)	Fruit	It detects volatile substances or gases that contribute to the maturation procedure (e.g., ethylene).
	Food fresh (www.vanprob.com)	Meat	This self-adhesive labeling indication is a nonreversible time-tracking device that may be set to expire inside a certain "consume within" period, which might be anything from a few days to many months.
Gas indicators	Ageless Eye (www.mgc.co.jp)	Meat	These sachets contain a tablet that functions as a marker for oxygen concentrations.

(*Continued*)

Table 12.3 (Continued)

Technology	Commercial Name	Application	Principle
	Tell-Tab (www.impakcorporation.com)	All products	This is a device contained within the packaging that signals the presence of oxygen.
	O_2Sense (www.evigence.com)	Fresh products	A visually discernible indicator for identifying leaking in modified atmosphere packaging (MAP).
	Novas Insignia Technologies (www.insigniatechnologies.com)	Products packed in a modified atmosphere	It indicates the occurrence of damage to the packaging. This enables producers and vendors to eliminate the item from the supply network.
	Shelf Life Guard (www.upm.com)	Meat	The consumer is notified when air has infiltrated the modified atmosphere gases inside the packaging as a result of a breach or leakage.
Sensors	O2xyDot (www.oxysense.com)	All products	Optical sensors are integrated into transparent or semitransparent packaging to facilitate the measurement of oxygen, enabling sensitive, fast, and nondestructive assessments.
	Flex Alert (www.flex-alert.com)	Coffee beans, dried nuts, seeds, wine barrels, and fresh fruit	A versatile biosensor designed for the detection of toxins in packaged foods across the entire supply chain. This formulation has been meticulously engineered to target *Escherichia coli*, *Listeria* spp., *Salmonella* spp., and aflatoxins.
Radio frequency identification systems (RFID) technology	Easy2log (www.environmental-expert.com)	Seafood, meat and poultry, milk-based products, and frozen food	This is an economical, semipassive device designed for the observation of temperature-sensitive items throughout their transit and storage phases. The tag possesses the capability to compute the mean kinetic temperature and allows for user-defined configurations regarding the remaining shelf-life duration. Additionally, it can trigger alerts if these parameters surpass the thresholds established by the user.
	CS8304 (www.convergence.com.hk)	Cold chain	The tag offers 10,000 items of logging memory for the preservation of temperature information. The LED indicator signals deviations in temperature parameters.
	TempTRIP (www.temptrip.com)	Cold chain	This temperature tag employs ultrahigh frequency for wireless communication with readers, which transmit the outcomes immediately to a web page on the Internet.

with color changes, other indicators. Photochromic ink and other materials are used in other systems, such as OnVu, CoolVu Food, and Smart dot, to demonstrate color changes at certain temperatures. TempDot and WarmMark are other systems that monitor temperature during shipping and storage to ensure that product transport meets the parameters of temperature. Freshtag/Checkpoint and OliTec use enzymatic reactions or multilayer labels to indicate changes in food quality caused by environmental factors. There are some systems that monitor cold-chain conditions, such as TOPCRYO and Traceo, which use microbial labels or color changes to indicate fluctuations in temperature. The FreshCode and Keep-it monitor product shelf life and gases associated with spoilage. Ageless Eye and Tell-Tab are gas indicators that indicate oxygen levels and packaging integrity. Novas Insignia Technologies and O2Sense detect leaks in MAP. O2xyDot and Flex Alert use sensors to detect oxygen or toxins, while RFID technologies such as Easy2log and TempTRIP monitor the temperature-sensitive products during storage and transportation. For fresh products, meats, dairy, seafood, beverages, and cold-chain management, these intelligent packaging systems are built for different applications to ensure optimal product quality and safety.

12.10 Challenges and future directions

The food packaging industry represents an attractive industry that significantly contributes to the current economic landscape. Food package design is important to make sure that food is safe every step of the entire supply chain, to make storage and shipping more efficient, and to make it easier for stores and customers to get information about products (Gustavo et al., 2018). Roughly 9% of the large number of research concerning food packaging (41,602 papers between 1960 and 2022) is dedicated to the development of packaging. The development of food packaging with sustainability in mind seeks to achieve the least possible environmental impact upon the disposal of both the food product and its container. As a result, the advancement and creativity in food packaging are rapidly accelerating. The literature review from the past 10 years reveals a notable a rise in annually publications related to sustainable food packaging, exhibiting an annual growth rate of 19%–35%. Concurrently, there is a significant rise in the domain of active and intelligent food packaging, about a yearly rise ranging from 13% to 33%. In the realm of food, paper for packing ranks as the second most utilized and researched material, comprising 34%, following plastic, which accounts for 37% in both rigid and flexible varieties (Muller et al., 2017). An ongoing rise in activities associated with innovations and novel food packaging has been noted, accompanied by a growing yet rapid escalation in research focused on bio-based materials, particularly bioplastic food packaging, as well as active and intelligent packaging solutions for food. Both areas have experienced significant growth over the past few years (Fig. 12.7). These developments illustrate the ongoing efforts to formulate green and sustainable packaging products for food that concentrate on (i) increasing the performance of biopolymers by means of physical, chemical, and enzymatic modifications, as well as composite formulations, (ii)

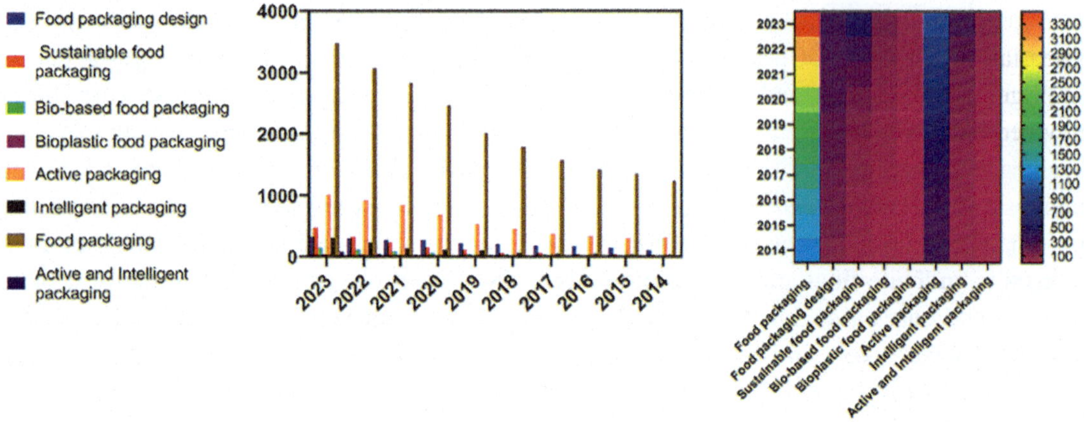

Figure 12.7

Published papers according to the search: S1: "Food packaging design," S2: "Sustainable food packaging," S3: "Bio-based food packaging," S4: "Bioplastic food packaging," S5: "Active packaging," S6: "Intelligent packaging," S7: "Food packaging," and S8: "Active" AND "Intelligent packaging."

Data source: Scopus.

creating natural-based nanocomposites for either active and intelligent environmentally friendly products, (iii) synthesizing novel renewable and biodegradable plastics, (iv) building up bioplastic production processes, (v) exploring new sustainable resources, particularly ignored raw materials, and (vi) enhancing the ability for material reusing and recycling (Jariyasakoolroj et al., 2020).

Intelligent packaging represents a significant advancement in food packaging technology, possessing the ability to assess and ensure the quality of foods for consumers. The intelligent packaging system remains in the developmental phase for numerous applications owing to the wide range of agricultural and food items. The gap exists among advanced intelligent packaging systems in the study phase and those in the marketing stage. Future investigations must consider certain obstacles to enhancing the commercial viability of the intelligent packaging systems. The complexity of food systems arises from numerous physical and chemical variables that exhibit significant variability, which, in turn, decreases the potential for commercializing intelligent packaging systems developed under in vitro or the laboratory and simplified experimental setups. Single intelligent packaging systems are anticipated to be incorporated into a cost-effective, compact unit with reduced sizes, aimed at delivering enhanced insights regarding the variables affecting quality of a food item. The synergistic investigation across electronics, mechanical, and food engineering technologies significantly contributes to the advancement of successful intelligent packaging developments. The integration of various sensors enhances the operational capabilities of intelligent packaging systems, enabling the observation of highly complicated product degradation processes. In this scenario, a thorough examination of the functional interactions among the sensors is essential. It is essential to develop more sophisticated systems to supersede the current intelligent packaging devices. Currently, the methods for IP are costly, primarily due to the elevated expenses associated with raw materials and highly complicated production processes. Therefore, the development of more economical intelligent packaging systems appears to be essential. Thus, enhancing the economic and technical viability of these novel systems is essential. Systems consisting of electronic chips integrated via artificial intelligence can help with enhanced communication with consumers, enabling more convenient alerts during both storage and consumption phases. The recent progress of nanosensors plays an essential part in the future improvement of intelligent packaging. At present, these systems are undergoing initial phases of research and development, with the anticipation that the commercialization of the gadgets will lead to a decrease in the expenses associated with the intelligent packaging systems. Nevertheless, emphasizing only developing systems is insufficient; it is imperative to also address the formulation of guidelines and the legal considerations necessary for effective intelligent packaging management. Another challenge involves the potential for counterfeiting within intelligent packaging systems; enhancing intelligent packaging security may require further investigation. The primary challenge faced by active packaging systems lies in having the capacity of the active materials to maintain their inherent characteristics. The development of novel production methodologies, along with the integration of new active materials exhibiting enhanced stability and superior physicochemical characteristics, will be the focus of current studies. The increasing market demand for secure antimicrobial materials is steering research trends toward the exploration of naturally produced antimicrobials and biopolymers. Nevertheless, several antimicrobial packaging technologies are being brought to market, influenced by negative impacts on the quality of food, customer acceptance, scalability, technological challenges, and economic factors. Additional investigation is required to address these issues along with the associated regulatory and financial issues. Normally, low molecular weight results in an immediate release number of active materials, consequently causing a decline in the efficacy of the active ingredient throughout the shelf life of food. The advancement of controlled-release packaging systems would be advantageous for applications that seek to prolong the delivery duration of active compounds while enhancing the consistency and reliability of release rates. Many studies focus only on the active features of packaging systems and overlook how these materials interact with real food. This gap often leads to poor success when these active packaging products reach the market. Companies also face technical and economic challenges during commercialization. For this reason, more interdisciplinary research is needed to make these technologies cost-effective, scalable, and suitable for real-world use.

A key research need is to determine the optimal level of active compounds for different foods and to study how these compounds are released in real-food conditions. Safety and regulatory issues, such as limits on how much of these active molecules can migrate into food, the use of food-grade additives, and toxicity concerns, also require further investigation. Nanotechnology offers promising new ways to solve many of these challenges. The movement (migration) of active materials is still one of the main limitations in active packaging. Nanomaterials and microfibers can provide slow and controlled release of active substances, showing strong potential for future development. Intelligent packaging can help monitor how well these systems work by providing real-time information about food quality. However, most research looks at either active or intelligent packaging separately. There is very little work exploring how these two systems can be combined. Future studies should focus on integrating both technologies to create complete packaging systems that work with real foods during storage, transport, and use. Combining active and intelligent functions could greatly improve food preservation and quality.

Intelligent packaging can detect food conditions and trigger the active components to respond when needed. Although this combined approach can be used for many food products, more research is required to design the most effective models for each type of food. It is also important to study how these systems affect the taste, smell, and appearance of the packaged food to ensure consumer acceptance. Overall, active intelligent hybrid packaging has strong potential, but cost considerations will be important for commercial use. Continued research in this area may significantly improve packaging for fresh-cut fruits and vegetables, meat, and seafood, providing better food safety, longer shelf life, real-time monitoring, and reduced food waste.

References

Ahmadi, A., Ahmadi, P., Sani, M.A., Ehsani, A., Ghanbarzadeh, B., 2021. Functional biocompatible nanocomposite films consisting of selenium and zinc oxide nanoparticles embedded in gelatin/cellulose nanofiber matrices. International Journal of Biological Macromolecules 175, 87–97. Available at https://doi.org/10.1016/j.ijbiomac.2021.01.135.

Almeida, C., Laso, J., de Sousa, D.B., Cooney, R., Quinteiro, P., Rowan, N., et al., 2023. Seafood consumers engagement in reducing environmental impacts from packaging. Science of the Total Environment 863, 160846. Available at https://doi.org/10.1016/j.scitotenv.2022.160846.

Al-Tayyar, N.A., Youssef, A.M., Al-Hindi, R.R., 2020. Antimicrobial packaging efficiency of ZnO-SiO2 nanocomposites infused into PVA/CS film for enhancing the shelf life of food products. Food Packaging and Shelf Life 25, 100523. Available at https://doi.org/10.1016/j.fpsl.2020.100523.

Alves, J., Gaspar, P.D., Lima, T.M., Silva, P.D., 2023. What is the role of active packaging in the future of food sustainability? A systematic review. Journal of the Science of Food and Agriculture 103 (3), 1004–1020. Available at https://doi.org/10.1002/jsfa.11880.

Asgher, M., Qamar, S.A., Bilal, M., Iqbal, H.M.N., 2020. Bio-based active food packaging materials: sustainable alternative to conventional petrochemical-based packaging materials. Food Research International 137, 109625. Available at https://doi.org/10.1016/j.foodres.2020.109625.

Azmin, S.N.H.M., Hayat, N.A.B.M., Nor, M.S.M., 2020. Development and characterization of food packaging bioplastic film from cocoa pod husk cellulose incorporated with sugarcane bagasse fibre. Journal of Bioresources and Bioproducts 5 (4), 248–255. Available at https://doi.org/10.1016/j.jobab.2020.10.003.

Balbinot-Alfaro, E., Craveiro, D.V., Lima, K.O., Costa, H.L.G., Lopes, D.R., Prentice, C., 2019. Intelligent packaging with pH indicator potential. Food Engineering Reviews 11 (4), 235–244. Available at https://doi.org/10.1007/s12393-019-09198-9.

Benitez, J.J., Florido-Moreno, P., Porras-Vázquez, J.M., Tedeschi, G., Athanassiou, A., Heredia-Guerrero, J.A., et al., 2024. Transparent, plasticized cellulose-glycerol bioplastics for food packaging applications. International Journal of Biological Macromolecules 273, 132956. Available at https://doi.org/10.1016/j.ijbiomac.2024.132956.

Bonilla, J., Atarés, L., Vargas, M., Chiralt, A., 2013. Properties of wheat starch film-forming dispersions and films as affected by chitosan addition. Journal of Food Engineering 114 (3), 303–312. Available at https://doi.org/10.1016/j.jfoodeng.2012.08.005.

Boura-Theodoridou, O., Giannakas, A., Katapodis, P., Stamatis, H., Ladavos, A., Barkoula, N.M., 2020. Performance of ZnO/chitosan nanocomposite films for antimicrobial packaging applications as a function of NaOH treatment and glycerol/PVOH blending. Food Packaging and Shelf Life 23, 100456. Available at https://doi.org/10.1016/j.fpsl.2019.100456.

Cao, Y., Xu, P., Lv, P., Lemstra, P.J., Cai, X., Yang, W., et al., 2020. Excellent UV resistance of polylactide by interfacial stereocomplexation with double-shell-structured TiO2 nanohybrids. ACS Applied Materials and Interfaces 12 (43), 49090–49100. Available at https://doi.org/10.1021/acsami.0c14423.

Chandrababu, V., Parameswaranpillai, J., Gopi, J.A., Pathak, C., Midhun Dominic, C.D., Feng, N.L., et al., 2024. Progress in food packaging applications of biopolymer-nanometal composites—a comprehensive review. Biomaterials Advances 162, 213921. Available at https://doi.org/10.1016/j.bioadv.2024.213921.

Chaudhary, P., Fatima, F., Kumar, A., 2020. Relevance of nanomaterials in food packaging and its advanced future prospects. Journal of Inorganic and Organometallic Polymers and Materials 30 (12), 5180–5192. Available at https://doi.org/10.1007/s10904-020-01674-8.

Chayavanich, K., Thiraphibundet, P., Imyim, A., 2020. Biocompatible film sensors containing red radish extract for meat spoilage observation. Spectrochimica Acta Part A: Molecular and Biomolecular Spectroscopy 226, 117601. Available at https://doi.org/10.1016/j.saa.2019.117601.

Chieng, B.W., Ibrahim, N.A., Yunus, W.M.Z.W., Hussein, M.Z., Then, Y.Y., Loo, Y.Y., 2014. Effects of graphene nanoplatelets and reduced graphene oxide on poly(lactic acid) and plasticized poly(lactic acid): a comparative study. Polymers 6 (8), 2232–2246. Available at https://doi.org/10.3390/polym6082232.

Corradini, M.G., 2018. Shelf life of food products: from open labeling to real-time measurements. Annual Review of Food Science and Technology 9, 251–269. Available at https://doi.org/10.1146/annurev-food-030117-012433.

Cruz, R.M.S., Krauter, V., Krauter, S., Agriopoulou, S., Weinrich, R., Herbes, C., et al., 2022. Bioplastics for food packaging: environmental impact, trends and regulatory aspects. Foods 11 (19), 3087. Available at https://doi.org/10.3390/foods11193087.

Cruz-Romero, M.C., Murphy, T., Morris, M., Cummins, E., Kerry, J.P., 2013. Antimicrobial activity of chitosan, organic acids and nano-sized solubilisates for potential use in smart antimicrobially-active packaging for potential food applications. Food Control 34 (2), 393–397. Available at https://doi.org/10.1016/j.foodcont.2013.04.042.

D'Almeida, A.P., de Albuquerque, T.L., 2024. Innovations in food packaging: from bio-based materials to smart packaging systems. Processes 12 (10), 2085. Available at https://doi.org/10.3390/pr12102085.

Das, P.P., Prathapan, R., Ng, K.W., 2024. Advances in biomaterials based food packaging systems: Current status and the way forward. Biomaterials Advances 164, 213988. Available at 10.1016/j.bioadv.2024.213988.

de Abreu, D.A.P., Cruz, J.M., Losada, P.P., 2012. Active and intelligent packaging for the food industry. Food Reviews International 28 (2), 146–187. Available at https://doi.org/10.1080/87559129.2011.595022.

Dey, A., Neogi, S., 2019. Oxygen scavengers for food packaging applications: a review. Trends in Food Science & Technology 90, 26–34. Available at https://doi.org/10.1016/j.tifs.2019.05.013.

Dilkes-Hoffman, L.S., Pratt, S., Lant, P.A., Levett, I., Laycock, B., 2018. Polyhydroxyalkanoate coatings restrict moisture uptake and associated loss of barrier properties of thermoplastic starch films. Journal of Applied Polymer Science 135 (25), 46379. Available at https://doi.org/10.1002/app.46379.

Elsabee, M.Z., Abdou, E.S., 2013. Chitosan based edible films and coatings: a review. Materials Science and Engineering C 33 (4), 1819–1841. Available at https://doi.org/10.1016/j.msec.2013.01.010.

European Commission, 2004. Regulation (EC) No 1935/2004 of the European Parliament and of the Council of 27 October 2004 on materials and articles intended to come into contact with food and repealing Directives 80/590/EEC and 89/109/EEC. EEC.

European Commission, 2009. Commission Regulation (EC) No 450/2009 of 29 May 2009 on active and intelligent materials and articles intended to come into contact with food. Official Journal of European Union 135, 3–11.

Farasati Far, B., Jahanbakhshi, M., Jameie, L., Zolfigol, F., Taromi, P., Ertas, Y.N., 2024. Chitosan-graft-pomegranate extract hydrogel: a dual-functional pad for antibacterial and antioxidant enhancement for shelf life extension in food packaging. ACS Applied Polymer Materials 6 (16), 9545–9558. Available at https://doi.org/10.1021/acsapm.4c01240.

Figueroa-Lopez, K.J., Prieto, C., Pardo-Figuerez, M., Cabedo, L., Lagaron, J.M., 2023. Development and characterization of electrospun poly(3-hydroxybutyrate-co-3-hydroxyvalerate) biopapers containing cerium oxide nanoparticles for active food packaging applications. Nanomaterials 13 (5), 823. Available at https://doi.org/10.3390/nano13050823.

Figueroa-Lopez, K.J., Villabona-Ortíz, Á., Ortega-Toro, R., 2024. Sustainable starch-based films from cereals and tubers: a comparative study on cherry tomato preservation. Polymers 16 (20), 2913. Available at https://doi.org/10.3390/polym16202913.

Fontecha-Umaña, F., Ríos-Castillo, A.G., Ripolles-Avila, C., Rodríguez-Jerez, J.J., 2020. Antimicrobial activity and prevention of bacterial biofilm formation of silver and zinc oxide nanoparticle-containing polyester surfaces at various concentrations for use. Foods 9 (4), 442. Available at https://doi.org/10.3390/foods9040442.

Gaikwad, K.K., Singh, S., Ajji, A., 2019. Moisture absorbers for food packaging applications. Environmental Chemistry Letters 17 (2), 609–628. Available at https://doi.org/10.1007/s10311-018-0810-z.

Gaikwad, K.K., Singh, S., Shin, J., Lee, Y.S., 2020. Novel polyisoprene based UV-activated oxygen scavenging films and their applications in packaging of beef jerky. LWT 117, 108643. Available at https://doi.org/10.1016/j.lwt.2019.108643.

Gamage, A., Thiviya, P., Liyanapathiranage, A., Wasana, M.L.D., Jayakodi, Y., Bandara, A., et al., 2024. Polysaccharide-based bioplastics: eco-friendly and sustainable solutions for packaging. Journal of Composites Science 8 (10), 413. Available at https://doi.org/10.3390/jcs8100413.

Gan, I., Chow, W.S., 2018. Antimicrobial poly(lactic acid)/cellulose bionanocomposite for food packaging application: a review. Food Packaging and Shelf Life 17, 150–161. Available at https://doi.org/10.1016/j.fpsl.2018.06.012.

Ghandehari-Alavijeh, S., Can Karaca, A., Akbari-Alavijeh, S., Assadpour, E., Farzaneh, P., Saidi, V., et al., 2024. Application of encapsulated flavors in food products: opportunities and challenges. Food Chemistry 436, 137743. Available at https://doi.org/10.1016/j.foodchem.2023.137743.

Gustavo, J.U., Pereira, G.M., Bond, A.J., Viegas, C.V., Borchardt, M., 2018. Drivers, opportunities and barriers for a retailer in the pursuit of more sustainable packaging redesign. Journal of Cleaner Production 187, 18–28. Available at https://doi.org/10.1016/j.jclepro.2018.03.197.

Hakeem, M.J., Feng, J., Nilghaz, A., Ma, L., Seah, H.C., Konkel, M.E., et al., 2020. Active packaging of immobilized zinc oxide nanoparticles controls Campylobacter jejuni in raw chicken meat. Applied and Environmental Microbiology 86 (22), e01195-20. Available at https://doi.org/10.1128/aem.01195-20.

Han, J.H., 2005. New technologies in food packaging: overiew. In: Han, J.H. (Ed.), Innovations in Food Packaging. Academic Press, pp. 3–11. Available at https://doi.org/10.1016/B978-012311632-1/50033-4.

Haylock, R., Rosentrater, K.A., 2018. Cradle-to-grave life cycle assessment and techno-economic analysis of polylactic acid composites with traditional and bio-based fillers. Journal of Polymers and the Environment 26 (4), 1484–1503. Available at https://doi.org/10.1007/s10924-017-1041-2.

Hossain, F., Follett, P., Salmieri, S., Vu, K.D., Harich, M., Lacroix, M., 2019. Synergistic effects of nanocomposite films containing essential oil nanoemulsions in combination with ionizing radiation for control of rice weevil Sitophilus oryzae in stored grains. Journal of Food Science 84 (6), 1439–1446. Available at https://doi.org/10.1111/1750-3841.14603.

Ibrahim, S., Fahmy, H., Salah, S., 2021. Application of interactive and intelligent packaging for fresh fish shelf-life monitoring. Frontiers in Nutrition 8, 677884. Available at https://doi.org/10.3389/fnut.2021.677884.

Jafarzadeh, S., Ariffin, F., Mahmud, S., Alias, A.K., Hosseini, S.F., Ahmad, M., 2017. Improving the physical and protective functions of semolina films by embedding a blend nanofillers (ZnO-nr and nano-kaolin). Food Packaging and Shelf Life 12, 66–75. Available at https://doi.org/10.1016/j.fpsl.2017.03.001.

Jafarzadeh, S., Rhim, J.W., Alias, A.K., Ariffin, F., Mahmud, S., 2019. Application of antimicrobial active packaging film made of semolina flour, nano zinc oxide and nano-kaolin to maintain the quality of low-moisture mozzarella cheese during low-temperature storage. Journal of the Science of Food and Agriculture 99 (6), 2716–2725. Available at https://doi.org/10.1002/jsfa.9439.

Jafarzadeh, S., Yildiz, Z., Yildiz, P., Strachowski, P., Forough, M., Esmaeili, Y., et al., 2024. Advanced technologies in biodegradable packaging using intelligent sensing to fight food waste. International Journal of Biological Macromolecules 261 (Pt 1), 129647. Available at https://doi.org/10.1016/j.ijbiomac.2024.129647.

Jariyasakoolroj, P., Leelaphiwat, P., Harnkarnsujarit, N., 2020. Advances in research and development of bioplastic for food packaging. Journal of the Science of Food and Agriculture 100 (14), 5032–5045. Available at https://doi.org/10.1002/jsfa.9497.

Jiang, A., Patel, R., Padhan, B., Palimkar, S., Galgali, P., Adhikari, A., et al., 2023. Chitosan based biodegradable composite for antibacterial food packaging application. Polymers 15 (10), 2235. Available at https://doi.org/10.3390/polym15102235.

Jieying, S., Tingting, L., Caie, W., Dandan, Z., Gongjian, F., Xiaojing, L., 2024. Paper-based material with hydrophobic and antimicrobial properties: advanced packaging materials for food applications. Comprehensive Reviews in Food Science and Food Safety 23 (3), e13373. Available at https://doi.org/10.1111/1541-4337.13373.

Jouki, M., Khazaei, N., Ghasemlou, M., HadiNezhad, M., 2013. Effect of glycerol concentration on edible film production from cress seed carbohydrate gum. Carbohydrate Polymers 96 (1), 39–46. Available at https://doi.org/10.1016/j.carbpol.2013.03.077.

Kachoei, M., Divband, B., Rahbar, M., Esmaeilzadeh, M., Ghanizadeh, M., Alam, M., et al., 2021. A novel developed bioactive composite resin containing silver/zinc oxide (ag/zno) nanoparticles as an antimicrobial material against Streptococcus mutans, Lactobacillus, and Candida albicans. Evidence-Based Complementary and Alternative Medicine 2021, 1–8. Available at https://doi.org/10.1155/2021/4743411.

Kale, G., Kijchavengkul, T., Auras, R., Rubino, M., Selke, S.E., Singh, S.P., 2007. Compostability of bioplastic packaging materials: an overview. Macromolecular Bioscience 7 (3), 255–277. Available at https://doi.org/10.1002/mabi.200600168.

Kanatt, S.R., Makwana, S.H., 2020. Development of active, water-resistant carboxymethyl cellulose-poly vinyl alcohol-aloe vera packaging film. Carbohydrate Polymers 227, 115303. Available at https://doi.org/10.1016/j.carbpol.2019.115303.

Kerry, J.P., O'Grady, M.N., Hogan, S.A., 2006. Past, current and potential utilisation of active and intelligent packaging systems for meat and muscle-based products: A review. Meat Science 74 (1), 113–130. Available at https://doi.org/10.1016/j.meatsci.2006.04.024.

Kim, J.M., Lee, M.H., Ko, J.A., Kang, D.H., Bae, H., Park, H.J., 2018. Influence of food with high moisture content on oxygen barrier property of polyvinyl alcohol (PVA)/vermiculite nanocomposite coated multilayer packaging film. Journal of Food Science 83 (2), 349–357. Available at https://doi.org/10.1111/1750-3841.14012.

Kim, I., Viswanathan, K., Kasi, G., Thanakkasaranee, S., Sadeghi, K., Seo, J., 2022. ZnO nanostructures in active antibacterial food packaging: preparation methods, antimicrobial mechanisms, safety issues, future prospects, and challenges. Food Reviews International 38 (4), 537–565. Available at https://doi.org/10.1080/87559129.2020.1737709.

Kumar, Y., Bist, Y., Thakur, D., Nagar, M., Saxena, D.C., 2024b. A review on the role of pH-sensitive natural pigments in biopolymers based intelligent food packaging films. International Journal of Biological Macromolecules 276 (Pt 1), 133869. Available at https://doi.org/10.1016/j.ijbiomac.2024.133869.

Kumar, P., Gautam, S., Bansal, D., Kaur, R., 2024a. Starch-based antibacterial food packaging with ZnO nanoparticle. Journal of Food Science and Technology 61 (1), 178–191. Available at https://doi.org/10.1007/s13197-023-05834-9.

Lei, Y., Mao, L., Yao, J., Zhu, H., 2021. Improved mechanical, antibacterial and UV barrier properties of catechol-functionalized chitosan/polyvinyl alcohol biodegradable composites for active food packaging. Carbohydrate Polymers 264, 117997. Available at https://doi.org/10.1016/j.carbpol.2021.117997.

Li, S., Tang, W., Chen, S., Si, Y., Liu, R., Guo, X., 2023. Flexible organic polymer gas sensor and system integration for smart packaging. Advanced Sensor Research 2 (11), 2300030. Available at https://doi.org/10.1002/adsr.202300030.

Liesenfeld, J., Jablonski, J.J., da Silva, J.R.F., Buenos, A.A., Scheuer, C.J., 2024. Exploring the influence of graphene incorporation on the characteristics of 3D-printed PLA. International Journal of Advanced Manufacturing Technology 130 (11–12), 5813–5835. Available at https://doi.org/10.1007/s00170-024-13032-w.

Machiels, J., Appeltans, R., Bauer, D.K., Segers, E., Henckens, Z., Van Rompaey, W., et al., 2021. Screen printed antennas on fiber-based substrates for sustainable HF RFID assisted E-fulfilment smart packaging. Materials 14 (19), 5500. Available at https://doi.org/10.3390/ma14195500.

Mane, K.A., 2016. A review on active packaging: an innovation in food packaging. International Journal of Environment, Agriculture and Biotechnology 1 (3), 544–549. Available at https://doi.org/10.22161/ijeab/1.3.35.

Martins, C., Vilarinho, F., Sanches Silva, A., Andrade, M., Machado, A.V., Castilho, M.C., et al., 2018. Active polylactic acid film incorporated with green tea extract: Development, characterization and effectiveness. Industrial Crops and Products 123, 100–110. Available at https://doi.org/10.1016/j.indcrop.2018.06.056.

Masmoudi, F., Bessadok, A., Dammak, M., Jaziri, M., Ammar, E., 2016. Biodegradable packaging materials conception based on starch and polylactic acid (PLA) reinforced with cellulose. Environmental Science and Pollution Research 23 (20), 20904–20914. Available at https://doi.org/10.1007/s11356-016-7276-y.

Mills, A., 2005. Oxygen indicators and intelligent inks for packaging food. Chemical Society Reviews 34 (12), 1003–1011. Available at https://doi.org/10.1039/b503997p.

Miltz, J., Perry, M., 2005. Evaluation of the performance of iron-based oxygen scavengers, with comments on their optimal applications. Packaging Technology and Science 18 (1), 21–27. Available at https://doi.org/10.1002/pts.671.

Mortezagholi, B., Movahed, E., Fathi, A., Soleimani, M., Forutan Mirhosseini, A., Zeini, N., et al., 2022. Plant-mediated synthesis of silver-doped zinc oxide nanoparticles and evaluation of their antimicrobial activity against bacteria cause tooth decay. Microscopy Research and Technique 85 (11), 3553–3564. Available at https://doi.org/10.1002/jemt.24207.

Muller, J., González-Martínez, C., Chiralt, A., 2017. Combination of poly(lactic) acid and starch for biodegradable food packaging. Materials 10 (8), 952. Available at https://doi.org/10.3390/ma10080952.

Narancic, T., Cerrone, F., Beagan, N., O'Connor, K.E., 2020. Recent advances in bioplastics: application and biodegradation. Polymers 12 (4), 920. Available at https://doi.org/10.3390/POLYM12040920.

Nguyen, H.L., Tran, T.H., Hao, L.T., Jeon, H., Koo, J.M., Shin, G., et al., 2021. Biorenewable, transparent, and oxygen/moisture barrier nanocellulose/nanochitin-based coating on polypropylene for food packaging applications. Carbohydrate Polymers 271, 118421. Available at https://doi.org/10.1016/j.carbpol.2021.118421.

Nikolic, M.V., Vasiljevic, Z.Z., Auger, S., Vidic, J., 2021. Metal oxide nanoparticles for safe active and intelligent food packaging. Trends in Food Science and Technology 116, 655–668. Available at https://doi.org/10.1016/j.tifs.2021.08.019.

Oluwasina, O.O., Awonyemi, I.O., 2021. Citrus peel extract starch-based bioplastic: effect of extract concentration on packed fish and bioplastic properties. Journal of Polymers and the Environment 29 (6), 1706–1716. Available at https://doi.org/10.1007/s10924-020-01990-7.

Ozdemir, M., Floros, J.D., 2004. Active food packaging technologies. Critical Reviews in Food Science and Nutrition 44 (3), 185–193. Available at https://doi.org/10.1080/10408690490441578.

Paidari, S., Esmaeili, Y., Ibrahim, S.A., Vahedi, S., Al-Hilif, S.A., Zamindar, N., 2024. Application of nanoparticles to enhance the microbial quality and shelf life of food products. Microbial Biotechnology in the Food Industry: Advances, Challenges, and Potential Solutions. Springer International Publishing, Iran, pp. 74–102. Available at https://doi.org/10.1007/978-3-031-51417-3_4.

Park, H.J., Chinnan, M.S., Shewfelt, R.L., 1994. Edible coating effects on storage life and quality of tomatoes. Journal of Food Science 59, 568–570. Available at 10.1111/j.1365-2621.1994.tb05563.x.

Păuşescu, I., Dreavă, D.-M., Bîtcan, I., Argetoianu, R., Dăescu, D., Medeleanu, M., 2022. Bio-based pH indicator films for intelligent food packaging applications. Polymers 14 (17), 3622. Available at https://doi.org/10.3390/polym14173622.

Pârvulescu, C., Anăstăsoaie, V., Tomescu, R., Aldrigo, M., Cristea, D., 2023. Multilayer smart holographic label with integrated RFID for product security and monitoring. Micromachines 14 (3), 692. Available at https://doi.org/10.3390/mi14030692.

Peelman, N., Ragaert, P., De Meulenaer, B., Adons, D., Peeters, R., Cardon, L., et al., 2013. Application of bioplastics for food packaging. Trends in Food Science & Technology 32 (2), 128–141. Available at https://doi.org/10.1016/j.tifs.2013.06.003.

Pereira, D.C.de S., Teixeira, N.S., Zin, I.M., Gonçalves, R.A.dos S., Melo, N.R.de, 2020. Active and intelligent packaging: security, legal aspects and global market. Brazilian Journal of Development 6 (8), 61766–61794. Available at https://doi.org/10.34117/bjdv6n8-164.

Reshmy, R., Madhavan, A., Philip, E., Paul, S.A., Sindhu, R., Binod, P., et al., 2021. Sugarcane bagasse derived nanocellulose reinforced with frankincense (Boswellia serrata): physicochemical properties, biodegradability and antimicrobial effect for controlling microbial growth for food packaging application. Environmental Technology & Innovation 21, 101335. Available at https://doi.org/10.1016/j.eti.2020.101335.

Rahman, L.F., Alam, L., Marufuzzaman, M., Sumaila, U.R., 2021. Traceability of sustainability and safety in fishery supply chain management systems using radio frequency identification technology. Foods 10 (10), 2265. Available at https://doi.org/10.3390/foods10102265.

Realini, C.E., Marcos, B., 2014. Active and intelligent packaging systems for a modern society. Meat Science 98 (3), 404–419. Available at https://doi.org/10.1016/j.meatsci.2014.06.031.

Restuccia, D., Spizzirri, U.G., Parisi, O.I., Cirillo, G., Curcio, M., Iemma, F., et al., 2010. New EU regulation aspects and global market of active and intelligent packaging for food industry applications. Food Control 21 (11), 1425–1435. Available at https://doi.org/10.1016/j.foodcont.2010.04.028.

Robertson, G.L., 2005. Food Packaging: Principles and Practice. CRC Press, Boca Raton.

Rossa, V., Monteiro Ferreira, L.E., da Costa Vasconcelos, S., Tai Shimabukuro, E.T., Gomes da Costa Madriaga, V., Carvalho, A.P., et al., 2022. Nanocomposites based on the graphene family for food packaging: historical perspective, preparation methods, and properties. RSC Advances 12 (22), 14084–14111. Available at https://doi.org/10.1039/d2ra00912a.

Rovera, C., Türe, H., Hedenqvist, M.S., Farris, S., 2020. Water vapor barrier properties of wheat gluten/silica hybrid coatings on paperboard for food packaging applications. Food Packaging and Shelf Life 26, 100561. Available at https://doi.org/10.1016/j.fpsl.2020.100561.

Siciliano, S., Lopresto, C.G., Lamonaca, F., 2024. From traditional packaging to smart bio-packaging for food safety: a review. Euro-Mediterranean Journal for Environmental Integration 9 (4), 1971–1986. Available at https://doi.org/10.1007/s41207-024-00627-8.

Siddiqui, A., Chand, K., 2022. Enhancement of shelf life of food using active packaging technologies. Innovative Approaches for Sustainable Development: Theories and Practices in Agriculture. Springer International Publishing, India, pp. 133–144. Available at https://doi.org/10.1007/978-3-030-90549-1_8.

Siddiqui, S.A., Sundarsingh, A., Bahmid, N.A., Nirmal, N., Denayer, J.F.M., Karimi, K., 2023. A critical review on biodegradable food packaging for meat: materials, sustainability, regulations, and perspectives in the EU. Comprehensive Reviews in Food Science and Food Safety 22 (5), 4147–4185. Available at https://doi.org/10.1111/1541-4337.13202.

Soltani Firouz, M., Mohi-Alden, K., Omid, M., 2021. A critical review on intelligent and active packaging in the food industry: research and development. Food Research International 141, 110113. Available at https://doi.org/10.1016/j.foodres.2021.110113.

Stoll, L., Domenek, S., Hickmann Flôres, S., Nachtigall, S.M.B., de Oliveira Rios, A., 2021. Polylactide films produced with bixin and acetyl tributyl citrate: functional properties for active packaging. Journal of Applied Polymer Science 138 (17), 50302. Available at https://doi.org/10.1002/app.50302.

Sulistyo, J., Winarno, P.S., Pratiwi, I.Y., Ridfan, L.P., Pranata, K.M., Chik, R.M.R., 2023. Preparation of active food packaging and coating material based on bacterial cellulose to increase food safety. Jurnal Teknologi dan Industri Pangan 34 (1), 48–61. Available at https://doi.org/10.6066/jtip.2023.34.1.48.

Supardi, Y.A., Karmini, M., 2023. Effect of using cassava and glycerol as food storage on the quality of bioplastic packaged food. Healthcare in Low-Resource Settings 11 (2), 11778. Available at https://doi.org/10.4081/hls.2023.11778.

Suryani, S., Rihayat, T., Nurhanifa, N., Riskina, S., 2020. Modification of poly lactid acid (PLA)/chitosan with cinnamon essential oil for antibacterial applications. IOP Conference Series: Materials Science and Engineering 830 (4), 042017. Available at https://doi.org/10.1088/1757-899x/830/4/042017.

Swetha, T.A., Bora, A., Mohanrasu, K., Balaji, P., Raja, R., Ponnuchamy, K., et al., 2023. A comprehensive review on polylactic acid (PLA) – synthesis, processing and application in food packaging. International Journal of Biological Macromolecules 234, 123715. Available at https://doi.org/10.1016/j.ijbiomac.2023.123715.

Taha, E., 2023. Shelf-life extension of fish fillets using gelatins edible coating extracted from the skins of nile tilapia and catfish. Egyptian Journal of Food Science 51 (1), 127–138. Available at https://doi.org/10.21608/ejfs.2023.182983.1152.

Tamjid, E., Najafi, P., Khalili, M.A., Shokouhnejad, N., Karimi, M., Sepahdoost, N., 2024. Review of sustainable, eco-friendly, and conductive polymer nanocomposites for electronic and thermal applications: current status and future prospects. Discover Nano 19 (1), 29. Available at https://doi.org/10.1186/s11671-024-03965-2.

Thakur, S., Chaudhary, J., Sharma, B., Verma, A., Tamulevicius, S., Thakur, V.K., 2018. Sustainability of bioplastics: opportunities and challenges. Current Opinion in Green and Sustainable Chemistry 13, 68–75. Available at https://doi.org/10.1016/j.cogsc.2018.04.013.

Upadhyay, P., Zubair, M., Roopesh, M.S., Ullah, A., 2024. An overview of advanced antimicrobial food packaging: emphasizing antimicrobial agents and polymer-based films. Polymers 16 (14), 2007. Available at https://doi.org/10.3390/polym16142007.

Vanderroost, M., Ragaert, P., Devlieghere, F., De Meulenaer, B., 2014. Intelligent food packaging: the next generation. Trends in Food Science & Technology 39 (1), 47–62. Available at https://doi.org/10.1016/j.tifs.2014.06.009.

Varghese, S.A., Siengchin, S., Parameswaranpillai, J., 2020. Essential oils as antimicrobial agents in biopolymer-based food packaging – a comprehensive review. Food Bioscience 38, 100785. Available at https://doi.org/10.1016/j.fbio.2020.100785.

Vicente, D., Proença, D.N., Morais, P.V., 2023. The role of bacterial polyhydroalkanoate (PHA) in a sustainable future: a review on the biological diversity. International Journal of Environmental Research and Public Health 20 (4), 2959. Available at https://doi.org/10.3390/ijerph20042959.

Vilela, C., Kurek, M., Hayouka, Z., Röcker, B., Yildirim, S., Antunes, M.D.C., et al., 2018. A concise guide to active agents for active food packaging. Trends in Food Science & Technology 80, 212–222. Available at https://doi.org/10.1016/j.tifs.2018.08.006.

Wagh, R.V., Priyadarshi, R., Rhim, J.W., 2023. Novel bacteriophage-based food packaging: an innovative food safety approach. Coatings 13 (3), 609. Available at https://doi.org/10.3390/coatings13030609.

Waldo-Mendoza, M.Á., Rivera-García, N.A., Robles-Martínez, M., Mayorga-Colunga, P.C., Martínez-Montejano, R.C., Pérez, E., 2024. Effect of the interlayer distribution of ZnO decorated with Ag nanoparticles on the antimicrobial activity of multilayer poly(methyl methacrylate) films. Journal of Vinyl and Additive Technology 30 (6), 1621–1634. Available at https://doi.org/10.1002/vnl.22146.

Wang, K., Lim, P.N., Tong, S.Y., Thian, E.S., 2019. Development of grapefruit seed extract-loaded poly(ε-caprolactone)/chitosan films for antimicrobial food packaging. Food Packaging and Shelf Life 22, 100396. Available at https://doi.org/10.1016/j.fpsl.2019.100396.

Yam, K.L., 2012. 8 - Intelligent packaging to enhance food safety and quality. In: Yam, K.L., Lee, D.S. (Eds.), Emerging Food Packaging Technologies. Woodhead Publishing, pp. 137–152. Available at https://doi.org/10.1533/9780857095664.2.137.

Yildirim, S., Röcker, B., Pettersen, M.K., Nilsen-Nygaard, J., Ayhan, Z., Rutkaite, R., Radusin, T., Suminska, P., Marcos, B., Coma, V., 2017. Active packaging applications for food. Comprehensive Reviews in Food Science and Food Safety 17, 165–199. Available at 10.1111/1541-4337.12322.

Yam, K.L., Lee, D.S., 2012. 1 - Emerging food packaging technologies: an overview. In: Yam, K.L., Lee, D.S. (Eds.), Emerging Food Packaging Technologies. Woodhead Publishing, pp. 1–9. Available at https://doi.org/10.1533/9780857095664.1.

Yin, Y., Woo, M.W., 2024. Transitioning of petroleum-based plastic food packaging to sustainable bio-based alternatives. Sustainable Food Technology 2 (3), 548–566. Available at https://doi.org/10.1039/d4fb00028e.

Yousefi, H., Su, H.M., Imani, S.M., Alkhaldi, K., Filipe, C.D., Didar, T.F., 2019. Intelligent food packaging: a review of smart sensing technologies for monitoring food quality. ACS Sensors 4 (4), 808–821. Available at https://doi.org/10.1021/acssensors.9b00440.

Zhai, Z., Du, X., Long, Y., Zheng, H., 2022. Biodegradable polymeric materials for flexible and degradable electronics. Frontiers in Electronics 3, 985681. Available at https://doi.org/10.3389/felec.2022.985681.

Zhang, H., Cheng, X., Liu, C., Liu, Z., Liu, L., Feng, C., et al., 2024. Ultraviolet-blocking polymers and composites: recent advances and future perspectives. Journal of Materials Chemistry A 12 (47), 32638–32664. Available at https://doi.org/10.1039/d4ta06335j.

Zhang, J., Zou, X., Zhai, X., Huang, X.W., Jiang, C., Holmes, M., 2019. Preparation of an intelligent pH film based on biodegradable polymers and roselle anthocyanins for monitoring pork freshness. Food Chemistry 272, 306–312. Available at https://doi.org/10.1016/j.foodchem.2018.08.041.

Zheng, H., Tang, H., Yang, C., Chen, J., Wang, L., Dong, Q., et al., 2022. Evaluation of the slow-release polylactic acid/polyhydroxyalkanoates active film containing oregano essential oil on the quality and flavor of chilled pufferfish (Takifugu obscurus) fillets. Food Chemistry 385, 132693. Available at https://doi.org/10.1016/j.foodchem.2022.132693.

Zheng, Y., Yanful, E.K., Bassi, A.S., 2005. A review of plastic waste biodegradation. Critical Reviews in Biotechnology 25 (4), 243–250. Available at https://doi.org/10.1080/07388550500346359.

Zimet, P., Mombrú, Á.W., Mombrú, D., Castro, A., Villanueva, J.P., Pardo, H., et al., 2019. Physico-chemical and antilisterial properties of nisin-incorporated chitosan/carboxymethyl chitosan films. Carbohydrate Polymers 219, 334–343. Available at https://doi.org/10.1016/j.carbpol.2019.05.013.

CHAPTER 13

Artificial intelligence and its role in bioplastic production

Yasaman Esmaeili[1], Colin J. Barrow[2], Minoo Naebe[2,3], Ali Najafi[4], and Shima Jafarzadeh[1,2]

[1]School of Engineering, Deakin University, Geelong Waurn Ponds Campus, Geelong, VIC, Australia, [2]Center for Sustainable Bioproducts, Deakin University, Geelong Waurn Ponds Campus, Geelong, VIC, Australia, [3]Institute for Frontier Materials, Deakin University, Geelong Waurn Ponds Campus, Geelong, VIC, Australia, [4]Department of food Science and Technology , Da.C, Islamic Azad University, Damghan, Iran

13.1 Introduction

Global plastic production has rocketed, growing from just 2 million tons in 1950 to 367 million tons in 2020. Each year, we produce over 300 million tons of plastic waste, but we recycle only 9% of it. This has led to a major problem: tiny plastic particles, called microplastics, are now found everywhere in our water, food, and even our blood. Furthermore, recent research confirms the widespread contamination of microplastics in drinking water, food, and even the human bloodstream, posing significant health risks as these particles can accumulate in organs and interfere with biological functions. Projections indicate that if current trends persist, conventional plastics could be responsible for up to 15% of global carbon emissions by 2050 (Phillip, 2024; Esmaeili et al., 2022; Paidari et al., 2024). Bioplastics are plastics produced from organic materials like plants, rather than petroleum (Senosha et al., 2025). Bioplastics are considered more eco-friendly alternative to traditional plastics, primarily due to their lower greenhouse gas emissions and the absence of harmful pollutants. While their primary application currently lies in packaging, they show significant potential for expansion into high-value industries like electronics and automotive components (Sidek and Draman, 2019; Esmaeili et al., 2022; Esmaeili et al., 2026). Beyond their reduced carbon footprint, some bioplastics offer superior properties such as enhanced mechanical strength, thermal resistance, and improved performance in oxygen permeability, gas barriers, and water vapor transmission, making them suitable for a wider range of applications (Jafarzadeh et al., 2025).

The agri-food sector increasingly requires new technologies and innovative approaches to address significant challenges like climate change, biodiversity loss, food waste, plastic

Bioplastics for Food Packaging. DOI: https://doi.org/10.1016/B978-0-443-41654-5.00005-6

pollution, and a shortage of skilled workers, all of which have been exacerbated by recent conflicts and pandemics (Ahmed et al., 2025). The Fifth Industrial Revolution, also known as Industry 5.0, represents a new scientific approach developed to tackle complex problems. This evolution focuses on integrating advanced automation and digital technologies. By combining information and communication tools, it enables better connections and real-time data sharing, thereby opening new possibilities for more accurate farming, smart agriculture, and intelligent food production systems (Hassoun et al., 2024). On the other hand, the use of artificial intelligence (AI) for process optimization is rapidly gaining importance. By harnessing machine learning (ML) and AI, producers of bioplastics can streamline the conversion of waste biomass, resulting in greater energy efficiency, lower production costs, and a reduced environmental footprint. AI can also identify the ideal conditions for enzymatic hydrolysis, fermentation, or polymerization, which makes the manufacturing process more efficient and scalable (Ghasemlou and Barrow, 2024; Syarifa et al., 2023; Sajadizadeh et al., 2023). Fig. 13.1 shows the main research themes related to sustainable development, circular economy, and biopolymers, with strong links among sustainability, biodegradation, packaging, and ML.

The application of ML and AI has profoundly transformed the development of bioplastics, aligning with principles of a circular economy and a sustainable future. Furthermore, these technologies provide capabilities for real-time monitoring, remote control, and predictive modeling in bioplastics manufacturing. Integrating ML and AI notably improves the sustainability and efficiency of bioplastic production (Li et al., 2020). Numerous studies have also documented the use of ML algorithms and AI-based systems for identifying, classifying, and optimizing the cultivation processes pertinent to bioplastic synthesis (Bin Abu Sofian et al., 2024; Ibarra-Pérez et al., 2023; Uddin et al., 2025).

By integrating cutting-edge AI and ML technologies into the production of bioplastics, this study demonstrates a significant advancement toward a sustainable circular economy. The chapter details the potential of these materials as an alternative to conventional plastics, focusing on their biodegradability and diverse applications.

13.2 Foundations and definitions

13.2.1 Industry 5.0: principles and paradigms

The food industry is undergoing a significant technological transformation, leading to a new paradigm defined by greater efficiency, sustainability, and superior consumer engagement. Innovations are being integrated throughout the supply chain to optimize all processes, including production, distribution, and consumption. Notably, the adoption of AI is revolutionizing agricultural practices by leveraging predictive analytics for advanced crop management and quality control (Martini et al., 2024). Industry 5.0 signifies a new direction

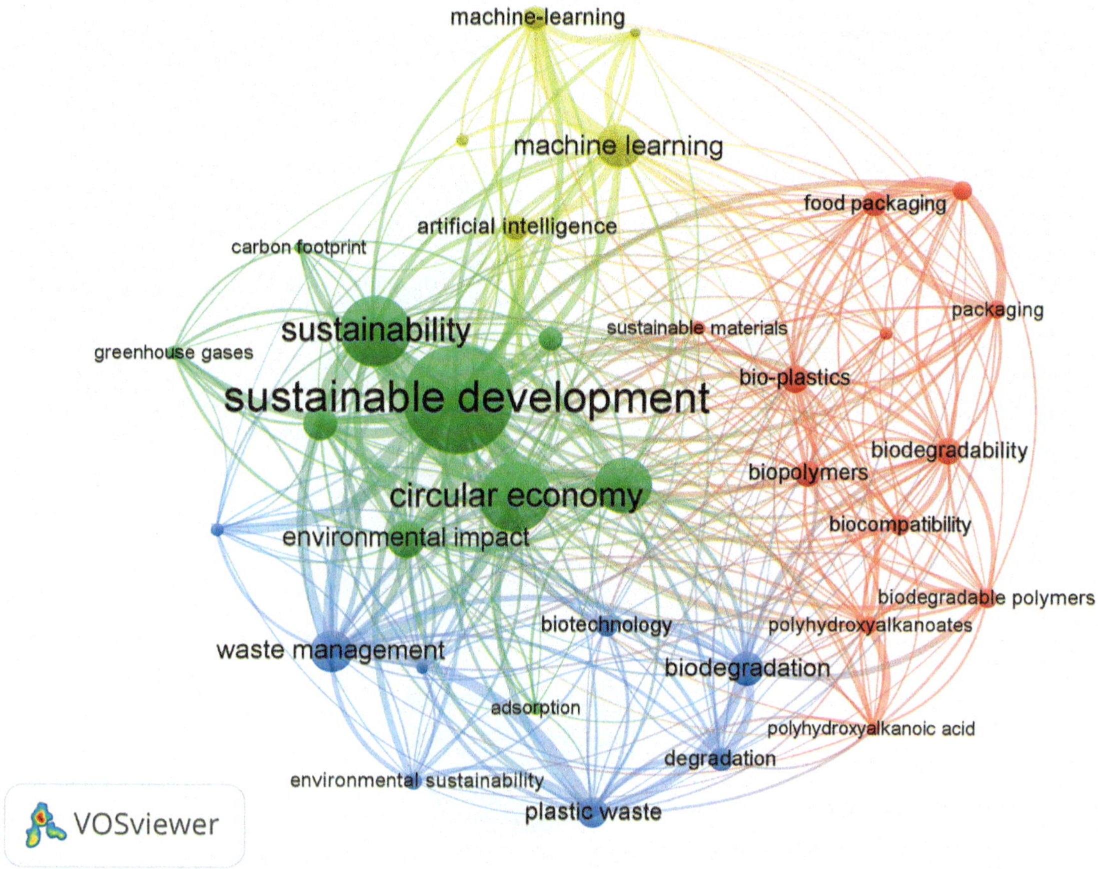

Figure 13.1

Keyword co-occurrence network of publications related to sustainable development, circular economy, and bio-based materials. The visualization, generated using VOSviewer, illustrates major research clusters linking concepts such as sustainable development, circular economy, biopolymers, biodegradation, and artificial intelligence. Node size represents keyword frequency, while link thickness indicates the strength of co-occurrence between terms.

that merges the efficiency of machines with the creative and personal touch that only humans can provide. It is not a separate revolution but rather an enhancement of Industry 4.0, aiming to create a balance between automation and the unique skills of people in the workplace (Fig. 13.2).

Looking back, the first Industrial Revolution used steam and waterpower to take over manual tasks. The second revolution introduced electricity, which allowed for mass production. The third was centered on automating tasks with programmable and digital technologies. The fourth, known as Industry 4.0, introduced smart manufacturing that relied on AI, cloud computing, the Internet of Things (IoT), robotics, and data analysis (Mezzina et al., 2022).

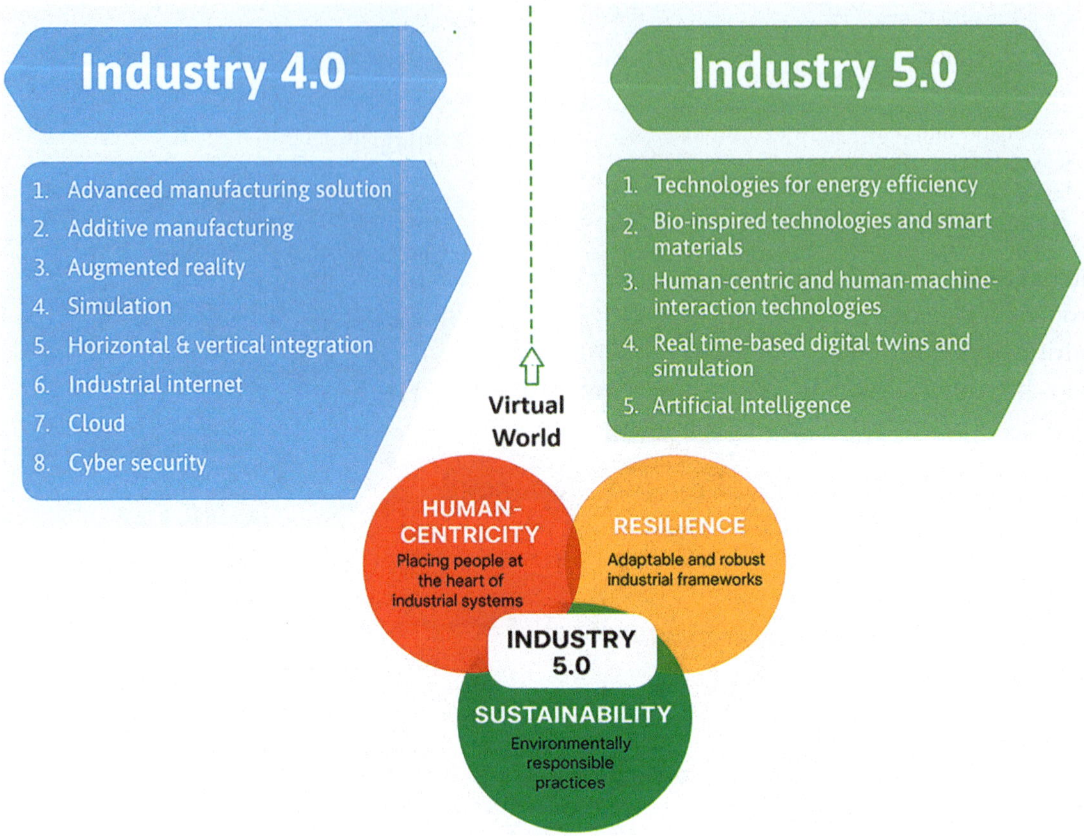

Figure 13.2
From Industry 4.0 to Industry 5.0: principles and paradigms.

Today, Industry 5.0 confronts important questions that were often ignored in past industrial eras (Vuppalapati et al., 2023):

- Does the push for full automation come at the cost of human creativity?
- Can we build a production system where people are vital and cannot be replaced, instead of eliminating their jobs?
- How can we make technological advancements serve both productivity and the greater good of society?

To address these questions, Industry 5.0 is founded on three fundamental pillars. The first is human-centric design, which places people at the heart of industrial systems by recognizing the value of their cognitive abilities and making their participation essential to operations. The second pillar is sustainability, which emphasizes environmentally responsible practices that conserve resources, reduce waste, and minimize ecological impact. The third is resilience, focusing on building adaptable and robust industrial frameworks capable of withstanding disruptions and responding effectively to global changes. Together, these pillars aim to create a more balanced, inclusive, and forward-thinking approach to industrial development (Wang et al., 2023).

13.3 Sustainable packaging: concepts, materials, and drivers

Packaging plays a crucial role in the modern economy, especially in the food industry, although it is often overlooked. In food sector, it serves important purposes such as protecting products, supporting transportation and storage, and providing essential marketing and nutritional information (Åkerman et al., 2025). Due to a growing global population and busy lifestyles, demand for packaged food and beverages has risen, which has led to the increased use of materials such as plastic, paper, cardboard, multimaterials (composed of separate components), and multilayer materials such as liquid packaging and multilayer plastics (Åkerman et al., 2025; Cordeiro et al., 2025).

However, today's world is moving quickly toward more eco-friendly packaging, thanks to both environmental issues and new regulations (Stabler et al., 2024). Significant environmental consequences stem from modern packaging, notably carbon emissions contributing to climate change, biodiversity loss caused by raw material extraction, and concerns over hazardous chemicals in plastics. At the same time, the rising use of packaging has strengthened the push for a circular economy. Among the most pressing issues is the growing amount of plastic waste in nature, which has led to reduce plastic consumption (Table 13.1). In contrast to less visible threats like greenhouse gas emissions, the visible presence of packaging waste as litter provides a sensible and powerful symbol of widespread environmental degradation (Åkerman et al., 2025). Significant efforts are being made to develop bioplastics for sustainable food packaging (Bandara and Indunil, 2022; Garrido-Romero et al., 2022; Hosen et al., 2022; Markevičiūtė and Varžinskas, 2022). Fig. 13.3 illustrates various resources used for production of biodegradable polymers. As an ideal alternative to traditional plastic packaging, cellulose packaging benefits from abundant and renewable raw materials and has been widely adopted in the express industry. Despite these advantages, its mechanical properties limit its reusability (Huang et al., 2025). In addition, plant-based bioplastics are now being commercialized. These include both biodegradable types, like polylactic acid (PLA), polyhydroxyalkanoates (PHAs), and polyhydroxybutyrate, and nondegradable versions, such as bio-PE and bio-polyethylene terephthalate (PET). Additionally, to reduce agriculture's environmental impact and land use, researchers are studying how to use by-products and waste from the agri-food industry as raw materials (De Moura et al., 2017; Jõgi and Bhat, 2020; Markevičiūtė and Varžinskas, 2022).

PET/cellulose@LDH membrane has been successfully fabricated by Huang et al. in a recent study. This advanced material, created through in situ growth of LDH nanosheets on cellulose, followed by hydrophobic modification and PET electrospinning, offers significant

Table 13.1 Comparative analysis of sustainable packaging materials.

Type of material	Applied food product	Application purpose	Characteristics	References
Corn starch/silver	–	Silver was used to reinforce bioplastics to increase its antimicrobial properties	Corn starch with silver-based samples increased antimicrobial properties and showed a 44% bacterial reduction rate against *Escherichia coli* and 50% against *Staphylococcus aureus*.	Kowser et al. (2025)
Chitosan/lignin-based bioplastics	–	Enhancement of mechanical and thermal properties	Incorporation of lignin improved the antioxidant properties and mechanical strength of the bioplastic Increase ultraviolet (UV)-blocking features Improvement of thermal stability and flame retardancy	Agustiany et al. (2025)
Starch-based bioplastic extracted from corn and potato	–	Enhancement of mechanical strength, flexibility, and biodegradability	Major improvement in tensile strength of 22.5% (6.08 MPa) and a 31.7% increase in Young's modulus (0.103 GPa) Faster degradation rate	Gurunathan et al. (2025)
Onion peel/potato starch bioplastic	–	Evaluation of influence of onion peel on film structure, mechanical performance, water resistance, gas permeability, and antioxidant activity	Improvement of mechanical properties and improved water resistance Water contact angles increased from 75° up to 87°, and water vapor permeability was reduced by up to 60%, improving moisture resistance	Vallejo et al. (2025)
Essential oil-infused nanocellulose	Meat and seafood	Controlling bacterial growth	Microbial reduction to 2–5 log CFU; shelf-life extension: 7–12 days	Firmanda et al. (2023)

Starch/bamboo cellulose	–	Developing a modified starch/cellulose film for enhanced performance in food packaging applications	Tensile strength enhanced Phosphorylated starch–cellulose composites showed improved mechanical properties Improved hydrophobicity and UV protection Potential for broader use in sustainable packaging applications	Omar et al. (2022)
Cellulose/chitosan	Kefir	Encapsulation of *Lactiplantibacillus plantarum* by emulsion technique	Enhancement of stability of encapsulated probiotics in kefir Cl–Ch hybrid encapsulation enables targeted probiotic delivery	Afzaal et al. (2022)
Carrot pomace/ cellulose nanofibers	Banana	Shelf-life extension	Enzymatic browning significantly delayed (up to 7 days)	Amoroso et al. (2022)
Cellulose nanocrystal/ mangosteen peel	–	Improvement of mechanical properties	Addition of cellulose nanocrystal (CNC) has a positive effect on tensile strength, tensile modulus, and elongation at break compared to pure bioplastic without CNC	Muhammad et al. (2019)

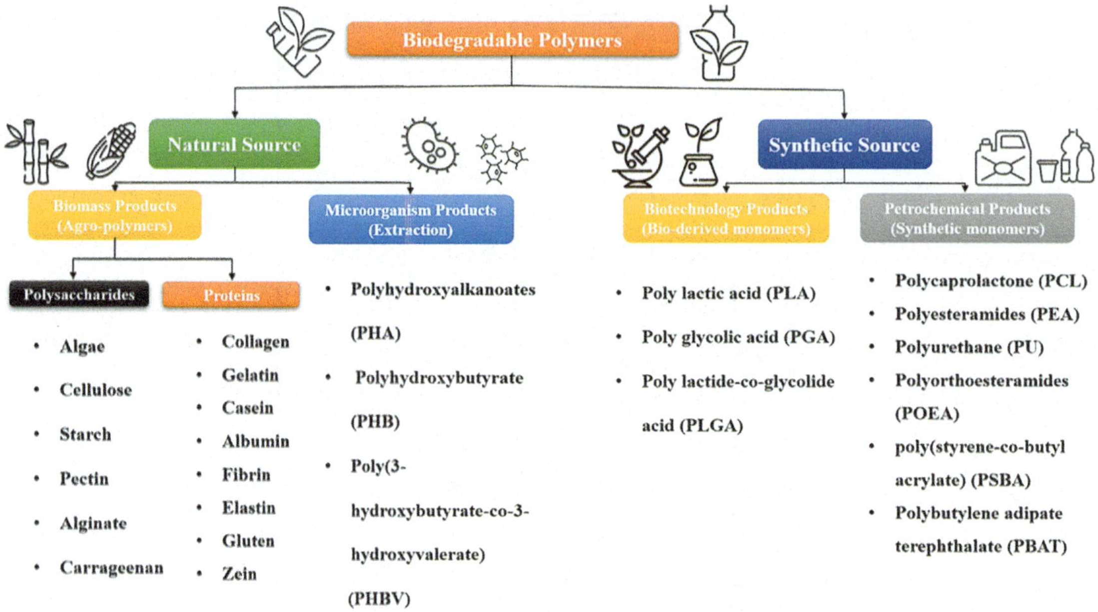

Figure 13.3

Classification of biodegradable polymers based on their origin and production source. Biodegradable polymers are categorized into two main groups: natural-source polymers, derived from biomass (e.g., polysaccharides and proteins) or microorganisms, and synthetic-source polymers, produced through biotechnological or petrochemical routes. Examples include polysaccharides such as starch and cellulose, microbial polymers like polyhydroxyalkanoates (PHAs) and polyhydroxybutyrate (PHB), and synthetic biopolymers such as polylactic acid (PLA), polyglycolic acid (PGA), and polybutylene terephthalate (PBAT).

advantages. It boasts high hydrophobic properties with a static water contact angle of 130 degrees, ensuring reduced contaminant adhesion during application. Moreover, its excellent breathability and mechanical strength make it a highly practical solution for the food packaging industry (Huang et al., 2025).

In another investigation, Riahi et al. created a multifunctional film for active packaging, Zn-doped avocado-derived carbon dots (Zn-ACDs) were added into a cellulose nanofiber (CNF)/pullulan (PUL) composite. This process resulted in a dense cross section and improved the interfacial compatibility of the film. Moreover, the addition of 5 wt.% Zn-ACDs to the CNF/PUL matrix significantly improved its performance. The new material provided 100.0% ultraviolet (UV) blocking and exhibited excellent antioxidant properties (100.0% of ABTS and 68.0% of DPPH). Furthermore, it eliminated the foodborne pathogens *Listeria monocytogenes* ATCC15313 and *Escherichia coli* O157:H7 after just 3 hours of exposure (Riahi et al., 2024).

As the most inexpensive renewable biopolymer, lignin offers significant potential in biorefineries, biofuels, and as an ideal material for developing environmentally friendly nanofilms for packaging application (Kirar et al., 2024). Research on lignin-derived films includes El-Nemr et al.'s (2020) development of a polyvinyl alcohol (PVA)/gelatin/lignin film for antibacterial packaging material. Zhang et al. (2020b) developed a PVA/chitosan/lignin film for antimicrobial packaging.

In a study by Kirar et al., the soda method was successfully employed to extract and characterize lignin from wheat straw. This purified lignin served as a precursor for synthesizing lignin nanoparticles (PNSL) and copper oxide nanoparticles (L@CuONPs), which were then incorporated into PVA-polyethylene glycol (PEG) films to form PVA-PEG-PNSL-L@CuONPs nanocomposite films. These nanocomposite films exhibited remarkable antimicrobial efficacy, achieving up to a 5-log reduction (approximately 99.999%) against *E. coli*, *Bacillus megaterium*, and *Candida albicans*. Moreover, the validation of their UV protective activity highlights their potential as eco-friendly, cost-effective, and sustainable biomaterials. Derived from renewable sources, these lignin-based films also offer significant promise for the food packaging industry due to their affordability, biocompatibility, flexibility, and transparency (Kirar et al., 2024).

Starch is a great candidate for eco-friendly packaging due to its low cost and wide availability. Starch's application is limited by its high-water sensitivity and poor barrier properties. Starch primarily consists of amylose and amylopectin. To overcome its poor water resistance and mechanical properties, starch is often blended with other polymers to make it comparable to synthetic ones (Falua et al., 2022). Dilkushi et al. introduced a novel biocomposite film consisting of banana pseudostem, cassava starch, and poly(vinyl alcohol). A composition of 30% pseudostem powder emerged as the optimal formula, marking a fundamental discovery. It offers an outstanding balance of tensile strength (TS), elongation, water absorption, and gas permeability, representing a significant breakthrough for eco-friendly biocomposite films in food packaging (Dilkushi et al., 2024).

To sum up, sustainable packaging is becoming more important due to environmental concerns, government rules, and new technologies. Many eco-friendly materials like cellulose, starch, lignin, and food industry waste are now being used instead of traditional plastics. These natural materials help reduce pollution and can also offer useful features like blocking UV light, killing bacteria, and being strong and flexible. Still, to make these materials more common in everyday packaging, we need more than just better materials.

This leads to the idea of packaging as a socio-technical system in Industry 5.0, where people, technology, and sustainability all play a part. The next section looks at how packaging is not just about the materials we use but also about the way we design, produce, and manage packaging in smarter and more responsible ways.

13.4 Packaging as a socio-technical system for Industry 5.0

Replacing plastic packaging with biodegradable alternatives offers a significant opportunity for change. However, the large-scale commercialization of bio-packaging remains challenging because conventional plastics benefit from a well-established infrastructure for their production and management. Even with strong efforts and the best possible plans, plastic pollution is still expected to grow in the coming years. These challenges are pushing changes in systems that involve both people and technology, leading to new ways to support sustainability. The goal is to tackle major problems like the loss of natural resources, growing food needs, and climate change by creating new materials that use fewer resources and produce less pollution. Socio-technical systems are made up of different networks that link people, organizations, technologies, and knowledge. While these parts are closely connected and rely on each other, each one has its own goals, values, ways of working, and resources (Ocicka et al., 2023).

Industry 5.0 is the next phase after Industry 4.0, moving beyond just automation. Its main goal is to create a more sustainable "circular economy" by finding ways to reuse, repurpose, and recycle products instead of discarding them. It also aims to make industries tougher and better prepared to handle crises. Central to this shift are key technologies that advance sustainable manufacturing using intelligent products and dynamic life cycle assessment (Gaikwad et al., 2025). Industry 5.0 builds on the technological advancements of Industry 4.0 by adding a human-centric focus to create a more sustainable and resilient food system. While Industry 4.0 introduced digital transformation through automation and real-time data, Industry 5.0 emphasizes the synergy between humans and machines to improve worker well-being and social responsibility. This collaborative approach uses cutting-edge technology to increase efficiency, transparency, and waste minimization, aligning with the principles of a circular economy (Binti Azhar et al., 2021). Fig. 13.4 shows potential applications of AI in plastic waste mitigation. This diagram highlights six key domains where AI can contribute: plastic sorting and recycling, plastic detection, supply chain optimization, predictive modeling, bioplastics research and development, and public education and engagement. Together, these applications demonstrate how AI technologies can improve efficiency, enable waste tracking, and support the transition toward sustainable alternatives.

On the other hand, Industry 5.0 technologies can enhance the entire food system, from postharvest management and logistics to processing, packaging, and retail. This adaptable approach, known as Food Industry 5.0, effectively addresses food security, consumer demand, and resource efficiency while aiding human decision-making (Gaikwad et al., 2025).

The circular economy conserves key resources like raw materials, energy, and water while minimizing waste, emissions, and environmental harm. To cut food waste, this approach uses new preservation technologies like cold chains and smart packaging and improves supply chain

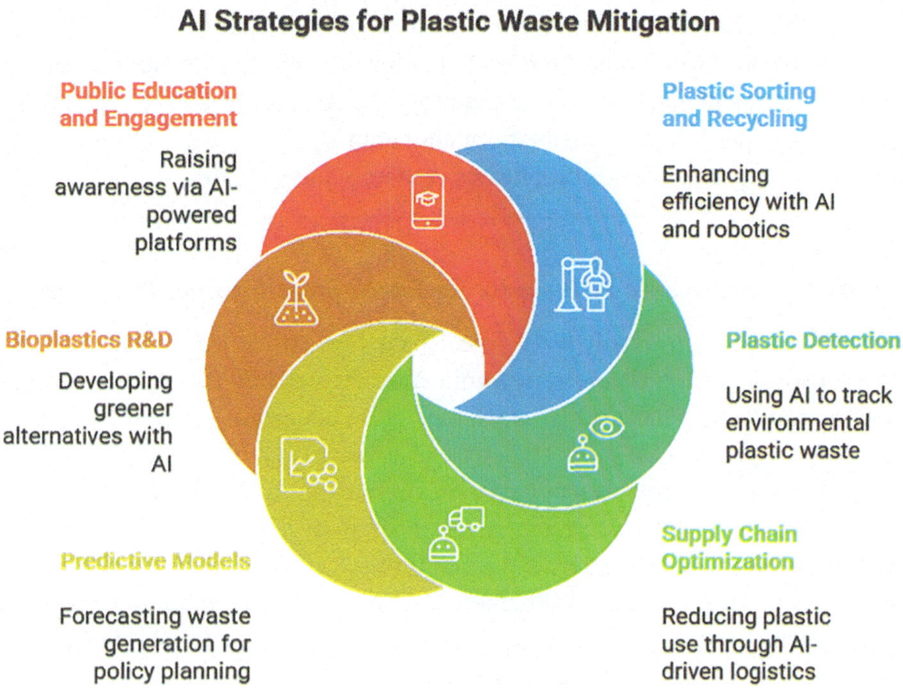

Figure 13.4

Artificial intelligence (AI) strategies for plastic waste mitigation. AI contributes to plastic waste reduction through multiple interconnected strategies, including public education and engagement, bioplastics research and development, predictive modeling for waste forecasting, supply chain optimization, plastic detection, and AI-assisted sorting and recycling. These integrated approaches enhance efficiency, promote sustainability, and support the transition toward a circular economy.

efficiency, inventory management, and the use of food by-products. AI and ML make food waste management more efficient, sustainable, and cost-effective. AI enhances food safety, delivery schedules, and logistics. Using predictive analytics, ML predicts how much waste will be created and what customers will buy, which helps manage stock, share extra food, and demand planning. Furthermore, AI-powered image recognition identifies subpar food throughout the supply chain to reduce waste. In cities, AI also improves waste management by predicting waste types, monitoring bin levels, sorting materials, and optimizing collection routes with Geographic Information System tools. These technologies work together to support effective treatment, recycling, and disposal in a circular economy (Anusha et al., 2024).

13.5 Artificial intelligence applications in bioplastic development

The world has changed to focus on technology, and we are all trying to adapt. This is happening because of ML and AI, which use computers to analyze large amounts of data and make decisions (Nayarisseri et al., 2021). This chapter explains how AI and ML are improving the entire life cycle of bioplastics. AI helps us find the best natural materials, make production more efficient, and predict the quality of the final product. Using AI leads to smarter, larger-scale, and more eco-friendly manufacturing. The next sections will show how AI is accelerating progress in both bioplastics research and industry.

13.5.1 Artificial intelligence in biomass feedstock identification and optimization

Reducing our reliance on traditional fossil-based plastics is crucial for sustainable development. Biomass provides a renewable alternative, but its feedstock composition and system parameters are highly variable and uncertain (Li et al., 2020). The development of bioplastics is supported by the availability of diverse biomass sources, many of which are abundant and not extensively utilized. Studies have confirmed the viability of producing bioplastics from a range of materials, including microorganisms like microalgae (Rahman and Miller, 2017), cellulose, and bio-derived polyethylene from sugarcane (Nandiyanto et al., 2020). Furthermore, research has encompassed starch-based sources (e.g., corn, banana peels, potato peels, and rice) (Shrirakshaya, 2020; Yamada et al., 2020), agricultural by-products like rice straw (Bilo et al., 2018) and corn stalks (Nandiyanto et al., 2020), and novel inputs such as soybeans, wood, food waste, and keratin from chicken feather (Senosha et al., 2025).

A study on PLA from corn stover in the United States used Monte Carlo simulations to assess life cycle global warming potential (GWP). Results showed that GWP per 1 kg of PLA ranged from 1.9 to 4.3 kg CO_2eq, depending on end-of-life treatment and biorefinery energy source. Incineration with wood pellets had the lowest emissions. Two AI models (random forest (RF) and neural networks) were trained on feedstock variations, with RF showing slightly better accuracy (Li et al., 2024). Integrating AL, ML, and the IoT into bioplastics production and cultivation marks a significant step toward an environmentally sustainable circular economy. Furthermore, emerging studies show that ML, specifically multivariate adaptive regression splines and artificial neural networks, can effectively predict the quality of torrefied biomass under various conditions, thus optimizing biofuel production (Chen et al., 2022). Ali et al. predicted the properties of synthesized bioplastics, demonstrating the potential of AI and ML in bioplastics research. Furthermore, utilizing microalgae in PLA production helps mitigate the risk of microplastic contamination (Ali et al., 2023). In another research, Wong et al. present an examination of integrating Industry 5.0 principles into macroalgae biorefineries. Their research underscores the potential for this approach to significantly enhance sustainability and ensure biofuel production aligns with the economic and human-centric pillars of Industry 5.0. The proposed strategy deploys human–robot collaboration, advanced automation, AI, IoT, and Big Data analytics to achieve superior sustainability outcomes, reduce operational expenditures, and maximize revenue generation (Wong et al., 2022).

PHAs are a class of biodegradable polyester synthesized by a variety of microorganisms. This synthesis occurs under specific culture conditions, namely, nutrient limitation in the presence of an excess carbon source. These materials are a promising substitute for petroleum-based plastics because they are fully biodegradable and have similar physical properties (Bin Abu Sofian et al., 2024). A mixed-integer nonlinear programming model by Ramos et al. optimizes PHA production plant design by maximizing net present value and finding ideal operating conditions. The method also facilitates the selection of the most efficient cell extraction techniques to maximize biopolymer recovery (Ramos et al., 2019).

Biodegradable starch films show significant promise for food packaging applications. Kathuria et al. employed the k-nearest neighbor (KNN) algorithm to classify these films based on parameters such as thickness, water vapor permeability (WVP), TS, and transparency. Their study involved producing 12 films from various botanical starch sources via the casting method, resulting in a database of 36 samples. Among these, the 5% cassava starch formulation emerged as optimal, characterized by a WVP of 1.21×10^{-10} g/m/s/Pa^{-1}, TS of 2.34 MPa, a thickness of 0.193 mm, and a water activity (Aw) of 0.408. The combined use of KNN and principal component analysis effectively classified and facilitated the selection of optimal biodegradable starch films (Kathuria et al., 2022).

Berger et al. employed a decision tree method to optimize the conversion of orange peels into biodegradable polymers. Their study analyzed key production variables, including orange peel powder particle size, starch types, cooling methods, and dehydration processes. This approach efficiently organized and analyzed the data, identifying optimal conditions: a 250 μm particle size, 100% corn starch, room-temperature cooling, and effective dehydration. The decision tree model proved highly effective in optimizing biopolymer production, enabling a structured exploration of the best ingredient and method combinations for high-quality bioplastics (Berger et al., 2020).

13.5.2 Sensor-integrated bioplastics

Freshness indicator packaging is a major improvement over traditional packaging because it can actively show freshness of food samples. This innovation enhances consumer convenience by eliminating complex procedures. At the same time, powerful digital technologies like ML, the IoT, and three-dimensional printing are creating new solutions for industrial challenges. The combination of digital technology and packaging design will lead to better materials and more intuitive ways to display freshness. Therefore, biodegradable freshness indicators represent a critical opportunity that demands urgent attention as the food industry evolves from Industry 4.0 to 5.0 (Zhao et al., 2025). Intelligent packaging is increasingly incorporating advanced technologies like time–temperature integrators, radiofrequency identification (RFID), novel freshness indicators, sophisticated biosensors, barcodes, the IoT, augmented reality, and AI (Nagaveni and Poosarla, 2024). Fig. 13.5 provides an overview of these trends.

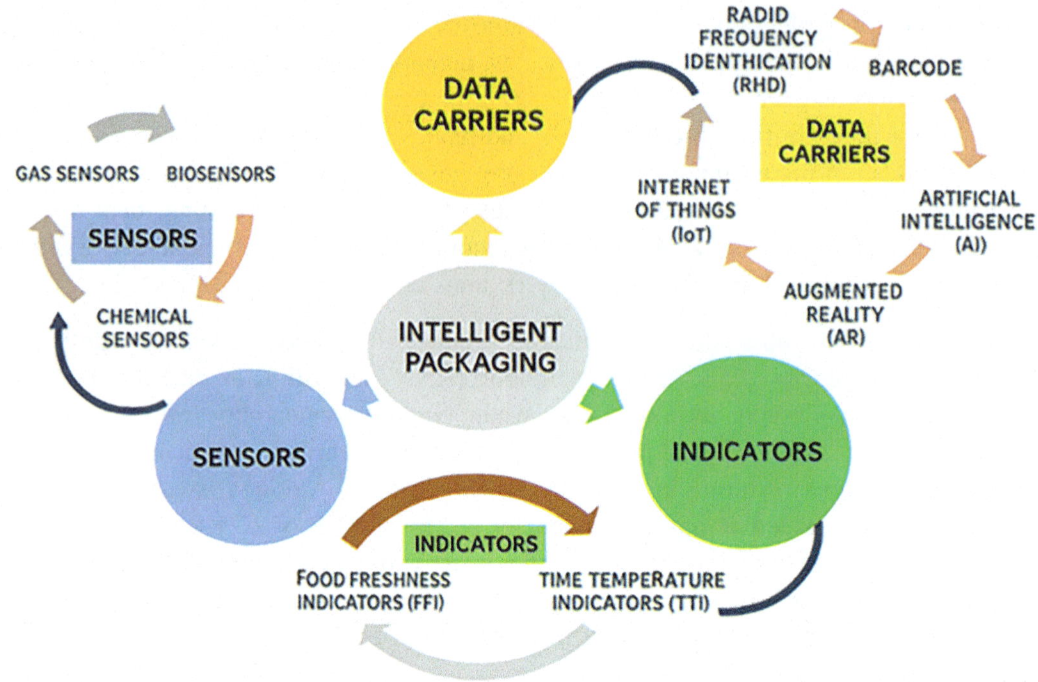

Figure 13.5

Overview of trends in intelligent packaging. Intelligent packages are currently trending toward the synthesis of time-temperature integrators (TTIs), radiofrequency identification (RFID), novel freshness indicators, sophisticated biosensor technologies, barcodes, Internet of things (IoT), augmented reality (AR), artificial intelligence (AI), and other groundbreaking technologies. *Reprinted with permission from Nagaveni, S., Poosarla, V.G., 2024. Chapter 14 - Future trends in intelligent packaging solutions. In: Bangar, S.P., Trif, M. (Eds.), Intelligent Packaging: Current Technologies and Applications. Academic Press, pp. 383–411. https://doi.org/10.1016/B978-0-443-15388-4.00015-8.*

To address the critical need for better food safety in the meat industry, Damdam et al. have developed an electronic nose system that uses IoT to monitor beef quality and detect spoilage. The research highlights that real-time monitoring is crucial for avoiding the health risks and financial losses caused by spoiled meat. This system works by combining IoT technology with electronic sensors to remotely and continuously track the volatile organic compounds released as beef spoils, allowing for early detection and action. Adopting such advanced systems could revolutionize quality control practices for beef and other perishable foods, leading to improved food safety and less waste across the entire supply chain (Damdam et al., 2023).

Sahu et al. present a nondestructive food quality monitoring system using ML and edge IoT on a Raspberry Pi, where ML algorithms analyze data from IoT sensors to assess parameters like freshness, ripeness, and spoilage. The authors' experiments confirm the system's effectiveness, highlighting the potential for this technology to improve quality control and reduce food waste. A specific application of these principles is seen in a system designed for the seafood industry, which assesses salmon freshness during cold storage. It combines environmental sensors that monitor temperature and humidity with an electronic nose that analyzes volatile compounds from the fish. The reliability of this combined approach was demonstrated through experimental trials, underscoring how these technologies can ensure food safety and quality in industries where freshness is paramount (Sahu et al., 2020). Using smart sensors and RFID technology, AI monitors temperature and humidity to enhance cold-chain management for biodegradable products, which cuts down on spoilage and guarantees product safety for consumers (Prakash et al., 2024). As an intelligent component within packaging, RFID uses a system of tag, reader, and recording units to communicate product information by sending and receiving digital electromagnetic signals (Prakash et al., 2024). RFID tags track the quality of food in real time, making them ideal for cold-chain products by monitoring location, time, and temperature. They are more powerful than other tracking tools because they can hold more data (1 MB), communicate faster, and send signals farther (100 m). Some advanced RFID tags can even check how fresh fish is by detecting temperature, ammonia, and hydrogen sulfide gas levels. However, since these tags create electronic waste, researchers are developing "Green RFID" alternatives. These include biodegradable tags made of paper or wood that decompose easily and edible tags that are safely digested, providing smart solutions for both the environment and our health (Palanisamy et al., 2025).

13.5.3 Artificial intelligence–driven bioplastic property prediction and material design

Creating new materials is key to advancing technology in areas like energy and electronics. Finding them the old way is slow and expensive, as it relies on trial and error. ML speeds everything up by figuring out the connection between a material's structure and its abilities, helping us design better materials more efficiently (Yakoubi, 2025; Yakoubi et al., 2023a, 2023b). Researchers have utilized optimization algorithms and ML techniques to determine optimal configurations of material properties and structural designs. This approach facilitates the engineering of sustainable polymer composites for specific applications, with ideal properties, minimized waste material, and enhanced recyclability (Ruan et al., 2023). ML models predict a bioplastic's mechanical, thermal, optical, or electrical properties based on its specific composition and structure, using approaches like regression, classification, or deep learning that are chosen according to the data's complexity (Champa-Bujaico et al., 2022). A study by Özkan et al. demonstrated that an artificial neural network (ANN) model was the most accurate method for predicting the mechanical properties of three-component biocomposite films. Their comparison, which also included RF and MLR models, found RF to be nearly as accurate as ANN, with MLR being the least effective. All models used mechanical properties from two-component films as input data (Özkan et al., 2019). Enhancing the thermal stability of bioplastics is critical for expanding their applicability in high-temperature environments, and AI offers powerful tools for predicting and optimizing thermal performance based on material composition and processing parameters (Yakoubi et al., 2023a). A study on chitosan's thermal decomposition by Araujo et al. revealed activation energies ranging from 98.1 to 183.3 kJ/mol. Using a multilayer perceptron (MLP) neural network to interpret thermogravimetric analysis data, they found the model closely matched experimental results with minimal errors. The analysis successfully identified a correlation between increasing activation energy and polymer dehydration, highlighting the MLP's capability to capture complex thermal behaviors (Araujo et al., 2022). Characterizing the thermal breakdown of biomass components like hemicellulose, cellulose, and lignin is a complex process that requires significant experimental resources. To address this, Kartal et al. created an ANN model that uses proximate analysis results to predict thermal degradation. The model, implemented in TensorFlow, showed outstanding accuracy ($R^2 > 0.998$) in generating derivative thermogravimetry (DTG) curves. This new approach enables the estimation of thermal degradation at any temperature and allows for the instant calculation of biopolymer fractions in degraded biomass (Kartal et al., 2023). By accurately predicting breakdown rates, AI improves deterioration models and makes reusable packaging more economically viable through enhanced supply chain and manufacturing efficiency. For instance, BASF (a European multinational company and the largest chemical producer in the world) used AI-driven manufacturing to cut its biodegradable plastic waste by 15% without increasing costs (Vaquerizo and Kiss, 2023). In a new study, Karaoğul et al. investigated the potential of Turkey's Chirich (*Asphodelus aestivus*) tuber for sustainable biohybrid film production. The team employed sophisticated computational algorithms, gradient boosting regression (GBR), random forest regression, and AdaBoost Regression, to model the films' properties and durability. The results highlighted the GBR algorithm as the most robust model, achieving an exceptional accuracy of 99.92% (R^2). This use of ML-driven optimization allows for precise property prediction and enhances the scalability of production. By pioneering an innovative, data-driven approach, this research advances the industrial application of bio-based polymers and contributes to global initiatives for mitigating plastic waste through responsible manufacturing (Vaquerizo and Kiss, 2023).

13.6 Challenges, risks, and ethical considerations

The integration of AI into recyclable packaging could transform the sector, several significant obstacles must first be addressed. A primary challenge is the considerable cost of implementation, making a particular challenge for small- and medium-sized enterprises. To mitigate this, public–private partnerships and government support may be essential. Moreover, AI in materials research raises critical concerns about data privacy, intellectual property, and the ownership and security of research findings. Cybersecurity threats not only reduce confidence in AI systems but also put private research data at risk. Regulatory and standardization gaps leave manufacturers and regulators uncertain, potentially resulting in contradictory biodegradability claims or even greenwashing. For successful deployment, public trust and acceptance are paramount, requiring focused educational efforts, clear communication, and broad stakeholder involvement (Ulfy et al., 2025). In addition, accurately modeling biopolymers is challenging due to their complex and varied molecular structures. For instance, the numerous bonds and interactions in materials like lignin make them hard to describe with standard math. Simple ML models often fail because they cannot grasp the nonlinear nature of these systems. Therefore, we need more advanced modeling tools that combine ML with physical and chemical simulations.

Another key problem is the lack of good, labeled data, especially in biochemistry. Labeling datasets is a major task, and without accurate labels, ML models cannot perform well (Malashin et al., 2024). The rapid growth of the bioplastics market also introduces environmental challenges related to feedstock farming and material processing. Comprehensive environmental evaluations, such as life cycle assessments, are hampered by the scarcity of robust datasets, reducing the reliability of sustainability claims (Grabowski et al., 2015).

Moving forward, progress will require upgrading data infrastructure, enhancing transparency and interpretability of AI models, and developing sustainable, ethically guided algorithms. Interdisciplinary collaboration among environmental scientists, biotechnologists, data analysts, and social scientists will be key to ensuring that technological innovation aligns with ecological stewardship. By prioritizing responsible and inclusive innovation, the integration of AI and ML into biopolymer research and recyclable packaging can drive the creation of advanced materials, accelerate the transition to a circular economy, and ultimately strengthen the foundation of a resilient and sustainable bioeconomy (Zhang et al., 2020a).

13.7 Conclusion and future outlook

The intersection of AI and bioplastic production represents a paradigm shift in addressing global plastic pollution while advancing sustainability. As demonstrated, AI enables precise process optimization, predictive modeling, and material innovation that traditional methods cannot achieve at scale. Its integration into Industry 5.0 frameworks reinforces a human-centric, resilient, and environmentally responsible production system. However, challenges remain, particularly in data availability, model interpretability, cost of implementation, and ethical considerations such as data privacy and greenwashing risks. Overcoming these barriers will require interdisciplinary collaboration, robust regulatory frameworks, and stronger public–private partnerships. Looking ahead, the future of bioplastics lies in harnessing AI not only for technical breakthroughs but also for fostering transparency, consumer trust, and circular economic practices. If guided responsibly, AI-driven bioplastic innovation can significantly reduce dependency on fossil-based plastics, enhance food security, and contribute to a more sustainable and resilient global bioeconomy.

References

Afzaal, M., Saeed, F., Ateeq, H., Shah, Y.A., Hussain, M., Javed, A., et al., 2022. Effect of cellulose–chitosan hybrid-based encapsulation on the viability and stability of probiotics under simulated gastric transit and in kefir. Biomimetics 7 (3), 109. Available at https://doi.org/10.3390/biomimetics7030109.

Agustiany, E.A., Nawawi, D.S., Fatriasari, W., Wahit, M.U., Vahabi, H., Kayla, D.S., et al., 2025. Mechanical, morphological, thermal, and fire-retardant properties of sustainable chitosan-lignin based bioplastics. International Journal of Biological Macromolecules 306, 141445. Available at https://doi.org/10.1016/j.ijbiomac.2025.141445.

Ahmed, M.S., Kareem, S.A., Mustafa, M.M., 2025. Emerging trends shaping the sustainable future of food Industry 5.0: challenges and opportunities. In: Dutta, P.K., Hamad, A., Haghi, A.K., Prabhakar, P.K. (Eds.), Food and Industry 5.0: Transforming the Food System for a Sustainable Future. Sustainable Development Goals Series. Springer, Cham, pp. 277–287. Available at https://doi.org/10.1007/978-3-031-76758-6_17.

Åkerman, M., Hakala, H., Harlin, A., Sundqvist, H., Leinonen, I., Varho, V., et al., 2025. Introduction: navigating the sustainability transition in food packaging: challenges, policies and interdisciplinary pathways. In: Tynkkynen, N., Hakala, H., Leinonen, I., Varho, V., Åkerman, M., Harlin, A. (Eds.), Sustainability in Food Packaging: Systems, Actors and Innovation. Springer, Cham, pp. 1–13. Available at https://doi.org/10.1007/978-3-031-87492-5_1.

Ali, W., Ali, H., Gillani, S., Zinck, P., Souissi, S., 2023. Polylactic acid synthesis, biodegradability, conversion to microplastics and toxicity: a review. Environmental Chemistry Letters 21 (3), 1761–1786. Available at https://doi.org/10.1007/s10311-023-01564-8.

Amoroso, L., De France, K.J., Milz, C.I., Siqueira, G., Zimmermann, T., Nyström, G., 2022. Sustainable cellulose nanofiber films from carrot pomace as sprayable coatings for food packaging applications. ACS Sustainable Chemistry and Engineering 10 (1), 342–352. Available at https://doi.org/10.1021/acssuschemeng.1c06345.

Anusha, K., Uma, K., Jayasri, K., Kambham, S., Dandamudi, S.D., 2024. IoT based food spoilage detection using machine learning techniques. 2024 International Conference on Advances in Computing, Communication and Applied Informatics (ACCAI), Chennai, India. IEEE, pp. 1–7. Available at https://doi.org/10.1109/ACCAI61061.2024.10602355.

Araujo, N.R.S., Sebastião, R.C.O., Freitas-Marques, M.B., da Nova Mussel, W., Yoshida, M.I., Virtuoso, L.S., 2022. Multilayer perceptron neural network applied to TG dynamic data of biopolymer chitosan – a robust tool to study the kinetics of solid thermal decomposition. Thermal Science and Engineering Progress 36, 101490. Available at https://doi.org/10.1016/j.tsep.2022.101490.

Bandara, R., Indunil, G.M., 2022. Food packaging from recycled papers: chemical, physical, optical properties and heavy metal migration. Heliyon 8 (10), e10959. Available at https://doi.org/10.1016/j.heliyon.2022.e10959.

Berger, P., Cachutt-Alvarado, C., Domínguez-Soberanes, J., Hernández-Lozano, L.C., Ortega-Fraustro, D., Macías-Ochoa, M.F., 2020. Biopolymers obtained from orange waste based on a decision tree. 2020 IEEE ANDESCON, Quito, Ecuador. IEEE, pp. 1–6. Available at https://doi.org/10.1109/ANDESCON50619.2020.9271999.

Bilo, F., Pandini, S., Sartore, L., Depero, L.E., Gargiulo, G., Bonassi, A., et al., 2018. A sustainable bioplastic obtained from rice straw. Journal of Cleaner Production 200, 357–368. Available at https://doi.org/10.1016/j.jclepro.2018.07.252.

Bin Abu Sofian, A.D.A., Lim, H.R., Manickam, S., Ang, W.L., Show, P.L., 2024. Towards a sustainable circular economy: algae-based bioplastics and the role of Internet-of-Things and machine learning. ChemBioEng Reviews 11 (1), 39–59. Available at https://doi.org/10.1002/cben.202300028.

binti Azhar, N.N., Yu, K.L., Ling, T.C., Show, P.L., 2021. Application of Industry 5.0 in the production of fine chemicals and biopolymers. In: Show, P.L., Chew, K.W., Ling, T.C. (Eds.), The Prospect of Industry 5.0 in Biomanufacturing. CRC Press, Boca Raton, pp. 229–256. Available at https://doi.org/10.1201/9781003080671-11-11.

Champa-Bujaico, E., García-Díaz, P., Díez-Pascual, A.M., 2022. Machine learning for property prediction and optimization of polymeric nanocomposites: a state-of-the-art. International Journal of Molecular Sciences 23 (18), 10712. Available at https://doi.org/10.3390/ijms231810712.

Chen, W.H., Aniza, R., Arpia, A.A., Lo, H.J., Hoang, A.T., Goodarzi, V., et al., 2022. A comparative analysis of biomass torrefaction severity index prediction from machine learning. Applied Energy 324. Available at https://doi.org/10.1016/j.apenergy.2022.119689.

Cordeiro, A., Hussain, M., Ramachandran, T., Beemkumar, N., Kumar, R., Karthikeyan, A., et al., 2025. Advancements in packaging materials: trends, sustainability, and future prospects. Circular Economy and Sustainability 5 (4), 2959–2990. Available at https://doi.org/10.1007/s43615-025-00586-4.

Damdam, A.N., Ozay, L.O., Ozcan, C.K., Alzahrani, A., Helabi, R., Salama, K.N., 2023. IoT-enabled electronic nose system for beef quality monitoring and spoilage detection. Foods 12 (11), 2227. Available at https://doi.org/10.3390/foods12112227.

De Moura, I.G., De Sa, A.V., Abreu, A.S.L.M., Machado, A.V.A., 2017. 7 - Bioplastics from agro-wastes for food packaging applications. In: Grumezescu, A.M. (Ed.), Food Packaging. Academic Press, pp. 223–263. Available at https://doi.org/10.1016/B978-0-12-804302-8.00007-8.

Dilkushi, H.A.S., Jayarathna, S., Manipura, A., Chamara, H.K.B.S., Edirisinghe, D., Vidanarachchi, J.K., et al., 2024. Development and characterization of biocomposite films using banana pseudostem, cassava starch and poly(vinyl alcohol): a sustainable packaging alternative. Carbohydrate Polymer Technologies and Applications 7, 100472. Available at https://doi.org/10.1016/j.carpta.2024.100472.

El-Nemr, K.F., Mohamed, H.R., Ali, M.A., Fathy, R.M., Dhmees, A.S., 2020. Polyvinyl alcohol/gelatin irradiated blends filled by lignin as green filler for antimicrobial packaging materials. International Journal of Environmental Analytical Chemistry 100 (14), 1578–1602. Available at https://doi.org/10.1080/03067319.2019.1657108.

Esmaeili, Y., Zamindar, N., Mohammadi, R., 2022. The effect of polypropylene film containing nano-hydroxyapatite on Physicochemical and microbiological properties of button mushrooms (Agaricus bisporus) under Modified atmosphere packaging. Journal of Food Measurement and Characterization 17, 773–786. Available at https://doi.org/10.1007/s11694-022-01613-w.

Esmaeili, Y., Timms, W., Barrow, C.J., Naebe, M., Jafarzadeh, S., 2026. Organic wastes to next-generation bioplastics through intelligent biomanufacturing of polyhydroxyalkanoates. npj Materials Sustainability 4. Available at https://doi.org/10.1038/s44296-026-00104-z.

Falua, K.J., Pokharel, A., Babaei-Ghazvini, A., Ai, Y., Acharya, B., 2022. Valorization of starch to biobased materials: a review. Polymers 14 (11), 2215. Available at https://doi.org/10.3390/polym14112215.

Firmanda, A., Fahma, F., Warsiki, E., Syamsu, K., Arnata, I.W., Sartika, D., et al., 2023. Antimicrobial mechanism of nanocellulose composite packaging incorporated with essential oils. Food Control 147, 109617. Available at https://doi.org/10.1016/j.foodcont.2023.109617.

Gaikwad, S., Anerao, K., Thakur, M., Khedkar, R., Sharma, S., Shankar, N., 2025. Industry 5.0 innovations for waste management and circular economy in sustainable food systems: a systematic review. BIO Web of Conferences 178, 01004. Available at https://doi.org/10.1051/bioconf/202517801004.

Garrido-Romero, M., Aguado, R., Moral, A., Brindley, C., Ballesteros, M., 2022. From traditional paper to nanocomposite films: analysis of global research into cellulose for food packaging. Food Packaging and Shelf Life 31, 100788. Available at https://doi.org/10.1016/j.fpsl.2021.100788.

Ghasemlou, M., Barrow, C.J., Adhikari, B., 2024. The future of bioplastics in food packaging: an industrial perspective. Food Packaging and Shelf Life 43, 101279. Available at https://doi.org/10.1016/j.fpsl.2024.101279.

Grabowski, A., Selke, S.E.M., Auras, R., Patel, M.K., Narayan, R., 2015. Life cycle inventory data quality issues for bioplastics feedstocks. International Journal of Life Cycle Assessment 20 (5), 584–596. Available at https://doi.org/10.1007/s11367-015-0853-3.

Gurunathan, M.K., Navasingh, R.J.H., Selvam, J.D.R., Čep, R., 2025. Development and characterization of starch bioplastics as a sustainable alternative for packaging. Scientific Reports 15 (1), 15264. Available at https://doi.org/10.1038/s41598-025-00221-0.

Hassoun, A., Jagtap, S., Trollman, H., Garcia-Garcia, G., Duong, L.N.K., Saxena, P., et al., 2024. From Food Industry 4.0 to Food Industry 5.0: identifying technological enablers and potential future applications in the food sector. Comprehensive Reviews in Food Science and Food Safety 23 (6), e370040. Available at https://doi.org/10.1111/1541-4337.70040.

Hosen, M.D., Hossain, M.S., Islam, M.A., Haque, A.N.M.A., Naebe, M., 2022. Utilisation of natural wastes: water-resistant semi-transparent paper for food packaging. Journal of Cleaner Production 364, 132665. Available at https://doi.org/10.1016/j.jclepro.2022.132665.

Huang, S., Wang, M., Dai, Y., Deng, C., Xue, S., Qiu, F., et al., 2025. Waste to treasure: upcycling waste express packing to sustainable packaging materials. Journal of Environmental Chemical Engineering 13 (1), 115016. Available at https://doi.org/10.1016/j.jece.2024.115016.

Ibarra-Pérez, D., Faba, S., Hernández-Muñoz, V., Smith, C., Galotto, M.J., Garmulewicz, A., 2023. Predicting the composition and mechanical properties of seaweed bioplastics from the scientific literature: a machine learning approach for modeling sparse data. Applied Sciences 13 (21), 11841. Available at https://doi.org/10.3390/app132111841.

Jafarzadeh, S., Qazanfarzadeh, Z., Parandi, E., Esmaeili, Y., Barrow, C.J., Timms, W., Naebe, M., 2025. Eco-friendly plastics from cereal-derived by-products and waste: A circular economy approach for sustainable packaging. Materials Today Chemistry 47, 102782. Available at https://doi.org/10.1016/j.mtchem.2025.102782.

Jõgi, K., Bhat, R., 2020. Valorization of food processing wastes and by-products for bioplastic production. Sustainable Chemistry and Pharmacy 18, 100326. Available at https://doi.org/10.1016/j.scp.2020.100326.

Kartal, F., Dalbudak, Y., Özveren, U., 2023. Prediction of thermal degradation of biopolymers in biomass under pyrolysis atmosphere by means of machine learning. Renewable Energy 204, 774–787. Available at https://doi.org/10.1016/j.renene.2023.01.017.

Kathuria, C., Mehrotra, D., Misra, N.K., 2022. A novel random forest approach to predict phase transition. International Journal of System Assurance Engineering and Management 13 (1), 494–503. Available at https://doi.org/10.1007/s13198-021-01302-9.

Kirar, S., Mohne, D., Singh, M., Sagar, V., Bhise, A., Goswami, S., et al., 2024. Eco-friendly lignin nanocomposite films as advanced UV protective and antimicrobial sustainable packaging materials. Sustainable Materials and Technologies 40, e00864. Available at https://doi.org/10.1016/j.susmat.2024.e00864.

Kowser, M.A., Mahmud, H., Chowdhury, M.A., Hossain, N., Mim, J.J., Islam, S., 2025. Fabrication and characterization of corn starch based bioplastic for packaging applications. Results in Materials 25, 100662. Available at https://doi.org/10.1016/j.rinma.2025.100662.

Li, J., Wang, Y., Xu, C., Liu, S., Dai, J., Lan, K., 2024. Bioplastic derived from corn stover: life cycle assessment and artificial intelligence-based analysis of uncertainty and variability. Science of The Total Environment 946, 174349. Available at https://doi.org/10.1016/j.scitotenv.2024.174349.

Li, X., Deng, Y., Li, X., Ma, X., Wang, J., Li, J., 2020. Integration of marine macroalgae (Chaetomorpha maxima) with a moving bed bioreactor for nutrient removal from maricultural wastewater. Archaea (Vancouver, B.C.) 2020, 8848120. Available at https://doi.org/10.1155/2020/8848120.

Malashin, I., Martysyuk, D., Tynchenko, V., Gantimurov, A., Semikolenov, A., Nelyub, V., et al., 2024. Machine learning-based process optimization in biopolymer manufacturing: a review. Polymers 16 (23), 3368. Available at https://doi.org/10.3390/polym16233368.

Markevičiūtė, Z., Varžinskas, V., 2022. Plant-origin feedstock applications in fully green food packaging: the potential for tree-free paper and plant-origin bio-plastics in the Baltic Sea region. Sustainability 14 (12), 7393. Available at https://doi.org/10.3390/su14127393.

Martini, B., Bellisario, D., Coletti, P., 2024. Human-centered and sustainable artificial intelligence in Industry 5.0: challenges and perspectives. Sustainability 16 (13), 5448. Available at https://doi.org/10.3390/su16135448.

Mezzina, G., Ciccarese, D., De Venuto, D., 2022. Preventing food waste by the shelf-life dynamical control in smart homes. IEEE Sensors Letters 6 (8), 1–4. Available at https://doi.org/10.1109/lsens.2022.3191410.

Muhammad, A., Roslan, A., Sanusi, S.N.A., Shahimi, M.Q., Nazari, N.Z., 2019. Mechanical properties of bioplastic form cellulose nanocrystal (CNC) mangosteen peel using glycerol as plasticizer. Journal of Physics: Conference Series 1349 (1), 012099. Available at https://doi.org/10.1088/1742-6596/1349/1/012099.

Nagaveni, S., Poosarla, V.G., 2024. Chapter 14 - Future trends in intelligent packaging solutions. In: Bangar, S.P., Trif, M. (Eds.), Intelligent Packaging: Current Technologies and Applications. Academic Press, pp. 383–411. Available at https://doi.org/10.1016/B978-0-443-15388-4.00015-8.

Nandiyanto, A.B.D., Fiandini, M., Ragadhita, R., Sukmafitri, A., Salam, H., Triawan, F., 2020. Mechanical and biodegradation properties of cornstarch-based bioplastic material. Materials Physics and Mechanics 44 (3), 380–391. Available at https://doi.org/10.18720/MPM.4432020_9.

Nayarisseri, A., Khandelwal, R., Tanwar, P., Madhavi, M., Sharma, D., Thakur, G., et al., 2021. Artificial intelligence, big data and machine learning approaches in precision medicine & drug discovery. Current Drug Targets 22 (6), 631–655. Available at https://doi.org/10.2174/1389450122999210104205732.

Ocicka, B., Kędzia, G., Brzeziński, J., 2023. State of the art and future scenarios for bio-packaging market transition: evidence from Poland. International Journal of Emerging Markets 20 (13), 201–222. Available at https://doi.org/10.1108/IJOEM-10-2022-1614.

Omar, F.N., Hafid, H.S., Zhu, J., Bahrin, E.K., Nadzri, F.Z.M., Wakisaka, M., 2022. Starch-based composite film reinforcement with modified cellulose from bamboo for sustainable packaging application. Materials Today Communications 33, 104392. Available at https://doi.org/10.1016/j.mtcomm.2022.104392.

Özkan, M., Karakoç, A., Borghei, M., Wiklund, J., Rojas, O.J., Paltakari, J., 2019. Machine learning assisted design of tailor-made nanocellulose films: a combination of experimental and computational studies. Polymer Composites 40 (10), 4013–4022. Available at https://doi.org/10.1002/pc.25262.

Palanisamy, Y., Kadirvel, V., Ganesan, N.D., 2025. Recent technological advances in food packaging: sensors, automation, and application. Sustainable Food Technology 3 (1), 161–180. Available at https://doi.org/10.1039/d4fb00296b.

Paidari, S., Esmaeili, Y., Ibrahim, S.A., Vahedi, S., Al-Hilifi, S.A., Zamindar, N., 2024. Microbial Biotechnology in the Food Industry. pp. 75–102. Available at https://doi.org/10.1007/978-3-031-51417-3_4.

Phillip, A., 2024. Bioplastics from waste biomass: paving the way for a sustainable future. International Journal for Research in Applied Science and Engineering Technology 12 (9), 518–533. Available at https://doi.org/10.22214/ijraset.2024.64225.

Prakash, S., Kumari, M., Chauhan, A.K., 2024. The intervention of nanotechnology in food packaging: a review. Journal of Materials Science 59 (7), 2585–2601. Available at https://doi.org/10.1007/s10853-024-09360-7.

Rahman, A., Miller, C.D., 2017. Chapter 6 - Microalgae as a source of bioplastics. In: Rastogi, R.P., Madamwar, D., Pandey, A. (Eds.), Algal Green Chemistry: Recent Progress in Biotechnology. Elsevier, pp. 121–138. Available at https://doi.org/10.1016/B978-0-444-63784-0.00006-0.

Ramos, F.D., Delpino, C.A., Villar, M.A., Diaz, M.S., 2019. Design and optimization of poly (hydroxyalkanoate) s production plants using alternative substrates. Bioresource Technology 289, 121699.

Riahi, Z., Khan, A., Rhim, J.W., Shin, G.H., Kim, J.T., 2024. Sustainable packaging film based on cellulose nanofibres/pullulan impregnated with zinc-doped carbon dots derived from avocado peel to extend the shelf life of chicken and tofu. International Journal of Biological Macromolecules 258, 129302. Available at https://doi.org/10.1016/j.ijbiomac.2024.129302.

Ruan, S.X., Zhang, X.B., Luo, Z.H., 2023. Investigation and optimization of polyolefin elastomers polymerization processes using multi-objective genetic algorithm. Chemical Engineering Research and Design 193, 383–393. Available at https://doi.org/10.1016/j.cherd.2023.03.047.

Sahu, B., Tiwari, A., Raheja, J.L., Kumar, S., 2020. Development of machine learning edge IoT based non-destructive food quality monitoring system using raspberry pi. 2020 IEEE International Conference on Computing, Power and Communication Technologies (GUCON), Greater Noida, India. IEEE, pp. 449–455. Available at https://doi.org/10.1109/GUCON48875.2020.9231061.

Sajadizadeh, H., Zamindar, N., Esmaeili, Y., Lorenzo, J.M., Pateiro, M., Goksen, G., Paidari, S., Liu, L., Doudi, M., 2023. Antimicrobial activity of Oliveria decumbens vent. Essential oil and Ferula asa-foetida extract on Turkey breast meat. Journal of Food Measurement and Characterization 18, 1218–1228. Available at https://doi.org/10.1007/s11694-023-02155-5.

Senosha, M.P., Matheri, A.N., Mohamed, B., 2025. Review on application of sustainability, circular and digital economy on bioplastics production. Circular Economy and Sustainability 5 (2), 1269–1289. Available at https://doi.org/10.1007/s43615-024-00453-8.

Shrirakshaya, S.S., 2020. Cellulose and starch as the source of bioplastic. International Journal for Research in Applied Science and Engineering Technology 8 (6), 562–567. Available at https://doi.org/10.22214/ijraset.2020.6089.

Sidek, I.S., Draman, S.F.S., Abdullah, S.R.S., Anuar, N., 2019. Current development on bioplastics and its future prospects: an introductory review. Inwascon Technology Magazine 1, 03–08. Available at https://doi.org/10.26480/itechmag.01.2019.03.08.

Stabler, D., Hakala, H., Huikkola, T., Mention, A.-L., 2024. Aligning servitization and circularity: the role of institutional confluence in sustainable business models. Journal of Cleaner Production 462, 142666. Available at https://doi.org/10.1016/j.jclepro.2024.142666.

Syarifa, R., Esmaeili, Y., Jafarzadeh, S., Garavand, F., Mahmud, S., Ariffin, F., 2023. An investigation of the morphological, thermal, mechanical, and barrier properties of an active packaging containing micro- and nano-sized ZnO particles. Food Science & Nutrition 11, 7373–7382. Available at https://doi.org/10.1002/fsn3.3665.

Uddin, M.H., Mulla, M.H., Abedin, T., Manap, A., Yap, B.K., Rajamony, R.K., et al., 2025. Advances in natural fiber polymer and PLA composites through artificial intelligence and machine learning integration. Journal of Polymer Research 32 (3), 76. Available at https://doi.org/10.1007/s10965-025-04282-7.

Ulfy, M.A., Haque, A., Huda, M.N., 2025. Integration of artificial intelligence in biodegradable plastic packaging design: exploring stakeholder attitudes. SSRN Electronic Journal IX (VI), 1728–1737. Available at https://doi.org/10.2139/ssrn.5367646.

Vallejo, M., Esteves, B., Carvalho, P., Coimbra, M.A., Oliveira, M., Ferreira, P., et al., 2025. Upcycling onion peels for enhanced gas barrier and antioxidant bioplastics based on starch from potato processing slurries. Waste Management 205, 115005. Available at https://doi.org/10.1016/j.wasman.2025.115005.

Vaquerizo, L., Kiss, A.A., 2023. Thermally self-sufficient process for cleaner production of e-methanol by CO2 hydrogenation. Journal of Cleaner Production 433, 139845. Available at https://doi.org/10.1016/j.jclepro.2023.139845.

Vuppalapati, C., Ilapakurti, A., Vissapragada, S., Kedari, S., Mamidi, V., Vuppalapati, R., et al., 2023. Ecuador Banana Production & Democratization of Climate Change Machine Learning Models to Mobile Edge Devices!. 2023 Congress in Computer Science, Computer Engineering, & Applied Computing (CSCE), Las Vegas, NV, USA. IEEE, pp. 405–412. Available at https://doi.org/10.1109/CSCE60160.2023.00072.

Wang, Y., Kang, M., Liu, Y., Li, J., Xue, K., Wang, X., et al., 2023. Can digital intelligence and cyber-physical-social systems achieve global food security and sustainability? IEEE/CAA Journal of Automatica Sinica 10 (11), 2070–2080. Available at https://doi.org/10.1109/JAS.2023.123951.

Wong, K.H., Tan, I.S., Foo, H.C.Y., Chin, L.M., Cheah, J.R.N., Sia, J.K., et al., 2022. Third-generation bioethanol and L-lactic acid production from red macroalgae cellulosic residue: prospects of Industry 5.0 algae. Energy Conversion and Management 253, 115155. Available at https://doi.org/10.1016/j.enconman.2021.115155.

Yakoubi, S., 2025. Sustainable revolution: AI-driven enhancements for composite polymer processing and optimization in intelligent food packaging. Food and Bioprocess Technology 18 (1), 82–107. Available at https://doi.org/10.1007/s11947-024-03449-2.

Yakoubi, S., Kobayashi, I., Uemura, K., Saidani-Tounsi, M., Nakajima, M., Isoda, H., et al., 2023. Nanoscale delivery system for improving Bacillus subtilis probiotic viability: a promising safety-enhanced nanoemulsion. Food Bioscience 56, 103184. Available at https://doi.org/10.1016/j.fbio.2023.103184.

Yakoubi, S., Kobayashi, I., Uemura, K., Tounsi, M.S., Nakajima, M., Hiroko, I., et al., 2023. Enhancing plantain epicarp active edible coating performance through investigation of optimal spray coating conditions. Colloids and Surfaces A: Physicochemical and Engineering Aspects 678, 132474. Available at https://doi.org/10.1016/j.colsurfa.2023.132474.

Yamada, M., Morimitsu, S., Hosono, E., Yamada, T., 2020. Preparation of bioplastic using soy protein. International Journal of Biological Macromolecules 149, 1077–1083. Available at https://doi.org/10.1016/j.ijbiomac.2020.02.025.

Zhang, J., Yin, J., Wang, R., Zhang, X., 2020. Basic framework and main methods of uncertainty quantification. Mathematical Problems in Engineering 2020, 1–18. Available at https://doi.org/10.1155/2020/6068203.

Zhang, X., Liu, W., Liu, W., Qiu, X., 2020. High performance PVA/lignin nanocomposite films with excellent water vapor barrier and UV-shielding properties. International Journal of Biological Macromolecules 142, 551–558. Available at https://doi.org/10.1016/j.ijbiomac.2019.09.129.

Zhao, L., Liu, Y., He, M., Zhang, D., Liu, H., Sun, B., 2025. From Industry 4.0 to 5.0: exploring the opportunity of biodegradable freshness indicator packaging. Comprehensive Reviews in Food Science and Food Safety 24 (4), e70242. Available at https://doi.org/10.1111/1541-4337.70242.

Index

Note: Page numbers followed by "*f*" and "*t*" refer to figures and tables, respectively.

D

E

F

R

S

Printed and bound by CPI Group (UK) Ltd, Croydon, CR0 4YY
29/09/2026
02224670-0002